Classics in Mathematics

George Pólya • Gabor Szegő

# Problems and Theorems in Analysis I

**Springer**
*Berlin*
*Heidelberg*
*New York*
*Barcelona*
*Budapest*
*Hong Kong*
*London*
*Milan*
*Paris*
*Santa Clara*
*Singapore*
*Tokyo*

This picture shows G. Pólya (r.) and G. Szegő (l.) delivering their original manuscript to Springer in Berlin in 1925 (courtesy of G. Alexanderson).

**George Pólya**

Born in Budapest, December 13, 1887, George Pólya initially studied law, then languages and literature in Budapest. He came to mathematics in order to understand philosophy, but the subject of his doctorate in 1912 was in probability theory and he promptly abandoned philosophy.

After a year in Göttingen and a short stay in Paris, he received an appointment at the ETH in Zürich. His research was multifaceted, ranging from series, probability, number theory and combinatorics to astronomy and voting systems. Some of his deepest work was on entire functions. He also worked in conformal mappings, potential theory, boundary value problems, and isoperimetric problems in mathematical physics, as well as heuristics late in his career. When Pólya left Europe in 1940, he first went to Brown University, then two years later to Stanford, where he remained until his death on September 7, 1985.

**Gabor Szegő**

Born in Kunhegyes, Hungary, January 20, 1895, Szegő studied in Budapest and Vienna, where he received his Ph. D. in 1918, after serving in the Austro-Hungarian army in the First World War. He became a privatdozent at the University of Berlin and in 1926 succeeded Knopp at the University of Königsberg. It was during his time in Berlin that he and Pólya collaborated on their great joint work, the Problems and Theorems in Analysis. Szegő's own research concentrated on orthogonal polynomials and Toeplitz matrices. With the deteriorating situation in Germany at that time, he moved in 1934 to Washington University, St. Louis, where he remained until 1938, when he moved to Stanford. As department head at Stanford, he arranged for Pólya to join the Stanford faculty in 1942. Szegő remained at Stanford until his death on August 7, 1985.

George Pólya • Gabor Szegő

# Problems and Theorems in Analysis I

## Series. Integral Calculus. Theory of Functions

Reprint of the 1978 Edition

Springer

George Pólya †
Gabor Szegő †

*Translator:*
Dorothee Aeppli
1414 Chelmsfordn Street
St. Paul, MN 55108
USA

Originally published as Vol. 193 of the
*Grundlehren der mathematischen Wissenschaften*

Mathematics Subject Classification (1991): 05-01, 28-01, 30-01, 40-01

CIP data applied for

Die Deutsche Bibliothek – CIP-Einheitsaufnahme
**Pólya, George:**
Problems and theorems in analysis / George Pólya; Gabor Szegő.- [Nachdr.].- Berlin; Heidelberg; New York; Barcelona; Budapest; Hong Kong; London; Milan; Paris; Santa Clara, Singapore; Tokyo: Springer
(Classics in mathematics)
1. Series, integral calculus, theory of functions.-Reprint [der Ausg.] Berlin, Springer, 1978.- 1998
ISBN 3-540-63640-4

ISSN 1431-0821
ISBN 3-540-63640-4 Springer-Verlag Berlin Heidelberg New York



SPIN 10651015 41/3143-5 4 3 2 1 0 – Printed on acid-free paper

G. Pólya G. Szegö

# Problems and Theorems in Analysis I

## Series Integral Calculus Theory of Functions

Translation by D. Aeppli

Corrected Printing

Springer-Verlag
Berlin Heidelberg New York 1978

George Pólya Gabor Szegö
Department of Mathematics, Stanford University
Stanford, CA 94305/USA

Revised and enlarged translation of "Aufgaben und Lehrsätze aus der Analysis I", 4th ed., 1970; Heidelberger Taschenbücher, Band 73

AMS Subject Classifications (1970): 05-01, 28-01, 30-01, 40-01

ISBN 3-540-05672-6 Springer-Verlag Berlin Heidelberg New York
ISBN 0-387-05672-6 Springer-Verlag New York Heidelberg Berlin


Printed in Germany

Printing and Bookbinding: Brühlsche Universitätsdruckerei, Giessen
2141/3140-5432

## Preface to the English Edition

The present English edition is not a mere translation of the German original. Many new problems have been added and there are also other changes, mostly minor. Yet all the alterations amount to less than ten percent of the text. We intended to keep intact the general plan and the original flavor of the work.

Thus we have not introduced any essentially new subject matter, although the mathematical fashion has greatly changed since 1924. We have restricted ourselves to supplementing the topics originally chosen.

Some of our problems first published in this work have given rise to extensive research. To include all such developments would have changed the character of the work, and even an incomplete account, which would be unsatisfactory in itself, would have cost too much labor and taken up too much space.

We have to thank many readers who, since the publication of this work almost fifty years ago, communicated to us various remarks on it, some of which have been incorporated into this edition. We have not listed their names; we have forgotten the origin of some contributions, and an incomplete list would have been even less desirable than no list.

The first volume has been translated by Mrs. Dorothee Aeppli, the second volume by Professor Claude Billigheimer. We wish to express our warmest thanks to both for the unselfish devotion and scrupulous conscientiousness with which they attacked their far from easy task. Our thanks are also due to Dr. Klaus Peters for his unflagging interest and wise advice and to Dr. Julius G. Baron for his kind help with the proofsheets.

Stanford, March 1972

G. Pólya · G. Szegö

# Preface to the First German Edition

> What is good education? Giving systematically opportunity to the student to discover things by himself.
>
> (Condensed from H. Spencer.)

In the mathematical literature there exist a number of excellent and comprehensive collections of problems, books of exercises, review texts etc. The present book, in our view, differs from all these, both in its aim and in the scope and arrangement of the material covered, as well as in the manner in which we envisage its use. Consequently each of these points requires some explanation.

---

The chief aim of this book, which we trust is not unrealistic, is to accustom advanced students of mathematics, through systematically arranged problems in some important fields of analysis, to the ways and means of independent thought and research. It is intended to serve the need for individual active study on the part of both the student and the teacher. The book may be used by the student to extend his own reading or lecture material, or he may work quite independently through selected portions of the book in detail. The instructor may use it as an aid in organizing tutorials or seminars.

This book is no mere collection of problems. Its most important feature is the systematic arrangement of the material which aims to stimulate the reader to independent work and to suggest to him useful lines of thought. We have devoted more time, care and detailed effort to devising the most effective presentation of the material than might be apparent to the uninitiated at first glance.

The imparting of factual knowledge is for us a secondary consideration. Above all we aim to promote in the reader a correct attitude, a certain discipline of thought, which would appear to be of even more essential importance in mathematics than in other scientific disciplines.

---

General rules which could prescribe in detail the most useful discipline of thought are not known to us. Even if such rules could be formulated, they would not be very useful. Rather than knowing the correct rules of thought theoretically, one must have them assimilated into one's flesh and blood ready for instant and instinctive use. Therefore, for the schooling of one's powers of thought only the practice of thinking is really useful. The independent solving of challenging problems will aid the reader far more than the aphorisms which follow, although as a start these can do him no harm.

One should try to understand everything: isolated facts by collating them with related facts, the newly discovered through its connection with the already assimilated. the unfamiliar by analogy with the accustomed, special results through generalization, general results by means of suitable specialization, complex situations by dissecting them into their constituent parts, and details by comprehending them within a total picture.

There is a similarity between knowing one's way about a town and mastering a field of knowledge; from any given point one should be able to reach any other point[1]. One is even better informed if one can immediately take the most convenient and quickest path from the one point to the other. If one is very well informed indeed, one can even execute special feats, for example, to carry out a journey by systematically avoiding certain forbidden paths which are customary—such things happen in certain axiomatic investigations.

There is an analogy between the task of constructing a well-integrated body of knowledge from acquaintance with isolated truths and the building of a wall out of unhewn stones. One must turn each new insight and each new stone over and over, view it from all sides, attempt to join it on to the edifice at all possible points, until the new finds its suitable place in the already established, in such a way that the areas of contact will be as large as possible and the gaps as small as possible, until the whole forms one firm structure.

A straight line is determined by two points. Similarly, many a new result is obtained by means of a kind of linear interpolation between

[1] See e.g. problem **92** and the neighboring problems in Part VI, also problem **64** in Part VIII.

two extreme special cases[1]. A straight line is also determined by a direction and one point. New results also frequently arise from the fortunate coincidence of the direction of one's work with a notable special case. Also the drawing of parallels is a valuable method by means of which new results may be derived[2].

An idea which can be used only once is a trick. If one can use it more than once it becomes a method. In mathematical induction the result to be obtained and the means available for its proof are proportional, they stand in the ratio of $n+1$ to $n$. Hence, strengthening the statement to be proved may also be advantageous, for we strengthen at the same time the means available for its proof. It is also found in other circumstances that the more general statement may be easier to prove than the more particular; in such cases the most important achievement consists precisely in setting up the more general statement, in extracting the essential, in realizing the complete picture[3].

"Qui nimium probat, nihil probat". One should examine every proof with suspicion to see if in fact all the assumptions stated have been used. One should attempt to obtain the same conclusion from fewer hypotheses, or a stronger conclusion from the same hypotheses, and should be satisfied only when one has found counter-examples which show that the limit of the possible has been attained.

However, one must not forget that there are two kinds of generalization, one facile and one valuable. One is generalization by dilution, the other is generalization by concentration. Dilution means boiling the meat in a large quantity of water into a thin soup; concentration means condensing a large amount of nutritive material into an essence. The unification of concepts which in the usual view appear to lie far removed from each other is concentration. Thus, for example, group theory has concentrated ideas which formerly were found scattered in algebra, number theory, geometry and analysis and which appeared to be very different. Examples of generalization by dilution would be still easier to quote, but this would be at the risk of offending sensibilities.

---

[1] See e.g. problem **139** in Part I.

[2] See e.g. the first section in Chapter 1 of Part IV, in particular problems **13** and **14**.

[3] It happens frequently that a hint which is appended to the problem, or the grouping of the neighboring problems, points to a strengthening or a generalization of the result which may be useful for the solution of the problem. Thus compare with each other problems **1** and **2**, **3** and **4**, **5** and **7**, **6** and **8** of Part I.

Not all the subject matter of analysis is suitable for problems. A collection in which all the more important fields of analysis were exhaustively dealt with would necessarily become too extensive and awkward. One can, of course, make a selection in many different ways. We have placed the greatest weight on the central field of modern analysis, the theory of functions of a complex variable. However, we have kept ourselves somewhat apart from the common highway travelled by the usual lectures, textbooks and collections of problems, and have, all things being equal, given preference to those fields which lie closest to our personal interests. We have also taken problems from more difficult fields and such as are still very much in the developmental stage, which have scarcely or not at all been considered as yet in the textbook literature. The table of contents will illustrate this in more detail. Certain chapters may also be made use of by the specialist. But we have nowhere attempted to attain the completeness of a monograph since we have subordinated the selection of the material to our chief aim, which is to present the material, to the best of our ability, in an arrangement that provides guidance and suggestions to the reader.

The origin of the material is highly varied. We have made selections from the classical body of knowledge of mathematics and also from treatises of more recent date. We collected problems which had in part already been published in various periodicals and in part communicated to us verbally by their authors. We have adapted the material to our purpose, completed, reformulated and substantially expanded it. In addition we have published here for the first time in the form of problems a number of our own original results. We thus hope to be able to offer something new even to the expert.

---

The material is arranged in two volumes. The first comprises three parts of a more fundamental nature, the second six parts which are devoted to more specialized questions and applications.

Each volume presents in its first half problems and in its second half their solutions. In the part containing the problems, especially at the beginning of the separate chapters, there are also some explanations, which recall the general notions and theorems needed as a background. Often there is added to a problem a method of attack or hint. The solutions are presented in as brief and concise a form as possible. Trivial deductions are omitted for they should become clear enough after a serious consideration of the problem. In exceptional cases the solution is merely sketched and the reader is referred to the relevant literature. On occasion, extensions, other applications and unsolved problems are touched upon.

The parts are divided into chapters and these are subdivided into sections. Whenever an explanation follows or a new train of thought is introduced, we indicate this by a space.

The arrangement of the problems within the chapters and sections is the point wherein the present book differs perhaps even more from the other similar books known to us than in the selection of the material. Practice problems in the narrow sense which aim to clarify newly learned theorems and concepts by means of suitable special cases occupy relatively little space. Isolated problems are rare. The individual problem is mostly incorporated into a longer series of problems which on the average cover a section, and the organic construction of these series has been the object of our greatest concern.

One may group problems from various points of view—according to required previous knowledge, difficulty, method, or result. We have not committed ourselves to any of these viewpoints, but have chosen varying arrangements, which may reflect the different situations that one meets in independent research. One section for example may be concerned with a method which is explained briefly at the beginning and is afterwards applied to the solution of as many problems of various forms as possible and which is thus developed further and further. Another section may deal similarly with a theorem which is stated at the beginning (or proved, if this can be done easily and quickly) and is then applied and specialized in several ways. Still other sections are constructed on an ascending pattern: The general theorem appears only after preceding special cases and small, fragmentary remarks which suggest the result or lead to its proof. Occasionally a proof which is more difficult is attained in several steps, through a sequence of problems; each problem yields an auxiliary lemma, an independent part of the proof, or some perspective, and thus forms a link in a chain of ideas, by means of which the reader finally reaches the theorem that is to be proved. Some sections bring "miscellaneous problems" and are more loosely knit. They recapitulate preceding material by means of more difficult applications or present isolated results which are of interest in themselves.

Now and then four consecutive problems form a "proportion", in which the fourth has the same relation to the third as the second to the first (generalization, converse, application). Some paragraphs are devoted to a more detailed presentation and examination of analogies[1]. Here the problems are taken in turn from the two subjects which are placed in parallel. They belong together in pairs and form what may be termed a "continued proportion". This arrangement seems to us to be particularly instructive.

---

One may approach the book with a view to finding in it opportunity for practice for oneself, or for one's students, or simply for reading. In each case a suitable way of using it for the particular purpose can be found quite naturally.

The initial chapters of each of the parts mostly require comparatively little previous knowledge. The different parts are, though not entirely, yet largely independent of each other, and also the connection between the sections of the same part is frequently loose, so that for example one is not required to keep scrupulously to the given sequence of topics.

The reader who wishes to solve a problem should think not only about *what* is asked, but also *how* and *where* it is asked. Many problems, which would be intractable even for an advanced student if set in isolation, are here surrounded by preparatory and explanatory problems and presented in such a context that with some perseverance and a little inventiveness it should be possible to master them. There occur of course also really difficult problems without any preparation. These are contained mostly in sections of a looser structure (miscellaneous problems) or else occur only as isolated problems.

The hints are at the disposal of the reader but they are not intended to be forced on him.

If you are unable to solve a problem, you should not despair. The "Socratic method of teaching" does not aim at drilling people in giving quick answers, but to educate by means of questions. If repeated efforts have been unsuccessful, the reader can afterwards analyze the solution which is to be found in the second half of the volume with more incisive attention, bring out the actual principle which is the salient point, assimilate it, and commit it to his memory as a permanent acquisition.

The book, already while in process of being written, was repeatedly made use of for the organization of practice sessions and problem-solving seminars for students in the middle and upper semesters. In such

---

[1] See Part II, Chapter 2; Part IV, Chapter 1, § 1; Part V, Chapter 1, § 1; Part VIII, Chapter 1, § 4.

sessions the easier problems were discussed in the classroom and the answers were given orally by the students, while the more difficult problems were answered in writing by an appropriately set deadline. Important problems serving as paradigms were solved by the instructor. Within one semester it was possible to cover approximately the material of one chapter. Several chapters have been tested in this manner and in part revised on the basis of the experience which we have gained. We believe that we may in good conscience recommend for the organization of practice sessions and seminars the method which we have followed: to pose not isolated problems, but carefully considered connected sequences of problems. Nearly all the chapters of this book can be used as a basis for such instruction. It is obvious that some care should be exercised. Especially, for homework or examinations, it is advisable to replace some problems by related ones.

A continuous reading of the work, in which immediately after each problem the solution is also read, can be recommended only to more experienced readers. On the whole this is not in the spirit of the book. Nevertheless certain chapters are suitable for such continuous reading and may be used essentially as a teaching text. However, for this purpose the presentation is rather condensed; it was intended to allow some time for thinking about the problem between its formulation and its solution, and about the proposition between its statement and its proof.

If our undertaking has not been successful in all respects we appeal to two extenuating circumstances: Firstly, as the plan of this work is essentially new, we had no models which we could have followed. Secondly, a more extensive treatment of the various chapters would have required so much space and the improvement of the presentation in some aspects so much time that the carrying out of the entire plan would have been placed in jeopardy. In the interests of the project we would be grateful to the critical reader if he would direct our attention to possible deficiencies which could be eliminated at a later opportunity.

Numerous friends and colleagues have made available to us unpublished items and others have assisted us by reading the manuscript or the proofsheets. We gratefully mention by name A. Aeppli (Zürich), P. Bernays (Göttingen), A. Cohn (Berlin), R. Courant (Göttingen), P. Csillag (Budapest), L. Fejér (Budapest), M. Fekete (Budapest), A. Fleck (Berlin), F. Gassmann (Zürich), A. Haar (Szeged), A. Hirsch (Zürich), E. Jacobsthal (Berlin), L. Kollros (Zürich), J. Kürschák (Budapest), E. Landau (Göttingen), E. Lasker (Berlin), K. Löwner (Berlin), A. Ostrowski (Göttingen), M. Plancherel (Zürich), H. Prüfer (Jena), T. Radó (Szeged), M. Riesz (Stockholm), A. Stoll (Zürich), O. Toeplitz (Kiel), A. Walther (Göttingen). We were also permitted to incorporate some results from unpublished papers in the estate of A. Hurwitz, and also in that of F. and Th. Lukács. In particular we would like to thank sincerely T. Carleman (Lund) and I. Schur (Berlin) for their valuable problems and also A. and R. Brauer (Berlin), H. Rademacher (Hamburg), and H. Weyl (Zürich) for their truly devoted co-operation. Our sincere thanks are also due to the publisher who accommodated us in every respect in spite of the present difficult times.

Zürich and Berlin, October 1924

G. Pólya · G. Szegö

# Contents

Part One

## Infinite Series and Infinite Sequences

Chapter 1

### Operations with Power Series



Chapter 2

### Linear Transformations of Series. A Theorem of Cesàro



Chapter 3

### The Structure of Real Sequences and Series



Chapter 4

### Miscellaneous Problems

Part Two

## Integration

Chapter 1

### The Integral as the Limit of a Sum of Rectangles



Chapter 2

### Inequalities



Chapter 3

### Some Properties of Real Functions



Chapter 4

### Various Types of Equidistribution



Chapter 5

### Functions of Large Numbers



Part Three

## Functions of One Complex Variable. General Part

Chapter 1

### Complex Numbers and Number Sequences

# Notation and Abbreviations

We tried to be as consistent as possible in regard to notation and abbreviations and to denote quantities of the same nature by the same symbol, at least within a section. A particular notation might be specified for one or two sections. Otherwise the meaning of every letter is explained anew in every problem except when a previous problem is referred to. A problem that is closely related to the preceding one is introduced by the remark "continued"; if it is related to some other problem the relevant number is mentioned, e.g. "continuation of **286**".

We denote Parts by roman, Chapters (if necessary) by arabic numerals. The numbering of the problems recommences with each Part. The problem numbers are in boldface. Within the same Part only the number of the problem is given; if, however, we refer to another Part its number is indicated also; e.g. if we refer to the problem (or solution) **123** of Part II in a problem (or solution) of Part II we write "**123**"; in a problem (or solution) of any other Part we write "II **123**".

Remarks in square brackets [] mean hints in the problems and quotations in the solutions (in particular at the beginning of the solutions) or references to other problems that are used in the proof. All other remarks are in ordinary parentheses. A problem number quoted refers to the problem as well as to the solution except when the one or the other is stressed, e.g. [solution **38**].

Almost always references to the sources are given only in the solution. Problems that appeared already in print are quoted as such. If the author but no bibliography is mentioned the problem has been communicated to us as new. Problems whose number is preceded by the sign * (as ***5** in Part II) or contains a decimal point (as **60.10** in Part I) are new (that is, either not contained in the original German edition, or contained there, but essentially modified in the present English version). If the problem is the same as in the original, but the solution has some essential new feature, the sign * is used only in the solution. The abbreviations of the names of journals are taken from the index of Mathematical Reviews and, if not listed there, from World List of Scientific Periodicals Published 1900–1960, Peter Brown, British Museum, Washington Butterworths, 1963.

The journals most often quoted are:

| | |
|---|---|
| Acta Math. | = Acta Mathematica, Stockholm |
| Amer. Math. Monthly | = The American Mathematical Monthly |
| Arch. Math. Phys. | = Archiv der Mathematik und Physik |
| Atti Accad. Naz. Lincei Rend. Cl. Sci. Fis. Mat. Natur. | = Atti dell' Accademia Nazionale dei Lincei Rendiconti. Classe di Scienze Fisiche, Matematiche e Naturali, Roma |
| C. R. Acad. Sci. (Paris) Sér. A–B | = Comptes rendus hebdomadaires des séances de l'Académie des Sciences, Paris, Séries A et B |

| | |
|---|---|
| Giorn. Mat. Battaglini | = Giornale di Matematiche di Battaglini |
| Jber. deutsch. Math. Verein. | = Jahresbericht der deutschen Mathematiker-Vereinigung |
| J. reine angew. Math. | = Journal für die reine und angewandte Mathematik |
| Math. Ann. | = Mathematische Annalen |
| Math. Z. | = Mathematische Zeitschrift |
| Nachr. Akad. Wiss. Göttingen | = Nachrichten der Gesellschaft der Wissenschaften Göttingen |
| Nouv. Annls Math. | = Nouvelles Annales de mathématiques |
| Proc. Lond. Math. Soc. | = Proceedings of the London Mathematical Society |

The following textbooks are quoted repeatedly and they are cited by the name of the author only or by a particular abbreviation (e.g. Hurwitz-Courant; MPR.):

E. Hecke: Vorlesungen über die Theorie der algebraischen Zahlen. New York: Chelsea Publishing 1948.

E. Hille: Analytic Function Theory, Vol. I: Boston-New York-Chicago-Atlanta-Dallas-Palo Alto-Toronto-London: Ginn & Co. 1959; Vol. II: Waltham/Mass.-Toronto-London: Blaisdell Publishing 1962.

A. Hurwitz-R. Courant: Vorlesungen über allgemeine Funktionentheorie und elliptische Funktionen, 4th Ed. Berlin-Göttingen-Heidelberg-New York: Springer 1964.

K. Knopp: Theory and Applications of Infinite Series, 2nd Ed. London-Glasgow: Blackie & Son 1964.

G. Kowalewski: Einführung in die Determinantentheorie, 4th Ed. Berlin: Walter de Gruyter 1954.

G. Pólya: How to Solve It, 2nd Ed. Princeton: Princeton University Press 1971. Quoted: HSI.

G. Pólya: Mathematics and Plausible Reasoning, Vols. 1 and 2, 2nd Ed. Princeton: Princeton University Press 1968. Quoted: MPR.

G. Pólya: Mathematical Discovery, Vols. 1 and 2, Cor. Ed. New York: John Wiley & Sons 1968. Quoted: MD.

E. T. Whittaker and G. N. Watson: A Course of Modern Analysis, 4th Ed. London: Cambridge University Press 1952.

The following notation and abbreviations are used throughout the book:

$a_n \to a$ means: $a_n$ tends to $a$ as $n \to \infty$.

$a_n \sim b_n$ (read: $a_n$ is asymptotically equal to $b_n$) means: $b_n \neq 0$ for sufficiently large $n$ and $\frac{a_n}{b_n} \to 1$ as $n \to \infty$.

$O(a_n)$, with $a_n > 0$, denotes a quantity that divided by $a_n$ remains bounded; $o(a_n)$ a quantity that divided by $a_n$ tends to 0 as $n \to \infty$.

Such notation is used analogously in limit processes other than $n \to \infty$.

$x \to a + 0$ means $x$ converges from the right ($x \to a - 0$ from the left) to $a$.

$\exp(x) = e^x$, $e$ is the base of natural logarithms.

Given $n$ real numbers $a_1, a_2, \ldots, a_n$, $\max(a_1, a_2, \ldots, a_n)$ denotes the largest (or one of the largest) and $\min(a_1, a_2, \ldots, a_n)$ the smallest (or one of the smallest) among the numbers $a_1, a_2, \ldots, a_n$. $\max f(x)$ and $\min f(x)$ have an analogous meaning for a real function defined on the interval $a, b$, provided $f(x)$ assumes a maximum or a minimum on $a$, $b$. Otherwise we retain the same notation for the least upper and the greatest lower bound resp. (similarly in the case of a complex variable).

sgn $x$ stands for the signum function:

$$\operatorname{sgn} x = \begin{cases} +1 \text{ for } x > 0 \\ \ \ 0 \text{ for } x = 0 \\ -1 \text{ for } x < 0 \end{cases}$$

$[x]$ denotes the largest integer that is not larger than $x$ $(x - 1 < [x] \leqq x)$. Square brackets, however, are also used instead of ordinary parentheses if no misunderstanding is expected. (Their use in a very special sense is restricted to Part I, Chap. 1, § 5.)

$\bar{z}$ is the conjugate to the complex number $z$.

For the determinant with general term $a_{\lambda,\mu}$, $\lambda, \mu = 1, 2, \ldots, n$, we use the abbreviated notation

$$\left| a_{\lambda\mu} \right|_1^n \text{ or } \left| a_{\lambda\mu} \right|_{\lambda,\mu=1,2,\ldots,n} \text{ or } \left| a_{\lambda 1}, a_{\lambda 2}, \ldots, a_{\lambda n} \right|_1^n.$$

A non-empty connected open set (containing only interior points) is called a *region*. The closure of a region (the union of the open set and of its boundary) is called a *domain*. (As this terminology is not the most frequently used, we shall sometimes overemphasize it and speak of "open region" and "closed domain".)

A *continuous curve* is defined as the single-valued continuous image of the interval $0 \leqq t \leqq 1$, i.e. the set of points $z = x + iy$, where $x = \varphi(t)$, $y = \psi(t)$, $\varphi(t)$ and $\psi(t)$ both continuous on the interval $0 \leqq t \leqq 1$. The curve is *closed* if $\varphi(0) = \varphi(1)$, $\psi(0) = \psi(1)$, *without double points* if $\varphi(t_1) = \varphi(t_2)$, $\psi(t_1) = \psi(t_2)$, $t_1 < t_2$, imply $t_1 = 0$, $t_2 = 1$. A curve without double points is also called a *simple curve*. A not-closed, simple, continuous curve is often referred to as *simple arc*.

A closed continuous curve without double points (Jordan curve) in a plane determines two regions of which it is the common boundary.

Paths of integration of line, or complex, integrals are assumed to be continuous and rectifiable.

$(a, b)$ denotes the open interval $a < x < b$, $[a, b)$ the half-open interval $a \leqq x < b$, $(a, b]$ the half-open interval $a < x \leqq b$, $[a, b]$ the closed interval $a \leqq x \leqq b$. When we need not distinguish between these four cases we use the term "interval $a$, $b$".

"Iff" is used now and then as an abbreviation for "if and only if".

# Problems

## Part One

## Infinite Series and Infinite Sequences

### Chapter 1

### Operations with Power Series

#### § 1. Additive Number Theory, Combinatorial Problems, and Applications

***1.** In how many different ways can you change one dollar? That is, in how many different ways can you pay 100 cents using five different kinds of coins, cents, nickels, dimes, quarters and half-dollars (worth 1, 5, 10, 25, and 50 cents, respectively)?

***2.** Let $n$ stand for a non-negative integer and let $A_n$ denote the number of solutions of the Diophantine equation

$$x + 5y + 10z + 25u + 50v = n$$

in non-negative integers. Then the series

$$A_0 + A_1\zeta + A_2\zeta^2 + \cdots + A_n\zeta^n + \cdots$$

represents a rational function of $\zeta$. Find it.

***3.** In how many ways can you put the necessary stamps *in one row* on an airmail letter sent inside the U.S., using 2, 4, 6, 8 cents stamps? The postage is 10 cents. (Different arrangements of the same values are regarded as different ways.)

**4.** We call $B_n$ the number of all possible sums with value $n$ ($n$ a positive integer) whose terms are 1, 2, 3, or 4. (Two sums consisting of the same terms but in different order are regarded as different.) The series

$$1 + B_1\zeta + B_2\zeta^2 + \cdots + B_n\zeta^n + \cdots$$

represents a rational function of $\zeta$. Which one?

**5.** Someone owns a set of eight weights of 1, 1, 2, 5, 10, 10, 20, 50 grams respectively. In how many different ways can 78 grams be composed of such weights? (Replacing one weight by an other one of the same value counts as a different way.)

**6.** In how many different ways can one weigh 78 grams if the weights may be placed on *both* pans of the scales and the same weights are used as in problem **5**?

**7.** We consider sums of the form

$$\varepsilon_1 + \varepsilon_2 + 2\varepsilon_3 + 5\varepsilon_4 + 10\varepsilon_5 + 10\varepsilon_6 + 20\varepsilon_7 + 50\varepsilon_8,$$

where $\varepsilon_1, \varepsilon_2, \ldots, \varepsilon_8$ assume the values 0 or 1. We call $C_n$ the number of different sums with value $n$. Write the polynomial

$$C_0 + C_1\zeta + C_2\zeta^2 + \cdots + C_{99}\zeta^{99}$$

as a product.

**8.** Let $\varepsilon_1, \varepsilon_2, \ldots, \varepsilon_8$ assume the values $-1$, 0, 1. Modify problem **7** accordingly. Let $D_n$ denote the number of different sums of value $n$ Find the factorization of the following expression (function of $\zeta$)

$$\sum_{n=-99}^{99} D_n\zeta^n.$$

**9.** Generalize the preceding examples by replacing the particular values of the coins, stamps and weights by $a_1, a_2, \ldots, a_l$.

**10.** An assembly of $p$ persons elects a committee consisting of $n$ of its members. How many different committees can they choose?

**11.** There are $p$ persons sharing $n$ dollars. In how many ways can they distribute the money?

**12.** There are $p$ persons sharing $n$ dollars, each getting at least one dollar. In how many ways can they do it?

**13.** Consider the general homogeneous polynomial of degree $n$ in the $p$ variables $x_1, x_2, \ldots, x_p$. How many terms does it have?

**14.** Any weight that is a positive integral multiple of a given unit can be weighed with the weights 1, 2, 4, 8, 16, ... on one pan of the scales, and this can be done in exactly one way. That is, any positive integer admits a unique representation in the binary system.

**15.** A set of weights 1, 3, 9, 27, 81, ... can be used to weigh any weight that is a positive integral multiple of a given unit if both pans of the scales are used, and this can be done in exactly one way.

**16.** Write

$$(1 + q\zeta)(1 + q\zeta^2)(1 + q\zeta^4)(1 + q\zeta^8)(1 + q\zeta^{16}) \cdots$$
$$= a_0 + a_1\zeta + a_2\zeta^2 + a_3\zeta^3 + \cdots.$$

Find the general formula for $a_n$.

**17.** Consider the expansion

$$(1 - a)(1 - b)(1 - c)(1 - d) \cdots$$
$$= 1 - a - b + ab - c + ac + bc - abc - d + \cdots.$$

What is the sign of the $n$-th term?

**18.** Prove the identity

$$(1 + \zeta + \zeta^2 + \zeta^3 + \cdots + \zeta^9)(1 + \zeta^{10} + \zeta^{20} + \cdots + \zeta^{90})$$
$$\times (1 + \zeta^{100} + \zeta^{200} + \cdots + \zeta^{900}) \cdots = \frac{1}{1 - \zeta}.$$

**18.1.** The first and the third problem considered in the solution of **9** (concerned with $A_n$ and $C_n$ respectively) are the extreme cases of a common generalization which, properly extended, includes also **18.** Formulate such a generalization.

**18.2.** In a legislative assembly there are $2n + 1$ seats and three parties. In how many different ways can the seats be distributed among the parties so that no party attains a majority against a coalition of the other two parties?

**19.**

$$(1 + \zeta)(1 + \zeta^2)(1 + \zeta^3)(1 + \zeta^4) \cdots = \frac{1}{(1 - \zeta)(1 - \zeta^3)(1 - \zeta^5)(1 - \zeta^7) \cdots}.$$

**20.** Each positive integer can be decomposed into a sum of *different* positive integers in as many ways as it can be decomposed into a sum of *equal or different odd* positive integers. E.g. the decompositions of 6 into sums with different terms are

$$6, \quad 1 + 5, \quad 2 + 4, \quad 1 + 2 + 3,$$

with odd terms

$$1 + 5, \quad 3 + 3, \quad 1 + 1 + 1 + 3, \quad 1 + 1 + 1 + 1 + 1 + 1.$$

**21.** It is possible to write the positive integer $n$ in $2^{n-1} - 1$ ways as a sum of *smaller* positive integers. Two sums that differ in the order of terms only are now regarded as different. E.g. only the seven following sums add up to 4:

$$\begin{array}{llll} 1+1+1+1, & 1+1+2, & 2+2, & 1+3, \\ & 1+2+1, & & 3+1. \\ & 2+1+1, & & \end{array}$$

**22.** The total number of non-negative integral solutions of the following Diophantine equations is $n + 1$:

$$x + 2y = n, \quad 2x + 3y = n - 1, \quad 3x + 4y = n - 2, \ldots,$$

$$nx + (n+1)\,y = 1, \quad (n+1)\,x + (n+2)\,y = 0.$$

**23.** The total number $N$ of non-negative integral solutions of the following Diophantine equations

$$x + 2y = n - 1, \quad 2x + 3y = n - 3, \quad 3x + 4y = n - 5, \ldots$$

is smaller than $n + 2$; moreover the difference $n + 2 - N$ is equal to the number of divisors of $n + 2$ (cf. VIII, Chap. 1, § 5).

**24.** Prove that the total number of non-negative integral solutions of the following Diophantine equations is $n$:

$$x + 4y = 3n - 1, \quad 4x + 9y = 5n - 4, \quad 9x + 16y = 7n - 9, \ldots$$

**25.** The number of non-negative solutions of the Diophantine equation

$$x + 2y + 3z = n$$

is equal to the integer closest to $\frac{(n+3)^2}{12}$.

**26.** Let $a$, $b$ and $n$ be positive integers, $a$ and $b$ relatively prime to each other. The number of non-negative integral solutions of the equation

$$ax + by = n$$

is equal to $\left[\frac{n}{ab}\right]$ or $\left[\frac{n}{ab}\right] + 1$. [More may be less: to prove a more general or a more precise theorem may be less trouble.]

**27.** Let $a_1, a_2, \ldots, a_l$ be positive integers without a common factor different from 1 and $A_n$ be the number of non-negative integral solutions of

$$a_1x_1 + a_2x_2 + a_3x_3 + \cdots + a_lx_l = n.$$

Then we have

$$\lim_{n\to\infty} \frac{A_n}{n^{l-1}} = \frac{1}{a_1a_2\cdots a_l(l-1)!}.$$

**27.1** (continued). We suppose more: we assume that $a_i$ and $a_j$ are relatively prime whenever $i \neq j$. Then we can assert more:

$$A_n = P(n) + Q_n,$$

where $P(x)$ is a polynomial with rational coefficients of degree $l-1$ and the sequence $Q_n$ is periodical with the period $a_1a_2\cdots a_l$:

$$Q_{n+a_1a_2\cdots a_l} = Q_n.$$

**27.2** (continued). In the particular case **26** where $l = 2$, $a_1 = a$, $a_2 = b$

$$A_n \leqq 1, \quad \text{when} \quad n < ab,$$
$$A_n \geqq 1, \quad \text{when} \quad n > ab - a - b,$$

$A_{ab} = 2$, $A_{ab-a-b} = 0$ and generally

$$A_{n+ab} = A_n + 1.$$

**28.** The points in three-dimensional space whose Cartesian coordinates $x, y, z$ are integers are called *lattice points* of this space. How many lattice points of the closed positive octant ($x \geqq 0$, $y \geqq 0$, $z \geqq 0$) lie on the plane $x + y + z = n$? How many lattice points of the open octant ($x > 0$, $y > 0$, $z > 0$) are contained in this plane?

**29.** Let $n$ be a positive integer. How many lattice points $(x_1, x_2, \ldots, x_p)$ of the $p$-dimensional space lie in the "octahedron"

$$|x_1| + |x_2| + |x_3| + \cdots + |x_p| \leqq n?$$

**30.** Consider those lattice points in the closed cube

$$-n \leqq x, y, z \leqq n$$

that satisfy the condition

$$-s \leqq x + y + z \leqq s$$

where $n$ and $s$ are positive integers. The number of such lattice points is equal to

$$\frac{1}{2\pi}\int_{-\pi}^{\pi}\left(\frac{\sin\frac{2n+1}{2}t}{\sin\frac{t}{2}}\right)^3 \frac{\sin\frac{2s+1}{2}t}{\sin\frac{t}{2}}\,dt.$$

**31.** Let $n \geqq 3$. The number of positive integral solutions of

$$x + y + z = n$$

that satisfy the additional conditions

$$x \leqq y + z, \quad y \leqq z + x, \quad z \leqq x + y$$

is given by

$$\frac{(n+8)(n-2)}{8} \text{ for } n \text{ even}, \qquad \frac{n^2-1}{8} \text{ for } n \text{ odd}.$$

## § 2. Binomial Coefficients and Related Problems

The binomial coefficients $\binom{\mu}{r}$ are defined as the coefficients in the expansion of

$$(1+z)^\mu = \binom{\mu}{0} + \binom{\mu}{1} z + \binom{\mu}{2} z^2 + \cdots + \binom{\mu}{r} z^r + \cdots, \quad \binom{\mu}{0} = 1,$$

$\mu$ denotes any number. If $\mu$ is a positive integer then $\binom{\mu}{r}$ can be interpreted as the number of combinations of $\mu$ objects taken $r$ at a time, see **10**. In the following problems we assume that $n$ is a non-negative integer.

**31.1.** Prove that

$$\binom{n}{r} = \binom{n-1}{r-1} + \binom{n-1}{r}$$

in several ways, especially with and without a combinatorial interpretation, with and without the binomial theorem.

**31.2.** Prove that

$$\binom{n}{0} - \binom{n}{1} + \binom{n}{2} - \cdots + (-1)^r \binom{n}{r} = (-1)^r \binom{n-1}{r}.$$

**32.** $\binom{n}{0}^2 + \binom{n}{1}^2 + \binom{n}{2}^2 + \cdots + \binom{n}{n}^2 = \binom{2n}{n}.$

**33.** $\binom{2n}{0}^2 - \binom{2n}{1}^2 + \binom{2n}{2}^2 - \cdots - \binom{2n}{2n-1}^2 + \binom{2n}{2n}^2 = (-1)^n \binom{2n}{n}.$

**34.** Put

$$\sum_{k=0}^{\infty} a_k z^k \sum_{l=0}^{\infty} b_l z^l = \sum_{n=0}^{\infty} c_n z^n,$$

$$\sum_{k=0}^{\infty} \frac{\alpha_k}{k!} z^k \sum_{l=0}^{\infty} \frac{\beta_l}{l!} z^l = \sum_{n=0}^{\infty} \frac{\gamma_n}{n!} z^n.$$

Deduce from these identities

$$c_n = a_0 b_n + a_1 b_{n-1} + a_2 b_{n-2} + \cdots + a_n b_0,$$

$$\gamma_n = \alpha_0 \beta_n + \binom{n}{1} \alpha_1 \beta_{n-1} + \binom{n}{2} \alpha_2 \beta_{n-2} + \cdots + \alpha_n \beta_0.$$

**34.1.** Given $a_0, a_1, a_2, \ldots$ Define, for $n = 0, 1, 2, \ldots,$

$$b_n = a_0 - \binom{n}{1} a_1 + \binom{n}{2} a_2 - \cdots + (-1)^n a_n.$$

Derive hence, for $n = 0, 1, 2, \ldots,$

$$a_n = b_0 - \binom{n}{1} b_1 + \binom{n}{2} b_2 - \cdots + (-1)^n b_n.$$

**35.** Defining

$$x^{n|h} = x(x-h)(x-2h)\cdots(x-(n-1)h)$$

we have the identity

$$(x+y)^{n|h} = x^{n|h} + \binom{n}{1} x^{n-1|h} y^{1|h} + \binom{n}{2} x^{n-2|h} y^{2|h} + \cdots + y^{n|h}.$$

**36** (continued). Prove the following generalization of the multinomial theorem

$$(x_1 + x_2 + x_3 + \cdots + x_l)^{n|h}$$
$$= \sum_{\nu_1+\nu_2+\nu_3+\cdots+\nu_l=n} \frac{n!}{\nu_1!\,\nu_2!\,\nu_3!\cdots\nu_l!} x_1^{\nu_1|h} x_2^{\nu_2|h} x_3^{\nu_3|h} \cdots x_l^{\nu_l|h}.$$

**37.** $\binom{n}{1} - 2\binom{n}{2} + 3\binom{n}{3} - \cdots + (-1)^{n-1} n \binom{n}{n} = \begin{cases} 0 \text{ for } n \neq 1, \\ 1 \text{ for } n = 1. \end{cases}$

**38.**

$$\binom{n}{1} - \frac{1}{2}\binom{n}{2} + \frac{1}{3}\binom{n}{3} - \cdots + (-1)^{n-1}\frac{1}{n}\binom{n}{n}$$
$$= 1 + \frac{1}{2} + \frac{1}{3} + \cdots + \frac{1}{n}.$$

**39.** $\sum_{k=0}^{n} (-1)^{n-k} 2^{2k} \binom{n+k+1}{2k+1} = n + 1.$

**40.** $\sum_{\nu=0}^{n} \left(\frac{\nu}{n} - \alpha\right)^2 \binom{n}{\nu} x^\nu (1-x)^{n-\nu} = (x-\alpha)^2 + \frac{x(1-x)}{n}.$

**41.** Put

$$\varphi(x) = a_0 + a_1 x + a_2 x(x-1) + a_3 x(x-1)(x-2) + \cdots,$$
$$\psi(x) = a_0 + \frac{a_1}{2} x + \frac{a_2}{2^2} x(x-1) + \frac{a_3}{2^3} x(x-1)(x-2) + \cdots,$$

then

$$\binom{n}{0}\varphi(0) + \binom{n}{1}\varphi(1) + \binom{n}{2}\varphi(2) + \cdots + \binom{n}{n}\varphi(n) = 2^n \psi(n)$$

and

$$\binom{n}{0}\varphi(0)-\binom{n}{1}\varphi(1)+\binom{n}{2}\varphi(2)-\cdots+(-1)^n\binom{n}{n}\varphi(n)=(-1)^n a_n n!.$$

**42.**

$$\binom{n}{0}(0-n)^2+\binom{n}{1}(2-n)^2+\binom{n}{2}(4-n)^2+\cdots$$
$$+\binom{n}{\nu}(2\nu-n)^2+\cdots=2^n n.$$

**43.**

$$\left.\begin{aligned}&\binom{n}{0}(0-n)^2-\binom{n}{1}(2-n)^2+\binom{n}{2}(4-n)^2-\cdots\\&\qquad+(-1)^\nu\binom{n}{\nu}(2\nu-n)^2+\cdots\end{aligned}\right\}$$
$$=\begin{cases}0 \text{ for } n \neq 2,\\ 8 \text{ for } n=2.\end{cases}$$

**43.1.** Verify the identity

$$\sum_1^\infty \frac{(-1)^{n-1}x^n}{n!\,n}=e^{-x}\sum_1^\infty\left(1+\frac{1}{2}+\frac{1}{3}+\cdots+\frac{1}{n}\right)\frac{x^n}{n!}.$$

## § 3. Differentiation of Power Series

Let $y$ be an arbitrarily often differentiable function of $z$. We define the operation $\left(z\frac{d}{dz}\right)^n y$ by the recursion formula

$$\left(z\frac{d}{dz}\right)^n y=z\frac{d}{dz}\left(z\frac{d}{dz}\right)^{n-1}y,\qquad z\frac{d}{dz}y=zy'.$$

E.g.

$$\left(z\frac{d}{dz}\right)^n z^k=k^n z^k.$$

Let $f(z)=c_0+c_1x+\cdots+c_nx^n$ be an arbitrary polynomial. We define

$$f\left(z\frac{d}{dz}\right)y=c_0y+c_1\left(z\frac{d}{dz}\right)y+\cdots+c_n\left(z\frac{d}{dz}\right)^n y.$$

**44.** We have

$$f\left(z\frac{d}{dz}\right)z^k=f(k)\,z^k.$$

**45.** If $f(x)$ is a polynomial with integral coefficients (cf. VIII, Chap. 2, § 1) then the sum of the following series is an integral multiple of $e$, the base of the natural logarithms,

$$f(0)+\frac{f(1)}{1!}+\frac{f(2)}{2!}+\cdots+\frac{f(k)}{k!}+\cdots.$$

**46.** Define $f_n(z)$ by

$$\left(z\frac{d}{dz}\right)^n \frac{1}{1-z} = 1^n z + 2^n z^2 + 3^n z^3 + \cdots = \frac{f_n(z)}{(1-z)^{n+1}}, \qquad n = 1, 2, 3, \ldots$$

Then $f_n(z)$ is a polynomial of degree $n$ with positive coefficients (except for the absolute term $f_n(0) = 0$). Furthermore

$$f_n(1) = n!.$$

**47.** Let $f(x)$ and $g(x)$ be two arbitrary polynomials, assume however that $g(x)$ does not have any non-negative integral zeroes. The series

$$y = \frac{f(0)}{g(0)} + \frac{f(1)}{g(1)} z + \frac{f(2)}{g(2)} z^2 + \frac{f(3)}{g(3)} z^3 + \cdots$$

satisfies the differential equation

$$g\left(z\frac{d}{dz}\right) y = f\left(z\frac{d}{dz}\right) \frac{1}{1-z}.$$

This is soluble by quadratures.

**48.** Suppose that $f(x)$ and $g(x)$ are two polynomials relatively prime, that the degree of $g(x)$ is not smaller than the degree of $f(x)$ and that $g(0) = 0$, $g(k) \neq 0$ for $k = 1, 2, \ldots$ The series

$$y = 1 + \frac{f(1)}{g(1)} z + \frac{f(1)\,f(2)}{g(1)\,g(2)} z^2 + \frac{f(1)\,f(2)\,f(3)}{g(1)\,g(2)\,g(3)} z^3 + \cdots$$

satisfies the linear homogeneous differential equation

$$g\left(z\frac{d}{dz}\right) y = f\left(z\frac{d}{dz}\right) zy.$$

**49.** The series

$$y = 1 + \left(\frac{1}{2}\right)^2 z + \left(\frac{1\cdot 3}{2\cdot 4}\right)^2 z^2 + \cdots + \left(\frac{1\cdot 3 \cdots (2n-1)}{2\cdot 4 \cdots 2n}\right)^2 z^n + \cdots$$

satisfies the differential equation

$$z(1-z)\frac{d^2y}{dz^2} + (1-2z)\frac{dy}{dz} - \frac{1}{4} y = 0.$$

## § 4. Functional Equations and Power Series

**50.** The function

$$F(z) = (1-qz)\,(1-q^2z)\,(1-q^3z)\cdots, \qquad |q| < 1,$$

can be expanded in a power series

$$F(z) = A_0 + A_1 z + A_2 z^2 + A_3 z^3 + \cdots.$$

Find the coefficients $A_k$ by using the functional equation

$$F(z) = (1 - qz)\, F(qz).$$

**51.** Let $F(z)$ be the function defined in **50**. Find the coefficients of the power series

$$\frac{1}{F(z)} = B_0 + B_1 z + B_2 z^2 + B_3 z^3 + \cdots.$$

**52.** Determine the coefficients $C_i$ in the identity

$$(1 + qz)(1 + qz^{-1})(1 + q^3 z)(1 + q^3 z^{-1}) \cdots (1 + q^{2n-1} z)(1 + q^{2n-1} z^{-1})$$
$$= C_0 + C_1(z + z^{-1}) + C_2(z^2 + z^{-2}) + \cdots + C_n(z^n + z^{-n}).$$

**53.** Deduce the following equation from the identity **52** by taking the limit:

$$\prod_{n=1}^{\infty} (1 + q^{2n-1} z)(1 + q^{2n-1} z^{-1})(1 - q^{2n}) = \sum_{n=-\infty}^{\infty} q^{n^2} z^n, \qquad |q| < 1.$$

**54.**

$$\prod_{n=1}^{\infty} (1 - q^n) = \sum_{n=-\infty}^{\infty} (-1)^n q^{\frac{3n^2+n}{2}}, \qquad |q| < 1.$$

**55.**

$$\frac{1 - q^2}{1 - q} \cdot \frac{1 - q^4}{1 - q^3} \cdot \frac{1 - q^6}{1 - q^5} \cdots = \sum_{n=0}^{\infty} q^{\frac{n(n+1)}{2}}, \qquad |q| < 1.$$

**56.** We have for $|q| < 1$ the relation

$$\frac{1 - q}{1 + q} \cdot \frac{1 - q^2}{1 + q^2} \cdot \frac{1 - q^3}{1 + q^3} \cdot \frac{1 - q^4}{1 + q^4} \cdots = 1 - 2q + 2q^4 - 2q^9 + 2q^{16} - \cdots.$$

**57.** Let $|q| < 1$. Define

$$G(z) = \frac{q}{1 - q}(1 - z) + \frac{q^2}{1 - q^2}(1 - z)(1 - qz)$$
$$+ \frac{q^3}{1 - q^3}(1 - z)(1 - qz)(1 - q^2 z) + \cdots.$$

This function satisfies the functional equation

$$1 + G(z) - G(qz) = (1 - qz)(1 - q^2 z)(1 - q^3 z) \cdots.$$

**58.** Find the coefficients of the power series of the function $G(z)$ defined in **57**:

$$G(z) = D_0 + D_1 z + D_2 z^2 + D_3 z^3 + \cdots.$$

**59.** Prove the identity

$$\sum_{k=1}^{n} \frac{(1 - a^n)(1 - a^{n-1}) \cdots (1 - a^{n-k+1})}{1 - a^k} = n, \qquad n = 1, 2, 3, \ldots$$

Use it to derive the power series of $-\log(1-x)$. [We have

$$G(q^{-n}) = -n, \qquad n = 0, 1, 2, \ldots,$$

where $G(z)$ is the function defined in **57**.]

**60.** The power series

$$f(z) = 1 + \frac{z^2}{3} + \frac{z^4}{5} + \frac{z^6}{7} + \cdots$$

satisfies the functional equation

$$f\left(\frac{2z}{1+z^2}\right) = (1+z^2)\,f(z) \qquad [\mathbf{39}].$$

## § 5. Gaussian Binomial Coefficients

Let $n$ and $k$ denote integers, and $q$ a variable. We define the *Gaussian binomial coefficient*[1] as

$$\begin{bmatrix} n \\ k \end{bmatrix} = \frac{1-q^n}{1-q}\,\frac{1-q^{n-1}}{1-q^2}\cdots\frac{1-q^{n-k+1}}{1-q^k}$$

for $1 \leqq k \leqq n$ and as $\begin{bmatrix} n \\ 0 \end{bmatrix} = 1$ when $k = 0$. If $k$ is not an integer or does not satisfy the inequality $0 \leqq k \leqq n$ we set

$$\begin{bmatrix} n \\ k \end{bmatrix} = 0.$$

We suppose initially that $q$ avoids the roots of the denominators (certain roots of unity). Sometimes we shall find it necessary to emphasize the dependence on $q$; then we shall use the more explicit notation $\begin{bmatrix} n \\ k \end{bmatrix}_q$ for $\begin{bmatrix} n \\ k \end{bmatrix}$.

**60.1.** Show that

$$\lim_{q\to 1} \begin{bmatrix} n \\ k \end{bmatrix} = \binom{n}{k}.$$

Cf. **10**.

**60.2.** Show that

$$\begin{bmatrix} n \\ k \end{bmatrix} = \begin{bmatrix} n \\ n-k \end{bmatrix}.$$

Pay attention to the case $k = 0$.

**60.3.** Prove the identity in $x$

$$\prod_{k=1}^{n} (1 + q^{k-1}x) = \sum_{k=0}^{n} \begin{bmatrix} n \\ k \end{bmatrix} q^{k(k-1)/2} x^k.$$

---

[1] Cf. C. F. Gauss: Summatio quarundam serierum singularium, Opera, Vol. 2, especially p. 16–17.

**60.4.**

$$\begin{bmatrix} n+1 \\ k \end{bmatrix} = \begin{bmatrix} n \\ k \end{bmatrix} + \begin{bmatrix} n \\ k-1 \end{bmatrix} q^{n-k+1}.$$

**60.5.** Show that $\begin{bmatrix} n \\ k \end{bmatrix}$ which we have defined as a rational function of $q$ is, in fact, a polynomial in $q$, of degree $k(n-k)$

$$\begin{bmatrix} n \\ k \end{bmatrix} = \sum_{\alpha=0}^{k(n-k)} c_{n,k,\alpha} q^{\alpha}.$$

The coefficients $c_{n,k,\alpha}$ are positive integers and "symmetric", that is

$$c_{n,k,\alpha} = c_{n,k,k(n-k)-\alpha}.$$

(We may regard henceforth $\begin{bmatrix} n \\ k \end{bmatrix}$ as a polynomial, defined for all values of $q$.)

**60.6.** According to the notation explained above $\begin{bmatrix} n \\ k \end{bmatrix}_{q^2}$ is the expression which we obtain from $\begin{bmatrix} n \\ k \end{bmatrix}$ by substituting $q^2$ for $q$. Prove the identity in $z$

$$\prod_{k=1}^{n} (1 + q^{2k-1}z)(1 + q^{2k-1}z^{-1}) = \begin{bmatrix} 2n \\ n \end{bmatrix}_{q^2} + \sum_{k=1}^{n} \begin{bmatrix} 2n \\ n-k \end{bmatrix}_{q^2} q^{k^2}(z^k + z^{-k}).$$

**60.7.** Let $m$, $r$, and $s$ denote non-negative integers. For $q = -1$ the Gaussian binomial coefficients assume simple values:

$$\begin{bmatrix} 2m \\ r \end{bmatrix}_{-1} = \begin{cases} 0 & \text{when } r \text{ odd} \\ \binom{m}{s} & \text{when } r = 2s, \end{cases}$$

$$\begin{bmatrix} 2m+1 \\ 2s \end{bmatrix}_{-1} = \begin{bmatrix} 2m+1 \\ 2s+1 \end{bmatrix}_{-1} = \binom{m}{s}.$$

**60.8.** Show that

$$\begin{bmatrix} n \\ 0 \end{bmatrix} - \begin{bmatrix} n \\ 1 \end{bmatrix} + \begin{bmatrix} n \\ 2 \end{bmatrix} - \cdots + (-1)^n \begin{bmatrix} n \\ n \end{bmatrix}$$

equals 0 when $n$ is odd but

$$= (1-q)(1-q^3)(1-q^5)\cdots(1-q^{n-1})$$

when $n$ is even. [Call $F(q, n)$ the proposed expression. Then you have to prove that

$$F(q, n) = (1 - q^{n-1}) F(q, n-2), \quad \text{for} \quad n \geqq 3.$$

You may try to start by passing from $n$ to $n-1$ in using **60.4**.]

We consider a plane with an attached system of rectangular coordinates $x$, $y$. A point of which both coordinates are integers is called a *lattice point* (of the plane; we consider now only the two dimensional space, cf. **28**, **29**). A line parallel to one or the other of the coordinate axes that passes through a lattice point (and therefore through infinitely many lattice points) is called "street". These streets constitute a "network of streets" in which the lattice points are street corners; the network divides the plane into square "blocks" with unit sides. Let $r$ and $s$ be non-negative integers. A shortest path in the network of streets between the origin $(0, 0)$ and the street corner $(r, s)$ is of length

$$r + s = n;$$

we call it a "zig-zag path". The "area under the zig-zag path" is included by the path, the "horizontal" coordinate axis $y = 0$ and the "vertical" line $x = r$.

**60.9.** The number of zig-zag paths between the street corners $(0, 0)$ and $(r, s)$ is

$$\binom{n}{r}.$$

**60.10.** The number of those zig-zag paths between the street corners $(0, 0)$ and $(r, s)$ the area under which is $\alpha$ equals $c_{n,r,\alpha}$ (notation **60.5**).

In order to specify one of the $\binom{n}{r}$ zig-zag paths considered in **60.9** we view in succession the unit segments of which it consists starting from $(0, 0)$ and we write $x$ or $y$ according as the segment viewed is parallel to the $x$-axis or the $y$-axis. Thus the zig-zag path specified by the sequence of letters

$$xxyxyyx$$

ends at the point $(4, 3)$ and the area under it is 4.

Take any two letters in such a sequence; they form an *inversion* generated by the zig-zag path if and only if they are different and $y$ comes before $x$. Thus, in our above example there are four inversions.

**60.11.** The number of inversions generated by a zig-zag path equals the area under the path. (Thus **60.10** determines the number of certain paths, or letter sequences, having a given number $\alpha$ of inversions.)

## § 6. Majorant Series

Let

$$a_0, a_1, a_2, \ldots, a_n, \ldots$$

be any complex numbers and

$$p_0, p_1, p_2, \ldots, p_n, \ldots$$

be non-negative real numbers. We define

$$a_0 + a_1 z + a_2 z^2 + \cdots + a_n z^n + \cdots = A(z),$$

$$p_0 + p_1 z + p_2 z^2 + \cdots + p_n z^n + \cdots = P(z).$$

If the inequalities

$$|a_0| \leqq p_0, \ |a_1| \leqq p_1, \ |a_2| \leqq p_2, \ldots, |a_n| \leqq p_n, \ldots$$

hold simultaneously for all $n$ we use the notation

$$A(z) \ll P(z),$$

in words: "$P(z)$ is a majorant of $A(z)$" or "$A(z)$ is a minorant of $P(z)$".

**61.** If $A(z) \ll P(z)$ and $A^*(z) \ll P^*(z)$ then we have also

$$A(z) + A^*(z) \ll P(z) + P^*(z)$$

and

$$A(z)\,A^*(z) \ll P(z)\,P^*(z).$$

**62.** If $n$ is a positive integer we have

$$\left(1 + \frac{z}{n}\right)^n \ll e^z.$$

**63.** Put

$$f(z) = z + a_2 z^2 + a_3 z^3 + \cdots + a_n z^n + \cdots.$$

From

$$z\frac{f'(z)}{f(z)} \ll \frac{1+z}{1-z}$$

deduce the inequalities

$$|a_n| \leqq n, \qquad n = 1, 2, 3, \ldots$$

**64.** Let $a_1, a_2, \ldots, a_l$ be positive integers. Prove the following relation twice, a) by applying, b) without applying, the results of **9**:

$$(1 + z^{a_1})(1 + z^{a_2}) \cdots (1 + z^{a_l})$$

$$\ll \frac{1}{(1 - z^{a_1})(1 - z^{a_2}) \cdots (1 - z^{a_l})} \ll \frac{1}{1 - z^{a_1} - z^{a_2} - z^{a_3} - \cdots - z^{a_l}}.$$

**64.1.** Let $z_1, z_2, \ldots, z_n$ denote the zeros of the polynomial $z^n + a_1 z^{n-1} + a_2 z^{n-2} + \cdots + a_n$ and define

$$s_k = z_1^k + z_2^k + \cdots + z_n^k$$

for $k = 1, 2, 3, \ldots$ Assume that

$$|s_k| \leqq 1 \quad \text{for} \quad k = 1, 2, \ldots, n.$$

Then

$$|a_k| \leqq 1$$

and [III **21**]

$$|z_k| < 2 \quad \text{for} \quad k = 1, 2, \ldots, n.$$

**64.2** (continued). Show by an example that the case of equality can be attained in the relation $|a_k| \leqq 1$.

## Chapter 2

# Linear Transformations of Series. A Theorem of Cesàro

## § 1. Triangular Transformations of Sequences into Sequences

**65.** Consider the infinite triangular array of numbers

$$\begin{array}{l} p_{00}, \\ p_{10}, p_{11}, \\ p_{20}, p_{21}, p_{22}, \\ \ldots\ldots\ldots\ldots \end{array}$$

Suppose that the numbers $p_{n\nu}$ are non-negative and that the sum of each row is 1:

$$p_{n\nu} \geqq 0,\ p_{n0} + p_{n1} + \cdots + p_{nn} = 1 \text{ for } \nu = 0, 1, \ldots,\ n;\ n = 0, 1, 2, \ldots$$

We transform any given sequence of numbers $s_0, s_1, s_2, \ldots, s_n, \ldots$ into a new sequence $t_0, t_1, t_2, \ldots, t_n, \ldots$ in setting

$$t_n = p_{n0}s_0 + p_{n1}s_1 + \cdots + p_{nn}s_n.$$

Assuming that the numbers $s_0, s_1, \ldots, s_n, \ldots$ are real, show that the value of $t_n$ is contained between their minimum and their maximum.

We have here an important particular case of a *linear transformation* of a sequence into a sequence. In defining $p_{n\nu} = 0$ for $\nu > n$ we extend our triangular array into an infinite square array, the *matrix of the transformation* $(p_{n\nu})$ in which $p_{n\nu}$ is the element in the $n$-th row and $\nu$-th column.

**66** (continued). We say that the transformation is *regular* if from the convergence of a sequence to a limit we can conclude the convergence of the transformed sequence to the same limit, that is, if necessarily

$$\lim_{n\to\infty} t_n = s$$

whenever

$$\lim_{n\to\infty} s_n = s.$$

The transformation defined in **65** is regular if and only if

$$\lim_{n\to\infty} p_{n\nu} = 0 \quad \text{for} \quad \nu = 0, 1, 2, \ldots$$

(This is a particular case of an important theorem of Toeplitz, see **80**, and III, Chap. 1, § 5.)

**67.** The existence of $\lim\limits_{n\to\infty} s_n$ implies

$$\lim_{n\to\infty} \frac{s_0 + s_1 + s_2 + \cdots + s_n}{n+1} = \lim_{n\to\infty} s_n.$$

**68.** If the sequence $p_1, p_2, \ldots, p_n, \ldots$ of positive numbers converges to the positive value $p$ then

$$\lim_{n\to\infty} \sqrt[n+1]{p_0 p_1 p_2 \cdots p_n} = p.$$

**68.1.** The numbers $a_0, a_1, a_2, \ldots$ are positive and

$$\lim_{n\to\infty} \frac{a_{n+1}}{a_n} = p.$$

Then $\lim\limits_{n\to\infty} \sqrt[n]{a_n}$ exists also and has the same value $p$.

**69.** Reduce the computation of $\lim\limits_{n\to\infty} \sqrt[n]{\frac{n^n}{n!}}$ to the computation of $\lim\limits_{n\to\infty} \left(1 + \frac{1}{n}\right)^n$.

**70.** Let the two given sequences

$$a_0, a_1, a_2, \ldots, a_n, \ldots$$

$$b_0, b_1, b_2, \ldots, b_n, \ldots$$

satisfy the conditions:

$$b_n > 0, \qquad n = 0, 1, 2, \ldots;$$

$$b_0 + b_1 + b_2 + \cdots + b_n + \cdots \qquad \text{diverges;}$$

$$\lim_{n\to\infty} \frac{a_n}{b_n} = s.$$

Then

$$\lim_{n\to\infty} \frac{a_0 + a_1 + a_2 + \cdots + a_n}{b_0 + b_1 + b_2 + \cdots + b_n} = s.$$

**71.** Assume $\alpha > 0$. Reduce the computation of

$$\lim_{n\to\infty} \frac{1^{\alpha-1} + 2^{\alpha-1} + 3^{\alpha-1} + \cdots + n^{\alpha-1}}{n^\alpha}$$

to the computation of

$$\lim_{n\to\infty} \frac{(n+1)^\alpha - n^\alpha}{n^{\alpha-1}}.$$

(The value of this limit is well known from calculus.)

**72.** Let $p_0, p_1, \ldots, p_n, \ldots$ be a sequence of positive numbers that satisfy the condition

$$\lim_{n\to\infty} \frac{p_n}{p_0 + p_1 + p_2 + \cdots + p_n} = 0.$$

The existence of $\lim_{n\to\infty} s_n = s$ implies

$$\lim_{n\to\infty} \frac{s_0 p_n + s_1 p_{n-1} + \cdots + s_n p_0}{p_0 + p_1 + \cdots + p_n} = s.$$

**73.** The two sequences of positive numbers

$$p_0, p_1, p_2, \ldots, p_n, \ldots; \qquad q_0, q_1, q_2, \ldots, q_n, \ldots$$

are assumed to satisfy the conditions

$$\lim_{n\to\infty} \frac{p_n}{p_0 + p_1 + p_2 + \cdots + p_n} = 0, \qquad \lim_{n\to\infty} \frac{q_n}{q_0 + q_1 + q_2 + \cdots + q_n} = 0.$$

Define a new sequence

$$r_n = p_0 q_n + p_1 q_{n-1} + p_2 q_{n-2} + \cdots + p_n q_0, \qquad n = 0, 1, 2, \ldots$$

This sequence satisfies again the condition

$$\lim_{n\to\infty} \frac{r_n}{r_0 + r_1 + r_2 + \cdots + r_n} = 0.$$

**74.** Let

$$p_0, p_1, p_2, \ldots, p_n, \ldots; \qquad q_0, q_1, q_2, \ldots, q_n, \ldots$$

be defined as in **73**, and let

$$s_0, s_1, s_2, \ldots, s_n, \ldots$$

be an arbitrary sequence. Consider

$$\lim_{n\to\infty} \frac{s_0 p_n + s_1 p_{n-1} + s_2 p_{n-2} + \cdots + s_n p_0}{p_0 + p_1 + p_2 + \cdots + p_n},$$

and

$$\lim_{n\to\infty} \frac{s_0 q_n + s_1 q_{n-1} + s_2 q_{n-2} + \cdots + s_n q_0}{q_0 + q_1 + q_2 + \cdots + q_n}.$$

If *both* these limits exist they are equal. (The proposition is of special interest if $\lim_{n\to\infty} s_n$ does *not* exist. If $\lim_{n\to\infty} s_n$ exists the proposition is a consequence of **72**.)

**75.** Let $\sigma > 0$. If the series

$$a_1 1^{-\sigma} + a_2 2^{-\sigma} + a_3 3^{-\sigma} + \cdots + a_n n^{-\sigma} + \cdots$$

is convergent, then

$$\lim_{n\to\infty} (a_1 + a_2 + a_3 + \cdots + a_n)\, n^{-\sigma} = 0.$$

(Series of this kind are called Dirichlet series. Cf. VIII, Chap. 1, § 5.)

**76.** Assume $p_1 > 0,\ p_2 > 0,\ p_3 > 0, \ldots$ and that the sequence $P_1, P_2, P_3, \ldots, P_n = p_1 + p_2 + p_3 + \cdots + p_n, \ldots$ is divergent, and $\lim_{n\to\infty} p_n P_n^{-1} = 0$. Then

$$\lim_{n\to\infty} \frac{p_1 P_1^{-1} + p_2 P_2^{-1} + \cdots + p_n P_n^{-1}}{\log P_n} = 1.$$

(Generalization of $1 + \frac{1}{2} + \frac{1}{3} + \cdots + \frac{1}{n} \sim \log n$.)

**77.** Let $p_1, p_2, p_3, \ldots, p_n, \ldots$ and $q_1, q_2, q_3, \ldots, q_n, \ldots$ be two sequences of positive numbers for which

$$\lim_{n\to\infty} \frac{p_1 + p_2 + p_3 + \cdots + p_n}{n p_n} = \alpha, \qquad \lim_{n\to\infty} \frac{q_1 + q_2 + q_3 + \cdots + q_n}{n q_n} = \beta,$$

$$\alpha + \beta > 0.$$

Then

$$\lim_{n\to\infty} \frac{p_1 q_1 + 2 p_2 q_2 + 3 p_3 q_3 + \cdots + n p_n q_n}{n^2 p_n q_n} = \frac{\alpha\beta}{\alpha + \beta}.$$

**78.** The series $a_1 + a_2 + a_3 + \cdots$ does not necessarily converge if

$$(a_1 - a_n) + (a_2 - a_n) + \cdots + (a_{n-1} - a_n)$$

is bounded as $n \to \infty$. If, however, the additional conditions

$$a_1 \geqq a_2 \geqq a_3 \geqq \cdots, \qquad \lim_{n\to\infty} a_n = 0$$

are satisfied the series $a_1 + a_2 + a_3 + \cdots$ must converge.

## § 2. More General Transformations of Sequences into Sequences

**79.** Consider the infinite matrix

$$\begin{matrix} p_{00}, p_{01}, p_{02}, \dots \\ p_{10}, p_{11}, p_{12}, \dots \\ p_{20}, p_{21}, p_{22}, \dots \\ \dots\dots\dots\dots\dots\dots \end{matrix}$$

Suppose that all the numbers $p_{n\nu}$ are non-negative and that the sum in each row is convergent and equal to 1 ($p_{n\nu} \geqq 0$; $\sum_{\nu=0}^{\infty} p_{n\nu} = 1$, for $n, \nu = 0, 1, 2, \dots$). Let $s_0, s_1, \dots, s_n, \dots$ form a bounded sequence. Define a new sequence $t_0, t_1, t_2, \dots, t_n, \dots$ by setting

$$t_n = p_{n0}s_0 + p_{n1}s_1 + p_{n2}s_2 + \cdots + p_{n\nu}s_\nu + \cdots.$$

Show that $t_n$ has a value between the upper and the lower bound of the sequence $s_0, s_1, \dots, s_n, \dots$ (whose terms are here supposed to be real).

**80** (continued). The convergence of the sequence $s_0, s_1, s_2, \dots$ to a limit $s$ implies the convergence of the transformed sequence $t_0, t_1, t_2, \dots$ to the same limit $s$ if and only if

$$\lim_{n\to\infty} p_{n\nu} = 0$$

for each fixed $\nu$. (This is the necessary and sufficient condition of the "regularity" of the transformation with matrix $(p_{n\nu})$; cf. **66**.)

**81.** Assume that the series

$$c_1 + 2c_2 + 3c_3 + 4c_4 + \cdots + nc_n + \cdots$$

converges. Then the series

$$c_n + 2c_{n+1} + 3c_{n+2} + 4n_{n+3} + \cdots = t_n$$

converges too and

$$\lim_{n\to\infty} t_n = 0.$$

**82.** Let the power series

$$f(x) = a_0 + a_1x + a_2x^2 + \cdots + a_nx^n + \cdots$$

be convergent for $x = 1$ and assume $0 < \alpha < 1$. Then the power series

$$f(\alpha) + \frac{f'(\alpha)}{1!}h + \frac{f''(\alpha)}{2!}h^2 + \cdots + \frac{f^{(n)}(\alpha)}{n!}h^n + \cdots$$

is convergent for $h = 1 - \alpha$.

## § 3. Transformations of Sequences into Functions. Theorem of Cesàro

**83.** Let the functions

$$\varphi_0(t), \varphi_1(t), \varphi_2(t), \ldots, \varphi_n(t), \ldots$$

be non-negative in the interval $0 < t < 1$ and assume that

$$\varphi_0(t) + \varphi_1(t) + \varphi_2(t) + \cdots + \varphi_n(t) + \cdots = 1$$

holds for all $0 < t < 1$. The sequence $s_0, s_1, s_2, \ldots, s_n, \ldots$ is supposed to be bounded. Construct the function

$$\Phi(t) = s_0\varphi_0(t) + s_1\varphi_1(t) + s_2\varphi_2(t) + \cdots + s_n\varphi_n(t) + \cdots.$$

The range of $\Phi(t)$ will fall into the interval between the upper and lower bounds of the sequence $s_0, s_1, s_2, \ldots, s_n, \ldots$

**84** (continued). Show that

$$\lim_{t\to 1-0} \left(s_0\varphi_0(t) + s_1\varphi_1(t) + s_2\varphi_2(t) + \cdots + s_n\varphi_n(t) + \cdots\right) = s$$

holds for every convergent sequence $s_0, s_1, s_2, \ldots$ for which $\lim\limits_{n\to\infty} s_n = s$ if and only if for each fixed $\nu$

$$\lim_{t\to 1-0} \varphi_\nu(t) = 0.$$

**85.** The infinite sequences

$$a_0, a_1, a_2, \ldots, a_n, \ldots; \qquad b_0, b_1, b_2, \ldots, b_n, \ldots$$

satisfy the following three conditions:

$$b_n > 0, \qquad n = 0, 1, 2, \ldots;$$

$$b_0 + b_1 t + b_2 t^2 + \cdots \begin{cases} \text{is convergent for } |t| < 1, \\ \text{divergent for } \quad t = 1; \end{cases}$$

$$\lim_{n\to\infty} \frac{a_n}{b_n} = s.$$

Then

$$a_0 + a_1 t + a_2 t^2 + \cdots + a_n t^n + \cdots \text{ converges for } |t| < 1$$

and

$$\lim_{t\to 1-0} \frac{a_0 + a_1 t + a_2 t^2 + \cdots + a_n t^n + \cdots}{b_0 + b_1 t + b_2 t^2 + \cdots + b_n t^n + \cdots} = s.$$

(This proposition is due to E. Cesàro. Several applications will be given in the sequel.)

**86.** If the series

$$a_0 + a_1 + a_2 + \cdots + a_n + \cdots = s$$

is convergent then

$$\lim_{t \to 1-0} (a_0 + a_1 t + a_2 t^2 + \cdots + a_n t^n + \cdots) = s.$$

**87.** Set

$$s_n = a_0 + a_1 + a_2 + \cdots + a_n, \qquad n = 0, 1, 2, 3, \ldots$$

If

$$\lim_{n \to \infty} \frac{s_0 + s_1 + s_2 + \cdots + s_n}{n+1} = s$$

exists, then

$$\lim_{t \to 1-0} (a_0 + a_1 t + a_2 t^2 + \cdots + a_n t^n + \cdots) = s.$$

(This proposition goes beyond **86** only if the series $a_0 + a_1 + a_2 + \cdots + a_n + \cdots$ diverges [**67**].)

**88.** If the following conditions are satisfied:

$$b_n > 0, \qquad \sum_{n=0}^{\infty} b_n \text{ divergent}, \qquad \lim_{n \to \infty} \frac{a_0 + a_1 + a_2 + \cdots + a_n}{b_0 + b_1 + b_2 + \cdots + b_n} = s,$$

then

$$\lim_{t \to 1-0} \frac{a_0 + a_1 t + a_2 t^2 + \cdots + a_n t^n + \cdots}{b_0 + b_1 t + b_2 t^2 + \cdots + b_n t^n + \cdots} = s,$$

provided that the series in the denominator converges for $|t| < 1$.

**89.** The following limit exists and is positive provided $\alpha$ is positive:

$$\lim_{t \to 1-0} (1-t)^\alpha (1^{\alpha-1} t + 2^{\alpha-1} t^2 + 3^{\alpha-1} t^3 + \cdots + n^{\alpha-1} t^n + \cdots).$$

**90.** If $0 < k < 1$ and if $k$ converges to 1, then

$$\int_0^1 \frac{dx}{\sqrt{(1-x^2)(1-k^2x^2)}} \sim \frac{1}{2} \log \frac{1}{1-k}. \qquad \text{[II 202.]}$$

**91.** Let $A_n$ and $B_n$ be the numerator and denominator, respectively, of the $n$-th convergent of the infinite continued fraction

$$\frac{a|}{|1} + \frac{a|}{|3} + \frac{a|}{|5} + \cdots \text{ and so } \frac{A_n}{B_n} = \frac{a|}{|1} + \frac{a|}{|3} + \frac{a|}{|5} + \cdots + \frac{a|}{|2n-3}, \quad a > 0.$$

Assume that this continued fraction converges. Find its value applying **85** and using the series

$$F(x) = \sum_{n=0}^{\infty} \frac{A_n}{n!} x^n, \qquad G(x) = \sum_{n=0}^{\infty} \frac{B_n}{n!} x^n.$$

[$F(x)$ and $G(x)$ satisfy a linear homogeneous differential equation of the second order by virtue of the recursion formulas for $A_n$ and $B_n$.]

**92.** Let $\sigma > 0$. If the series

$$a_1 1^{-\sigma} + a_2 2^{-\sigma} + a_3 3^{-\sigma} + \cdots + a_n n^{-\sigma} + \cdots$$

is convergent then we have

$$\lim_{t \to 1-0} (1-t)^\sigma (a_1 t + a_2 t^2 + a_3 t^3 + \cdots + a_n t^n + \cdots) = 0 \qquad [\mathbf{75}].$$

**93.** Show that

$$\lim_{t \to 1-0} \sqrt{1-t} \sum_{n=1}^{\infty} (t^{n^2} - 2t^{2n^2})$$

exists and is negative.

**94.** The two given sequences

$$a_0, a_1, a_2, \ldots, a_n, \ldots; \qquad b_0, b_1, b_2, \ldots, b_n, \ldots$$

satisfy the conditions

$$b_n > 0; \qquad \sum_{n=0}^{\infty} b_n t^n \text{ converges for all values of } t;$$

$$\lim_{n \to \infty} \frac{a_n}{b_n} = s.$$

Then $a_0 + a_1 t + a_2 t^2 + \cdots + a_n t^n + \cdots$ converges too for all values of $t$ and in addition

$$\lim_{t \to \infty} \frac{a_0 + a_1 t + a_2 t^2 + \cdots + a_n t^n + \cdots}{b_0 + b_1 t + b_2 t^2 + \cdots + b_n t^n + \cdots} = s. \qquad \text{(Cf. IV } \mathbf{72}.)$$

**95.** If $\lim_{n \to \infty} s_n = s$ exists then

$$\lim_{t \to \infty} \left( s_0 + s_1 \frac{t}{1!} + s_2 \frac{t^2}{2!} + \cdots + s_n \frac{t^n}{n!} + \cdots \right) e^{-t} = s.$$

**96.** Assume that the sum

$$a_0 + a_1 + a_2 + \cdots + a_n + \cdots = s$$

exists. Define

$$g(t) = a_0 + a_1 \frac{t}{1!} + a_2 \frac{t^2}{2!} + \cdots + a_n \frac{t^n}{n!} + \cdots.$$

Then

$$\int_0^\infty e^{-t} g(t)\, dt = s.$$

**97.** The Bessel function of order 0 is defined as

$$J_0(x) = 1 - \frac{1}{1!\,1!}\left(\frac{x}{2}\right)^2 + \frac{1}{2!\,2!}\left(\frac{x}{2}\right)^4 - \cdots + \frac{(-1)^m}{m!\,m!}\left(\frac{x}{2}\right)^{2m} + \cdots.$$

We have

$$\int_0^\infty e^{-t} J_0(t)\,dt = \frac{1}{\sqrt{2}}.$$

# Chapter 3

# The Structure of Real Sequences and Series

## § 1. The Structure of Infinite Sequences

**98.** Let the terms of the sequence $a_1, a_2, a_3, \ldots$ satisfy the condition

$$a_{m+n} \leqq a_m + a_n, \qquad m, n = 1, 2, 3, \ldots;$$

then the sequence

$$\frac{a_1}{1}, \frac{a_2}{2}, \frac{a_3}{3}, \ldots, \frac{a_n}{n}, \ldots$$

either converges to its lower bound or diverges properly to $-\infty$.

**99.** Assume that the terms of the sequence $a_1, a_2, a_3, \ldots$ satisfy the condition

$$a_m + a_n - 1 < a_{m+n} < a_m + a_n + 1.$$

Then

$$\lim_{n\to\infty} \frac{a_n}{n} = \omega$$

exists; $\omega$ is finite and we have

$$\omega n - 1 < a_n < \omega n + 1.$$

**100.** If the general term of a series which is neither convergent nor properly divergent tends to 0 the partial sums are everywhere dense between their lowest and their highest limit points.

**101.** Let $a_n > 0$, $\lim\limits_{n\to\infty} a_n = 0$ and the series $a_1 + a_2 + \cdots + a_n + \cdots$ be divergent. Put $a_1 + a_2 + \cdots + a_n = s_n$ and denote by $[s_n]$ the largest integer $\leqq s_n$. Find the limit points of the sequence

$$s_1 - [s_1],\ s_2 - [s_2], \ldots,\ s_n - [s_n], \ldots$$

**102.** Assume that there exists for the sequence $t_1, t_2, \ldots, t_n, \ldots$ a sequence of positive numbers $\varepsilon_1, \varepsilon_2, \ldots, \varepsilon_n, \ldots$, converging to 0, for which

$$t_{n+1} > t_n - \varepsilon_n \quad \text{for all } n.$$

Then the numbers $t_1, t_2, \ldots, t_n, \ldots$ are everywhere dense between their lowest and highest limit points.

**103.** Let $\nu_1, \nu_2, \ldots, \nu_n, \ldots$ be positive integers, $\nu_1 \leqq \nu_2 \leqq \nu_3 \leqq \cdots$. The set of limit points of the sequence

$$\frac{\nu_1}{1 + \nu_1}, \frac{\nu_2}{2 + \nu_2}, \ldots, \frac{\nu_n}{n + \nu_n}, \ldots$$

consists of a closed interval (of length 0 if the limit exists).

**104.** A subsequence whose terms are the successive partial sums of an absolutely convergent series can be picked out from every convergent sequence.

**105.** A sequence $t_1, t_2, \ldots, t_n, \ldots$ that diverges to $+\infty$ contains a minimum (i.e. there exists a $t_n$ such that $t_m \geqq t_n$ for all $m$).

**106.** A convergent sequence has either a maximum or a minimum or both.

The following propositions show that even the most extravagant sequences behave occasionally like good mannered sequences, i.e. they show some feature of monotone sequences.

**107.** Let $l_1, l_2, l_3, \ldots, l_m, \ldots$ be a sequence of positive numbers (positive in the sense of $> 0$) and let $\liminf_{m\to\infty} l_m = 0$. Then there are infinitely many subscripts $n$ for which $l_n$ is smaller than all the terms $l_1, l_2, l_3, \ldots, l_{n-1}$ preceding $l_n$, ($l_n < l_k$, $k = 1, 2, 3, \ldots, n - 1$).

**108.** Let $l_1, l_2, l_3, \ldots, l_m, \ldots$ be a sequence of positive numbers (positive in the sense of $> 0$) and let $\lim_{m\to\infty} l_m = 0$. Then there are infinitely many subscripts $n$ for which $l_n$ is larger than all the terms $l_{n+1}, l_{n+2}, \ldots$ following $l_n$, ($l_n > l_{n+k}$, $k = 1, 2, 3, \ldots$). (Not only the conclusion but also the hypothesis is different from the one in **107**.)

**109.** Given two sequences

$$l_1, l_2, l_3, \ldots, l_m, \ldots; \qquad l_m > 0;$$

$$s_1, s_2, s_3, \ldots, s_m, \ldots; \qquad s_1 > 0, \quad s_{m+1} > s_m, \qquad m = 1, 2, 3, \ldots$$

satisfying the conditions

$$\lim_{m\to\infty} l_m = 0, \qquad \limsup_{m\to\infty} l_m s_m = +\infty.$$

Then there are infinitely many subscripts $n$ such that two different kinds of inequalities hold simultaneously:

$$l_n > l_{n+1}, \quad l_n > l_{n+2}, \quad l_n > l_{n+3}, \ldots,$$

$$l_n s_n > l_{n-1}s_{n-1}, \quad l_n s_n > l_{n-2}s_{n-2}, \ldots, \quad l_n s_n > l_1 s_1. \qquad [\mathbf{107}, \mathbf{108}.]$$

**110.** If the sequence $\frac{L_1}{1}, \frac{L_2}{2}, \ldots, \frac{L_m}{m}, \ldots$ tends to $+\infty$ and if $A$ is larger than its minimum [**105**] then there exists a subscript $n$ (or several subscripts $n$), $n \geqq 1$, so that the quotients

$$\frac{L_n - L_{n-1}}{1}, \quad \frac{L_n - L_{n-2}}{2}, \quad \frac{L_n - L_{n-3}}{3}, \ldots, \quad \frac{L_n}{n}$$

are $\leqq A$ and the infinitely many quotients

$$\frac{L_{n+1} - L_n}{1}, \quad \frac{L_{n+2} - L_n}{2}, \quad \frac{L_{n+3} - L_n}{3}, \ldots$$

are all $\geqq A$. [The quantities in question can be interpreted as the slopes of certain connecting lines between the points with cartesian coordinates

$$(0, L_0), (1, L_1), (2, L_2), \ldots, (m, L_m), \ldots; \qquad L_0 = 0.$$

This interpretation leads to a geometric proof of the statement.]

**111.** Assume that the sequence $l_1, l_2, l_3, \ldots, l_m, \ldots$ satisfies the sole condition

$$\lim_{m\to\infty} l_m = +\infty.$$

Let $A$ be larger than $l_1$ ($A > l_1$). Then there exists a subscript $n$, $n \geqq 1$, such that all these inequalities hold simultaneously:

$$\frac{l_{n-\mu+1} + \cdots + l_{n-1} + l_n}{\mu} \leqq A \leqq \frac{l_{n+1} + l_{n+2} + \cdots + l_{n+\nu}}{\nu},$$
$$\mu = 1, 2, \ldots, n; \; \nu = 1, 2, 3, \ldots.$$

If $A$ tends to infinity then so does $n$.

**112.** Let the sequence $l_1, l_2, \ldots, l_m, \ldots$ satisfy the two conditions

$$\limsup_{m\to\infty} (l_1 + l_2 + \cdots + l_m) = +\infty, \qquad \lim_{m\to\infty} l_m = 0,$$

and assume $l_1 > A > 0$. Then there exists a subscript $n$, $n \geqq 1$, such that the inequalities

$$\frac{l_{n-\mu+1} + \cdots + l_{n-1} + l_n}{\mu} \geqq A \geqq \frac{l_{n+1} + l_{n+2} + \cdots + l_{n+\nu}}{\nu},$$
$$\mu = 1, 2, \ldots, n; \; \nu = 1, 2, 3, \ldots$$

hold simultaneously. If $A$ tends to 0, then $n$ tends to infinity.

## § 2. Convergence Exponent

The *convergence exponent* of the sequence $r_1, r_2, r_3, \ldots, r_m, \ldots$ where $0 < r_1 \leqq r_2 \leqq \cdots$, $\lim_{m\to\infty} r_m = \infty$, is defined as the number $\lambda$ having the following property:

$$r_1^{-\sigma} + r_2^{-\sigma} + r_3^{-\sigma} + \cdots + r_m^{-\sigma} + \cdots$$

converges for $\sigma > \lambda$ and diverges for $\sigma < \lambda$. (For $\sigma = \lambda$ it may converge or diverge.) For $\sigma = 0$ the series is divergent, therefore $\lambda \geqq 0$. If the series does not converge for any $\sigma$ then $\lambda = \infty$.

**113.** Show that

$$\limsup_{m\to\infty} \frac{\log m}{\log r_m} = \lambda.$$

**114.** Let $x_1, x_2, x_3, \ldots, x_m, \ldots$ be arbitrary real numbers, $x_m \neq 0$. If there exists a positive distance $\delta$ such that $|x_l - x_k| > \delta$, $l < k$, $l, k = 1, 2, 3, \ldots$, then the convergence exponent of $|x_1|, |x_2|, |x_3|, \ldots, |x_m|, \ldots$ is at most 1.

**115.** Let $\beta$ be larger than the convergence exponent of the sequence $r_1, r_2, \ldots$ Then there exist infinitely many subscripts $n$ for which the $n-1$ inequalities

$$\frac{r_1}{r_n} < \left(\frac{1}{n}\right)^{\frac{1}{\beta}}, \quad \frac{r_2}{r_n} < \left(\frac{2}{n}\right)^{\frac{1}{\beta}}, \quad \ldots, \quad \frac{r_{n-1}}{r_n} < \left(\frac{n-1}{n}\right)^{\frac{1}{\beta}}$$

are satisfied. [**107.**]

**116.** Assume that the convergence exponent $\lambda$ of the sequence $r_1, r_2, r_3, \ldots, r_m, \ldots$ is positive and $0 < \alpha < \lambda < \beta$. Then there are infinitely many subscripts such that the two types of inequalities

$$\frac{r_\mu}{r_n} > \left(\frac{\mu}{n}\right)^{\frac{1}{\alpha}} \quad \text{for } \mu = n-1,\ n-2, \ldots, 1,$$

$$\frac{r_\nu}{r_n} > \left(\frac{\nu}{n}\right)^{\frac{1}{\beta}} \quad \text{for } \nu = n+1,\ n+2,\ n+3, \ldots$$

hold simultaneously. [**109.**]

## § 3. The Maximum Term of a Power Series

**117.** Suppose $0 < r_1 < r_2 < r_3 < \cdots$. For what values of $x$, $x \geqq 0$, is the $m$-th term of the series

$$1 + \frac{x}{r_1} + \frac{x^2}{r_1 r_2} + \cdots + \frac{x^m}{r_1 r_2 \cdots r_m} + \cdots$$

larger than all the other terms, $m = 0, 1, 2, \ldots$?

**118.** Assume

$$0 < r_1 \leqq r_2 \leqq r_3 \leqq \cdots, \qquad 0 < s_1 \leqq s_2 \leqq s_3 \leqq \cdots,$$

$$\lim_{m\to\infty} \frac{r_m}{s_m} = \infty.$$

Then arbitrarily large values of $n$ and $r$ can be found for which the following inequalities hold simultaneously [**111**]:

$$\frac{r^k}{r_1 r_2 \cdots r_k} \cdot \frac{r_1 r_2 \cdots r_n}{r^n} \leqq \frac{s_n^k}{s_1 s_2 \cdots s_k} \frac{s_1 s_2 \cdots s_n}{s_n^n} \leqq 1, \qquad k = 0, 1, 2, 3, \ldots$$

(At the root of this fact, and of **122**, lies the comparison of two power series

$$1 + \frac{x}{r_1} + \frac{x^2}{r_1 r_2} + \cdots + \frac{x^m}{r_1 r_2 \cdots r_m} + \cdots,$$

$$1 + \frac{y}{s_1} + \frac{y^2}{s_1 s_2} + \cdots + \frac{y^m}{s_1 s_2 \cdots s_m} + \cdots.)$$

Suppose that $p_0 \geqq 0$, $p_1 \geqq 0, \ldots, p_m \geqq 0$, and that $p_i \neq 0$ for at least one subscript $i$. Let $\varrho$, $\varrho > 0$, possibly $\varrho = \infty$, be the radius of convergence of the power series

$$p_0 + p_1 x + p_2 x^2 + \cdots + p_m x^m + \cdots.$$

The sequence

$$p_0, p_1 x, p_2 x^2, \ldots, p_m x^m, \ldots$$

converges to 0 if $0 < x < \varrho$. Therefore there exists [**105**] a *maximum term* whose value is denoted by $\mu(x)$. I.e.

$$p_m x^m \leqq \mu(x), \qquad m = 0, 1, 2, \ldots$$

The *central subscript* $\nu(x)$ is the subscript of the maximum term, i.e. $\mu(x) = p_{\nu(x)} x^{\nu(x)}$. If several of the terms $p_m x^m$ are equal to $\mu(x)$ we call $\nu(x)$ the largest of the corresponding subscripts. More details in IV, Chap. 1.

**119.** For an everywhere convergent power series in $x$ which is not merely a polynomial the central subscript $\nu(x)$ tends to $\infty$ with $x$.

**120.** The subscript of the maximum term increases as $x$ increases. (One might consider this situation as somewhat unusual: in the course of successive changes the position of maximum importance is held by more and more capable individuals.)

**121.** The series

$$p_0 + p_1 x + p_2 x^2 + \cdots + p_m x^m + \cdots$$

with positive coefficients and finite radius of convergence $\varrho$ ($p_m > 0$, $\varrho > 0$) is such that one term after the other, all terms in turn, become maximum term. Then $\frac{1}{\varrho}$ is the radius of convergence of the series

$$\frac{1}{p_0} + \frac{x}{p_1} + \frac{x^2}{p_2} + \cdots + \frac{x^m}{p_m} + \cdots.$$

**122.** The dominance of the maximum term is more pronounced in an always convergent series than in one that does not always converge (it is strongest in a polynomial). More exactly: let the radius of convergence of the power series

$$a_0 + a_1x + a_2x^2 + \cdots + a_mx^m + \cdots$$

be infinite and that of the power series

$$b_0 + b_1y + b_2y^2 + \cdots + b_my^m + \cdots$$

be finite. Suppose $a_m \geqq 0$, $b_m > 0$, $m = 0, 1, 2, \ldots$ The coefficients $b_0, b_1, b_2, \ldots$ be such that all the terms $b_my^m$ become in turn maximum term [**120**]. Then a value $\bar{y}$ can be determined for certain arbitrarily large positive $\bar{x}$ such that for these corresponding values the respective series have the same central subscript. Let the common central subscript be $n$. Then all the following inequalities hold simultaneously:

$$\frac{a_k\bar{x}^k}{a_n\bar{x}^n} \leqq \frac{b_k\bar{y}^k}{b_n\bar{y}^n} \leqq 1, \qquad k = 0, 1, 2, \ldots$$

[Consider the maximum term of $\sum_{m=0}^{\infty} \frac{a_m}{b_m} z^m$.]

**123.** If there are values $x^*$ to which no $y$ corresponds in the sense of **122** then they are "rare". They have a finite logarithmic measure, i.e. the set of points $\log x^*$, $x^*$ exceptional value, may be covered by countably many intervals of finite total length.

## § 4. Subseries

Let $t_1, t_2, t_3, \ldots, t_n, \ldots$ be integers, $0 < t_1 < t_2 < t_3 < \cdots$. The series $a_{t_1} + a_{t_2} + a_{t_3} + \cdots + a_{t_n} + \cdots$ is called *a subseries* of the series $a_1 + a_2 + a_3 + \cdots + a_n + \cdots$.

**124.** From the harmonic series

$$\frac{1}{1} + \frac{1}{2} + \frac{1}{3} + \cdots + \frac{1}{n} + \cdots$$

remove all terms that contain the digit 9 in the decimal representation of the denominator. The resulting subseries is convergent.

**125.** If all the subseries of a series converge then the series is absolutely convergent.

**126.** Let $k$ and $l$ denote positive integers. Must the convergent series $a_1 + a_2 + a_3 + \cdots$ be absolutely convergent if all its subseries of the form

$$a_k + a_{k+l} + a_{k+2l} + a_{k+3l} + \cdots$$

(subscripts in arithmetic progression) converge?

**127.** Let $k$ and $l$ be integers, $k \geqq 1$, $l \geqq 2$. Must the convergent series $a_1 + a_2 + a_3 + \cdots$ be absolutely convergent if all its subseries of the form

$$a_k + a_{kl} + a_{kl^2} + a_{kl^3} + \cdots$$

(subscripts in geometric progression) are convergent?

**128.** Let $\varphi(x)$ denote a polynomial assuming integral values for integral $x$, $\varphi(x) = c_0x^l + c_1x^{l-1} + \cdots$ [VIII, Chap. 2]. Assume that the degree is $l \geqq 1$ and that the coefficient $c_0$ of $x^l$ is positive ($c_0 > 0$). The values $\varphi(0), \varphi(1), \varphi(2), \ldots$ form a generalized arithmetic progression of order $l$; since $c_0 > 0$ only a finite number of terms of the progression can be negative. Must a convergent series $a_1 + a_2 + a_3 + \cdots$ converge absolutely if all its subseries whose subscripts form a generalized arithmetic progression

$$a_{\varphi(0)} + a_{\varphi(1)} + a_{\varphi(2)} + a_{\varphi(3)} + \cdots$$

(omitting the terms with negative subscripts) converge?

**129.** If the series $a_1 + a_2 + a_3 + \cdots$ converges absolutely and if every subseries

$$a_l + a_{2l} + a_{3l} + \cdots, \qquad l = 1, 2, 3, \ldots$$

has the sum 0 then $a_1 = a_2 = \cdots = 0$.

**130.** Consider the set of points determined by all the subseries of

$$\frac{2}{3} + \frac{2}{9} + \frac{2}{27} + \cdots + \frac{2}{3^n} + \cdots.$$

This set is perfect and nowhere dense (closed and dense in itself, but nowhere dense in the set of all real numbers). (We have to consider all the subseries, finite and infinite, including the "empty" subseries to which we attribute the sum 0.)

**131.** Let the terms of the convergent series

$$p_1 + p_2 + p_3 + \cdots + p_n + \cdots = s$$

satisfy the inequalities

$$p_1 \geqq p_2 \geqq p_3 \geqq \cdots,$$

$$0 < p_n \leqq p_{n+1} + p_{n+2} + p_{n+3} + \cdots, \qquad n = 1, 2, 3, \ldots$$

Then it is possible to represent any number $\sigma$ in the half-closed interval

$0 < \sigma \leqq s$ by an infinite subseries:

$$p_{t_1} + p_{t_2} + p_{t_3} + \cdots = \sigma.$$

**132.** Find the series $p_1 + p_2 + p_3 + \cdots$ that satisfies the conditions

$$p_1 = \frac{1}{2}, \quad p_n = p_{n+1} + p_{n+2} + p_{n+3} + \cdots, \quad n = 1, 2, 3, \ldots,$$

and verify that in this case every $\sigma$ mentioned in **131** can be represented by one infinite subseries only.

## § 5. Rearrangement of the Terms

**132.1.** By rearranging the factors of the infinite product

$$\left(1+\frac{1}{2}\right)\left(1-\frac{1}{3}\right)\left(1+\frac{1}{4}\right)\left(1-\frac{1}{5}\right)\left(1+\frac{1}{6}\right)\cdots = P_{1,1}$$

we obtain the infinite product

$$P_{p,q} = \left(1+\frac{1}{2}\right)\left(1+\frac{1}{4}\right)\cdots\left(1+\frac{1}{2p}\right)$$
$$\times\left(1-\frac{1}{3}\right)\left(1-\frac{1}{5}\right)\cdots\left(1-\frac{1}{2q+1}\right)\left(1+\frac{1}{2p+2}\right)\cdots$$

in which blocks of $p$ factors greater than 1 alternate with blocks of $q$ factors smaller than 1. (Factors of the same kind remain in the "natural" order.) Show that

$$P_{p,q} = \sqrt{\frac{p}{q}}. \qquad \text{[II 202.]}$$

**132.2.** By rearranging the terms of the infinite series

$$\frac{1}{2} - \frac{1}{3} + \frac{1}{4} - \frac{1}{5} + \cdots = S_{1,1} = 1 - \log 2$$

we obtain the infinite series

$$S_{p,q} = \frac{1}{2} + \frac{1}{4} + \frac{1}{6} + \cdots + \frac{1}{2p} - \frac{1}{3} - \frac{1}{5} - \cdots - \frac{1}{2q+1} + \frac{1}{2p+2} + \cdots$$

in which blocks of $p$ positive terms alternate with blocks of $q$ negative terms. (Terms of the same kind remain in the "natural" order, steadily decreasing in absolute value.) Show that

$$S_{p,q} - S_{1,1} = \frac{1}{2}\log\frac{p}{q}.$$

Let $r_1, r_2, r_3, \ldots, s_1, s_2, s_3, \ldots$ be two sequences of steadily increasing natural numbers without common terms. Suppose furthermore that all positive integers appear in one or the other of the two sequences

($r_m < r_{m+1}$, $s_n < s_{n+1}$, $r_m \gtrless s_n$ for $m, n = 1, 2, 3, \ldots$). The two series

$$a_{r_1} + a_{r_2} + a_{r_3} + \cdots, \qquad a_{s_1} + a_{s_2} + a_{s_3} + \cdots$$

(the "reds" and the "blacks") are *complementary* subseries of the series $a_1 + a_2 + a_3 + \cdots$. Let $\nu_1, \nu_2, \nu_3, \ldots$ be a sequence of integers such that each natural number $1, 2, 3, \ldots$ appears once and only once in it (a permutation of the natural numbers). The series

$$a_{\nu_1} + a_{\nu_2} + a_{\nu_3} + \cdots + a_{\nu_n} + \cdots$$

is obtained from the series

$$a_1 + a_2 + a_3 + \cdots + a_n + \cdots$$

by *rearrangement*. We call special attention to the rearrangements where $a_{r_m}$ ($m = 1, 2, 3, \ldots$) is the $r_m$-th term before as well as after the rearrangement, i.e. which preserve the subseries $a_{r_1} + a_{r_2} + a_{r_3} + \cdots$. If before as well as after the rearrangement $a_{r_m}$ precedes $a_{r_n}$ and $a_{s_m}$ precedes $a_{s_n}$ for all number pairs $m, n$, $m < n$, we say that the rearrangement *shifts* the two complementary subseries *relatively to each other* (and it leaves each in its original order).

**133.** If one of the two to each other complementary subseries of a convergent series is convergent the other is convergent too. A rearrangement which only shifts the two subseries relatively to each other does not change the sum of such a series.

**134.** If one of the two to each other complementary subseries of a conditionally convergent series diverges to $+\infty$ then the other diverges to $-\infty$. Provided that all the terms of one of these two subseries are of the same sign it is possible to obtain an arbitrary sum for the whole series by shifting the two subseries relatively to each other.

**135.** It is not possible to accelerate by rearrangement the divergence of a divergent series with positive monotone decreasing terms.

**136.** By rearranging the series we can slow down arbitrarily the divergence of a divergent series with positive terms which tend to 0. More explicitely: Assume

$$p_n > 0, \qquad \lim_{n\to\infty} p_n = 0, \qquad \lim_{n\to\infty} (p_1 + p_2 + \cdots + p_n) = \infty,$$

$$0 < Q_1 < Q_2 < \cdots < Q_n < \cdots, \qquad \lim_{n\to\infty} Q_n = \infty.$$

Then there exists a rearranged series $p_{\nu_1} + p_{\nu_2} + p_{\nu_3} + \cdots + p_{\nu_n} + \cdots$ such that $\quad p_{\nu_1} + p_{\nu_2} + \cdots + p_{\nu_n} \leqq Q_n \quad$ for $\quad n = 1, 2, 3 \ldots$

**137.** Assume that

$$a_1 + a_2 + a_3 + \cdots + a_n + \cdots = s \text{ is convergent,}$$
$$\text{and } |a_1| + |a_2| + |a_3| + \cdots + |a_n| + \cdots \quad \text{is divergent.}$$

Let $s' < s < s''$. By a rearrangement that leaves all the negative terms at their places the sum $s'$ can be realized; by a rearrangement that leaves all the positive terms at their places the sum $s''$ can be realized [**136**].

### § 6. Distribution of the Signs of the Terms

**138.** Assume that $p_n > 0$, $p_1 \geqq p_2 \geqq p_3 \geqq \cdots$ and that the series

$$p_1 + p_2 + p_3 + \cdots + p_n + \cdots$$

is divergent and the series

$$\varepsilon_1 p_1 + \varepsilon_2 p_2 + \varepsilon_3 p_3 + \cdots + \varepsilon_n p_n + \cdots,$$

where $\varepsilon_k$ is $-1$ or $+1$, is convergent. If under these conditions a certain percentage of terms is positive then it is 50%. More precisely:

$$\liminf_{n\to\infty} \frac{\varepsilon_1 + \varepsilon_2 + \cdots + \varepsilon_n}{n} \leqq 0 \leqq \limsup_{n\to\infty} \frac{\varepsilon_1 + \varepsilon_2 + \cdots + \varepsilon_n}{n}.$$

**139.** Suppose $p_n > 0$, $p_1 \geqq p_2 \geqq p_3 \geqq \cdots$ and that the series

$$\varepsilon_1 p_1 + \varepsilon_2 p_2 + \varepsilon_3 p_3 + \cdots + \varepsilon_n p_n + \cdots,$$

where $\varepsilon_k$ is $-1$ or $+1$, is convergent. Then

$$\lim_{n\to\infty} (\varepsilon_1 + \varepsilon_2 + \varepsilon_3 + \cdots + \varepsilon_n)\, p_n = 0.$$

(Notice the two wellknown extreme cases

$$\varepsilon_1 = \varepsilon_2 = \varepsilon_3 = \cdots \quad \text{and} \quad \varepsilon_1 = -\varepsilon_2 = \varepsilon_3 = -\varepsilon_4 = \cdots.)$$

# Chapter 4

# Miscellaneous Problems

## § 1. Enveloping Series

We say that the series $a_0 + a_1 + a_2 + \cdots$ *envelops* the number $A$ if the relations

$$|A - (a_0 + a_1 + a_2 + \cdots + a_n)| < |a_{n+1}|, \qquad n = 0, 1, 2, \ldots$$

are satisfied. The enveloping series may be convergent or divergent; if it converges its sum is $A$. Assume that $A, a_0, a_1, a_2, \ldots$ are all real. If we have

$$A - (a_0 + a_1 + a_2 + \cdots + a_n) = \theta_n a_{n+1}, \quad \text{for all } n = 0, 1, 2, \ldots$$
$$\text{and } 0 < \theta_n < 1,$$

the number $A$ is enveloped by the series $a_0 + a_1 + a_2 + \cdots$, and in fact it lies between two consecutive partial sums. In this situation we say the series is *enveloping* $A$ in the *strict sense.* G. A. Scott and G. N. Watson [Quart. J. pure appl. Math. (London) Vol. 47, p. 312 (1917)] use the expression "arithmetically asymptotic" for a ciosely related concept. The terms of a strictly enveloping series have necessarily alternating signs.

**140.** Suppose that $f(x)$ is a real function of the real variable $x$. If the functions $|f'(x)|, |f''(x)|, \ldots$ are steadily and strictly decreasing in the interval $[0, x]$, $x > 0$, then $f(x)$ is enveloped in the strict sense by its Maclaurin series.

**141.** The functions

$$e^{-x}; \log(1 + x); (1 + x)^{-p}, \qquad p > 0;$$

are strictly enveloped by their Maclaurin series for $x > 0$.

**142** (continued). Prove the same for the functions[1]

$$\cos x, \sin x.$$

**143** (continued). Prove the same for the functions

$$\text{arc tan } x, \quad J_0(x) = 1 - \frac{1}{1!\,1!}\left(\frac{x}{2}\right)^2 + \frac{1}{2!\,2!}\left(\frac{x}{2}\right)^4 - \cdots \qquad [\mathbf{141, 142}].$$

**144.** Suppose that the terms of the series $a_0 + a_1 + a_2 + \cdots$ are alternately positive and negative and that there exists a number $A$ such that

$$A - (a_0 + a_1 + a_2 + \cdots + a_n)$$

assumes always the same sign as the next term, $a_{n+1}$. Then the series envelops $A$ in the strict sense.

**145.** If the series $a_0 + a_1 + a_2 + \cdots$, $a_n$ real, $n = 0, 1, 2, \ldots$, envelops the real number $A$ and if in addition $|a_1| > |a_2| > |a_3| > \cdots$

[1] Obviously only the non-vanishing terms of the Maclaurin series are to be considered. E.g. the $n$-th partial sum of the Maclaurin series for $\cos x$ is

$$1 - \frac{x^2}{2!} + \frac{x^4}{4!} - \cdots + (-1)^n \frac{x^{2n}}{(2n)!}, \qquad n = 0, 1, 2, \ldots$$

then the terms $a_1, a_2, a_3, \ldots$ have alternating signs and $A$ is enveloped in the strict sense.

**146.** Let the function $f(x)$ assume real values for real $x$, $x > R > 0$. If $f(x)$ is enveloped for $x > R$ by the real series $a_0 + \frac{a_1}{x} + \frac{a_2}{x^2} + \frac{a_3}{x^3} + \cdots$ then the numbers $a_1, a_2, a_3, \ldots$ have alternating signs and the series is strictly enveloping.

**147.** Suppose that the real valued function $f(t)$ is infinitely often differentiable for $t \geqq 0$ and that all its derivatives $f^{(n)}(t)$ $(n = 0, 1, 2, \ldots)$ have decreasing absolute values and converge to 0 for $t \to \infty$. Then the integral

$$\int_0^\infty f(t) \cos xt \, dt$$

is, for real $x$, strictly enveloped by the series

$$-\frac{f'(0)}{x^2} + \frac{f'''(0)}{x^4} - \frac{f^{\mathrm{V}}(0)}{x^6} + \frac{f^{\mathrm{VII}}(0)}{x^8} - \cdots.$$

(Example: $f(t) = e^{-t}$.)

**148.** The number $\frac{2}{3}$ is enveloped by the series

$$\frac{3}{4} + \frac{1}{4} - \frac{3}{8} - \frac{1}{8} + \frac{3}{16} + \frac{1}{16} - \frac{3}{32} - \frac{1}{32} + \cdots,$$

but not in the strict sense.

**149**[1]. Plot the first seven terms of the series

$$e^i = 1 + \frac{i}{1!} - \frac{1}{2!} - \frac{i}{3!} + \frac{1}{4!} + \frac{i}{5!} - \frac{1}{6!} - \cdots$$

successively as complex numbers and compute so the value of $e^i$ to three decimals.

**150.** Let $\mathfrak{H}$ denote a ray with origin $z = 0$. Assume that along $\mathfrak{H}$ all the derivatives of the function $f(z)$ attain the maximum of their absolute values at the origin and only there; i.e.

$$|f^{(n)}(0)| > |f^{(n)}(z)|$$

whenever $z$ is on $\mathfrak{H}$ and $|z| > 0$. Then:

a) The function $f(t)$ is enveloped for every $z$ on $\mathfrak{H}$ by the Maclaurin series

$$f(0) + \frac{f'(0)}{1!} z + \frac{f''(0)}{2!} z^2 + \cdots \qquad \textbf{[140]}.$$

[1] In **149**–**155** the terms of the series are complex numbers; they are regarded as points in the Gaussian plane (complex plane) [III **1** et seqq.].

b) the function $F(z) = \int_0^\infty e^{-t} f\left(\frac{t}{z}\right) dt$ is for every $z$ on $\overline{\mathfrak{H}}$ enveloped by the series

$$f(0) + \frac{f'(0)}{z} + \frac{f''(0)}{z^2} + \frac{f'''(0)}{z^3} + \cdots \qquad [\mathbf{147}];$$

$\overline{\mathfrak{H}}$ is the mirror image of $\mathfrak{H}$ with respect to the $x$-axis; the integral is taken along the positive $t$-axis and converges under our assumptions on $f(z)$.

**151.** The Maclaurin series of $e^{-z}$, $\log(1+z)$ and $(1+z)^{-p}$, $p > 0$, envelop the respective functions for $\Re z \geqq 0$, $z \neq 0$.

**152.** Let $z$ be restricted by the following conditions:

$$-\frac{\pi}{4} \leqq \arg z \leqq \frac{\pi}{4}, \qquad \frac{3\pi}{4} \leqq \arg z \leqq \frac{5\pi}{4}, \qquad z \neq 0.$$

Then the function $e^{\frac{z^2}{2}} \int_z^\infty e^{-\frac{t^2}{2}} dt$ is enveloped by the series

$$\frac{1}{z} - \frac{1}{z^3} + \frac{1 \cdot 3}{z^5} - \frac{1 \cdot 3 \cdot 5}{z^7} + \cdots$$

(strictly enveloped if $z$ is real).

**153.** Suppose that $a_n$ and $b_n$ are arbitrary complex numbers that have the same argument, i.e. $\frac{a_n}{b_n}$ is real and positive. If at a certain point $z \neq 0$ the two series

$$a_0 + a_1 z + a_2 z^2 + \cdots + a_n z^n + \cdots, \qquad b_0 + b_1 z + b_2 z^2 + \cdots + b_n z^n + \cdots$$

envelop the values $\varphi(z)$ and $\psi(z)$ then the combined series

$$a_0 + b_0 + (a_1 + b_1) z + (a_2 + b_2) z^2 + \cdots + (a_n + b_n) z^n + \cdots$$

envelops the value $\varphi(z) + \psi(z)$ for this $z$. (The same is true for enveloping in the strict sense if all the coefficients and $z$ are real.)

**154.** If $z$ lies in the sectors described in **152** the function $z \coth z$ is enveloped by its power series

$$z \coth z = z \frac{e^z + e^{-z}}{e^z - e^{-z}} = 1 + B_1 \frac{(2z)^2}{2!} - B_2 \frac{(2z)^4}{4!} + B_3 \frac{(2z)^6}{6!} + \cdots$$

(strictly enveloped if $z$ is real). The coefficients $B_1, B_2, B_3, \ldots$ are called the Bernoulli numbers.

**155.** The function

$$\omega(z) = \log \Gamma(1+z) - (z + \tfrac{1}{2}) \log z + z - \tfrac{1}{2} \log(2\pi) \qquad [\text{II } \mathbf{31}]$$

can, for $\Re z > 0$, be written as an integral

$$\omega(z) = 2\int_0^\infty \frac{\arctan\frac{t}{z}}{e^{2\pi t} - 1}\, dt.$$

[Cf. E. T. Whittaker and G. N. Watson, pp. 251—252.] Prove that the resulting (divergent) Stirling series

$$\frac{B_1}{1\cdot 2\cdot z} - \frac{B_2}{3\cdot 4\cdot z^3} + \frac{B_3}{5\cdot 6\cdot z^5} - \cdots$$

envelops the function $\omega(z)$ if $\Re z > 0$ and $-\frac{\pi}{4} \leqq \arg z \leqq \frac{\pi}{4}$.

## § 2. Various Propositions on Real Series and Sequences

**156.** Assume that $\varphi(x)$ is defined for positive $x$ and that for $x$ large enough it can be represented in the form

$$\varphi(x) = a_0 + \frac{a_1}{x} + \frac{a_2}{x^2} + \cdots + \frac{a_k}{x^k} + \cdots$$

$a_k$ real, $k = 0, 1, 2, \ldots$ The infinite series

$$\varphi(1) + \varphi(2) + \varphi(3) + \cdots + \varphi(n) + \cdots$$

converges if and only if $a_0 = a_1 = 0$.

**157** (continued). Suppose $\varphi(n) \neq 0$. The infinite product

$$\varphi(1)\, \varphi(2)\, \varphi(3) \cdots \varphi(n) \cdots$$

converges if and only if $a_0 = 1$, $a_1 = 0$.

**158** (continued). Under what conditions does the following infinite series converge

$$\varphi(1) + \varphi(1)\,\varphi(2) + \varphi(1)\,\varphi(2)\,\varphi(3) + \cdots + \varphi(1)\,\varphi(2)\,\varphi(3)\cdots\varphi(n) + \cdots?$$

**159.** For which positive values of $\alpha$ does the following series converge

$$\sum_{n=1}^{\infty} (2 - e^{\alpha})\,(2 - e^{\alpha/2}) \cdots (2 - e^{\alpha/n})\,?$$

**160.**

$$\int_0^1 x^{-x}\, dx = \sum_{n=1}^{\infty} n^{-n}.$$

**161.** Considering positive square roots only we find

$$\sqrt{1 + \sqrt{1 + \sqrt{1 + \cdots}}} = 1 + \cfrac{1}{1 + \cfrac{1}{1 + \ddots}}.$$

**162.** Let $a_1, a_2, \ldots, a_n, \ldots$ be positive numbers and put

$$t_n = \sqrt{a_1 + \sqrt{a_2 + \cdots + \sqrt{a_n}}},$$

where only positive square roots are considered. The convergence of the sequence

$$(t) \qquad t_1, t_2, t_3, \ldots, t_n, \ldots$$

is related to

$$\limsup_{n\to\infty} \frac{\log\log a_n}{n} = \alpha$$

in the following way:

if $\alpha < 2$ then $(t)$ converges,

if $\alpha > 2$ then $(t)$ diverges.

**163** (continued). The sequence $(t)$ is certainly convergent if the series

$$\sum_{n=1}^{\infty} 2^{-n} a_n (a_1 a_2 \cdots a_n)^{-\frac{1}{n}} \text{ converges.}$$

**164.** We have for $0 < q < 1$

$$\frac{1-q}{1+q}\left(\frac{1-q^2}{1+q^2}\right)^{\frac{1}{2}}\left(\frac{1-q^4}{1+q^4}\right)^{\frac{1}{4}}\left(\frac{1-q^8}{1+q^8}\right)^{\frac{1}{8}}\cdots = (1-q)^2.$$

**165.** The series

$$\cdots + f''(x) + f'(x) + f(x) + \int_0^x f(\xi_1)\, d\xi_1 + \int_0^x d\xi_1 \int_0^{\xi_1} f(\xi_2)\, d\xi_2 + \cdots$$

is infinite in two directions. Suppose it does converge uniformly on some interval. Which function does the series represent?

**166.** Let $\varphi_n(x)$ and $\psi_n(x)$ denote the polynomials of degree $n$, $n = 0, 1, 2, \ldots$ defined by the formulas

$$\varphi_0(x) = 1, \quad \varphi_n'(x) = \varphi_{n-1}(x), \quad \varphi_n(0) = 0,$$
$$\psi_0(x) = 1, \quad \psi_n(x+1) - \psi_n(x) = \psi_{n-1}(x), \quad \psi_n(0) = 0, \quad n = 1, 2, 3, \ldots$$

Find the two sums $\varphi(x)$ and $\psi(x)$,

$$\varphi(x) = \varphi_0(x) + \varphi_1(x) + \cdots + \varphi_n(x) + \cdots,$$
$$\psi(x) = \psi_0(x) + \psi_1(x) + \cdots + \psi_n(x) + \cdots.$$

**167.** We define

$$x_n = y_n e^{-\frac{1}{12n}}, \qquad y_n = n!\, n^{-n-\frac{1}{2}} e^n, \qquad n = 1, 2, 3, \ldots$$

Then each interval $(x_k, y_k)$, $k = 1, 2, 3, \ldots$, contains the interval $(x_{k+1}, y_{k+1})$ as a subinterval.

**168.** The sequence

$$a_n = \left(1 + \frac{1}{n}\right)^{n+p}, \qquad n = 1, 2, 3, \ldots$$

is monotone decreasing if and only if $p \geqq \frac{1}{2}$.

**169.** The sequence

$$a_n = \left(1 + \frac{1}{n}\right)^n \left(1 + \frac{x}{n}\right), \qquad n = 1, 2, 3, \ldots$$

is monotone decreasing if and only if $x \geqq \frac{1}{2}$.

**170.** Let $n$ be a positive integer. Then we have

$$\frac{e}{2n+2} < e - \left(1 + \frac{1}{n}\right)^n < \frac{e}{2n+1}.$$

**171.** As is well-known, the number $e = \lim_{n\to\infty} \left(1 + \frac{1}{n}\right)^n$ is contained in the interval

$$\left(1 + \frac{1}{n}\right)^n < e < \left(1 + \frac{1}{n}\right)^{n+1} \qquad [168]$$

In which quarter of the interval is it contained?

**172.** The sequence

$$a_n = \left(1 + \frac{x}{n}\right)^{n+1}, \qquad n = 1, 2, 3, \ldots$$

is monotone decreasing if and only if $0 < x \leqq 2$.

**173.** Prove that the $n$-times iterated sine function

$$\sin_n x = \sin(\sin_{n-1} x), \qquad \sin_1 x = \sin x,$$

converges to 0 as $n \to \infty$ if $\sin x > 0$ and that, moreover,

$$\lim_{n\to\infty} \sqrt{\frac{n}{3}} \sin_n x = 1.$$

**174.** Assume that $0 < f(x) < x$ and

$$f(x) = x - ax^k + bx^l + x^l \varepsilon(x), \qquad \lim_{x\to 0} \varepsilon(x) = 0$$

for $0 < x < x_0$, where $1 < k < l$ and $a$, $b$ positive. Put

$$v_0 = x, \quad v_1 = f(v_0), \quad v_2 = f(v_1), \ldots, \quad v_n = f(v_{n-1}), \ldots$$

Then we have for $n \to \infty$

$$n^{\frac{1}{k-1}} v_n \to [(k-1)\,a]^{-\frac{1}{k-1}}.$$

**175.** Discuss the convergence of the series

$$v_1^s + v_2^s + v_3^s + \cdots,$$

where

$$v_1 = \sin x, \quad v_2 = \sin \sin x, \ldots, \quad v_n = \sin v_{n-1}, \ldots$$

We can obviously assume that $v_1 > 0$.

**176.** Prove the formula

$$e^x - 1 = u_1 + u_1 u_2 + u_1 u_2 u_3 + \cdots$$

where

$$u_1 = x \gtrless 0, \quad u_{n+1} = \log \frac{e^{u_n} - 1}{u_n}, \qquad n = 1, 2, 3, \ldots$$

**177.** Compute

$$s = \cos^3 \varphi - \frac{1}{3} \cos^3 3\varphi + \frac{1}{3^2} \cos^3 3^2\varphi - \frac{1}{3^3} \cos^3 3^3 \varphi + \cdots.$$

**178.** Let $a_n$, $b_n$, $b_n \neq 0$, $n = 0, 1, 2, 3, \ldots$, be two sequences that satisfy the conditions:

a) the power series $f(x) = \sum_{n=0}^{\infty} a_n x^n$ has a positive radius of convergence $r$.

b) the limit

$$\lim_{n \to \infty} \frac{b_n}{b_{n+1}} = q$$

exists and $|q| < r$.

We define

$$c_n = a_0 b_n + a_1 b_{n-1} + \cdots + a_n b_0, \qquad n = 0, 1, 2, \ldots$$

Then $\frac{c_n}{b_n}$ converges to $f(q)$ as $n \to \infty$.

**179.** Let

$$f_n(x) = a_{n1} x + a_{n2} x^2 + a_{n3} x^3 + \cdots, \qquad n = 1, 2, 3, \ldots$$

be arbitrary functions, $|a_{nk}| < A$ for all positive integral values of $n$ and $k$ and

$$\lim_{n \to \infty} f_n(x) = 0 \quad \text{if} \quad 0 < x < 1.$$

Then we have for fixed $k$, $k = 1, 2, 3, \ldots$

$$\lim_{n \to \infty} a_{nk} = 0.$$

**180.** Suppose that the series

$$a_{n0} + a_{n1} + a_{n2} + \cdots + a_{nk} + \cdots = s_n, \quad n = 0, 1, 2, \ldots,$$

have a common convergent majorant

$$A_0 + A_1 + A_2 + \cdots + A_k + \cdots = S,$$

i.e. for each $k$ the inequalities $|a_{nk}| \leqq A_k$ hold simultaneously for all $n$. Assume furthermore that

$$\lim_{n\to\infty} a_{nk} = a_k$$

exists for $k = 0, 1, 2, \ldots$ Then the series

$$a_0 + a_1 + a_2 + \cdots + a_k + \cdots = s$$

is convergent and

$$\lim_{n\to\infty} s_n = s.$$

**181.** Justify the limits in **53** and **59**.

**181.1.** Assume that

$$a_{i1} + a_{i2} + \cdots + a_{in} + \cdots = s_i$$

converges for $i = 1, 2, 3, \ldots$, define $U_i$ as the least upper bound of

$$|a_{i1} + a_{i2} + \cdots + a_{in}|, \qquad n = 1, 2, 3, \ldots,$$

and assume that

$$U_1 + U_2 + \cdots + U_n + \cdots$$

converges. Then the series

$$(*) \quad a_{11} + a_{12} + a_{21} + a_{13} + a_{22} + a_{31} + \cdots + a_{1n} + a_{2,n-1} + \cdots,$$

which you obtain by arranging the numbers in the array

$$\begin{array}{l} a_{11}\ a_{12}\ a_{13} \ldots \ldots a_{1n} \ldots \\ a_{21}\ a_{22}\ a_{23} \ldots \ldots a_{2,n-1} \ldots \\ a_{31}\ a_{32}\ a_{33} \ldots \ldots \ldots \\ \ldots \ldots \ldots \ldots \ldots \ldots \end{array}$$

"diagonally", converges and its sum is

$$s_1 + s_2 + \cdots + s_n + \cdots.$$

(The interesting point is that the absolute convergence of the double series $\Sigma\Sigma a_{ik}$ is not assumed.)

**182.** If $\alpha$ is fixed, $\alpha > 0$, and $n$ an integer increasing to $+\infty$, then

$$\sum_{\nu=1}^{\infty}{}' \nu^{\alpha-1}(n^\alpha - \nu^\alpha)^{-2} \sim \frac{n^{1-\alpha}}{3}\left(\frac{\pi}{\alpha}\right)^2.$$

The meaningless term with subscript $\nu = n$ has to be omitted, which is indicated by the comma at the summation sign. Notice the case $\alpha = 1$.

**183.** Let the numbers $\varepsilon_0, \varepsilon_1, \varepsilon_2, \ldots, \varepsilon_n, \ldots$ assume one of the three values $-1, 0, 1$. Then we have

$$\varepsilon_0\sqrt{2+\varepsilon_1\sqrt{2+\varepsilon_2\sqrt{2+\cdots}}} = 2\sin\left(\frac{\pi}{4}\sum_{n=0}^{\infty}\frac{\varepsilon_0\varepsilon_1\varepsilon_2\cdots\varepsilon_n}{2^n}\right).$$

(The left hand side must be interpreted as the limit of

$$\varepsilon_0\sqrt{2+\varepsilon_1\sqrt{2+\varepsilon_2\sqrt{2+\cdots+\varepsilon_n\sqrt{2}}}}, \qquad n = 0, 1, 2, \ldots$$

for $n \to \infty$. These expressions are well defined for all $n$. Non-negative square roots are used throughout.)

**184.** Every value $x$ of the interval $[-2, 2]$ can be written in the form

$$x = \varepsilon_0\sqrt{2+\varepsilon_1\sqrt{2+\varepsilon_2\sqrt{2+\cdots}}}$$

where $\varepsilon_k$, $k = 0, 1, 2, \ldots$, is either $-1$ or $+1$. The representation is unique if $x$ is not of the form $2\cos\frac{p}{2^q}\pi$, $p, q$ integers, $0 < p < 2^q$. These are the only numbers that may be written in the finite form

$$\varepsilon_0\sqrt{2+\varepsilon_1\sqrt{2+\varepsilon_2\sqrt{2+\cdots+\varepsilon_n\sqrt{2}}}}.$$

The finite representation can be extended to an infinite one in two ways: by putting either

$$\varepsilon_{n+1} = 1, \qquad \varepsilon_{n+2} = -1, \qquad \varepsilon_{n+3} = \varepsilon_{n+4} = \cdots = 1,$$

or

$$\varepsilon_{n+1} = -1, \qquad \varepsilon_{n+2} = -1, \qquad \varepsilon_{n+3} = \varepsilon_{n+4} = \cdots = 1.$$

**185** (continued). The number $x$ is of the form $x = 2\cos k\pi$, $k$ rational, if and only if the sequence $\varepsilon_0, \varepsilon_1, \varepsilon_2, \ldots$ is periodical after a certain term.

**185.1.** Construct a sequence of real numbers $a_1, a_2, \ldots, a_n, \ldots$ so that the series

$$a_1^l + a_2^l + \cdots + a_n^l + \cdots$$

diverges for $l = 5$, but converges when $l$ is any odd positive integer different from 5.

**185.2** (continued). Let the set of all odd positive integers be divided (arbitrarily) into two complementary subsets $C$ and $D$ (having no element in common). Show that there exists a sequence of real numbers $a_1, a_2, \ldots, a_n, \ldots$ such that the series mentioned converges when $l$ belongs to $C$ and diverges when $l$ belongs to $D$.

## § 3. Partitions of Sets, Cycles in Permutations

A *partition* of a set $S$ is formed by disjoint subsets of $S$ the union of which is $S$. "Disjoint" (or non-overlapping) means that the intersection of any two subsets involved is the empty set. If $k$ subsets are involved in the partition none of which is the empty set we speak of a *partition into k classes.*

We let $S_k^n$ stand for the number of different partitions of a set of $n$ elements into $k$ classes and $T_n$ for the total number of its different partitions (into any number of classes):

$$T_n = S_1^n + S_2^n + \cdots + S_n^n.$$

The $S_k^n$ are called the *Stirling numbers of the second kind.*

***186.** Tabulate the numbers $S_k^n$ for $n \leqq 8, 1 \leqq k \leqq n$.

***187.** Obviously

$$S_1^n = S_n^n = 1.$$

Show that

$$S_k^{n+1} = S_{k-1}^n + kS_k^n.$$

***188.** Show that

$$\frac{S_k^k}{z^{k+1}} + \frac{S_k^{k+1}}{z^{k+2}} + \cdots + \frac{S_k^n}{z^{n+1}} + \cdots = \frac{1}{z(z-1)(z-2)\cdots(z-k)}.$$

***189.** Show that, for $n \geqq 1$,

$$S_k^n = \frac{1}{k!}\left[k^n - \binom{k}{1}(k-1)^n + \binom{k}{2}(k-2)^n - \cdots + (-1)^k 0^n\right].$$

***190.** Show that

$$\sum_{n=k}^{\infty} \frac{S_k^n z^n}{n!} = \frac{(e^z - 1)^k}{k!}.$$

***191.** Prove the identity in $x$

$$x^n = S_1^n x + S_2^n x(x-1) + \cdots + S_n^n x(x-1)\cdots(x-n+1).$$

[**189**, III **221**.]

***192.** Prove the identity of **191** independently of **189** [by a combinatorial argument] and derive hence a new proof for **189**.

***193.** Define $T_0 = 1$. Then

$$T_0 + \frac{T_1 z}{1!} + \frac{T_2 z^2}{2!} + \cdots + \frac{T_n z^n}{n!} + \cdots = e^{e^z - 1}.$$

***194.** Show [by a combinatorial argument] that

$$T_{n+1} = \binom{n}{0}T_n + \binom{n}{1}T_{n-1} + \binom{n}{2}T_{n-2} + \cdots + \binom{n}{n}T_0.$$

***195.** Use **194** to prove **193** [34].

***196.** Show that for $n \geqq 1$

$$T_n = \frac{1}{e}\left(\frac{1^n}{1!} + \frac{2^n}{2!} + \frac{3^n}{3!} + \cdots\right).$$

We let $s_k^n$ stand for the number of those permutations of $n$ elements that are the products of $k$ disjoint cycles[1]. Obviously

$$s_1^n + s_2^n + s_3^n + \cdots + s_n^n = n!.$$

The $s_k^n$ are called the *Stirling numbers of the first kind.*

There are $n$ persons seated around $k$ round tables (where all seats are equal) so that at least one person sits at each table. We may regard as essential

(1) who sits next to whom, and whether $A$ is the left-hand or right-hand neighbor of $B$.

Or we regard as essential only

(2) which people sit at the same table, no matter in which order.

The number of different seatings is $s_k^n$ in case (1) and $S_k^n$ in case (2). Obviously

$$s_k^n \geqq S_k^n.$$

***197.** Tabulate the numbers $s_k^n$ for $n \leqq 8$, $1 \leqq k \leqq n$.

***198.** Obviously

$$s_1^n = (n-1)!, \qquad s_n^n = 1.$$

Show that

$$s_k^{n+1} = s_{k-1}^n + n s_k^n.$$

***199.** Show that

$$x(x+1)(x+2)\cdots(x+n-1) = s_1^n x + s_2^n x^2 + \cdots + s_n^n x^n$$

or, which is the same, that

$$x(x-1)(x-2)\cdots(x-n+1) = (-1)^{n-1} s_1^n x + \cdots - s_{n-1}^n x^{n-1} + s_n^n x^n.$$

***200.** Show that

$$\sum_{n=k}^{\infty} \frac{s_k^n z^n}{n!} = \frac{1}{k!}\left(\log\frac{1}{1-z}\right)^k.$$

(Compare **200** with **190**, **199** with **188**, and again **199** with **191**. See also VII **54.2** and VIII **247.1**.)

Define $\tilde{S}_k^n$ as the number of partitions of a set of $n$ elements into $k$ classes each of which contains more than one element.

---

[1] See, e.g., Garrett Birkhoff and Saunders MacLane: A Survey of Modern Algebra, 3rd Ed. New York: Macmillan 1965, p. 137.

***201.** Tabulate the numbers $\tilde{S}_k^n$ for $n \leqq 8, 1 \leqq k \leqq n$.

***202.** Obviously

$$\tilde{S}_1^n = 1 \text{ when } n \geqq 2, \qquad \tilde{S}_k^n = 0 \text{ when } n < 2k.$$

Show that

$$\tilde{S}_k^{n+1} = n\tilde{S}_{k-1}^{n-1} + k\tilde{S}_k^n.$$

***203.** Show that

$$\tilde{S}_2^n = 2^{n-1} - n - 1, \qquad \tilde{S}_n^{2n} = 1 \cdot 3 \cdot 5 \cdots (2n-1).$$

***204.** Returning to $S_k^n$, show that

$$S_{n-1}^n = \binom{n}{2}, \qquad S_{n-2}^n = \binom{n}{3}\frac{3n-5}{4}, \qquad S_{n-3}^n = \binom{n}{4}\frac{(n-2)(n-3)}{2}$$

and, generally, that $S_{n-a}^n$ is a polynomial in $n$ of degree $2a$ divisible by $n(n-1)(n-2)\cdots(n-a)$; of course, $a \geqq 1$.

***205.** Define $\tilde{T}_0 = 1$,

$$\tilde{T}_n = \tilde{S}_1^n + \tilde{S}_2^n + \cdots + \tilde{S}_n^n$$

for $n \geqq 1$. Compute $T_n$ and $\tilde{T}_n$ for $n \leqq 8$.

***206.** $\tilde{T}_{n+1} = \binom{n}{1}\tilde{T}_{n-1} + \binom{n}{2}\tilde{T}_{n-2} + \cdots + \binom{n}{n-1}\tilde{T}_1 + \binom{n}{n}\tilde{T}_0$.

***207.** $\tilde{T}_0 + \frac{\tilde{T}_1 z}{1!} + \frac{\tilde{T}_2 z^2}{2!} + \cdots + \frac{\tilde{T}_n z^n}{n!} + \cdots = e^{e^z - 1 - z}$.

***208.** $\tilde{T}_n = \binom{n}{0}T_n - \binom{n}{1}T_{n-1} + \binom{n}{2}T_{n-2} - \cdots + (-1)^n\binom{n}{n}T_0$.

($\tilde{T}_n = \Delta^n T_0$, if we use the notation of the calculus of finite differences, see introduction to III **220**.)

***209.** If the function $F(t)$ has an $n$-th derivative

$$\left(\frac{d}{dx}\right)^n F(e^x) = S_1^n F'(e^x)\, e^x + S_2^n F''(e^x)\, e^{2x} + \cdots + S_n^n F^{(n)}(e^x)\, e^{nx}.$$

***210.** Derive the identities in the variables $z$ and $w$:

$$\sum_{n=0}^{\infty} \sum_{k \leqq n} \binom{n}{k} \frac{w^k z^n}{n!} = e^{z(w+1)}, \tag{1}$$

$$\sum_{n=0}^{\infty} \sum_{k \leqq n} S_k^n \frac{w^k z^n}{n!} = e^{(e^z-1)w}, \tag{2}$$

$$\sum_{n=0}^{\infty} \sum_{k \leqq n} s_k^n \frac{w^k z^n}{n!} = (1-z)^{-w} \tag{3}$$

either on the basis of the foregoing [**10, 190, 200**] or independently of the foregoing.

(On the other hand, we could regard (1), (2) and (3) as defining $\binom{n}{k}$, $S_k^n$ and $s_k^n$, respectively, and then take them as basis and starting point for establishing the propositions, especially the combinatorial propositions, discussed in the section herewith concluded.)

Part Two

# Integration

## Chapter 1

## The Integral as the Limit of a Sum of Rectangles

### § 1. The Lower and the Upper Sum

Let $f(x)$ be a bounded function on the finite interval $[a, b]$. The points with abscissae $x_0, x_1, x_2, \ldots, x_{n-1}, x_n$, where

$$a = x_0 < x_1 < x_2 < \cdots < x_{n-1} < x_n = b,$$

constitute a subdivision of this interval. Denote by $m_\nu$ and $M_\nu$ the greatest lower and the least upper bound, respectively, of $f(x)$ on the $\nu$-th subinterval $[x_{\nu-1}, x_\nu]$, $\nu = 1, 2, \ldots, n$. We call

$$L = \sum_{\nu=1}^{n} m_\nu (x_\nu - x_{\nu-1}) \quad \text{the } \textit{lower sum},$$

$$U = \sum_{\nu=1}^{n} M_\nu (x_\nu - x_{\nu-1}) \quad \text{the } \textit{upper sum}$$

belonging to the subdivision $x_0, x_1, x_2, \ldots, x_{n-1}, x_n$. Any upper sum is always larger (not smaller) than any lower sum, regardless of the subdivision considered. If there exists *only one* number which is neither smaller than any lower sum nor larger than any upper sum, then this number, denoted by the symbol

$$\int_a^b f(x)\, dx,$$

is called the *definite integral* of $f(x)$ over the interval $[a, b]$ and $f(x)$ is called (properly) integrable over $[a, b]$ in the sense of Riemann.

Example:

$$f(x) = \frac{1}{x^2}, \qquad a > 0.$$

$$\frac{1}{x_\nu^2} < \frac{1}{x_{\nu-1}x_\nu} < \frac{1}{x_{\nu-1}^2},$$

consequently

$$\sum_{\nu=1}^{n} \frac{x_\nu - x_{\nu-1}}{x_\nu^2} < \sum_{\nu=1}^{n} \frac{x_\nu - x_{\nu-1}}{x_{\nu-1}x_\nu} < \sum_{\nu=1}^{n} \frac{x_\nu - x_{\nu-1}}{x_{\nu-1}^2}.$$

We have

$$\sum_{\nu=1}^{n} \frac{x_\nu - x_{\nu-1}}{x_{\nu-1}x_\nu} = \sum_{\nu=1}^{n} \left(\frac{1}{x_{\nu-1}} - \frac{1}{x_\nu}\right) = \frac{1}{a} - \frac{1}{b}.$$

The number $\frac{1}{a} - \frac{1}{b}$ is therefore larger than any lower sum and smaller than any upper sum. If we can prove that no other number with this property exists we can conclude that

$$\frac{1}{a} - \frac{1}{b} = \int_a^b \frac{1}{x^2}\,dx.$$

Since $\frac{1}{x^2}$ is monotone the desired proof is easy.

Cf. e.g. G. B. Thomas: Calculus and Analytic Geometry, 2nd Ed. Reading/Mass.: Addison-Wesley Publishing 1958, pp. 140–141.

**1.** Suppose that $a > 0$, $r$ integer, $r \geqq 2$. Show in a similar way as in the previous example that

$$\sum_{\nu=1}^{n} \frac{1}{r-1}\left(\frac{1}{x_\nu^{r-1}x_{\nu-1}} + \frac{1}{x_\nu^{r-2}x_{\nu-1}^2} + \cdots + \frac{1}{x_\nu x_{\nu-1}^{r-1}}\right)(x_\nu - x_{\nu-1})$$

$$= \frac{1}{r-1}\left(\frac{1}{a^{r-1}} - \frac{1}{b^{r-1}}\right) = \int_a^b \frac{dx}{x^r}.$$

**2.** Assume that $a > 0$ and that $r$ is a positive integer. Show that $\int_a^b x^r\,dx = \frac{b^{r+1} - a^{r+1}}{r+1}$, i.e. the number $\frac{b^{r+1} - a^{r+1}}{r+1}$ is larger than all lower sums and smaller than all upper sums.

The points of division $x_0, x_1, x_2, \ldots, x_{n-1}, x_n$ form an arithmetic progression if

$$x_{\nu+1} - x_\nu = x_\nu - x_{\nu-1}$$

for $\nu = 1, 2, \ldots, n-1$. They form a geometric progression if

$$\frac{x_{\nu+1}}{x_\nu} = \frac{x_\nu}{x_{\nu-1}}$$

for $\nu = 1, 2, \ldots, n-1$. In the second case we assume $a > 0$.

**3.** Work out the lower and the upper sum for the function $e^x$ on the interval $[a, b]$ with the points of division in arithmetic progression. Find the limit as $n$ becomes infinite.

**4.** Construct the lower and the upper sum for the function $\frac{1}{x}$ on the interval $[a, b]$ with the points of division in geometric progression, $a > 0$. Find the limit as $n$ becomes infinite.

***5.** We divide the interval $[1, 2]$ into $n$ subintervals by the $n+1$ points

$$\frac{n}{n}, \quad \frac{n+1}{n}, \quad \frac{n+2}{n}, \ldots, \frac{n+n}{n},$$

and consider the lower sum $L_n$ and the upper sum $U_n$ for the function $f(x) = \frac{1}{x}$ that belong to this subdivision. Show that

$$U_1 = 1, \qquad L_1 = 1 - \frac{1}{2}, \qquad U_2 = 1 - \frac{1}{2} + \frac{1}{3}, \cdots$$

and generally that the sequence $U_1, L_1, U_2, L_2, \ldots, U_n, L_n, \ldots$ is identical with the sequence of the partial sums of the series

$$1 - \frac{1}{2} + \frac{1}{3} - \frac{1}{4} + \cdots + \frac{1}{2n-1} - \frac{1}{2n} + \cdots$$

**6.** Consider the infinite sequence whose $n$-th term is the $n$-th partial sum of the series

$$\frac{\sin x}{1} + \frac{\sin 2x}{2} + \cdots + \frac{\sin nx}{n} + \cdots$$

at the point $x = \frac{\pi}{n+1}$. The sequence converges to a limit different from zero. (This fact shows that the series in question cannot converge uniformly in the neighbourhood of $x = 0$.)

**7.** Assume that the function $f(x)$ mentioned at the beginning of this chapter is the derivative of the function $F(x)$. Denote any lower sum of $f(x)$ by $L$ and any upper sum by $U$. Then we have

$$L \leqq F(b) - F(a) \leqq U.$$

(But $F(b) - F(a)$ is not necessarily the only number satisfying this double inequality for all $L$ and $U$.)

## § 2. The Degree of Approximation

**8.** Assume that $0 < \xi < 1$, that the function $f(x)$ is monotone increasing on the interval $[0, \xi]$ and monotone decreasing on the interval $[\xi, 1]$ and that $f(\xi) = M$. Then the difference

$$\Delta_n = \int_0^1 f(x)\,dx - \frac{1}{n}\left[f\left(\frac{1}{n}\right) + f\left(\frac{2}{n}\right) + \cdots + f\left(\frac{n}{n}\right)\right]$$

tends to zero like $\frac{1}{n}$ as $n \to \infty$. We find

$$-\frac{M - f(0)}{n} \leqq \Delta_n \leqq \frac{M - f(1)}{n}.$$

**9.** Suppose that the function $f(x)$ is of bounded variation on the interval $[0, 1]$. The difference

$$\Delta_n = \int_0^1 f(x)\,dx - \frac{1}{n}\left[f\left(\frac{1}{n}\right) + f\left(\frac{2}{n}\right) + \cdots + f\left(\frac{n}{n}\right)\right]$$

tends to zero like $\frac{1}{n}$ as $n \to \infty$. In fact, calling the total variation $V$ we have

$$|\Delta_n| \leqq \frac{V}{n}.$$

**10.** Suppose that the function $f(x)$ has a bounded and integrable derivative in the interval $[a, b]$. We write

$$\Delta_n = \int_a^b f(x)\,dx - \frac{b-a}{n} \sum_{\nu=1}^{n} f\left(a + \nu\frac{b-a}{n}\right).$$

Find $\lim\limits_{n\to\infty} n\,\Delta_n$.

**11.** Assume that $f(x)$ is twice differentiable and that $f''(x)$ is properly integrable over $[a, b]$. Then the difference

$$\Delta_n' = \int_a^b f(x)\,dx - \frac{b-a}{n} \sum_{\nu=1}^{n} f\left(a + (2\nu - 1)\frac{b-a}{2n}\right)$$

tends to zero like $\frac{1}{n^2}$ as $n \to \infty$. More precisely, $\lim\limits_{n\to\infty} n^2 \Delta_n'$ exists; determine its value.

**12** (continued). The difference

$$\Delta_n'' = \int_a^b f(x)\,dx - \frac{b-a}{2n+1}\left[f(a) + 2\sum_{\nu=1}^{n} f\left(a + 2\nu\frac{b-a}{2n+1}\right)\right]$$

converges to zero like $\frac{1}{n^2}$ as $n \to \infty$. More precisely, $\lim\limits_{n\to\infty} n^2 \Delta_n''$ exists; find its value. Show, in addition, that $\Delta_n'' \geqq 0$ if $f'(a) \geqq 0$ and $f''(x) \geqq 0$, $a \leqq x \leqq b$.

**13.** We write

$$U_n = \frac{1}{n+1} + \frac{1}{n+2} + \cdots + \frac{1}{2n}, \quad V_n = \frac{2}{2n+1} + \frac{2}{2n+3} + \cdots + \frac{2}{4n-1}.$$

Prove that

$$\lim_{n\to\infty} U_n = \log 2, \qquad \lim_{n\to\infty} V_n = \log 2,$$

$$\lim_{n\to\infty} n\,(\log 2 - U_n) = \frac{1}{4}, \qquad \lim_{n\to\infty} n^2\,(\log 2 - V_n) = \frac{1}{32}.$$

**14.** The expression

$$\frac{1}{\sin\frac{\pi}{n}} + \frac{1}{\sin\frac{2\pi}{n}} + \cdots + \frac{1}{\sin\frac{(n-1)\pi}{n}} - \frac{2n}{\pi}\,(\log 2n + C - \log \pi)$$

is bounded for increasing $n$; $C$ is Euler's constant [solution **18**].

**15.**

$$\lim_{n\to\infty} e^{\frac{n}{4}}\, n^{-\frac{n+1}{2}}\, (1^1\, 2^2\, 3^3 \cdots n^n)^{\frac{1}{n}} = 1.$$

**16.** Suppose that $\alpha$ is positive and that $x_n$ is the (only) root of the equation

$$\frac{1}{2x} + \frac{1}{x-1} + \frac{1}{x-2} + \cdots + \frac{1}{x-n} = \alpha$$

in the interval $(n, \infty)$. Prove that

$$\lim_{n\to\infty} \left(x_n - \frac{n+\frac{1}{2}}{1-e^{-\alpha}}\right) = 0. \qquad [\mathbf{12}.]$$

**17.** Assume that $\alpha$ is positive and that $x_n'$ is the (only) root in the interval $(n, \infty)$ of the equation

$$\frac{1}{x} + \frac{2x}{x^2-1^2} + \frac{2x}{x^2-2^2} + \cdots + \frac{2x}{x^2-n^2} = \alpha.$$

Verify that

$$\lim_{n\to\infty} \left(x_n' - \left(n + \frac{1}{2}\right)\frac{1+e^{-\alpha}}{1-e^{-\alpha}}\right) = 0. \qquad [\mathbf{12}.]$$

**18.** Suppose that $f(x)$ is differentiable and that $f'(x)$ is monotone increasing or decreasing to zero as $x \to \infty$. Then the following limit exists:

$$\lim_{n\to\infty} \left(\tfrac{1}{2}f(1) + f(2) + f(3) + \cdots + f(n-1) + \tfrac{1}{2}f(n) - \int_1^n f(x)\,dx\right) = s.$$

Assume that $f'(x)$ increases. Then the two inequalities

$$\tfrac{1}{8}f'(n) < \tfrac{1}{2}f(1) + f(2) + f(3) + \cdots + f(n-1) + \tfrac{1}{2}f(n) - \int_1^n f(x)\,dy - s < 0$$

can be established. Note the particular cases $f(x) = \frac{1}{x}$, $f(x) = -\log x$.

**19.** Assume that $f(x)$ is differentiable for $x \geqq 1$ and that $f'(x)$ is monotone increasing to $\infty$ as $x \to \infty$. Then

$$\tfrac{1}{2}f(1) + f(2) + f(3) + \cdots + f(n-1) + \tfrac{1}{2}f(n) = \int_1^n f(x)\,dx + O[f'(n)].$$

More precisely,

$$0 < \tfrac{1}{2}f(1) + f(2) + f(3) + \cdots + f(n-1) + \tfrac{1}{2}f(n) - \int_1^n f(x)\,dx < \tfrac{1}{8}f'(n) - \tfrac{1}{8}f'(1).$$

**19.1.** We may regard the relation [**18**]

$$\lim_{n\to\infty}\left(1 + \frac{1}{2} + \frac{1}{3} + \cdots + \frac{1}{n} - \log n\right) = C$$

as the definition of Euler's constant $C$. Derive hence that

$$1 - \frac{1}{2} + \frac{1}{3} - \cdots + \frac{1}{2n-1} - \frac{1}{2n} + \cdots = \log 2.$$

**19.2.** The definition of $C$ given in **19.1** is convenient. Yet it would be desirable to approximate $C$ by rational numbers, to represent $C$ as the sum of a series whose terms are rational. Prove that the following series fulfills this desideratum:

$$\begin{aligned}
&\frac{1}{1} - \\
&-\frac{1}{2} - \frac{1}{3} + \\
&+\frac{1}{4} + \frac{1}{4} + \frac{1}{4} - \\
&-\frac{1}{5} - \frac{1}{6} - \frac{1}{7} - \frac{1}{8} + \\
&+\frac{1}{9} + \frac{1}{9} + \frac{1}{9} + \frac{1}{9} + \frac{1}{9} - \\
&-\frac{1}{10} - \frac{1}{11} - \frac{1}{12} - \frac{1}{13} - \frac{1}{14} - \frac{1}{15} + \\
&+\cdots.
\end{aligned}$$

Take the terms in the order as you read a book: from left to right and from top to bottom. This is essential since the series is not absolutely convergent. Its terms are non increasing in absolute value.

## § 3. Improper Integrals Between Finite Limits

Let $f(x)$ be defined on the finite interval $[a, b]$ except at the point $x = c$, $a \leqq c \leqq b$, in the neighbourhood of which $f(x)$ assumes arbitrarily large values. Furthermore let $f(x)$ be properly integrable over each closed subinterval of $[a, b]$ that does not contain $c$. Then the integral $\int_a^b f(x)\,dx$ is defined as the limit

$$\int_a^b f(x)\,dx = \lim_{\varepsilon,\varepsilon'\to+0}\left(\int_a^{c-\varepsilon} f(x)\,dx + \int_{c+\varepsilon'}^b f(x)\,dx\right).$$

(If $c$ coincides with $a$ or $b$ there is only one integral to consider.) If $f(x)$ becomes infinite at several (finitely many) points of the interval $[a, b]$ the integral is defined in a similar way.

Assume that $f(x)$ is defined for $x \geqq a$, furthermore that it is properly integrable over any finite interval $[a, \omega]$. Then we define

$$\int_0^\infty f(x)\,dx = \lim_{\omega\to\infty}\int_a^\omega f(x)\,dx.$$

One type of improper integral may be transformed into the other type of improper integral by an appropriate substitution.

**20.** Assume that the function $f(x)$ is monotone on the interval $(0, 1)$. It need not be bounded at the points $x = 0$, $x = 1$, we assume however that the improper integral $\int_0^1 f(x)\,dx$ exists. Under these conditions

$$\lim_{n\to\infty}\frac{f\left(\frac{1}{n}\right)+f\left(\frac{2}{n}\right)+\cdots+f\left(\frac{n-1}{n}\right)}{n} = \int_0^1 f(x)\,dx.$$

**21** (continued). If $\varphi(x)$ is properly integrable over $[0, 1]$ we have

$$\lim_{n\to\infty}\frac{\varphi\left(\frac{1}{n}\right)f\left(\frac{1}{n}\right)+\varphi\left(\frac{2}{n}\right)f\left(\frac{2}{n}\right)+\cdots+\varphi\left(\frac{n-1}{n}\right)f\left(\frac{n-1}{n}\right)}{n} = \int_0^1 \varphi(x)\,f(x)\,dx.$$

**22.** Prove in a way different from I **71** that for $\alpha > 0$

$$\lim_{n\to\infty}\frac{1^{\alpha-1}+2^{\alpha-1}+\cdots+n^{\alpha-1}}{n^\alpha} = \frac{1}{\alpha}.$$

**23.** We put

$$\sum_{k=1}^\infty k^{\alpha-1}z^k \sum_{l=1}^\infty l^{\beta-1}z^l = \sum_{n=1}^\infty a_n z^n, \qquad \alpha, \beta > 0.$$

Then

$$\lim_{n\to\infty} n^{1-\alpha-\beta}a_n$$

exists and is non-zero. (If $0 < \alpha < 1$, $0 < \beta < 1$, $\alpha + \beta \geqq 1$, $z = -1$, the product series is divergent although the two factors are convergent.)

**24.** The converse of the statement of problem **20** is not valid: There are functions, monotone in the interval (0, 1), for which $\lim_{n\to\infty} \sum_{k=1}^{n-1} \frac{1}{n} f\left(\frac{k}{n}\right)$ exists but not the integral.

**25.** The integral $\int_0^1 f(x)\,dx$ exists if $f(x)$ is monotone in the interval (0, 1), finite at $x = 0$ or $x = 1$ and if the following limit is finite

$$\lim_{n\to\infty} \frac{f\left(\frac{1}{n}\right) + f\left(\frac{2}{n}\right) + \cdots + f\left(\frac{n-1}{n}\right)}{n}.$$

**26.** Let the monotone function $f(x)$ be defined on (0, 1). Then the equation

$$\lim_{n\to\infty} \frac{1}{n} \sum_{\nu=1}^{n} f\left(\frac{2\nu-1}{2n}\right) = \int_0^1 f(x)\,dx$$

holds under the condition that the integral at right exists.

**27.** If $\alpha > 0$

$$\lim_{n\to\infty} \frac{1^{\alpha-1} - 2^{\alpha-1} + 3^{\alpha-1} - \cdots + (-1)^{n-1} n^{\alpha-1}}{n^\alpha} = 0.$$

**28.** If $f(x)$ is properly integrable over $[0, 1]$ we have obviously

$$\lim_{n\to\infty} \frac{f\left(\frac{1}{n}\right) - f\left(\frac{2}{n}\right) + f\left(\frac{3}{n}\right) - \cdots + (-1)^n f\left(\frac{n-1}{n}\right)}{n} = 0.$$

Show that this is true also if $f(x)$ is improperly integrable but monotone.

**29.** If $f(x)$ is monotone for $x > 0$, $\lim_{n\to\infty} \varepsilon_n = 0$, $c > 0$, $\varepsilon_n > \frac{c}{n}$ we find

$$\lim_{n\to\infty} \frac{f(\varepsilon_n) + f\left(\varepsilon_n + \frac{1}{n}\right) + f\left(\varepsilon_n + \frac{2}{n}\right) + \cdots + f\left(\varepsilon_n + \frac{n-1}{n}\right)}{n} = \int_0^1 f(x)\,dx,$$

provided that the integral at right exists and $f(x)$ is finite at $x = 1$.

## § 4. Improper Integrals Between Infinite Limits

**30.** Assume that the monotone function $f(x)$ is defined for $x \geqq 0$ and that the improper integral $\int_0^\infty f(x)\,dx$ exists. Then we have

$$\lim_{h\to+0} h\left(f(h) + f(2h) + f(3h) + \cdots\right) = \lim_{h\to+0} h \sum_{n=1}^{\infty} f(nh) = \int_0^\infty f(x)\,dx.$$

**31.** The $\Gamma$-function is defined for $\alpha > 0$ (or $\Re\alpha > 0$) by the integral

$$\Gamma(\alpha) = \int_0^\infty e^{-x} x^{\alpha-1}\, dx.$$

Using I **89** prove that

$$\Gamma(\alpha) = \lim_{n\to\infty} \frac{n^{\alpha-1}\, n!}{\alpha(\alpha+1)\cdots(\alpha+n-1)}, \qquad \alpha > 0.$$

**32.** As it is well known [cf. E. T. Whittaker and G. N. Watson, p. 246] Euler's constant $C$ can be written as an integral

$$C = \int_0^\infty e^{-x}\left(\frac{1}{1-e^{-x}} - \frac{1}{x}\right) dx.$$

Show that

$$C = \lim_{t\to 1-0}\left[(1-t)\left(\frac{t}{1-t} + \frac{t^2}{1-t^2} + \frac{t^3}{1-t^3} + \cdots + \frac{t^n}{1-t^n} + \cdots\right) - \log\frac{1}{1-t}\right].$$

**33.**

$$\lim_{t\to 1-0}(1-t)\left(\frac{t}{1+t} + \frac{t^2}{1+t^2} + \frac{t^3}{1+t^3} + \cdots + \frac{t^n}{1+t^n} + \cdots\right) = \log 2.$$

**34.**

$$\lim_{t\to 1-0}(1-t)^2\left(\frac{t}{1-t} + 2\frac{t^2}{1-t^2} + 3\frac{t^3}{1-t^3} + \cdots + n\frac{t^n}{1-t^n} + \cdots\right) = \frac{\pi^2}{6}.$$

**35.** We have

$$\lim_{t\to 1-0}\sqrt{1-t}\,(1 + t + t^4 + t^9 + \cdots + t^{n^2} + \cdots) = \frac{\sqrt{\pi}}{2}$$

and more generally for $\alpha > 0$

$$\lim_{t\to 1-0}\sqrt[\alpha]{1-t}\,(1 + t^{1^\alpha} + t^{2^\alpha} + t^{3^\alpha} + \cdots + t^{n^\alpha} + \cdots) = \frac{1}{\alpha}\Gamma\left(\frac{1}{\alpha}\right).$$

**36.** Compute

$$\lim_{t\to+\infty}\left(\frac{1}{t} + \frac{2t}{t^2+1^2} + \frac{2t}{t^2+2^2} + \cdots + \frac{2t}{t^2+n^2} + \cdots\right).$$

**37.** Let $\alpha > 1$ and put $g(t) = \prod_{n=1}^{\infty}\left(1 + \frac{t}{n^\alpha}\right)$. Show that

$$\lim_{t\to+\infty}\frac{\log g(t)}{t^{\frac{1}{\alpha}}} = \frac{\pi}{\sin\frac{\pi}{\alpha}}.$$

**38.** Establish the equation

$$\int_0^\infty \log(1 - 2x^{-2}\cos 2\varphi + x^{-4})\, dx = 2\pi\sin\varphi, \qquad 0 \leqq \varphi \leqq \pi,$$

with help of the following identity, valid for all complex $t$:

$$\frac{\sin t}{t} = \frac{e^{it} - e^{-it}}{2it} = \prod_{n=1}^{\infty}\left(1 - \frac{t^2}{n^2\pi^2}\right).$$

**39.** Compute the integral $\int_0^a \log x\,dx$ as the limit of an *infinite* sum of rectangles corresponding to the points of division

$$a,\ aq,\ aq^2,\ aq^3,\ \ldots, \qquad 0 < q < 1.$$

**40.** Let $k$ be a fixed positive number and $n$ an integer increasing to infinity. Then

$$\sum_{\nu=0}^{n}\binom{n}{\nu}^k \sim \frac{2^{kn}}{\sqrt{k}}\left(\frac{2}{\pi n}\right)^{\frac{k-1}{2}}$$

[**58**]. Observe the particular cases $k = 1$, $k = 2$.

## § 5. Applications to Number Theory

**41.** Divide the integer $n$ by $\nu$, $\nu = 1, 2, 3, \ldots, n$, and call the resulting remainder $n_\nu$. E.g. $17_3 = 2$, $10_{20} = 10$; obviously $n \equiv n_\nu \pmod{\nu}$, $0 \leqq n_\nu < \nu$. Find the probability that $n_\nu \geqq \frac{\nu}{2}$.

Solution:

$$n = \nu\left[\frac{n}{\nu}\right] + n_\nu,$$

consequently

$$0 \leqq \frac{2n}{\nu} - 2\left[\frac{n}{\nu}\right] = \frac{2n_\nu}{\nu} < 2.$$

In the favorable case we have

$$n_\nu \geqq \frac{\nu}{2}, \text{ thus } \left[\frac{2n}{\nu}\right] - 2\left[\frac{n}{\nu}\right] = 1;$$

in the adverse case

$$n_\nu < \frac{\nu}{2}, \text{ thus } \left[\frac{2n}{\nu}\right] - 2\left[\frac{n}{\nu}\right] = 0 \qquad \text{[VIII 3]}.$$

The probability in question is therefore

$$w_n = \frac{1}{n}\sum_{\nu=1}^{n}\left(\left[\frac{2n}{\nu}\right] - 2\left[\frac{n}{\nu}\right]\right).$$

As $n \to \infty$ this sum approaches the proper integral

$$\int_0^1 \left(\left[\frac{2}{x}\right] - 2\left[\frac{1}{x}\right]\right) dx = \lim_{n\to\infty} \sum_{\nu=1}^{n-1} \int_{\frac{1}{\nu+1}}^{\frac{1}{\nu}} \left(\left[\frac{2}{x}\right] - 2\left[\frac{1}{x}\right]\right) dx$$

$$= \lim_{n\to\infty} \sum_{\nu=1}^{n-1} \left(\frac{1}{\nu+\frac{1}{2}} - \frac{1}{\nu+1}\right)$$

$$= 2\left(\frac{1}{3} - \frac{1}{4} + \frac{1}{5} - \frac{1}{6} + \cdots\right) = 2\log 2 - 1 = 0{,}38629\ldots$$

**42** (continued). Compute

$$\lim_{n\to\infty} \frac{1}{n}\left(\frac{n_1}{1} + \frac{n_2}{2} + \frac{n_3}{3} + \cdots + \frac{n_n}{n}\right).$$

**43** (continued). Compute

$$\lim_{n\to\infty} \frac{n_1 + n_2 + n_3 + \cdots + n_n}{n^2}.$$

**44** (continued). Let $A_\alpha$ be the number of those fractions of the form $\frac{n_k}{k}$, $k = 1, 2, \ldots, n$, that are smaller than a given number $\alpha$, $0 \leqq \alpha \leqq 1$. The quotient $\frac{A_\alpha}{n}$ tends with increasing $n$ towards the limit

$$\int_0^1 \frac{1-x^\alpha}{1-x}\, dx. \qquad \text{[VIII 4.]}$$

**45.** Let $\sigma_\alpha(n)$ denote the sum of the $\alpha$-th powers of all the divisors of $n$ (VIII, Chap. 1, § 5) and

$$\Sigma_\alpha(n) = \sigma_\alpha(1) + \sigma_\alpha(2) + \sigma_\alpha(3) + \cdots + \sigma_\alpha(n) = \sum_{\nu=1}^{n} \left[\frac{n}{\nu}\right] \nu^\alpha \qquad \text{[VIII 81].}$$

Then we have for $\alpha > 0$

$$\lim_{n\to\infty} \frac{\Sigma_\alpha(n)}{n^{\alpha+1}} = \frac{\zeta(\alpha+1)}{\alpha+1},$$

where $\zeta(s)$ denotes Riemann's $\zeta$-function ($\zeta(z) = \sum_{n=1}^{\infty} n^{-z}$, cf. VIII, Chap. 1, § 5). For $\alpha > 1$ we even have the inequality

$$\left|\frac{\Sigma_\alpha(n)}{n^{\alpha+1}} - \frac{\zeta(\alpha+1)}{\alpha+1}\right| \leqq \frac{2\zeta(\alpha)-1}{n}, \qquad n = 1, 2, 3, \ldots$$

[$\left(\frac{1}{x} - \left[\frac{1}{x}\right]\right) x^\alpha$ is properly integrable over the interval [0, 1] if $\alpha > 0$, [**107**]; $\left[\frac{1}{x}\right] x^\alpha$ is of bounded variation if $\alpha > 1$.]

**46.** Let $\tau(n) = \sigma_0(n)$ denote the number of divisors of $n$. Then

$$\tau(1) + \tau(2) + \tau(3) + \cdots + \tau(n) = \sum_{\nu=1}^{n} \left[\frac{n}{\nu}\right]$$
$$= n(\log n + 2C - 1) + O(\sqrt{n}) \qquad \text{[VIII 79]},$$

$C$ is Euler's constant. [Apply the idea of **9** to $\frac{1}{x} - \left[\frac{1}{x}\right]$ in the interval $\left(\frac{1}{m}, 1\right)$, $m = [\sqrt{n}] + 1$; solution **18**.]

**47.** Denote by $O_n$ the number of odd divisors and by $E_n$ the number of even divisors of the integer $n$. E.g. $O_{20} = 2$, $E_{20} = 4$. Prove that

$$\lim_{n\to\infty} \frac{O_1 - E_1 + O_2 - E_2 + \cdots + O_n - E_n}{n} = \log 2.$$

## § 6. Mean Values and Limits of Products

The *arithmetic*, *geometric* and *harmonic means* of the numbers $a_1, a_2, \ldots, a_n$ are

$$\frac{a_1 + a_2 + a_3 + \cdots + a_n}{n}, \qquad \sqrt[n]{a_1 a_2 a_3 \cdots a_n}, \qquad \frac{n}{\frac{1}{a_1} + \frac{1}{a_2} + \frac{1}{a_3} + \cdots + \frac{1}{a_n}},$$

respectively. For the last two expressions all the $a_i$'s are supposed to be positive. (More details in Chap. 2.)

**48.** Suppose that the function $f(x)$ is defined on $[a, b]$ and properly integrable over this interval. Define

$$f_{\nu n} = f(a + \nu\delta_n), \qquad \delta_n = \frac{b - a}{n}.$$

Then

$$\lim_{n\to\infty} \frac{f_{1n} + f_{2n} + f_{3n} + \cdots + f_{nn}}{n} = \frac{1}{b - a}\int_a^b f(x)\,dx,$$

$$\lim_{n\to\infty} \sqrt[n]{f_{1n} f_{2n} f_{3n} \cdots f_{nn}} = e^{\frac{1}{b-a}\int_a^b \log f(x)\,dx},$$

$$\lim_{n\to\infty} \frac{n}{\frac{1}{f_{1n}} + \frac{1}{f_{2n}} + \frac{1}{f_{3n}} + \cdots + \frac{1}{f_{nn}}} = \frac{b - a}{\int_a^b \frac{dx}{f(x)}}.$$

These three limits are called the *arithmetic*, *geometric* and *harmonic mean* of the *function* $f(x)$. In the last two relations the greatest lower bound of $f(x)$ should be positive. (To improper integrals apply with caution!)

**49.** Prove the existence of

$$\lim_{n\to\infty} \frac{\sqrt[n]{n!}}{n}$$

in a way different from I **69** and that the limit is equal to the geometric mean of $f(x) = x$ on the interval $[0, 1]$, i.e. $= \frac{1}{e}$.

**50.** Let $a$ and $d$ be positive numbers and call $A_n$ the arithmetic, and $G_n$ the geometric, mean of the numbers $a, a + d, a + 2d, \ldots, a + (n - 1)d$. Then we obtain

$$\lim_{n\to\infty} \frac{G_n}{A_n} = \frac{2}{e}.$$

**51.** Let $A_n$ denote the arithmetic, and $G_n$ the geometric, mean of the binomial coefficients

$$\binom{n}{0}, \binom{n}{1}, \binom{n}{2}, \ldots, \binom{n}{n}.$$

Show that

$$\lim_{n\to\infty} \sqrt[n]{A_n} = 2, \qquad \lim_{n\to\infty} \sqrt[n]{G_n} = \sqrt{e}.$$

**52.** Prove

$$\frac{1}{2\pi}\int_0^{2\pi} \log(1 - 2r\cos x + r^2)\, dx = \begin{cases} 2\log r & \text{for } r \geqq 1, \\ 0 & \text{for } 0 \leqq r \leqq 1. \end{cases}$$

**53.** Let $r$ be positive and smaller than 1; let $x$ assume any value in the interval $[0, 2\pi]$ and $\xi$ denote the number closest to $x$ for which

$$\sin(x - \xi) = r \sin x.$$

Then we find

$$\frac{1}{2\pi}\int_0^{2\pi} \log(1 - 2r\cos\xi + r^2)\, dx = \log(1 - r^2).$$

[Interpret $e^{ix}$, $e^{i\xi}$, $r$ in the complex plane.]

**54.** Assume that $f(x)$ is properly integrable over $[a, b]$. Using the same notation as in **48** establish

$$\lim_{n\to\infty} (1 + f_{1n}\delta_n)(1 + f_{2n}\delta_n)\cdots(1 + f_{nn}\delta_n) = e^{\int_a^b f(x)dx}.$$

**55.** Compute

$$\lim_{n\to\infty} \frac{(n^2 + 1)(n^2 + 2)\cdots(n^2 + n)}{(n^2 - 1)(n^2 - 2)\cdots(n^2 - n)}.$$

**56.** Prove the identity

$$\left(1+\frac{1}{\alpha-1}\right)\left(1-\frac{1}{2\alpha-1}\right)\left(1+\frac{1}{3\alpha-1}\right)\left(1-\frac{1}{4\alpha-1}\right)\cdots$$
$$\cdots\left(1+\frac{1}{(2n-1)\alpha-1}\right)\left(1-\frac{1}{2n\alpha-1}\right)$$
$$=\frac{(n+1)\alpha}{(n+1)\alpha-1}\cdot\frac{(n+2)\alpha}{(n+2)\alpha-1}\cdots\frac{(n+n)\alpha}{(n+n)\alpha-1}.$$

Whence there follows that the product on the left hand side extended to infinity has the limit $2^{\frac{1}{\alpha}}$, provided that $\alpha \neq 0, 1, \frac{1}{2}, \frac{1}{3}, \frac{1}{4}, \ldots$ $[(2n)! = 2^n n!\, 1\cdot 3\cdot 5 \cdots (2n-1).]$

**57.** Let $\alpha, \beta, \delta$ be fixed, $\delta > 0$ and

$$a = 1+\frac{\alpha}{n}, \qquad b = 1+\frac{\beta}{n}, \qquad d = \frac{\delta}{n}.$$

Show that

$$\lim_{n\to\infty}\frac{a}{b}\cdot\frac{a+d}{b+d}\cdot\frac{a+2d}{b+2d}\cdots\frac{a+(n-1)d}{b+(n-1)d} = (1+\delta)^{\frac{\alpha-\beta}{\delta}}.$$

**58.** Let $n$ and $\nu$ be integers, $0<\nu<n$. If $n$ and $\nu$ increase to infinity in such a way that

$$\lim_{n\to\infty}\frac{\nu-\frac{n}{2}}{\sqrt{n}} = \lambda,$$

then

$$\lim_{n\to\infty}\frac{\sqrt{n}}{2^n}\binom{n}{\nu} = \sqrt{\frac{2}{\pi}}\, e^{-2\lambda^2}.$$

**59.** Let $t$ be a fixed real number. Define $z = 2ne^{\frac{it}{\sqrt{n}}}$. Then we have

$$\lim_{n\to\infty}\left|\frac{2n-1}{z-1}\cdot\frac{2n-2}{z-2}\cdot\frac{2n-3}{z-3}\cdots\frac{2n-n}{z-n}\right| = \left(\frac{2}{e}\right)^{t^2}.$$

## § 7. Multiple Integrals

**60.** Suppose that the function $f(x, y)$ is the second mixed derivative of a function $F(x, y)$,

$$\frac{\partial^2 F(x,y)}{\partial x\,\partial y} = f(x,y),$$

in the rectangle $R$

$$a \leqq x \leqq b, \qquad c \leqq y \leqq d.$$

The points $x_\mu, y_\nu, \mu = 0, 1, 2, \ldots, m; \nu = 0, 1, 2, \ldots, n;$

$$a = x_0 < x_1 < x_2 < \cdots < x_{m-1} < x_m = b,$$

$$c = y_0 < y_1 < y_2 < \cdots < y_{n-1} < y_n = d$$

determine a subdivision of the given rectangle $R$ into "subrectangles" $R_{\mu\nu}$: $x_{\mu-1} \leqq x \leqq x_\mu$, $y_{\nu-1} \leqq y \leqq y_\nu$, $\mu = 1, 2, \ldots, m$; $\nu = 1, 2, \ldots, n$. We denote by $M_{\mu\nu}$, $m_{\mu\nu}$ the least upper and the greatest lower bound resp. of $f(x, y)$ in the rectangle $R_{\mu\nu}$. Defining the

$$\text{upper sum } U = \sum_{\mu=1}^{m} \sum_{\nu=1}^{n} M_{\mu\nu}(x_\mu - x_{\mu-1})(y_\nu - y_{\nu-1}),$$

$$\text{and the lower sum } L = \sum_{\mu=1}^{m} \sum_{\nu=1}^{n} m_{\mu\nu}(x_\mu - x_{\mu-1})(y_\nu - y_{\nu-1})$$

we find that

$$L \leqq F(b, d) - F(b, c) - F(a, d) + F(a, c) \leqq U.$$

**61.**

$$\iint\limits_{0 \leqq x \leqq y \leqq \pi} \log|\sin(x - y)|\, dx\, dy = -\frac{\pi^2}{2} \log 2.$$

[Compute the square of the absolute value of the following determinant in two ways

$$\begin{vmatrix} 1 & 1 & 1 & \cdots 1 \\ 1 & \varepsilon & \varepsilon^2 & \cdots \varepsilon^{n-1} \\ 1 & \varepsilon^2 & \varepsilon^4 & \cdots \varepsilon^{2(n-1)} \\ \cdots & \cdots & \cdots & \cdots \\ 1 & \varepsilon^{n-1} & \varepsilon^{2(n-1)} & \cdots \varepsilon^{(n-1)(n-1)} \end{vmatrix}, \qquad \varepsilon = e^{\frac{2\pi i}{n}}.]$$

**62.** Let $f(x, y)$ be properly integrable over the square $0 \leqq x \leqq 1$, $0 \leqq y \leqq 1$. Show that

$$\lim_{n\to\infty} \prod_{\mu=1}^{n} \prod_{\nu=1}^{n} \left[1 + \frac{1}{n^2} f\left(\frac{\mu}{n}, \frac{\nu}{n}\right)\right] = e^{\int_0^1 \int_0^1 f(x,y)\,dx\,dy}.$$

**63.** Let $f(x, y)$ be properly integrable over the square $0 \leqq x \leqq 1$, $0 \leqq y \leqq 1$. Compute

$$\lim_{n\to\infty} \prod_{\nu=1}^{n} \left\{1 + \frac{1}{n^2}\left[f\left(\frac{1}{n}, \frac{\nu}{n}\right) + f\left(\frac{2}{n}, \frac{\nu}{n}\right) + \cdots + f\left(\frac{n}{n}, \frac{\nu}{n}\right)\right]\right\}.$$

**64.** The three-dimensional domain $\mathfrak{D}$ is defined by the inequalities

$$-1 \leqq x, y, z \leqq 1, \qquad -\sigma \leqq x + y + z \leqq \sigma.$$

Using I **30** show that the volume of $\mathfrak{D}$ is

$$\iiint\limits_{\mathfrak{D}} dx\, dy\, dz = \frac{2^3}{\pi} \int\limits_{-\infty}^{\infty} \left(\frac{\sin t}{t}\right)^3 \frac{\sin \sigma t}{t}\, dt.$$

**65.** Let $\alpha_1, \alpha_2, \ldots, \alpha_p$ be arbitrary positive numbers, define

$$f_\nu(z) = 1^{\alpha_\nu - 1} z + 2^{\alpha_\nu - 1} z^2 + \cdots + n^{\alpha_\nu - 1} z^n + \cdots, \qquad \nu = 1, 2, \ldots, p$$

and

$$f_1(z)\, f_2(z) \cdots f_p(z) = \sum_{n=1}^{\infty} a_n z^n.$$

Show that

$$\lim_{n\to\infty} \frac{a_n}{n^{\alpha_1+\alpha_2+\cdots+\alpha_p - 1}}$$

$$= \iint \cdots \int x_1^{\alpha_1 - 1} x_2^{\alpha_2 - 1} \cdots x_{p-1}^{\alpha_{p-1} - 1} (1 - x_1 - x_2 - \cdots - x_{p-1})^{\alpha_p - 1}$$

$$\times\, dx_1\, dx_2 \cdots dx_{p-1},$$

where the integral is taken over the domain described by the $p$ inequalities $x_1 \geqq 0,\ x_2 \geqq 0, \ldots, x_{p-1} \geqq 0,\ x_1 + x_2 + \cdots + x_{p-1} \leqq 1$ ($p - 1$ dimensional simplex).

**66** (continued).

$$\iint \cdots \int x_1^{\alpha_1 - 1} x_2^{\alpha_2 - 1} \cdots x_{p-1}^{\alpha_{p-1} - 1} (1 - x_1 - x_2 - \cdots - x_{p-1})^{\alpha_p - 1}$$

$$\times\, dx_1\, dx_2 \cdots dx_{p-1} = \frac{\Gamma(\alpha_1)\, \Gamma(\alpha_2) \cdots \Gamma(\alpha_p)}{\Gamma(\alpha_1 + \alpha_2 + \cdots + \alpha_p)}.$$

**67.** Work out the product $\prod_{k=1}^{n} (1 + f_{kn}\, \delta_n)$ (same notation as in **48**, **54**) as a polynomial of degree $n$ in $\delta_n$. Prove that the term containing $\delta_n^p$ converges to the limit

$$\iint\limits_{a \leqq x_1 \leqq x_2 \leqq \cdots \leqq x_p \leqq b} \cdots \int f(x_1)\, f(x_2) \cdots f(x_p)\, dx_1\, dx_2 \cdots dx_p = \frac{1}{p!} \left( \int\limits_a^b f(x)\, dx \right)^p,$$

when $p$ is fixed and $n$ increases to infinity.

**68.** Suppose that the $2m$ functions

$$f(x_1), f(x_2), \ldots, \ f_m(x),$$

$$\varphi(x_1), \varphi(x_2), \ldots, \ \varphi_m(x)$$

are properly integrable over the interval $[a, b]$. Then we have

$$\begin{vmatrix} \int_a^b f_1(x)\,\varphi_1(x)\,dx & \int_a^b f_1(x)\,\varphi_2(x)\,dx & \cdots & \int_a^b f_1(x)\,\varphi_m(x)\,dx \\ \int_a^b f_2(x)\,\varphi_1(x)\,dx & \int_a^b f_2(x)\,\varphi_2(x)\,dx & \cdots & \int_a^b f_2(x)\,\varphi_m(x)\,dx \\ \cdots & \cdots & \cdots & \cdots \\ \int_a^b f_m(x)\,\varphi_1(x)\,dx & \int_a^b f_m(x)\,\varphi_2(x)\,dx & \cdots & \int_a^b f_m(x)\,\varphi_m(x)\,dx \end{vmatrix}$$

$$= \frac{1}{m!}\underbrace{\int_a^b\int_a^b\cdots\int_a^b}_{m} \begin{vmatrix} f_1(x_1) & f_1(x_2) & \cdots & f_1(x_m) \\ f_2(x_1) & f_2(x_2) & \cdots & f_2(x_m) \\ \cdots & \cdots & \cdots & \cdots \\ f_m(x_1) & f_m(x_2) & \cdots & f_m(x_m) \end{vmatrix} \cdot \begin{vmatrix} \varphi_1(x_1) & \varphi_1(x_2) & \cdots & \varphi_1(x_m) \\ \varphi_2(x_1) & \varphi_2(x_2) & \cdots & \varphi_2(x_m) \\ \cdots & \cdots & \cdots & \cdots \\ \varphi_m(x_1) & \varphi_m(x_2) & \cdots & \varphi_m(x_m) \end{vmatrix}$$

$$\times\, dx_1\, dx_2 \cdots dx_m.$$

[Compute in two different ways the product of the two matrices

$$\|f_{\nu n}^{(\lambda)}\|_{\substack{\lambda=1,2,\ldots,m \\ \nu=1,2,\ldots,n}} \cdot \|\varphi_{\nu n}^{(\mu)}\|_{\substack{\mu=1,2,\ldots,m \\ \nu=1,2,\ldots,n}};$$

$$f_{\nu n}^{(\lambda)} = f_\lambda\left(a + \nu\frac{b-a}{n}\right), \qquad \varphi_{\nu n}^{(\mu)} = \varphi_\mu\left(a + \nu\frac{b-a}{n}\right).]$$

Chapter 2

# Inequalities

## § 1. Inequalities

Let $a_1, a_2, \ldots, a_n$ be arbitrary real numbers. Their *arithmetic mean* $\mathfrak{A}(a)$ is defined as the expression

$$\mathfrak{A}(a) = \frac{a_1 + a_2 + \cdots + a_n}{n}.$$

If all the numbers $a_1, a_2, \ldots, a_n$ are positive we define their *geometric* and *harmonic means* as

$$\mathfrak{G}(a) = \sqrt[n]{a_1 a_2 \cdots a_n}, \qquad \mathfrak{H}(a) = \frac{n}{\dfrac{1}{a_1} + \dfrac{1}{a_2} + \cdots + \dfrac{1}{a_n}},$$

respectively. If $m$ denotes the smallest and $M$ the largest of the numbers $a_i$ then

$$m \leqq \mathfrak{A}(a) \leqq M, \qquad m \leqq \mathfrak{G}(a) \leqq M, \qquad m \leqq \mathfrak{H}(a) \leqq M.$$

For $\mathfrak{G}(a)$ and $\mathfrak{H}(a)$ we assume $m > 0$. The three numbers represent *mean values* of $a_1, a_2, \ldots, a_n$. The inequalities become equalities only if all the $a_i$'s are equal. The mean values have the following properties:

$$\frac{1}{\mathfrak{G}(a)} = \mathfrak{G}\left(\frac{1}{a}\right), \qquad \frac{1}{\mathfrak{H}(a)} = \mathfrak{A}\left(\frac{1}{a}\right),$$

$$\mathfrak{A}(a + b) = \mathfrak{A}(a) + \mathfrak{A}(b), \quad \mathfrak{G}(ab) = \mathfrak{G}(a)\,\mathfrak{G}(b), \quad \log \mathfrak{G}(a) = \mathfrak{A}\,(\log a).$$

Let the function $f(x)$ be defined and properly integrable on the interval $[x_1, x_2]$. We define the *arithmetic mean* $\mathfrak{A}(f)$ of $f(x)$ as

$$\mathfrak{A}(f) = \frac{1}{x_2 - x_1} \int_{x_1}^{x_2} f(x)\,dx.$$

If $f(x)$ is strictly positive, i.e. if there exists a positive constant $k$ such that $f(x) > k$ for $x$ in $[x_1, x_2]$, the *geometric* and *harmonic means* are defined as

$$\mathfrak{G}(f) = e^{\frac{1}{x_2 - x_1}\int_{x_1}^{x_2} \log f(x)\,dx}, \qquad \mathfrak{H}(f) = \frac{x_2 - x_1}{\displaystyle\int_{x_1}^{x_2} \frac{dx}{f(x)}}.$$

[**48.**] If $m$ denotes the greatest lower and $M$ the least upper bound of $f(x)$ on $[x_1, x_2]$ then

$$m \leqq \mathfrak{A}(f) \leqq M, \qquad m \leqq \mathfrak{G}(f) \leqq M, \qquad m \leqq \mathfrak{H}(f) \leqq M.$$

It is understood that $m > 0$ for $\mathfrak{G}(f)$ and $\mathfrak{H}(f)$.

The following relations are obvious:

$$\mathfrak{A}(f + g) = \mathfrak{A}(f) + \mathfrak{A}(g), \qquad \mathfrak{G}(fg) = \mathfrak{G}(f)\,\mathfrak{G}(g), \qquad \log \mathfrak{G}(f) = \mathfrak{A}(\log f).$$

Let $a_1, a_2, \ldots, a_n$ be arbitrary positive numbers which are not all equal. Then

$$\frac{n}{\frac{1}{a_1} + \frac{1}{a_2} + \cdots + \frac{1}{a_n}} < \sqrt[n]{a_1 a_2 \cdots a_n} < \frac{a_1 + a_2 + \cdots + a_n}{n},$$

i.e.

$$\mathfrak{H}(a) < \mathfrak{G}(a) < \mathfrak{A}(a).$$

(Theorem of the means, arithmetic, geometric and harmonic.) A very beautiful proof[1] was given by Cauchy in his Analyse algébrique (Note 2; Oeuvres Complètes, Ser. 2, Vol. 3; Paris: Gauthier-Villars 1897, pp. 375–377).

---

[1] It is obviously sufficient to prove $\mathfrak{A}(a) > \mathfrak{G}(a)$. Here is the passage referred to: "*La moyenne géométrique entre plusieurs nombres A, B, C, D, ... est toujours inférieure à leur moyenne arithmétique.*

Démonstration. — Soit $n$ le nombre des lettres $A, B, C, D, \ldots$ Il suffira de prouver, qu'on a généralement

$$\sqrt[n]{ABCD\cdots} < \frac{A+B+C+D+\cdots}{n} \tag{1}$$

ou, ce qui revient au même,

$$ABCD\cdots < \left(\frac{A+B+C+D+\cdots}{n}\right)^n. \tag{2}$$

Or, en premier lieu, on aura évidemment, pour $n = 2$,

$$AB = \left(\frac{A+B}{2}\right)^2 - \left(\frac{A-B}{2}\right)^2 < \left(\frac{A+B}{2}\right)^2,$$

et l'on en conclura, en prenant successivement $n = 4$, $n = 8$, ..., enfin $n = 2^m$

$$ABCD < \left(\frac{A+B}{2}\right)^2 \left(\frac{C+D}{2}\right)^2 < \left(\frac{A+B+C+D}{4}\right)^4,$$

$$ABCDEFGH < \left(\frac{A+B+C+D}{4}\right)^4 \left(\frac{E+F+G+H}{4}\right)^4$$

$$< \left(\frac{A+B+C+D+E+F+G+H}{8}\right)^8,$$

.........................................................................

$$ABCD\cdots < \left(\frac{A+B+C+D+\cdots}{2^m}\right)^{2^m}. \tag{3}$$

En second lieu, si $n$ n'est pas un terme de la progression géométrique

$$2, 4, 8, 16, \ldots,$$

on désignera par $2^m$ un terme de cette progression supérieur à $n$, et l'on fera

$$K = \frac{A+B+C+D+\cdots}{n};$$

puis, en revenant à la formule (3), et supposant dans le premier membre de cette formule les $2^m - n$ derniers facteurs égaux à $K$, on trouvera

$$ABCD\cdots K^{2^m-n} < \left[\frac{A+B+C+D+\cdots+(2^m-n)K}{2^m}\right]^{2^m}$$

ou, en d'autres termes,

$$ABCD\ldots K^{2^m-n} < K^{2^m}.$$

On aura donc par suite

$$ABCD\cdots < K^n = \left(\frac{A+B+C+D+\cdots}{n}\right)^n,$$

ce qu'il fallait démontrer."

**69.** Let the function $f(x)$ be defined and properly integrable on the interval $[x_1, x_2]$ and let $f(x)$ have a positive lower bound. Then

$$\frac{x_2 - x_1}{\int\limits_{x_1}^{x_2} \frac{dx}{f(x)}} \leqq e^{\frac{1}{x_2 - x_1}\int\limits_{x_1}^{x_2} \log f(x)\,dx} \leqq \frac{1}{x_2 - x_1}\int\limits_{x_1}^{x_2} f(x)\,dx,$$

or, with the notation just defined,

$$\mathfrak{H}(f) \leqq \mathfrak{G}(f) \leqq \mathfrak{A}(f).$$

**70.** Suppose that the (not necessarily differentiable) function $\varphi(t)$ satisfies for arbitrary values $t_1, t_2, t_1 \neq t_2$ the inequality

$$\varphi\left(\frac{t_1 + t_2}{2}\right) < \frac{\varphi(t_1) + \varphi(t_2)}{2}.$$

Then the more general inequality

$$\varphi\left(\frac{t_1 + t_2 + \cdots + t_n}{n}\right) < \frac{\varphi(t_1) + \varphi(t_2) + \cdots + \varphi(t_n)}{n}$$

holds, where the $t_i$'s are arbitrary but $t_i \neq t_j$ for at least one pair $i, j$.

A function $\varphi(t)$ defined on the interval $m \leqq t \leqq M$ is called *convex* if for each pair $t_1, t_2$ on $[m, M]$, $t_1 \neq t_2$ the inequality

$$\varphi\left(\frac{t_1 + t_2}{2}\right) \leqq \frac{\varphi(t_1) + \varphi(t_2)}{2}$$

is satisfied. (By the solution of **70** we have then generally

$$\varphi\left(\frac{t_1 + t_2 + \cdots + t_n}{n}\right) \leqq \frac{\varphi(t_1) + \varphi(t_2) + \cdots + \varphi(t_n)}{n}$$

for arbitrary points $t_1, t_2, \ldots, t_n$ of the interval.) If instead of the inequality with $\leqq$ strict inequality is supposed ($<$) then $\varphi(t)$ is termed *strictly convex.* If $-\varphi(t)$ is convex, $\varphi(t)$ is termed *concave.* (A more intuitive but somewhat clumsier terminology would be "convex from below" for convex and "convex from above" for concave.) In the sequel we will consider bounded convex functions only; these are continuous [cf. **124**; **110** often useful].

**71.** Suppose that the function $f(x)$ is properly integrable on the interval $[x_1, x_2]$ and $m \leqq f(x) \leqq M$ and, furthermore, that $\varphi(t)$ is defined and convex on the interval $[m, M]$. Then we have the inequality

$$\varphi\left(\frac{1}{x_2 - x_1}\int\limits_{x_1}^{x_2} f(x)\,dx\right) \leqq \frac{1}{x_2 - x_1}\int\limits_{x_1}^{x_2} \varphi[f(x)]\,dx.$$

**72.** Suppose that the function $\varphi(t)$ is defined on the interval $[m, M]$ and that $\varphi''(t)$ exists and $\varphi''(t) > 0$ on $[m, M]$. In this case $\varphi(t)$ is strictly convex. If we have $\varphi''(t) \geqq 0$ only, then $\varphi(t)$ is convex. (A function can be convex in an interval where its second derivative does not exist at all points.)

**73.** The functions

$$t^k \quad (0 < k < 1) \qquad \text{and} \qquad \log t$$

are concave on any positive interval;

$$t^k \quad (k < 0 \text{ or } k > 1) \qquad \text{and} \qquad t \log t$$

are convex on any positive interval;

$$\log(1 + e^t) \qquad \text{and} \qquad \sqrt{c^2 + t^2} \qquad (c > 0)$$

are everywhere convex.

**74.** Assume that $\varphi(t)$ is a convex function defined on $[m, M]$, that $p_1, p_2, \ldots, p_n$ are arbitrary positive numbers and that $t_1, t_2, \ldots, t_n$ are arbitrary points of the interval $[m, M]$. Then we have the inequality

$$\varphi\left(\frac{p_1t_1 + p_2t_2 + \cdots + p_nt_n}{p_1 + p_2 + \cdots + p_n}\right) \leqq \frac{p_1\varphi(t_1) + p_2\varphi(t_2) + \cdots + p_n\varphi(t_n)}{p_1 + p_2 + \cdots + p_n}.$$

**75.** Assume that $f(x)$ and $p(x)$ are two functions which are properly integrable over $[x_1, x_2]$ and that $m \leqq f(x) \leqq M$, $p(x) \geqq 0$ and $\int_{x_1}^{x_2} p(x)\,dx > 0$. Let $\varphi(t)$ denote a convex function defined on the interval $m \leqq t \leqq M$. Then we have

$$\varphi\left(\frac{\int_{x_1}^{x_2} p(x) f(x)\,dx}{\int_{x_1}^{x_2} p(x)\,dx}\right) \leqq \frac{\int_{x_1}^{x_2} p(x)\,\varphi[f(x)]\,dx}{\int_{x_1}^{x_2} p(x)\,dx}.$$

**76.** Suppose that on the interval $[m, M]$ the first and the second derivative of $\varphi(t)$ exist and that $\varphi''(t) > 0$. Then we find for positive $p_1, p_2, \ldots, p_n$

$$\varphi\left(\frac{p_1t_1 + p_2t_2 + \cdots + p_nt_n}{p_1 + p_2 + \cdots + p_n}\right) \leqq \frac{p_1\varphi(t_1) + p_2\varphi(t_2) + \cdots + p_n\varphi(t_n)}{p_1 + p_2 + \cdots + p_n}.$$

There is equality if and only if $t_1 = t_2 = \cdots = t_n$.

**77.** The functions $f(x)$ and $p(x)$ are assumed to be continuous on the interval $[x_1, x_2]$, $p(x)$ is strictly positive and $m \leqq f(x) \leqq M$; the function $\varphi(t)$ is defined on the interval $[m, M]$, $\varphi(t)$ can be differentiated twice and $\varphi''(t) > 0$. Then

$$\varphi\left(\frac{\int_{x_1}^{x_2} p(x) f(x)\,dx}{\int_{x_1}^{x_2} p(x)\,dx}\right) \leq \frac{\int_{x_1}^{x_2} p(x)\,\varphi[f(x)]\,dx}{\int_{x_1}^{x_2} p(x)\,dx}.$$

There is equality if and only if $f(x)$ is a constant.

**78.** Prove the following generalization of the proposition on the arithmetic, geometric and harmonic means: Let $p_1, p_2, \ldots, p_n, a_1, a_2, \ldots, a_n$ denote arbitrary positive numbers, $a_i \neq a_j$ for at least one pair $i \neq j$, $i, j = 1, 2, \ldots, n$; then the inequalities

$$\frac{p_1 + p_2 + \cdots + p_n}{\frac{p_1}{a_1} + \frac{p_2}{a_2} + \cdots + \frac{p_n}{a_n}} < e^{\frac{p_1 \log a_1 + p_2 \log a_2 + \cdots + p_n \log a_n}{p_1 + p_2 + \cdots + p_n}} < \frac{p_1 a_1 + p_2 a_2 + \cdots + p_n a_n}{p_1 + p_2 + \cdots + p_n}$$

are satisfied. Furthermore

$$e^{\frac{\frac{p_1}{a_1}\log a_1 + \frac{p_2}{a_2}\log a_2 + \cdots + \frac{p_n}{a_n}\log a_n}{\frac{p_1}{a_1} + \frac{p_2}{a_2} + \cdots + \frac{p_n}{a_n}}} < \frac{p_1 + p_2 + \cdots + p_n}{\frac{p_1}{a_1} + \frac{p_2}{a_2} + \cdots + \frac{p_n}{a_n}},$$

$$\frac{p_1 a_1 + p_2 a_2 + \cdots + p_n a_n}{p_1 + p_2 + \cdots + p_n} < e^{\frac{p_1 a_1 \log a_1 + p_2 a_2 \log a_2 + \cdots + p_n a_n \log a_n}{p_1 a_1 + p_2 a_2 + \cdots + p_n a_n}}$$

**79.** Let $f(x)$ and $p(x)$ be continuous and positive on the interval $[x_1, x_2]$; further, $f(x)$ is not a constant. Then we have the inequalities

$$\frac{\int_{x_1}^{x_2} p(x)\,dx}{\int_{x_1}^{x_2} \frac{p(x)}{f(x)}\,dx} < e^{\frac{\int_{x_1}^{x_2} p(x) \log f(x)\,dx}{\int_{x_1}^{x_2} p(x)\,dx}} < \frac{\int_{x_1}^{x_2} p(x) f(x)\,dx}{\int_{x_1}^{x_2} p(x)\,dx};$$

moreover

$$e^{\frac{\int_{x_1}^{x_2} \frac{p(x)}{f(x)} \log f(x)\,dx}{\int_{x_1}^{x_2} \frac{p(x)}{f(x)}\,dx}} < \frac{\int_{x_1}^{x_2} p(x)\,dx}{\int_{x_1}^{x_2} \frac{p(x)}{f(x)}\,dx}, \qquad \frac{\int_{x_1}^{x_2} p(x) f(x)\,dx}{\int_{x_1}^{x_2} p(x)\,dx} < e^{\frac{\int_{x_1}^{x_2} p(x) f(x) \log f(x)\,dx}{\int_{x_1}^{x_2} p(x) f(x)\,dx}}.$$

**80.** Let $a_1, a_2, \ldots, a_n, b_1, b_2, \ldots, b_n$ be arbitrary real numbers. Show that they satisfy the inequality

$$(a_1b_1 + a_2b_2 + \cdots + a_nb_n)^2 \leqq (a_1^2 + a_2^2 + \cdots + a_n^2)(b_1^2 + b_2^2 + \cdots + b_n^2).$$

The case of equality arises if and only if the numbers $a_\nu$ and $b_\nu$ are proportional to each other, i.e. $\lambda a_\nu + \mu b_\nu = 0$ for $\nu = 1, 2, \ldots, n$, $\lambda^2 + \mu^2 > 0$. (Cauchy's inequality.)

**81.** Suppose that $f(x)$ and $g(x)$ denote two functions that are properly integrable over the interval $[x_1, x_2]$. Then

$$\left(\int_{x_1}^{x_2} f(x)\, g(x)\, dx\right)^2 \leqq \int_{x_1}^{x_2} [f(x)]^2\, dx \int_{x_1}^{x_2} [g(x)]^2\, dx.$$

(Schwarz's inequality.)

**81.1.** The numbers $a_1, a_2, \ldots, a_n, b_1, b_2, \ldots, b_n, \alpha$ and $\beta$ are positive, $\alpha + \beta = 1$. Then

$$a_1^\alpha b_1^\beta + a_2^\alpha b_2^\beta + \cdots + a_n^\alpha b_n^\beta \leqq (a_1 + a_2 + \cdots + a_n)^\alpha (b_1 + b_2 + \cdots + b_n)^\beta.$$

(Hölder's inequality; **80** follows from the particular case where $\alpha = \beta$.)

**81.2.** The functions $f(x)$ and $g(x)$ and the numbers $\alpha$ and $\beta$ are positive,

$$\alpha + \beta = 1,$$

and the functions are integrable in the interval $[x_1, x_2]$. Then

$$\int_{x_1}^{x_2} [f(x)]^\alpha [g(x)]^\beta\, dx \leqq \left[\int_{x_1}^{x_2} f(x)\, dx\right]^\alpha \left[\int_{x_1}^{x_2} g(x)\, dx\right]^\beta.$$

(For the particular case where $\alpha = \beta$ see **81**.)

**81.3.** The numbers

$$a_1, a_2, \ldots, a_n; \quad b_1, b_2, \ldots, b_n; \quad \ldots; \quad l_1, l_2, \ldots, l_n,$$
$$\alpha, \beta, \ldots, \lambda$$

are positive,

$$\alpha + \beta + \cdots + \lambda = 1.$$

Then

$$a_1^\alpha b_1^\beta \cdots l_1^\lambda + a_2^\alpha b_2^\beta \cdots l_2^\lambda + \cdots + a_n^\alpha b_n^\beta \cdots l_n^\lambda$$
$$\leqq (a_1 + a_2 + \cdots + a_n)^\alpha (b_1 + b_2 + \cdots + b_n)^\beta \cdots (l_1 + l_2 + \cdots l_n)^\lambda.$$

**81.4.** The functions $f(x), g(x), \ldots, h(x)$ and the numbers $\alpha, \beta, \ldots, \lambda$ are positive,

$$\alpha + \beta + \cdots + \lambda = 1$$

and the functions integrable in the interval $[x_1, x_2]$. Then

$$\int_{x_1}^{x_2} [f(x)]^{\alpha} [g(x)]^{\beta} \cdots [h(x)]^{\lambda} dx \leqq \left[\int_{x_1}^{x_2} f(x)\, dx\right]^{\alpha} \left[\int_{x_1}^{x_2} g(x)\, dx\right]^{\beta} \cdots \left[\int_{x_1}^{x_2} h(x)\, dx\right]^{\lambda}.$$

**82.** Let $a_1, a_2, \ldots, a_n$ be arbitrary positive numbers, not all equal. Then the function

$$\psi(t) = \left(\frac{a_1^t + a_2^t + \cdots + a_n^t}{n}\right)^{\frac{1}{t}}$$

is monotone increasing with $t$. Find the values of

$$\psi(-\infty), \quad \psi(-1), \quad \psi(0), \quad \psi(1), \quad \psi(+\infty).$$

(Define $\psi(0)$ so that $\psi(t)$ is continuous at the point $t = 0$.)

**83.** Assume that the function $f(x)$, defined on $[x_1, x_2]$ is properly integrable and that it has a positive lower bound. The function

$$\psi(t) = \left(\frac{1}{x_2 - x_1} \int_{x_1}^{x_2} [f(x)]^t\, dx\right)^{\frac{1}{t}}$$

is non-decreasing for all $t$. Compute

$$\psi(-\infty), \quad \psi(-1), \quad \psi(0), \quad \psi(1), \quad \psi(+\infty).$$

For $\psi(0)$ see **82**. In computing $\psi(-\infty)$ and $\psi(\infty)$ assume that $f(x)$ is continuous.

**84.** Let $a_\nu, b_\nu, \nu = 1, 2, \ldots, n$, be arbitrary positive numbers. Prove the inequality

$$\sqrt[n]{(a_1 + b_1)(a_2 + b_2) \cdots (a_n + b_n)} \geqq \sqrt[n]{a_1 a_2 \cdots a_n} + \sqrt[n]{b_1 b_2 \cdots b_n},$$

i.e.

$$\mathfrak{G}(a + b) \geqq \mathfrak{G}(a) + \mathfrak{G}(b).$$

The relation becomes an equality if and only if $a_\nu = \lambda b_\nu$, $\nu = 1, 2, \ldots, n$.

**85.** The functions $f(x)$ and $g(x)$ are properly integrable over the interval $[x_1, x_2]$ and strictly positive. Then

$$e^{\frac{1}{x_2 - x_1}\int_{x_1}^{x_2} \log[f(x) + g(x)]dx} \geqq e^{\frac{1}{x_2 - x_1}\int_{x_1}^{x_2} \log f(x)dx} + e^{\frac{1}{x_2 - x_1}\int_{x_1}^{x_2} \log g(x)dx},$$

i.e.

$$\mathfrak{G}(f + g) \geqq \mathfrak{G}(f) + \mathfrak{G}(g).$$

**86.** The functions $f_1(x), f_2(x), \ldots, f_m(x)$ are defined on the interval $[x_1, x_2]$, properly integrable and strictly positive (there exists a constant

$K$ such that $0 < K \leqq f_\nu(x)$, $\nu = 1, 2, \ldots, m$, $x_1 \leqq x \leqq x_2$). Let $p_1, p_2, \ldots, p_m$ denote arbitrary positive numbers. Then

$$\mathfrak{G}(p_1 f_1 + p_2 f_2 + \cdots + p_m f_m) \geqq p_1 \mathfrak{G}(f_1) + p_2 \mathfrak{G}(f_2) + \cdots + p_m \mathfrak{G}(f_m).$$

**87.** Suppose that the functions $f_\nu(x)$, $\nu = 1, 2, \ldots, m$, are of bounded variation on the interval $[x_1, x_2]$ and that $p_1, p_2, \ldots, p_m$ are arbitrary positive numbers. Define

$$F(x) = \frac{p_1 f_1(x) + p_2 f_2(x) + \cdots + p_m f_m(x)}{p_1 + p_2 + \cdots + p_m}.$$

The lengths of the arcs of $f_1(x), f_2(x), \ldots, f_m(x)$ are denoted by $l_1, l_2, \ldots, l_m$, the length of the arc of $F(x)$ by $L$. (At a point of discontinuity the jump must be included in the length of the arc.) Then we have

$$L \leqq \frac{p_1 l_1 + p_2 l_2 + \cdots + p_m l_m}{p_1 + p_2 + \cdots + p_m}.$$

**88.** Let $f(x)$ be a positive continuous and periodic function with period $2\pi$ and let $p(x)$ be a non-negative and properly integrable function on $[0, 2\pi]$ with positive integral. Then

$$F(x) = \frac{\int_0^{2\pi} p(\xi) f(\xi + x)\, d\xi}{\int_0^{2\pi} p(\xi)\, d\xi}$$

is positive and continuous; furthermore

$$e^{\frac{1}{2\pi}\int_0^{2\pi} \log F(x) dx} \geqq e^{\frac{1}{2\pi}\int_0^{2\pi} \log f(x) dx}$$

i.e.

$$\mathfrak{G}(F) \geqq \mathfrak{G}(f).$$

**89.** Assume that $f(x)$ is a periodic function with period $2\pi$ and that $p(x)$ is non-negative and properly integrable over the interval $[0, 2\pi]$ and that its integral is positive. If $f(x)$ is of bounded variation then this is true also for

$$F(x) = \frac{\int_0^{2\pi} p(\xi) f(\xi + x)\, d\xi}{\int_0^{2\pi} p(\xi)\, d\xi}.$$

If $l$, $L$ denote the lengths of the arcs of $f(x)$ and $F(x)$ on $[0, 2\pi]$ then $l$ and $L$ satisfy the inequality

$$L \leqq l.$$

**90.** The arbitrary numbers $a_1, a_2, \ldots, a_n$ and $b_1, b_2, \ldots, b_n$ are positive. We define

$$\mathfrak{M}_\varkappa(a) = (a_1^\varkappa + a_2^\varkappa + \cdots + a_n^\varkappa)^{\frac{1}{\varkappa}}.$$

Then

$$\mathfrak{M}_\varkappa(a + b) \leqq \quad \text{or} \quad \geqq \mathfrak{M}_\varkappa(a) + \mathfrak{M}_\varkappa(b),$$

according as $\varkappa \geqq 1$ or $\varkappa \leqq 1$. Equality is attained only for $a_\nu = \lambda b_\nu$, $\nu = 1, 2, \ldots, n$, or if $\varkappa = 1$. (What does the proposition mean in the case $\varkappa = 2$?) (Minkowski's inequality.)

**91.** The function $f(x)$ is defined on $[x_1, x_2]$, properly integrable and strictly positive. We introduce

$$\mathfrak{M}_\varkappa(f) = \left(\int_{x_1}^{x_2} [f(x)]^\varkappa \, dx\right)^{\frac{1}{\varkappa}}.$$

Let $g(x)$ be a function with the same properties as $f(x)$. Then we have

$$\mathfrak{M}_\varkappa(f + g) \leqq \quad \text{or} \quad \geqq \mathfrak{M}_\varkappa(f) + \mathfrak{M}_\varkappa(g),$$

according as $\varkappa \geqq 1$ or $\varkappa \leqq 1$.

**92.** Let $a, A, b, B$ be positive numbers, $a < A$, $b < B$. If the $n$ numbers $a_1, a_2, \ldots, a_n$ lie between $a$ and $A$, and the $n$ numbers $b_1, b_2, \ldots, b_n$ between $b$ and $B$ we can prove that

$$1 \leqq \frac{(a_1^2 + a_2^2 + \cdots + a_n^2)(b_1^2 + b_2^2 + \cdots + b_n^2)}{(a_1 b_1 + a_2 b_2 + \cdots + a_n b_n)^2} \leqq \left(\frac{\sqrt{\frac{AB}{ab}} + \sqrt{\frac{ab}{AB}}}{2}\right)^2.$$

The first inequality is identical with **80**. The second inequality becomes an equality if and only if

$$k = \frac{\frac{A}{a}}{\frac{A}{a} + \frac{B}{b}} n, \quad l = \frac{\frac{B}{b}}{\frac{A}{a} + \frac{B}{b}} n$$

are integers and if $k$ of the numbers $a_\nu$ coincide with $a$ and the remaining $l \,(= n - k)$ of the $a_\nu$'s coincide with $A$, while the corresponding $b_\nu$'s coincide with $B$ and $b$ resp.

**93.** Let $a, A, b, B$ be positive numbers $a < A$, $b < B$. If the two functions $f(x)$ and $g(x)$ are properly integrable over the interval $[x_1, x_2]$ and if $a \leqq f(x) \leqq A$, $b \leqq g(x) \leqq B$ on $[x_1, x_2]$ then

$$1 \leqq \frac{\int_{x_1}^{x_2} [f(x)]^2 \, dx \int_{x_1}^{x_2} [g(x)]^2 \, dx}{\left(\int_{x_1}^{x_2} f(x) \, g(x) \, dx\right)^2} \leqq \left(\frac{\sqrt{\frac{AB}{ab}} + \sqrt{\frac{ab}{AB}}}{2}\right)^2.$$

The first inequality is identical with Schwarz's inequality.

**93.1.** The numbers $a_1, a_2, \ldots, a_n$, $r$, and $s$ are positive, $r < s$, $n > 1$ and $\Sigma$ stands for $\sum_{\nu=1}^{n}$. Then

$$(\Sigma a_\nu^s)^{\frac{1}{s}} < (\Sigma a_\nu^r)^{\frac{1}{r}}.$$

**94.** The function $f(x)$ is defined on $(0, 1)$, non-decreasing, $f(x) \geqq 0$, but not identically zero. Let $0 < a < b$. If all the integrals occurring exist we find the inequalities

$$1 - \left(\frac{a-b}{a+b+1}\right)^2 \leqq \frac{\left(\int_0^1 x^{a+b} f(x)\,dx\right)^2}{\int_0^1 x^{2a} f(x)\,dx \int_0^1 x^{2b} f(x)\,dx} < 1.$$

The inequality on the right hand side is well known. The inequality on the left becomes an equality if and only if $f(x)$ is a constant.

## § 2. Some Applications of Inequalities

**94.1.** A solid is so located in a rectangular coordinate system that its intersection with any straight line that is parallel to one of the three coordinate axes is either empty or consists of just one point or just one line segment. (Such a solid need not be convex.) Let $S$ be the surface area of the boundary of the solid and $P$, $Q$ and $R$ the areas of its orthogonal projections onto the three coordinate planes respectively. Show that

$$2(P^2 + Q^2 + R^2)^{\frac{1}{2}} \leqq S \leqq 2(P + Q + R)$$

and point out simple polyhedra for which the case of equality is attained on one or the other side.

**94.2.** Let $E$ denote the area of the surface of an ellipsoid with semi-axes $a$, $b$, $c$ and prove that

$$\frac{4\pi(bc + ca + ab)}{3} \leqq E \leqq \frac{4\pi(a^2 + b^2 + c^2)}{3}.$$

[Derive, or take for granted, that

$$E = \iint (b^2c^2\xi^2 + c^2a^2\eta^2 + a^2b^2\zeta^2)^{1/2}\,d\omega:$$

the integration is extended over the surface of the unit sphere of which $(\xi, \eta, \zeta)$ is a point,

$$\xi^2 + \eta^2 + \zeta^2 = 1,$$

and $d\omega$ the surface element,

$$\iint d\omega = 4\pi.]$$

**94.3** (continued). The surface area of the ellipsoid is larger than the surface area of the sphere with the same volume, that is

$$E > 4\pi(abc)^{\frac{2}{3}}$$

unless $a = b = c$.

**95.** Let us call the ratio of the electrostatic capacity to the volume of a conductor its "specific capacity". Show that the specific capacity of an ellipsoid with three axes is always between the harmonic and the arithmetic means of the specific capacities of the three spheres whose radii are equal to the three semiaxes of the ellipsoid.

In analytic terms: we have to prove the inequalities

$$\frac{3}{\frac{bc}{a} + \frac{ca}{b} + \frac{ab}{c}} < \frac{1}{2}\int_0^\infty \frac{du}{\sqrt{(a^2+u)(b^2+u)(c^2+u)}} < \frac{\frac{a}{bc} + \frac{b}{ca} + \frac{c}{ab}}{3},$$

which hold for all positive number triples $(a, b, c)$ unless $a = b = c$.

**95.1** (continued). The capacity of the ellipsoid is larger than the capacity of the sphere with the same volume. That is, the upper bound in the double inequality **95** can be replaced by the sharper $(abc)^{-1/3}$.

**95.2.** If all the roots of the equation of degree $n$

$$x^n - a_1x^{n-1} + a_2x^{n-2} - \cdots = 0$$

are real, they are all contained in the interval with the endpoints

$$\frac{a_1}{n} \pm \frac{n-1}{n}\left(a_1^2 - \frac{2n}{n-1}a_2\right)^{\frac{1}{2}}.$$

**95.3.** We consider the non-decreasing sequence of positive numbers $\gamma_1, \gamma_2, \gamma_3, \ldots$

$$0 < \gamma_1 \leqq \gamma_2 \leqq \gamma_3 \leqq \cdots.$$

We set $\gamma_1 = \gamma$,

$$\gamma_1^{-n} + \gamma_2^{-n} + \gamma_3^{-n} + \cdots = s_n$$

and assume that this series is convergent for $n = 1$ (and so also for $n \geqq 1$). Prove that

$$\frac{1}{s_1} < \left(\frac{1}{s_2}\right)^{\frac{1}{2}} < \left(\frac{1}{s_3}\right)^{\frac{1}{3}} < \cdots < \gamma < \cdots < \frac{s_3}{s_4} < \frac{s_2}{s_3} < \frac{s_1}{s_2}$$

and that

$$\lim_{n\to\infty}\left(\frac{1}{s_n}\right)^{1/n} = \gamma = \lim_{n\to\infty}\frac{s_n}{s_{n+1}}.$$

(If $s_1, s_2, s_3, \ldots$ are given, we have here a "perfect" scheme for computing $\gamma$. The $n$-th step yields the lower bound $s_n^{-1/n}$ and the upper bound $s_n/s_{n+1}$ for $\gamma$, the next step improves both bounds and both bounds converge to the desired $\gamma$ as $n$ tends to $\infty$.

Observe that

$$s_1 z + s_2 z^2 + s_3 z^3 + \cdots = \sum_{\nu=1}^{\infty} \frac{z}{\gamma_\nu - z} = -\frac{zG'(z)}{G(z)}$$

where

$$G(z) = \prod_{\nu=1}^{\infty}\left(1 - \frac{z}{\gamma_\nu}\right)$$

and $\gamma$ is the zero of $G(z)$ nearest to the origin. Compare **197**, III **342**.)

**95.4** (continued). Even if $s_n$ is not given for all values of $n$ but only for $n = 1, 2, 4, 8, \ldots, 2^m, \ldots$, we can devise a scheme for computing $\gamma$. Prove that

$$\frac{1}{s_1} < \left(\frac{1}{s_2}\right)^{\frac{1}{2}} < \left(\frac{1}{s_4}\right)^{\frac{1}{4}} < \cdots < \gamma < \cdots < \left(\frac{s_4}{s_8}\right)^{\frac{1}{4}} < \left(\frac{s_2}{s_4}\right)^{\frac{1}{2}} < \frac{s_1}{s_2}$$

and that

$$\lim_{n\to\infty}\left(\frac{s_n}{s_{2n}}\right)^{1/n} = \gamma.$$

**95.5.** A wire which forms a closed plane curve $C$ carries a unit electric current and exerts a force $F$ on a unit magnetic pole in the plane of, and interior to, $C$. Given $A$, the area enclosed by $C$, prove that $F$ is a minimum when $C$ is a circle and the magnetic pole is at its center.—[Assume that $C$ is star-shaped with respect to the magnetic pole [III **109**] which is located at the origin of a system of polar coordinates $r$, $\varphi$. Then

$$F = \int_0^{2\pi} \frac{d\varphi}{r}.$$

Express $A$.]

**96.** Assume that

$$a_{\mu\nu} \geqq 0, \qquad \sum_{\mu=1}^{n} a_{\mu\nu} = \sum_{\nu=1}^{n} a_{\mu\nu} = 1, \qquad x_\nu \geqq 0$$

and

$$y_\mu = a_{\mu 1}x_1 + a_{\mu 2}x_2 + \cdots + a_{\mu n}x_n, \qquad \mu, \nu = 1, 2, \ldots, n.$$

Then we have

$$y_1 y_2 \cdots y_n \geqq x_1 x_2 \cdots x_n.$$

**97.** Let $a_1, a_2, \ldots, a_n$ be positive numbers, $M$ be their arithmetic, $G$ their geometric mean and let $\varepsilon$ denote a proper fraction. Show that the inequality

$$\frac{M-G}{G} \leqq \varepsilon$$

implies the inequalities

$$1 + \varrho < \frac{a_i}{M} < 1 + \varrho', \qquad i = 1, 2, \ldots, n$$

where $\varrho$ and $\varrho'$ denote the only negative and the only positive root respectively of the transcendental equation

$$(1 + x)\, e^{-x} = (1 - \varepsilon)^n.$$

## Chapter 3

## Some Properties of Real Functions

### § 1. Proper Integrals

**98.** We define

$$g(x) = \sin^2 \pi x + \sin^2 \pi x \cos^2 \pi x + \sin^2 \pi x \cos^4 \pi x + \cdots + \sin^2 \pi x \cos^{2k} \pi x + \cdots$$

and

$$G(x) = \lim_{n \to \infty} g(n!\, x).$$

Is $G(x)$ integrable?

**99.** Let the function $f(x)$ (cf. also **169**, VIII **240**) be given by

$$f(x) = \begin{cases} 0 & \text{if} \quad x \text{ is irrational} \\ \dfrac{1}{q} & \text{if} \quad x \text{ is rational}, \quad x = \dfrac{p}{q}, \qquad (p, q) = 1, q \geqq 1. \end{cases}$$

Show that $f(x)$ is continuous for every irrational $x$, discontinuous for every rational $x$ and properly integrable over any interval.

**100.** The two functions $f(x)$ and $\varphi(x)$ are properly integrable over $[a, b]$. Subdivide the interval:

$$a = x_0 < x_1 < x_2 < \cdots < x_{n-1} < x_n = b,$$

$$x_{\nu-1} < y_\nu < x_\nu, \quad x_{\nu-1} < \eta_\nu < x_\nu, \quad \nu = 1, 2, \ldots, n.$$

If $\max(x_\nu - x_{\nu-1}) \to 0$ (the subinterval of maximal length converges to 0) we obtain the relation

$$\lim_{n\to\infty} \sum_{\nu=1}^{n} f(y_\nu)\,\varphi(\eta_\nu)\,(x_\nu - x_{\nu-1}) = \int_a^b f(x)\,\varphi(x)\,dx.$$

**101.** Suppose that $f(x)$ is properly integrable over $[a, b]$ and $\varphi(x)$ properly integrable over $[a, b + d]$, $d > 0$. Then

$$\lim_{\delta\to+0} \int_a^b f(x)\,\varphi(x+\delta)\,dx = \int_a^b f(x)\,\varphi(x)\,dx.$$

**102.** Let $f(x)$ denote a properly integrable function on $[a, b]$. There exist to every positive number $\varepsilon$ two *step-functions*, $\psi(x)$ and $\Psi(x)$, such that for the entire interval $[a, b]$

$$\psi(x) \leqq f(x) \leqq \Psi(x)$$

and

$$\int_a^b \Psi(x)\,dx - \int_a^b \psi(x)\,dx < \varepsilon.$$

It is even possible to choose $\psi(x)$ and $\Psi(x)$ so that their points of discontinuity are equidistant.

**103** (continued). If $f(x)$ is of bounded variation $\psi(x)$ and $\Psi(x)$ may be chosen so that the total variation of neither exceeds the total variation of $f(x)$.

**104.** We define

$$4[x] - 2[2x] + 1 = s(x).$$

Then we have ($n$ integer) the limit relation

$$\lim_{n\to\infty} \int_0^1 f(x)\,s(nx)\,dx = 0$$

for any properly integrable function $f(x)$ on $[0, 1]$. [Sketch $s(nx)$, VIII **3**.]

**105.** Let $f(x)$ be properly integrable over $[a, b]$. Then we can prove that

$$\lim_{n\to\infty} \int_a^b f(x)\sin nx\,dx = 0.$$

**106** (continued). Yet

$$\lim_{n\to\infty} \int_a^b f(x)\,|\sin nx|\,dx = \frac{2}{\pi}\int_a^b f(x)\,dx.$$

Suppose that the function $f(x)$ is bounded on the interval $[a, b]$ and that this interval is subdivided by the points $x_0, x_1, x_2, \ldots, x_{n-1}, x_n$, whereby

$$a = x_0 < x_1 < x_2 < \cdots < x_{n-1} < x_n = b.$$

The greatest lower and the least upper bounds of $f(x)$ in the subinterval $[x_{\nu-1}, x_\nu]$ are denoted by $m_\nu$ and $M_\nu$ respectively. Then $M_\nu - m_\nu$ is called the *oscillation* of $f(x)$ on the $\nu$-th subinterval. The function $f(x)$ is properly integrable over $[a, b]$ if and only if to every pair $\varepsilon$, $\eta$ of positive numbers a subdivision of the interval can be found so that the total length of the subintervals for which the oscillation is larger than $\varepsilon$ is smaller than $\eta$. [Riemann's criterion, cf. l.c. **105**, Riemann, p. 226.]

**107.** The function $\left(\frac{1}{x} - \left[\frac{1}{x}\right]\right) x^\alpha$ is properly integrable over $[0, 1]$ for $\alpha \geqq 0$.

**108.** If $f(x)$ is properly integrable over $[a, b]$ the points of continuity of $f(x)$ are everywhere dense on this interval.

**109.** The function $f(x)$ is properly integrable over $[a, b]$. The equation

$$\int_a^b (f(x))^2\, dx = 0$$

is satisfied if and only if $f(\xi) = 0$ at every point of continuity $\xi$ of $f(x)$, $a < \xi < b$.

**110.** Assume that the function $y = f(x)$ is properly integrable over $[a, b]$, that $m \leqq f(x) \leqq M$ on this interval and that the function $\varphi(y)$ is continuous on $[m, M]$. Then also $\varphi[f(x)]$ is properly integrable over $[a, b]$.

**111.** If $f(x)$ and $\varphi(y)$ are properly integrable $\varphi[f(x)]$ need not be properly integrable. [**98**, **99**.]

## § 2. Improper Integrals

**112.** If $f(x)$ is a monotone function on the interval $(0, 1]$ and if $\int_0^1 x^a f(x)\, dx$ exists then

$$\lim_{x \to 0} x^{a+1} f(x) = 0.$$

**113.** If $f(x)$ is monotone on the interval $[1, \infty)$ and if $\int_1^\infty x^a f(x)\, dx$ exists then

$$\lim_{x \to \infty} x^{a+1} f(x) = 0.$$

**114.** Determine those pairs of real values $\alpha, \beta$ for which the integral

$$\int_0^\infty x^\alpha \left|\cos x\right|^{x^\beta} dx$$

is convergent.

**114.1.** Construct a function $f(x)$ that takes positive values and is bounded and integrable in any finite subinterval of $[0, \infty)$ and such that

$$\int_0^\infty [f(x)]^\alpha \, dx$$

converges for $\alpha = 1$ but diverges for any real value of $\alpha$ different from 1.

**115.** The functions

$$f_1(x), \quad f_2(x), \quad f_3(x), \quad \ldots, \quad f_n(x), \quad \ldots$$

are properly integrable over any finite interval and they satisfy the following conditions:

On every finite interval we have $\lim_{n\to\infty} f_n(x) = f(x)$ uniformly in $x$.

There is a function $F(x)$ such that $|f_n(x)| \leq F(x)$ and $\int_{-\infty}^{\infty} F(x)\,dx$ exists.

Then the limit and the integral can be interchanged:

$$\lim_{n\to\infty} \int_{-\infty}^{\infty} f_n(x)\,dx = \int_{-\infty}^{\infty} f(x)\,dx.$$

[Analogous to I **180**.]

**116.** Prove **58** using VI **31**.

**117.** If the Dirichlet series [VIII, Chap. 1, § 5]

$$a_1 1^{-s} + a_2 2^{-s} + a_3 3^{-s} + \cdots + a_n n^{-s} + \cdots = D(s)$$

converges for $s = \sigma$, $\sigma > 0$ we obtain for $s > \sigma$

$$D(s)\,\Gamma(s) = \int_0^\infty P(y)\,y^{s-1}\,dy,$$

where

$$a_1 e^{-y} + a_2 e^{-2y} + a_3 e^{-3y} + \cdots + a_n e^{-ny} + \cdots = P(y).$$

**118.** Suppose that $f(x)$ is properly integrable over every finite interval and that $\int_{-\infty}^{\infty} |f(x)|\,dx$ exists. Then we have

$$\lim_{n\to\infty} \int_{-\infty}^{+\infty} f(x) \sin nx\,dx = 0, \quad \lim_{n\to\infty} \int_{-\infty}^{+\infty} f(x)\,|\sin nx|\,dx = \frac{2}{\pi} \int_{-\infty}^{+\infty} f(x)\,dx.$$

**118.1.** There are rational functions $R(x)$ such that for an arbitrary function $f(x)$ of the real variable $x$

$$\int_{-\infty}^{\infty} f(R(x))\,dx = \int_{-\infty}^{\infty} f(x)\,dx$$

provided that the integral on the right hand side exists. Show that this property belongs to those, and only to those rational functions that are of the form

$$R(x) = \varepsilon\left(x - \alpha - \frac{p_1}{x-\alpha_1} - \frac{p_2}{x-\alpha_2} - \cdots - \frac{p_n}{x-\alpha_n}\right)$$

where $\varepsilon = +1$ or $-1$, $\alpha, \alpha_1, \alpha_2, \ldots, \alpha_n$ are real, and $p_1, p_2, \ldots, p_n$ are positive, numbers; $n \geqq 0$. (The case $n = 0$ must be interpreted as meaning $R(x) = \varepsilon(x - \alpha)$.)

## § 3. Continuous, Differentiable, Convex Functions

**119.** Are there actually functions of three variables?

More precisely: Is it possible to write every real function $f(x, y, z)$ of three variables with the help of two functions $\varphi(x, y)$ and $\psi(u, z)$ in the form of

$$f(x, y, z) = \psi(\varphi(x, y), z)\,?$$

Discuss the question:

(1) if $f(x, y, z)$, $\varphi(x, y)$, $\psi(u, z)$ are defined for all real values of the variables,

(2) if these functions are defined for all real values of the variables and are *continuous*.

**119a.** Setting $x + y = S(x, y)$, $xy = P(x, y)$ we can write

$$yz + zx + xy = S\{P(x, y), P[S(x, y), z]\};$$

in this formula $xy + yz + zx$ is composed of four functions of two variables "boxed" in each other. Prove that it is impossible to express $f(x, y, z)$ using only three functions of two variables boxed in one another if these functions are defined for all pairs of real values and if they are arbitrarily often differentiable. [We have to show that $f(x, y, z) = yz + zx + xy$ cannot be represented by any of the combinations $\varphi\{\psi[\chi(x, y), z], z\}$, $\varphi[\psi(x, z), \chi(y, z)]$, $\varphi\{\psi[\chi(x, y), z], x\}$.]

**120.** Let the function $f(x)$ have a continuous derivative on $(a, b)$. Decide whether it is possible to find for each point $\xi$ of this interval two points $x_1, x_2$, $x_1 < \xi < x_2$, such that

$$\frac{f(x_2) - f(x_1)}{x_2 - x_1} = f'(\xi).$$

**121.** Assume that the function $f(x)$ is differentiable on $[a, b]$, but not a constant and that $f(a) = f(b) = 0$. Then there exists at least one point $\xi$ on $(a, b)$ for which

$$|f'(\xi)| > \frac{4}{(b-a)^2} \int_a^b f(x)\,dx.$$

**122.** The function $f(x)$ is twice differentiable. Then there exists a point $\xi$ in $(x_0 - r, x_0 + r)$ where

$$f''(\xi) = \frac{3}{r^3} \int_{x_0-r}^{x_0+r} [f(x) - f(x_0)]\,dx.$$

**122.1.** When is the mean value of a function in an interval a simple mean of those two values that the function takes at the endpoints of the interval?

Assume that $f(x)$ is defined, bounded and integrable in the interval $[a, b]$. Introduce the abbreviations

$$f(u) = U, \qquad f(v) = V,$$

$$\frac{1}{v-u} \int_u^v f(x)\,dx = W$$

and determine the most general function $f(x)$ satisfying

$$W = \frac{U+V}{2}$$

for all $u$ and $v$ subject to the condition

$$a \leqq u < v \leqq b.$$

**122.2** (continued). Analogous question for

$$W = \sqrt{UV};$$

assume that $f(x) > 0$ for $a \leqq x \leqq b$.

**122.3** (continued). Analogous question for

$$W = \frac{2UV}{U+V};$$

assume that $f(x) > 0$ for $a \leqq x \leqq b$.

**123.** The numbers $p_0, p_1, p_2, \ldots, p_n, \ldots$ are non-negative, at least two of the $p_\nu$'s do not vanish. Then the logarithm of the series

$$p_0 + p_1 e^x + p_2 e^{2x} + \cdots + p_n e^{nx} + \cdots$$

is a function of $x$, convex on every interval where the series converges.

**124.** A bounded convex function [p. 65] is everywhere continuous and it is even everywhere differentiable from the left and from the right.

**125.** Suppose that the real function $f(x)$ is defined on a finite or infinite interval and that it has there a continuous derivative $f'(x)$. Consider the points of intersection of all the horizontal tangents of the curve $y = f(x)$ with the $y$-axis, i.e. the set $M$ of all points

$$y = f(x) \quad \text{for which} \quad f'(x) = 0.$$

Prove that the set $M$ cannot fill out an entire interval. (This proposition admits a farreaching generalization.)

**126.** If a monotone sequence of continuous functions converges on a closed interval to a continuous function it converges uniformly. (Theorem of Dini.)

**127.** Prove the following counterpart to **126**: If a sequence of monotone (continuous or discontinuous) functions converges on a closed interval to a continuous function it converges uniformly.

## § 4. Singular Integrals. Weierstrass' Approximation Theorem

**128.** If the functions

$$p_1(t), \quad p_2(t), \quad \ldots, \quad p_n(t), \quad \ldots$$

are continuous on the interval $[a, b]$ and if they satisfy the conditions

$$p_n(t) \geqq 0, \qquad \int_a^b p_n(t)\, dt = 1, \qquad n = 1, 2, 3, \ldots,$$

then the terms of the sequence

$$\int_a^b p_1(t) f(t)\, dt, \qquad \int_a^b p_2(t) f(t)\, dt, \ldots, \qquad \int_a^b p_n(t) f(t)\, dt, \ldots$$

are between the minimum and the maximum of the continuous function $f(t)$. (Cf. I **65**, I **79**, I **83**.)

**129.** Let $x$ be a fixed point of the interval $[a, b]$ considered in **128**. In order that

$$\lim_{n\to\infty} \int_a^b p_n(t) f(t)\, dt = f(x)$$

holds for all functions $f(t)$ continuous on $[a, b]$ it is necessary and sufficient that

$$\lim_{n\to\infty} \left( \int_a^{x-\varepsilon} p_n(t)\, dt + \int_{x+\varepsilon}^b p_n(t)\, dt \right) = 0$$

for all positive values of $\varepsilon$ for which $a < x - \varepsilon < x + \varepsilon < b$ (if $x = a$ or $x = b$ the first or second integral resp. under the limit sign has to be omitted.) (Cf. I **66**, I **80**, I **84**.)

**130.** We have

$$\lim_{\varepsilon\to+0} \varepsilon \int_0^\infty e^{-\varepsilon t} f(t)\, dt = \lim_{t\to\infty} f(t),$$

provided that the integral on the left and the limit on the right hand side exist.

**131.** If the integral

$$\int_0^\infty t^\lambda f(t)\, dt$$

converges for $\lambda = \alpha$ and for $\lambda = \beta$, $\alpha < \beta$, it converges for $\alpha \leqq \lambda \leqq \beta$ and it represents a continuous function of $\lambda$ on that interval.

**132.** We assume that

$$p_1(x, t), \quad p_2(x, t), \ldots, \quad p_n(x, t), \ldots$$

are continuous functions of $x$ and $t$, $a \leqq {x \atop t} \leqq b$, and that for each $n$

$$p_n(x, t) \geqq 0, \qquad \int_a^b p_n(x, t)\, dt = 1.$$

Let $f(t)$ denote a continuous function. The functions

$$f_n(x) = \int_a^b p_n(x, t)\, f(t)\, dt, \qquad n = 1, 2, 3, \ldots$$

lie between the minimum and the maximum of $f(t)$ on $[a, b]$ for any $x$, $a \leqq x \leqq b$; i.e. $\min_{a\leqq x\leqq b} f(x) \leqq f_n(x) \leqq \max_{a\leqq x\leqq b} f(x)$. Furthermore

$$\lim_{n\to\infty} f_n(x) = f(x) \quad \text{for} \quad a < x < b,$$

provided

$$\lim_{n\to\infty} \left( \int_a^{x-\varepsilon} p_n(x, t)\, dt + \int_{x+\varepsilon}^b p_n(x, t)\, dt \right) = 0$$

uniformly for $a + \varepsilon \leqq x \leqq b - \varepsilon$, $\varepsilon$ fixed and positive. The convergence is uniform on any closed subinterval of $(a, b)$.

**133.** Let $f(x)$ denote a continuous function on $[0, 1]$. The convergence of

$$\lim_{n\to\infty} \frac{1}{2} \cdot \frac{3}{2} \cdot \frac{5}{4} \cdots \frac{2n+1}{2n} \int_0^1 f(t)\, [1 - (x - t)^2]^n\, dt = f(x)$$

is uniform for $\varepsilon \leqq x \leqq 1 - \varepsilon$, $0 < \varepsilon < \frac{1}{2}$, $\varepsilon$ fixed.

**134.** Let $f(x)$ be a continuous periodic function with period $2\pi$. Then

$$\lim_{n\to\infty} \frac{1}{2n\pi} \int_0^{2\pi} f(t) \left( \frac{\sin n \dfrac{x-t}{2}}{\sin \dfrac{x-t}{2}} \right)^2 dt = f(x),$$

the convergence is uniform for all $x$.

**135.** Every function defined and continuous on the finite interval $[a, b]$ can be approximated uniformly on $[a, b]$ by polynomials to any degree of accuracy. (Weierstrass' approximation theorem.)

**136.** Every continuous function that is periodic with period $2\pi$ can be uniformly approximated by trigonometric polynomials [VI, § 2] to any assigned degree of accuracy. (Weierstrass' approximation theorem.)

**137.** Let $f(x)$ denote a function that is properly integrable over $[a, b]$ ($[0, 2\pi]$). Two polynomials (trigonometric polynomials), $p(x)$ and $P(x)$, can then be found for any positive $\varepsilon$ so that for $a \leqq x \leqq b$ ($0 \leqq x \leqq 2\pi$)

$$p(x) \leqq f(x) \leqq P(x)$$

and

$$\int_a^b P(x)\,dx - \int_a^b p(x)\,dx < \varepsilon \qquad \left(\int_0^{2\pi} P(x)\,dx - \int_0^{2\pi} p(x)\,dx < \varepsilon\right).$$

**138.** The $n$-th moment of a function $f(x)$ is given by

$$\int_a^b f(t)\,t^n\,dt.$$

If all the moments of a function that is defined and continuous on the finite interval $[a, b]$ vanish then the function vanishes identically.

**139.** If all the moments

$$\int_a^b f(t)\,t^n\,dt, \qquad n = 0, 1, 2, \ldots,$$

of a function that is properly integrable over the interval $[a, b]$ vanish then the function $f(x)$ vanishes at every point of continuity.

**140.** If the first $n$ moments vanish,

$$\int_a^b f(x)\,dx = \int_a^b f(x)\,x\,dx = \int_a^b f(x)\,x^2\,dx = \cdots = \int_a^b f(x)\,x^{n-1}\,dx = 0,$$

of a function $f(x)$ defined and continuous on the finite or infinite interval $(a, b)$ then the function changes sign (V, Chap. 1, § 2) at least $n$ times in the interval $(a, b)$ unless it is identically 0.

**141.** The $2n$-th and $(2n+1)$-th trigonometric moment (Fourier constants, cf. VI, § 4) of a function with period $2\pi$ are defined as

$$\int_0^{2\pi} \cos nx\, f(x)\,dx \quad \text{and} \quad \int_0^{2\pi} \sin nx\, f(x)\,dx.$$

If the first $2n+1$ trigonometric moments of a continuous function $f(x)$ with period $2\pi$ vanish then $f(x)$ changes sign at least $2n+2$ times in any interval of length $> 2\pi$ (V, Chap. 1, § 2) unless $f(x)$ is identically 0.

**142.** Let the function $\varphi(x)$ be defined and continuous for $x \geqq 0$. Suppose that the integral

$$J(k) = \int_0^\infty e^{-kx} \varphi(x)\, dx$$

converges for $k = k_0$ and that it vanishes for a sequence of $k$'s increasing in arithmetic progression:

$$J(k_0) = J(k_0 + \alpha) = J(k_0 + 2\alpha) = \cdots = J(k_0 + n\alpha) = \cdots = 0, \quad \alpha > 0.$$

Then $\varphi(x)$ vanishes identically.

**143.** The $\Gamma$-function

$$\Gamma(s) = \lim_{n\to\infty} \frac{n^s\, n!}{s(s+1)\cdots(s+n)}$$

can be written as an integral [**31**]. Use this fact to prove that $\Gamma(s)$ does not have any zeroes. [$\Gamma(s+1) = s\Gamma(s)$, **142**].

We associate with each function that is defined on [0, 1] the polynomials

$$K_n(x) = \sum_{\nu=0}^{n} f\left(\frac{\nu}{n}\right)\binom{n}{\nu} x^\nu (1-x)^{n-\nu}, \qquad n = 0, 1, 2, \ldots$$

This polynomial is bounded on [0, 1] from below by the greatest lower bound, and from above by the least upper bound, of $f(x)$ and it coincides with $f(x)$ at the endpoints.

**144.** Work out the polynomials $K_n(x)$, $n = 0, 1, 2, \ldots$ for

$$f(x) = 1, \quad f(x) = x, \quad f(x) = x^2, \quad f(x) = e^x.$$

**145.** Let $x$ be any point on [0, 1] and

$$1 = \sum_{\nu=0}^{n} \binom{n}{\nu} x^\nu (1-x)^{n-\nu} = \Sigma^{\mathrm{I}} + \Sigma^{\mathrm{II}},$$

where $\Sigma^{\mathrm{I}}$ refers to the subscripts for which $|\nu - nx| \leqq n^{3/4}$ and $\Sigma^{\mathrm{II}}$ to those for which $|\nu - nx| > n^{3/4}$, $n \geqq 1$. Then

$$\Sigma^{\mathrm{II}} < \frac{1}{4} n^{-\frac{1}{2}}.$$

**146.** Let $f(x)$ be continuous on [0, 1]. The polynomials $K_n(x)$ converge uniformly to $f(x)$ on [0, 1]. (New proof of Weierstrass' theorem, **135**.)

## Chapter 4

# Various Types of Equidistribution

## § 1. Counting Function. Regular Sequences

In the sequel we are considering monotone sequences of positive numbers. The *counting function* $N(r)$ of such a sequence $r_1, r_2, \ldots, r_n, \ldots,$ $0 < r_1 \leqq r_2 \leqq r_3 \leqq \cdots \leqq r_n \leqq \cdots$, is defined as the number of those $r_n$'s that are not larger than $r$, $r \geqq 0$:

$$N(r) = \sum_{r_n \leqq r} 1.$$

(If $f(t)$ is a function of $t$ then $\sum_{r_n \leqq r} f(r_n)$ denotes the sum $f(r_1) + f(r_2) + \cdots + f(r_m)$, $r_m \leqq r < r_{m+1}$.) E.g. if $r_1 = 1$, $r_2 = 2$, $r_3 = 3, \ldots$ then $N(r) = [r]$.

$N(r)$ is a piecewise constant, non-decreasing function whose jumps are integers and which is everywhere continuous on the right.

**147.** If $f(t)$ is differentiable and $f'(t)$ properly integrable, $t > 0$, then

$$\sum_{r_n \leqq r} f(r_n) = N(r)\, f(r) - \int_0^r N(t)\, f'(t)\, dt.$$

**148.** Let $N(r)$ denote the counting function of the sequence $r_1, r_2, r_3, \ldots, r_n, \ldots$, which increases to infinity. Then

$$\limsup_{r\to\infty} \frac{N(r)}{r} = \limsup_{n\to\infty} \frac{n}{r_n}, \qquad \liminf_{r\to\infty} \frac{N(r)}{r} = \liminf_{n\to\infty} \frac{n}{r_n},$$

$$\limsup_{r\to\infty} \frac{\log N(r)}{\log r} = \limsup_{n\to\infty} \frac{\log n}{\log r_n}, \qquad \liminf_{r\to\infty} \frac{\log N(r)}{\log r} = \liminf_{n\to\infty} \frac{\log n}{\log r_n}.$$

**149.** The counting function $N(r)$ and the convergence exponent $\lambda$ of the sequence $r_1, r_2, r_3, \ldots, r_n, \ldots$ [I, Chap. 3, § 2] are connected by the relation

$$\limsup_{r\to\infty} \frac{\log N(r)}{\log r} = \lambda.$$

**150.** A function $L(r)$ defined and positive for $r > 0$ is termed *slowly increasing* if it is monotone increasing and satisfies the condition

$$\lim_{r\to\infty} \frac{L(2r)}{L(r)} = 1.$$

Show that

$$\lim_{r\to\infty} \frac{L(cr)}{L(r)} = 1, \qquad c > 0.$$

**151.** Suppose that $L(r)$ is positive for $r > 0$, monotone increasing and that for $r$ sufficiently large

$$L(r) = (\log r)^{\alpha_1} (\log_2 r)^{\alpha_2} \cdots (\log_k r)^{\alpha_k}, \qquad \alpha_1 > 0.$$

[$\log_k x = \log_{k-1} (\log x)$.] Then $L(r)$ is slowly increasing.

**152.** If $L(r)$ is slowly increasing then

$$\lim_{r\to\infty} \frac{\log L(r)}{\log r} = 0.$$

**153.** If $N(r)$ denotes the counting function of the sequence $r_1, r_2, r_3, \ldots, r_n, \ldots$ and if

$$N(r) \sim r^\lambda L(r),$$

where $L(r)$ is slowly increasing, $0 < \lambda < \infty$, then $\lambda$ is the convergence exponent of the sequence $r_1, r_2, r_3, \ldots, r_n, \ldots$

A sequence $r_1, r_2, r_3, \ldots, r_n, \ldots$ of the type considered in **153** will be called a ***regular sequence*** in the sequel, **154**—**159**. Later on (e.g. IV **59**—IV **65**) sequences for which $N(r) \sim \frac{r^\lambda}{L(r)}$ will also be termed regular. If we take the term in this broader sense also the prime numbers 2, 3, 5, 7, 11, ... form a regular sequence and the propositions **153**—**159** remain valid without alteration.

**154.** The counting function of a regular sequence with convergence exponent $\lambda$ satisfies the relation

$$\lim_{r\to\infty} \frac{N(cr)}{N(r)} = c^\lambda, \qquad c > 0.$$

**155.** Let $N(r)$ be the counting function of the regular sequence $r_1, r_2, r_3, \ldots, r_n, \ldots$ with convergence exponent $\lambda$ and $f(x)$ be a piecewise constant function on the interval $(0, c]$, $c > 0$. Then

$$\lim_{r\to\infty} \frac{1}{N(r)} \sum_{r_n \leq cr} f\left(\frac{r_n}{r}\right) = \int_0^{c^\lambda} f\left(x^{\frac{1}{\lambda}}\right) dx.$$

**156.** The limit relation in **155** is also valid if $f(x)$ denotes a properly integrable function on $[0, c]$.

**157.** Let $N(r)$ denote the counting function of the regular sequence $r_1, r_2, \ldots, r_n, \ldots$ with convergence exponent $\lambda$ and let $\alpha > 0$. Then

$$\lim_{r\to\infty} \frac{1}{N(r)} \sum_{r_n \leq r} \left(\frac{r_n}{r}\right)^{\alpha-\lambda} = \int_0^1 x^{\frac{\alpha-\lambda}{\lambda}} dx = \frac{\lambda}{\alpha}.$$

**158** (continued).

$$\lim_{r\to\infty}\frac{1}{N(r)}\sum_{r_n>r}\left(\frac{r_n}{r}\right)^{-\alpha-\lambda}=\int_1^\infty x^{\frac{-\alpha-\lambda}{\lambda}}\,dx=\frac{\lambda}{\alpha}.$$

**159.** Assume that $N(r)$ is the counting function of the regular sequence $r_1, r_2, \ldots, r_n, \ldots$ with convergence exponent $\lambda$, that $f(x)$ is defined for $x > 0$ and properly integrable over every finite interval $[a, b]$, $0 < a < b$ and that furthermore

$$|f(x)| < x^{\alpha-\lambda} \quad \text{in a neighbourhood of } x = 0$$

and

$$|f(x)| < x^{-\alpha-\lambda} \quad \text{in a neighbourhood of } x = \infty, \qquad \alpha > 0.$$

Show that

$$\lim_{r\to\infty}\frac{1}{N(r)}\sum_{n=1}^{\infty} f\left(\frac{r_n}{r}\right)=\int_0^\infty f\left(x^{\frac{1}{\lambda}}\right)dx.$$

**160.** Suppose that the function $f(x)$ is defined and monotone on the interval $(0, 1]$ and that it satisfies in the neighbourhood of $x = 0$ the inequality

$$|f(x)| < x^{\alpha-\lambda}, \qquad \alpha > 0.$$

The counting function and the convergence exponent of the positive sequence $r_1, r_2, \ldots, r_n, \ldots$ are called $N(r)$ and $\lambda$ resp.; let $0 < \lambda < \infty$. Then

$$\liminf_{r\to\infty}\frac{1}{N(r)}\sum_{r_n\leqq r} f\left(\frac{r_n}{r}\right)\leqq\int_0^1 f\left(x^{\frac{1}{\lambda}}\right)dx\leqq\limsup_{r\to\infty}\frac{1}{N(r)}\sum_{r_n\leqq r} f\left(\frac{r_n}{r}\right). \qquad [\text{I } \mathbf{115}.]$$

The sequence $r_n$ need not be regular.

**161.** The function $f(x)$ is defined for $x > 0$, is positive and decreasing and satisfies the inequalities

$$f(x) < x^{\alpha-\lambda} \quad \text{in the neighbourhood of } x = 0$$

$$f(x) < x^{-\alpha-\lambda} \quad \text{in the neighbourhood of } x = \infty, \qquad \alpha > 0.$$

The sequence $r_1, r_2, r_3, \ldots, r_n, \ldots$ is defined as in **160**. Then

$$\liminf_{r\to\infty}\frac{1}{N(r)}\sum_{n=1}^{\infty} f\left(\frac{r_n}{r}\right)\leqq\int_0^\infty f\left(x^{\frac{1}{\lambda}}\right)dx. \qquad [\text{I } \mathbf{116}.]$$

## § 2. Criteria of Equidistribution

A sequence of the form

$$x_1, x_2, x_3, \ldots, x_n, \ldots$$

is called *equidistributed* in the interval [0, 1] if all the $x_1, x_2, x_3, \ldots, x_n, \ldots$ are on [0, 1] and if for every function that is properly integrable over [0, 1] the following equation holds:

$$(*) \qquad \lim_{n\to\infty} \frac{f(x_1) + f(x_2) + \cdots + f(x_n)}{n} = \int_0^1 f(x)\,dx.$$

The term "equidistribution" is explained by the following criterion:

**162.** A sequence $x_1, x_2, x_3, \ldots, x_n, \ldots$, $0 \leqq x_n \leqq 1$, is equidistributed on [0, 1] if and only if the "probability" of a term $x_n$ to fall into a certain subinterval of [0, 1] is equal to the length of that subinterval. More precisely, if the sequence has the following property: Let $[\alpha, \beta]$ be an arbitrary subinterval of [0, 1] and $\nu_n(\alpha, \beta)$ denote the number of $x_\nu$'s, $\nu = 1, 2, \ldots, n$, on $[\alpha, \beta]$, then

$$\lim_{n\to\infty} \frac{\nu_n(\alpha, \beta)}{n} = \beta - \alpha. \qquad [\mathbf{102.}]$$

**163.** Let $[\alpha, \beta]$ be an arbitrary subinterval of [0, 1] and $s_n(\alpha, \beta)$ denote the sum of the $x_\nu$'s, $\nu = 1, 2, , \ldots, n$, that fall into $[\alpha, \beta]$. A sequence $x_1, x_2, x_3, \ldots, x_n, \ldots$, $0 \leqq x \leqq 1$, is equidistributed if and only if

$$\lim_{n\to\infty} \frac{s_n(\alpha, \beta)}{n} = \frac{\beta^2 - \alpha^2}{2}.$$

**164.** A sequence $x_1, x_2, x_3, \ldots, x_n, \ldots$, $0 \leqq x_n \leqq 1$, is equidistributed on [0, 1] if and only if for every positive integer $k$

$$\lim_{n\to\infty} \frac{x_1^k + x_2^k + x_3^k + \cdots + x_n^k}{n} = \frac{1}{k+1}. \qquad [\mathbf{137.}]$$

**165.** A sequence $x_1, x_2, x_3, \ldots, x_n, \ldots$, $0 \leqq x_n \leqq 1$, is equidistributed on [0, 1] if and only if the two equations

$$\lim_{n\to\infty} \frac{\cos 2\pi k x_1 + \cos 2\pi k x_2 + \cdots + \cos 2\pi k x_n}{n} = 0,$$

$$\lim_{n\to\infty} \frac{\sin 2\pi k x_1 + \sin 2\pi k x_2 + \cdots + \sin 2\pi k x_n}{n} = 0$$

hold for every positive integer $k$. [**137.**]

## § 3. Multiples of an Irrational Number

**166.** Let $\theta$ be an irrational number. The numbers

$$x_n = \theta n - [\theta n]$$

are equidistributed on the interval [0, 1].

**167.** Let $\theta$ denote an irrational number. Put $\varepsilon_n = 1$ or $\varepsilon_n = 0$ according as the integer next to $n\theta$ is larger or smaller than $n\theta$. Let $a$ and $d$, $a \geqq 0$, $d > 0$, stand for two integers. Then we find that

$$\lim_{n\to\infty} \frac{\sum_{k=0}^{n-1} \varepsilon_{a+kd}}{n} = \frac{1}{2}.$$

**168.** Let $\theta$ be an irrational number and $\alpha$ be defined as $\alpha = q\theta$, $q$ integer, $q \neq 0$. As $z$ converges to $e^{2\pi i\alpha}$ along the ray $\arg z = 2\pi\alpha$, the function

$$f(z) = \sum_{n=1}^{\infty} (n\theta - [n\theta])\, z^n, \qquad z \text{ arbitrary complex, } |z| < 1,$$

increases to $\infty$ in such a way that

$$\lim_{r\to 1-0} (1-r)\, f(re^{2\pi i\alpha}) = \frac{1}{2\pi i q} \qquad \text{[I 88]}.$$

**169.** Determine for real $x$ the function

$$f(x) = \lim_{n\to\infty} \frac{\cos^2 \pi x + \cos^4 2\pi x + \cos^6 3\pi x + \cdots + \cos^{2n} n\pi x}{n}.$$

**170.** The decimal fraction

$$\theta = 0.12345678910111213\ldots$$

(the natural numbers listed consecutively) represents an irrational number. According to **166** the numbers

$$n\theta - [n\theta], \qquad n = 1, 2, 3, \ldots$$

are everywhere dense on the interval $[0, 1]$. Show that this is already the case for the subset

$$10^n\theta - [10^n\theta], \qquad n = 0, 1, 2, 3, \ldots$$

**171.** The number

$$e = 1 + \frac{1}{1!} + \frac{1}{2!} + \frac{1}{3!} + \cdots + \frac{1}{n!} + \cdots$$

is irrational. [VIII **258**.] Prove that the only limit point of the set

$$n!\,e - [n!\,e], \qquad n = 1, 2, 3, \ldots$$

is zero.

**172.** Suppose that the polynomial $P(x) = a_1x + a_2x^2 + \cdots + a_rx^r$ has at least one irrational coefficient. Then the numbers

$$P(n) - [P(n)], \qquad n = 1, 2, 3, \ldots$$

have infinitely many limit points.

**173.** Let $\theta$ be an irrational number, $x_n = n\theta - [n\theta]$, $n = 1, 2, 3, \ldots$ and let $\alpha_1, \alpha_2, \alpha_3, \ldots, \alpha_n, \ldots$ be a monotone decreasing sequence of positive numbers whose sum diverges. Then we find for any properly integrable function $f(x)$ on $[0, 1]$ that

$$\lim_{n\to\infty} \frac{\alpha_1 f(x_1) + \alpha_2 f(x_2) + \cdots + \alpha_n f(x_n)}{\alpha_1 + \alpha_2 + \cdots + \alpha_n} = \int_0^1 f(x)\,dx.$$

## § 4. Distribution of the Digits in a Table of Logarithms and Related Questions

**174.** The function $g(t)$ has the following properties for $t \geqq 1$:

(1) $g(t)$ is continuously differentiable;

(2) $g(t)$ is monotone increasing to $\infty$ as $t \to \infty$;

(3) $g'(t)$ is monotone decreasing to 0 as $t \to \infty$;

(4) $tg'(t)$ tends to $\infty$ as $t \to \infty$.

Then the numbers

$$x_n = g(n) - [g(n)], \qquad n = 1, 2, 3, \ldots$$

are equidistributed on the interval $[0, 1]$.

**175.** Suppose that $a > 0$, $0 < \sigma < 1$. The sequence

$$x_n = an^\sigma - [an^\sigma]$$

is equidistributed on the interval $[0, 1]$.

**176.** Let $a > 0$, $\sigma > 1$. The numbers

$$x_n = a(\log n)^\sigma - [a(\log n)^\sigma]$$

are equidistributed on $[0, 1]$.

**177.** For $0 < \sigma < 1$, $\xi \neq 0$ the series

$$\frac{\sin 1^\sigma \xi}{1^\varrho} + \frac{\sin 2^\sigma \xi}{2^\varrho} + \frac{\sin 3^\sigma \xi}{3^\varrho} + \cdots + \frac{\sin n^\sigma \xi}{n^\varrho} + \cdots$$

is absolutely convergent if and only if $\varrho > 1$.

**178.** Suppose that the square roots of the natural numbers $1, 2, 3, \ldots$ are written up one below the other in an infinite array. Examine the digits at the $j$-th decimal place (to the right of the decimal point), $j \geqq 1$. Each digit $0, 1, 2, \ldots, 9$ appears on the average equally often. More precisely: let $\nu_g(n)$ denote the number of those integers $\leqq n$ whose square roots show a $g$ at the $j$-th decimal place. Then

$$\lim_{n\to\infty} \frac{\nu_g(n)}{n} = \frac{1}{10}, \qquad g = 0, 1, 2, \ldots, 9.$$

**179.** Assume $a > 0$ and $x_n = a \log n - [a \log n]$, $n = 1, 2, 3, \ldots$ and that the arbitrary function $f(x)$ is defined and properly integrable over [0, 1]. Then the limit relation

$$\lim_{n\to\infty} \frac{f(x_1) + f(x_2) + \cdots + f(x_n)}{n} = \int_0^1 f(x)\, K(x, \xi)\, dx$$

holds provided that $n$ increases to infinity in such a manner that $x_n \to \xi$, $0 \leqq \xi \leqq 1$. The function $K(x, \xi)$ is given by

$$K(x, \xi) = \begin{cases} \dfrac{\log q}{q-1} q^{x-\xi+1} & \text{if } 0 \leqq x < \xi \\ \dfrac{\log q}{q-1} q^{x-\xi} & \text{if } \xi < x \leqq 1, \quad q = e^{1/a}, \quad 0 < \xi < 1; \end{cases}$$

$$K(x, 0) = K(x, 1) = \frac{\log q}{q-1} q^x.$$

**180** (continued). The limit points of

$$\frac{f(x_1) + f(x_2) + \cdots + f(x_n)}{n}, \qquad n = 1, 2, 3, \ldots$$

cover an entire interval $J = J(a, f)$ which depends on $a$ and $f$ only. This interval degenerate into a point if and only if $f(x) = c$, $c$ a constant, at each point of continuity. What can you say about $J(a, f)$ if $a$ is a very large or a very small positive number?

**181.** Suppose that the common logarithms (to the base 10) of the natural numbers 1, 2, 3, 4, ... are listed below each other in an infinite table of logarithms. Consider the digits at the $j$-th decimal place (to the right of the decimal point), $j \geqq 1$. There exists no definite probability for the distribution of the digits 0, 1, 2, ..., 9 in this sequence. More exactly: let $\nu_g(n)$ denote the number of those integers $\leqq n$ whose logarithms show the digit $g$ at their $j$-th decimal place. Then the quotients $\frac{\nu_g(n)}{n}$ do not have a limit as $n \to \infty$: Their limit points fill out an entire interval of positive length.

**182.** The function $g(t)$ has the following properties for $t \geqq 1$:

(1) $g(t)$ is continuously differentiable;

(2) $g(t)$ is monotone increasing to $\infty$ as $t \to \infty$;

(3) $g'(t)$ is monotone decreasing to 0 as $t \to \infty$;

(4) $tg'(t) \to 0$ as $t \to \infty$.

(Cf. **174**.) Then the numbers

$$x_n = g(n) - [g(n)], \qquad n = 1, 2, 3, \ldots,$$

are everywhere dense on the interval [0, 1] but they are not equidistributed. Their distribution is characterized by the following limit theorem: Let the function $f(x)$ be properly integrable over the interval [0, 1]. If $n$

increases to infinity so that $x_n \to \xi$, $0 < \xi < 1$, then

$$\lim_{n\to\infty} \frac{f(x_1) + f(x_2) + \cdots + f(x_n)}{n} = f(\xi)$$

holds provided that $f(x)$ is continuous at $x = \xi$. If $f(x)$ has a simple discontinuity (jump) at the point $x = \xi$, the set of limit points of

$$\frac{f(x_1) + f(x_2) + \cdots + f(x_n)}{n}$$

covers the interval $[f(\xi - 0), f(\xi + 0)]$. The statement is true also for $\xi = 0$ or $\xi = 1$ if $f(1) = f(0)$ and if $f(x)$ is extended so that it becomes a periodic function with period 1. [Then $f(1+0) = f(+0)$, $f(1-0) = f(-0)$.]

**183.** The sequence

$$x_n = a (\log n)^\sigma - [a (\log n)^\sigma], \qquad n = 1, 2, 3, \ldots$$

is for $0 < \sigma < 1$ everywhere dense on the interval $[0, 1]$ but not equidistributed. [**176, 179.**]

**184.** Assume that the square roots of the logarithms of the natural numbers 1, 2, 3, 4, ... are tabulated below each other in an infinite array. Consider the digits at the $j$-th decimal place (to the right of the decimal point) $j \geqq 1$. There exists no definite probability for the distribution of the digits 0, 1, 2, ..., 9 at the $j$-th decimal place. More exactly: Let $\nu_g(n)$ denote the number of integers $k$ among the first $n$ integers for which $\sqrt{\log k}$ has the digit $g$ at the $j$-th decimal place. Then the quotients $\frac{\nu_g(n)}{n}$, $n = 1, 2, 3, \ldots$ are everywhere dense between 0 and 1.

## § 5. Other Types of Equidistribution

**185.** Imagine in the $p$-dimensional space a rectilinear uniform motion described by the equations $x_\nu(t) = a_\nu + \theta_\nu t$, $a_\nu$, $\theta_\nu$ constants, $\nu = 1, 2, \ldots, p$, $t$ time. If the numbers $\theta_1, \theta_2, \ldots, \theta_p$ are rationally independent (i.e. if $n_1\theta_1 + n_2\theta_2 + \cdots + n_p\theta_p = 0$, $n_1, n_2, \ldots, n_p$ rational, has the only solution $n_1 = n_2 = \cdots = n_p = 0$) any function $f(x_1, x_2, \ldots, x_p)$, that is periodic in $x_1, x_2, \ldots, x_p$ with period 1 and properly integrable over the unit cube $0 \leqq x_\nu \leqq 1$, $\nu = 1, 2, \ldots, p$, satisfies the relation

$$\lim_{t\to\infty} \frac{1}{t} \int_0^t f(x_1(t), x_2(t), \ldots, x_p(t))\, dt = \int_0^1 \int_0^1 \cdots \int_0^1 f(x_1, x_2, \ldots, x_p)\, dx_1\, dx_2 \cdots dx_p.$$

**186.** Let $\alpha_1$, $\alpha_2$, $\beta_1$, $\beta_2$ be arbitrary constants, $0 \leqq \alpha_1 < \alpha_2 < 1$, $0 \leqq \beta_1 < \beta_2 < 1$. The conditions

$$\alpha_1 \leqq x - [x] \leqq \alpha_2, \quad \beta_1 \leqq y - [y] \leqq \beta_2$$

determine an infinite number of rectangles with sides parallel to the axes and congruent mod 1, i.e. they are congruent by translations parallel to the axes through integral lengths. The equations $x = a + \theta_1 t$, $y = b + \theta_2 t$, $a, \theta_1, b, \theta_2$ constants, $t$ time, define a linear uniform motion. Let $T(t)$ denote the sum of the time intervals up to time $t$ the moving point is spending in one of the above mentioned rectangles. In the case where $\theta_1 : \theta_2$ is irrational we can establish the relation

$$\lim_{t\to\infty} \frac{T(t)}{t} = (\alpha_2 - \alpha_1)(\beta_2 - \beta_1).$$

**187.** A billiard ball is moving rectilinearly with constant speed on a smooth square table with surface $\mathfrak{F}$. The ball is reflected by the cushion each time according to the law of reflection (angle of incidence = angle of reflection). Suppose that the tangent of the angle between the direction of the motion and a side of the billiard table is irrational. We denote by $T(t)$ the sum of the time intervals up to the time $t$ that the moving ball spends in a certain subregion of size $\mathfrak{f}$. Then

$$\lim_{t\to\infty} \frac{T(t)}{t} = \frac{\mathfrak{f}}{\mathfrak{F}}.$$

The numbers

$$\frac{1}{n}, \quad \frac{2}{n}, \quad \frac{3}{n}, \ldots, \quad \frac{n}{n}, \qquad n = 1, 2, 3, \ldots$$

which appear in the construction of sums of rectangles (subdivision according to an arithmetic progression) are in a certain sense equidistributed. A similar type of equidistribution comes up in the next two problems.

**188.** Let $r_{1n}, r_{2n}, r_{3n}, \ldots, r_{\varphi n}$ denote the positive integers that are smaller than $n$ and relative prime to $n$; their number is $\varphi = \varphi(n)$ [VIII **25**]. Then

$$\lim_{n\to\infty} \frac{f\left(\frac{r_{1n}}{n}\right) + f\left(\frac{r_{2n}}{n}\right) + f\left(\frac{r_{3n}}{n}\right) + \cdots + f\left(\frac{r_{\varphi n}}{n}\right)}{\varphi(n)} = \int_0^1 f(x)\,dx$$

holds for any properly integrable function $f(x)$ on [0, 1]. [VIII **35**.]

**189.** We write down in increasing order all reduced fractions $\leqq 1$ whose numerators and denominators are among the numbers $1, 2, 3, \ldots, n$:

$$w_1, \quad w_2, \quad w_3, \ldots, \quad w_N$$

(Farey series

$$w_1 = \frac{1}{n}, \ldots, \quad w_N = \frac{1}{1}, \quad N = N(n) = \varphi(1) + \varphi(2) + \cdots + \varphi(n)).$$

Then the relation

$$\lim_{n\to\infty} \frac{f(w_1) + f(w_2) + f(w_3) + \cdots + f(w_N)}{N} = \int_0^1 f(x)\, dx$$

holds for any properly integrable function $f(x)$ on $[0, 1]$. [I **70**.]

Some of the number sequences occurring in the preceding problems were equidistributed, i.e. the probability that a number would fall into a certain interval was proportional to its length, e.g. **166**, **175**, **188**. This will not be the case in the following examples: For these sequences there exists a certain probability density according to which the density of points may be different in different subintervals. A similar case appeared already in **159**.

**190.** Suppose that $f(x)$ is a properly integrable function on $\left[0, \sqrt{\frac{2}{\pi}}\right]$ and that there exists a positive number $p$ such that $x^{-p}f(x)$ is bounded on this interval. We set

$$\frac{\sqrt{n}\binom{n}{\nu}}{2^n} = s_{\nu n}, \qquad \nu = 0, 1, \ldots, n; \qquad n = 1, 2, 3, \ldots$$

Then

$$\lim_{n\to\infty} \frac{f(s_{0n}) + f(s_{1n}) + f(s_{2n}) + \cdots + f(s_{nn})}{\sqrt{n}} = \int_{-\infty}^{+\infty} f\left(\sqrt{\frac{2}{\pi}}\, e^{-2x^2}\right) dx.$$

**191.** Let

$$x_{1n}, x_{2n}, \ldots, x_{nn}, \qquad -1 < x_{\nu n} < 1, \qquad \nu = 1, 2, \ldots, n,$$

be the zeros of the $n$-th Legendre polynomial $P_n(x)$ [VI **97**] and $\lambda$ be real, $\lambda > 1$. Then

$$\lim_{n\to\infty} \frac{\log\left(1 + \frac{x_{1n}}{\lambda}\right) + \log\left(1 + \frac{x_{2n}}{\lambda}\right) + \cdots + \log\left(1 + \frac{x_{nn}}{\lambda}\right)}{n} = \log \frac{\lambda + \sqrt{\lambda^2 - 1}}{2\lambda},$$

where the positive value of the square root is considered. [Use **203**.]

**192** (continued). Let $k$ be any positive integer. Show that

$$\lim_{n\to\infty} \frac{x_{1n}^k + x_{2n}^k + \cdots + x_{nn}^k}{n} = \frac{1}{\pi} \int_0^\pi \cos^k \vartheta\, d\vartheta. \qquad \text{[I 179.]}$$

**193.** Let

$$x_{1n}, x_{2n}, \ldots, x_{nn}, \qquad -1 < x_{\nu n} < 1, \qquad \nu = 1, 2, \ldots, n,$$

denote the zeros of the $n$-th Legendre polynomial $P_n(x)$, and $f(x)$ be a properly integrable function on $[-1, 1]$. Then

$$\lim_{n\to\infty} \frac{f(x_{1n}) + f(x_{2n}) + \cdots + f(x_{nn})}{n} = \frac{1}{\pi} \int_0^\pi f(\cos\vartheta)\, d\vartheta.$$

**194.** Assume that $\alpha \leqq x \leqq \beta$ is an arbitrary subinterval of $[-1, 1]$ and that $\nu_n(\alpha, \beta)$ is the number of zeros in $[\alpha, \beta]$ of the $n$-th Legendre polynomial. Then

$$\lim_{n\to\infty} \frac{\nu_n(\alpha, \beta)}{n} = \frac{\arccos \alpha - \arccos \beta}{\pi}.$$

The points $x_{\nu n}$ are not equidistributed on the interval $[-1, 1]$ but the values $\arccos x_n$ are equidistributed on $[0, \pi]$. We may interpret the interval $[-1, 1]$ as the horizontal diameter of a circle and each point $x$ as the normal projection of two points of the circumference onto the diameter. We are facing here an equidistribution on the circumference but not on the diameter.

# Chapter 5

# Functions of Large Numbers

## § 1. Laplace's Method

**195.** Let $p_1, p_2, \ldots, p_l$, $a_1, a_2, \ldots, a_l$ be arbitrary positive numbers. Then

$$\lim_{n\to\infty} \sqrt[n]{p_1 a_1^n + p_2 a_2^n + \cdots + p_l a_l^n}$$

exists and it is equal to the largest among the numbers $a_1, a_2, \ldots, a_l$.

**196.** Under the same hypotheses as in **195**

$$\lim_{n\to\infty} \frac{p_1 a_1^{n+1} + p_2 a_2^{n+1} + \cdots + p_l a_l^{n+1}}{p_1 a_1^n + p_2 a_2^n + \cdots + p_l a_l^n} = \max(a_1, a_2, \ldots, a_l).$$

**197.** Let $f(x)$ be an arbitrary polynomial whose zeros are all real and positive and for which

$$-\frac{f'(x)}{f(x)} = c_0 + c_1 x + c_2 x^2 + \cdots + c_n x^n \cdots.$$

Show that

$$\lim_{n\to\infty} \frac{1}{\sqrt[n]{c_n}} = \lim_{n\to\infty} \frac{c_{n-1}}{c_n}$$

exists and that it is equal to the smallest zero of $f(x)$.

**198.** The two functions $\varphi(x)$ and $f(x)$ are continuous and positive on the interval $[a, b]$. Then

$$\lim_{n\to\infty} \sqrt[n]{\int_a^b \varphi(x)\,[f(x)]^n\,dx}$$

exists and is equal to the maximum of $f(x)$ on $[a, b]$.

**199.** Under the same hypothesis as in **198**

$$\lim_{n\to\infty} \frac{\int_a^b \varphi(x)\,[f(x)]^{n+1}\,dx}{\int_a^b \varphi(x)\,[f(x)]^n\,dx} = \max f(x).$$

**200.** Let $k$ be a positive constant and $a < \xi < b$. Show that for $a, b, \xi, k$ fixed and $n \to \infty$

$$\int_a^b e^{-kn(x-\xi)^2}\,dx \sim \sqrt{\frac{\pi}{kn}}.$$

**201.** The functions $\varphi(x)$, $h(x)$ and $f(x) = e^{h(x)}$ are defined on the finite or infinite interval $[a, b]$ and satisfy the following conditions:

(1) $\varphi(x)\,[f(x)]^n = \varphi(x)\,e^{nh(x)}$ is absolutely integrable over $[a, b]$; $n = 0, 1, 2, \ldots$

(2) The function $h(x)$ attains its maximum only at the point $\xi$ in $(a, b)$; moreover, the least upper bound of $h(x)$ is smaller than $h(\xi)$ on any closed interval that does not contain $\xi$; there is, furthermore, a neighbourhood of $\xi$ where $h''(x)$ exists and is continuous; finally $h''(\xi) < 0$,

(3) $\varphi(x)$ is continuous at $x = \xi$, $\varphi(\xi) \neq 0$. Then the following asymptotic formula holds as $n \to \infty$[1]

$$\int_a^b \varphi(x)\,[f(x)]^n\,dx \sim \varphi(\xi)\,[f(\xi)]^{n+1/2}\sqrt{-\frac{2\pi}{nf''(\xi)}} = \varphi(\xi)\,e^{nh(\xi)}\sqrt{-\frac{2\pi}{nh''(\xi)}}.$$

[1] On the use of such integrals Laplace has this to say: ... On est souvent conduit à des expressions qui contiennent tant de termes et de facteurs, que les substitutions numériques y sont impraticables. C'est ce qui a lieu dans les questions de probabilité, lorsque l'on considère un grand nombre d'événements. Cependant il importe alors d'avoir la valeur numérique des formules, pour connaître avec quelle probabilité les résultats que les événements développent en se multipliant sont indiqués. Il importe surtout d'avoir la loi suivant laquelle cette probabilité approche sans cesse de la certitude qu'elle finirait par atteindre, si le nombre des événements devenait infini. Pour y parvenir, je considérai que les intégrales définies de différentielles multipliées par des facteurs élevés à de grandes puissances, donnaient par l'intégration, des formules composées d'un grand nombre de termes et de facteurs ...

He adds the following remark on his method of which **201** describes the first step: ... un procédé qui fait converger la série avec d'autant plus de rapidité, que la formule qu'elle représente est plus compliquée; en sorte qu'il est d'autant plus exact, qu'il devient plus nécessaire ... (Essai philosophique sur les probabilités' Oeuvres, Vol. 7. Paris: Gauthier-Villars 1886, p. XXXVIII.)

[We consider only a neighbourhood of $\xi$ and expand $h(x)$ in powers of $(x-\xi)$ up to terms of the second order.]

**202.** Let $n$ be an integer, $n \to +\infty$. Using the fact that

$$\int_0^{\frac{\pi}{2}} \sin^{2n} x\,dx = \int_0^{\frac{\pi}{2}} \cos^{2n} x\,dx = \frac{1\cdot 3 \cdots (2n-1)}{2\cdot 4 \cdots 2n}\,\frac{\pi}{2},$$

prove that

$$\frac{1\cdot 3 \cdots (2n-1)}{2\cdot 4 \cdots 2n} \sim \frac{1}{\sqrt{n\pi}}.$$

**203.** We assume that $\lambda$ is real, $\lambda > 1$; $P_n(x)$ denotes the $n$-th Legendre polynomial. As $n \to \infty$

$$P_n(\lambda) \sim \frac{1}{\sqrt{2n\pi}}\,\frac{(\lambda + \sqrt{\lambda^2-1})^{n+1/2}}{\sqrt[4]{\lambda^2-1}}.$$

The positive value of the roots must be used. [VI **86**.]

**204.** The Bessel function $J_\nu(t)$ can be defined by Hansen's expansion

$$e^{it\cos x} = J_0(t) + 2\sum_{\nu=1}^{\infty} i^\nu J_\nu(t)\cos \nu x.$$

Derive the following asymptotic formula:

$$J_\nu(it) \sim i^\nu \frac{e^t}{\sqrt{2\pi t}}, \qquad t \to +\infty, \quad \nu = 0, 1, 2, \ldots$$

**205.** Show that for positive $n$, $n \to +\infty$,

$$\Gamma(n+1) = \int_0^\infty e^{-x} x^n\,dx \sim \left(\frac{n}{e}\right)^n \sqrt{2\pi n}$$

and, more accurately,

$$\left(\frac{e}{n}\right)^n \Gamma(n+1) = \sqrt{2\pi n} + O\left(\frac{1}{\sqrt{n}}\right). \qquad \textbf{[18, I 167.]}$$

**206.** Let $k$ and $l$ be real numbers, $k > 1$. Prove that for $n \to +\infty$

$$\binom{nk+l}{n} \sim \frac{(k-1)^n}{\sqrt{2\pi n}}\left(\frac{k}{k-1}\right)^{nk+l+\frac{1}{2}}.$$

**207.** Assume that $\alpha$ is real and that $t$ is positive and increases to infinity. Then

$$\int_1^\infty x^\alpha \left(\frac{te}{x}\right)^x \frac{dx}{x} \sim \sqrt{2\pi}\, t^{\alpha-\frac{1}{2}} e^t.$$

**208.** Let $0 < \alpha < 1$. The following approximation is justified as $\tau \to +0$:

$$\int_0^\infty \exp\left(\frac{x^\alpha}{\alpha} - \tau x\right) dx \sim \sqrt{\frac{2\pi}{1-\alpha}}\,\tau^{-\frac{\alpha}{2(1-\alpha)}-1} \exp\left(\frac{1-\alpha}{\alpha}\,\tau^{-\frac{\alpha}{1-\alpha}}\right).$$

**209.** Let $\alpha > 0$. As $t \to \infty$ we obtain

$$\int_0^\infty x^{-\alpha x} t^x \, dx \sim \sqrt{\frac{2\pi}{e\alpha}}\, t^{\frac{1}{2\alpha}} \exp\left(e^{-1}\alpha t^{\frac{1}{\alpha}}\right).$$

## § 2. Modifications of the Method

**210.** Let $\alpha$ and $\beta$ be two real constants. Then the relation

$$\frac{1}{n!}\int_0^{n+\alpha\sqrt{n}+\beta} e^{-x} x^n \, dx = A + \frac{B}{\sqrt{n}} + o\left(\frac{1}{\sqrt{n}}\right)$$

holds where

$$A = \frac{1}{\sqrt{2\pi}} \int_{-\infty}^{\alpha} e^{-\frac{t^2}{2}} dt, \qquad B = \frac{1}{\sqrt{2\pi}}\left(\beta - \frac{\alpha^2+2}{3}\right) e^{-\frac{\alpha^2}{2}}.$$

**211.** We denote by $\lambda$ a positive proper fraction and by $x_n$ the only positive root of the transcendental equation

$$1 + \frac{x}{1!} + \frac{x^2}{2!} + \cdots + \frac{x^n}{n!} = \lambda e^x \qquad [\text{V } \mathbf{42}].$$

As $n \to \infty$ the root is given by

$$x_n = n + \alpha\sqrt{n} + \beta + o(1),$$

where $\alpha$ and $\beta$ satisfy the equations:

$$\frac{1}{\sqrt{2\pi}} \int_\alpha^\infty e^{-\frac{t^2}{2}} dt = \lambda, \qquad \beta = \frac{\alpha^2+2}{3}.$$

**212** (Continuation of **201**). Let $\alpha$ denote a real constant. Then, for $n \to \infty$,

$$\int_a^{\xi+\frac{\alpha}{\sqrt{n}}} \varphi(x)\,[f(x)]^n \, dx \sim \varphi(\xi)\,[f(\xi)]^{n+1/2} \frac{1}{\sqrt{-nf''(\xi)}} \int_{-\infty}^{\alpha c} e^{-\frac{t^2}{2}} dt$$

$$= \varphi(\xi)\, e^{nh(\xi)} \frac{1}{\sqrt{-nh''(\xi)}} \int_{-\infty}^{\alpha c} e^{-\frac{t^2}{2}} dt, \qquad c = \sqrt{-\frac{f''(\xi)}{f(\xi)}} = \sqrt{-h''(\xi)}.$$

**213.** The functions $\varphi(x)$, $h(x)$ and $f(x) = e^{h(x)}$ are defined on the finite or infinite interval $[a, b]$ and satisfy the following conditions:

(1) $\varphi(x)\,[f(x)]^n = \varphi(x)\,e^{nh(x)}$ is absolutely integrable over $[a, b]$, $n = 0, 1, 2, \ldots$

(2) The value of the function $h(x)$ at a point $\xi$ of $(a, b)$ is larger than its least upper bound in any closed interval to the left of $\xi$ which does not contain $\xi$. Moreover there is a neighbourhood of $\xi$ in which $h''(x)$ exists and is bounded. Finally $h'(\xi) > 0$.

(3) $\varphi(x)$ is continuous at $x = \xi$, $\varphi(\xi) \neq 0$.

Prove, for $n \to \infty$, the asymptotic formula

$$\int_a^{\xi + \frac{\alpha \log n}{n} + \frac{\beta}{n}} \varphi(x)\,[f(x)]^n\,dx \sim \frac{\varphi(\xi)}{h'(\xi)}\,e^{\beta h'(\xi)} \cdot n^{\alpha h'(\xi) - 1} \cdot e^{nh(\xi)},$$

where $\alpha$ and $\beta$ stand for real constants.

**214.** Let $\xi$ denote the only real root of the transcendental equation $e^{1+\xi}\xi = 1$. Then we have for $n \to \infty$

$$\frac{1}{n!}\int_0^{\xi n + \alpha \log n + \beta} e^x x^n\,dx \sim n^A B,$$

where $\alpha$ and $\beta$ are real constants and

$$A = \alpha\frac{1+\xi}{\xi} - \frac{1}{2}, \qquad B = \frac{1}{\sqrt{2\pi}}\,\frac{\xi}{1+\xi}\,e^{\beta\frac{1+\xi}{\xi}}.$$

**215.** Suppose that $n$ is odd and let $-x_n$ denote the only real root of the equation

$$1 + \frac{x}{1!} + \frac{x^2}{2!} + \cdots + \frac{x^n}{n!} = 0 \qquad [\text{V } \mathbf{74}].$$

As $n \to \infty$ $x_n$ is asymptotically given by

$$x_n = \xi n + \alpha \log n + \beta + o(1),$$

where $\xi$ is the only real root of the equation $e^{1+\xi}\xi = 1$ and $\alpha$ and $\beta$ are given by

$$\alpha = \frac{1}{2}\,\frac{\xi}{1+\xi}, \qquad \beta = \frac{\xi}{1+\xi}\log\left(\sqrt{2\pi}\,\frac{1+\xi}{\xi}\right).$$

**216.** Assume that the function $g(x)$ is monotone increasing for positive $x$ and that

$$\lim_{x\to+\infty} g(x) = +\infty, \qquad \lim_{x\to+\infty}\frac{g(x)}{x} = 0.$$

We define

$$a_n = \frac{1}{n!}\int_0^\infty e^{-x+g(x)} x^n\,dx.$$

If there is a positive number $\gamma$ such that

$$\lim_{x\to+\infty} \frac{g(\alpha x)}{g(x)}$$

exists and is a continuous function of $\alpha$ for $1-\gamma \leqq \alpha \leqq 1+\gamma$ then

$$\lim_{n\to\infty} \frac{\log a_n}{g(n)} = 1.$$

The method of problem **201** to evaluate functions of large numbers can be generalized in the following way: We have to estimate an integral of the form

$$\int_a^b \varphi(x)\, f_1(x)\, f_2(x) \cdots f_n(x)\, dx = \int_a^b \varphi(x)\, e^{h_1(x)+h_2(x)+\cdots+h_n(x)}\, dx$$

where the functions $h_1(x), h_2(x), \ldots, h_n(x)$ are positive on $(a, b)$ and attain their maximum at the same interior point $\xi$. Then we approximate

$$h_\nu(x) = h_\nu(\xi) + \tfrac{1}{2} h_\nu''(\xi)\,(x-\xi)^2 + \cdots \quad \text{by} \quad h_\nu(\xi) + \tfrac{1}{2} h_\nu''(\xi)\,(x-\xi)^2$$

and the integral by

$$\int_{-\infty}^{\infty} \varphi(\xi)\, e^{h_1(\xi)+h_2(\xi)+\cdots+h_n(\xi)-\frac{s}{2}t^2}\, dt.$$

We have supposed that $\varphi(\xi) \neq 0$, moreover $h_\nu'(\xi) = 0$, $h_\nu''(\xi) < 0$ as condition for the maximum at the point $\xi$, and $-h_1''(\xi) - h_2''(\xi) - \cdots - h_n''(\xi) = s$. The method can be justified in many instances and it is capable of adaptation and refinement.

**217.**

$$\lim_{n\to\infty} \int_{-\pi}^{\pi} \frac{n!\, 2^{2n\cos\vartheta}}{|(2ne^{i\vartheta}-1)(2ne^{i\vartheta}-2)(2ne^{i\vartheta}-3)\cdots(2ne^{i\vartheta}-n)|}\, d\vartheta = 2\pi.$$

[Put $\vartheta = \frac{x}{\sqrt{n}}$ and recall **59**, **115**.]

**217.1.** Analogy to **201** suggests sufficient conditions under which for $n \to +\infty$

$$\iint_{\mathfrak{R}} \varphi(x, y)\, e^{nh(x,y)}\, dx\, dy \sim \varphi(\xi, \eta)\, e^{nh(\xi,\eta)} \frac{2\pi}{n\sqrt{h_{xx}h_{yy} - h_{xy}^2}}$$

where the partial derivatives of second order $h_{xx}$, $h_{yy}$, and $h_{xy}$ are taken at the point $(\xi, \eta)$. Give a full statement and a proof.

## § 3. Asymptotic Evaluation of Some Maxima

**218.** The function

$$\sqrt{x}\,(x-1)(x-2)\cdots(x-n)\,a^{-x},$$

where $a > 1$, has the maximum $M_n$ on the interval $(n, +\infty)$. It can be approximated by

$$\frac{M_n}{n!} \sim \frac{1}{\sqrt{2\pi}} \frac{1}{(a-1)^{n+1/2}}. \qquad [16.]$$

**219.** The function

$$x(x^2 - 1^2)(x^2 - 2^2) \cdots (x^2 - n^2)\, a^{-x},$$

where $a > 1$, has the maximum $M_n$ on the interval $(n, +\infty)$. It can be approximated by

$$\frac{M_n}{n!^2} \sim \frac{1}{2\pi} \left(\frac{2\sqrt{a}}{a-1}\right)^{2n+1}. \qquad [17.]$$

**220.** We define $\sqrt{x} = Q_0(x)$,

$$\sqrt{x}\left(1 - \frac{x}{1}\right)\left(1 - \frac{x}{2}\right) \cdots \left(1 - \frac{x}{n}\right) = Q_n(x), \qquad n = 1, 2, 3, \ldots$$

The sequence of functions

$$Q_1(x)\, a^{-x}, \quad Q_2(x)\, a^{-x}, \quad \ldots, \quad Q_n(x)\, a^{-x}, \quad \ldots$$

is uniformly bounded for $x > 0$ if $a \geqq 2$; it is not uniformly bounded if $0 < a < 2$.

**221.** We define $x = P_0(x)$,

$$x\left(1 - \frac{x^2}{1}\right)\left(1 - \frac{x^2}{4}\right)\left(1 - \frac{x^2}{9}\right) \cdots \left(1 - \frac{x^2}{n^2}\right) = P_n(x), \qquad n = 1, 2, 3, \ldots$$

The sequence of functions

$$P_1(x)\, a^{-x}, \quad P_2(x)\, a^{-x}, \quad \ldots, \quad P_n(x)\, a^{-x}, \quad \ldots$$

is uniformly bounded for $x > 0$ if $a \geqq 3 + \sqrt{8}$; it is not uniformly bounded if $0 < a < 3 + \sqrt{8}$.

**222.** Assume that $a > 0$, $0 < \mu < 1$ and that $M_n$ is the maximum of $e^{-(x + ax^\mu)} x^n$ in the interval $(0, +\infty)$. We find

$$\lim_{n \to \infty} \left(\frac{M_n}{n!}\right)^{n^{-\mu}} = e^{-a}.$$

## § 4. Minimax and Maximin

***223.** The function $f(x, y)$ is continuous in the rectangle

$$a \leqq x \leqq a', \qquad b \leqq y \leqq b'.$$

Then its maximum for a given $x$ and $b \leqq y \leqq b'$ (along a segment parallel to the $y$-axis)

$$\max_y f(x, y) = \varphi(x)$$

is a continuous function of $x$.

We interpret the surface

$$z = f(x, y)$$

in a rectangular coordinate system $x, y, z$ with vertical $z$-axis as a topographical surface in a mountainous region. Then the curve

$$z = \varphi(x)$$

(in the $x, z$-plane) is the skyline of the range (as it appears when seen from a faraway point of the $y$-axis) and the minimax

$$\min_x \varphi(x) = \min_x \max_y f(x, y)$$

(the minimum of the maxima) refers to the lowest point in the skyline.

***224** (continued). Show that

$$\max_y \min_x f(x, y) \leqq \min_x \max_y f(x, y).$$

***225** (continued). Which one of the two signs, $<$ and $=$, is valid in the example

$$f(x, y) = 1 - (x - y + 1)^2,$$

$a = 0,\ a' = 2,\ b = 0,\ b' = 4$?

***226.** Add to the assumptions of **223** that $f(x, y) > 0$. Then

$$\lim_{n\to\infty} \left[ \int_a^{a'} \left( \int_b^{b'} [f(x, y)]^n \, dy \right)^{-1} dx \right]^{-\frac{1}{n}} = \min_x \max_y f(x, y).$$

Part Three

# Functions of One Complex Variable

## General Part

## Chapter 1

## Complex Numbers and Number Sequences

### § 1. Regions and Curves. Working with Complex Variables

The complex variable $z$ is written in the form

$$z = x + iy = re^{i\vartheta} \quad (x, y, r, \vartheta \text{ real}, r \geqq 0, \vartheta \text{ taken mod } 2\pi).$$

We call

$x = \Re z$ the real part of $z$, $y = \Im z$ the imaginary part of $z$,

$r = |z|$ the absolute value of $z$ (also modulus),

$\vartheta = \arg z$ the argument or amplitude of $z$.

The number $\bar{z} = x - iy = re^{-i\vartheta}$ is the conjugate of $z$.

**1.** The number $z + \bar{z}$ is real, $z - \bar{z}$ is purely imaginary, $z\bar{z}$ is real and not negative.

**2.** What sets of points in the $z$-plane are characterized by the conditions:

$$\Re z > 0;\quad \Re z \geqq 0;\quad a < \Im z < b;\quad \alpha \leqq \arg z \leqq \beta;\quad \Re z = 0;$$

$$|z - z_0| = R;\quad |z - z_0| < R;\quad |z - z_0| \leqq R;\quad R \leqq |z| \leqq R';\quad \Re \frac{1}{z} = \frac{1}{R}$$

($a, b, \alpha, \beta, R, R'$ real, $z_0$ complex, $a < b$, $\alpha < \beta < \alpha + 2\pi$, $0 < R < R'$)?

**3.** What sets of points in the $z$-plane are characterized by the conditions

$$|z - a| + |z - b| = k;\quad |z - a| + |z - b| \leqq k,\ k > 0?$$

**4.** What open set of the $z$-plane is characterized by the condition

$$|z^2 + az + b| < R^2?$$

For what values of $R$ is this set connected, for what values of $R$ is it not connected?

**5.** Assume $|a| < 1$. For any point of the complex $z$-plane

$$\left|\frac{z - a}{1 - \bar{a}z}\right|$$

is either $< 1$, or $= 1$, or $> 1$, and so the whole plane is divided into three subsets. Describe them.

**6.** Suppose $\Re a > 0$. For any point of the $z$-plane

$$\left|\frac{a - z}{\bar{a} + z}\right|$$

is either $< 1$, or $= 1$, or $> 1$, and so the whole plane is divided into three subsets. Describe them.

**7.** Let $\alpha$, $\beta$ be real; $a$ complex; $\alpha$, $\beta$ and $a$ are fixed. Suppose that the complex variables $z_1$ and $z_2$ satisfy the relation

$$\alpha z_1\bar{z}_1 + \bar{a}z_1\bar{z}_2 + a\bar{z}_1z_2 + \beta z_2\bar{z}_2 = 0.$$

If $\alpha\beta - a\bar{a} < 0$ the points $\frac{z_1}{z_2}$ lie on a circle, possibly a line segment. (The left hand side of the equation is called a Hermitian form of the variables $z_1$ and $z_2$.)

**8.** Let $a$ and $b$ be positive constants and the real variable $t$ signify time. Describe the curves given by the three equations

$$z_1 = ia + at,\ z_2 = -ibe^{-it},\ z = ia + at - ibe^{-it}.$$

**9.** Describe the motion of the point

$$z = (a + b)e^{it} - be^{i\frac{a+b}{b}t},$$

where $a$, $b$ are positive constants and $t$ denotes time.

**10.** Let the radius vector $r$ and the argument $\vartheta$ be functions of the time. The complex function $z = re^{i\vartheta}$ of the real variable $t$ is represented by the motion of a point in a plane. Compute the components of velocity and of acceleration parallel and perpendicular to the radius vector. [Differentiate $z$ twice with respect to $t$.]

**11.** For what values of $z$ is the absolute value of the $n$-th term $\frac{z^n}{n!}$ of

$$1 + \frac{z}{1!} + \frac{z^2}{2!} + \cdots + \frac{z^n}{n!} + \cdots$$

(exponential series in the complex plane) larger than the absolute value of any other term? $n = 0, 1, 2, \ldots$

**12.** For what values of $z$ is the absolute value of the $n$-th term of the series

$$1 + \frac{z}{1} + \frac{z(z-1)}{1 \cdot 2} + \frac{z(z-1)(z-2)}{1 \cdot 2 \cdot 3} + \cdots +$$

$$+ \frac{z(z-1)\cdots(z-n+1)}{1 \cdot 2 \cdots n} + \cdots = \sum_{n=0}^{\infty} \binom{z}{n}$$

(binomial series for $(1+t)^z$ with $t = 1$ and complex $z$) larger than the absolute value of any other term of this series? $n = 0, 1, 2, \ldots$

**13.** We put

$$P_0(z) = z, \quad P_n(z) = z\left(1 - \frac{z^2}{1^2}\right)\left(1 - \frac{z^2}{2^2}\right)\left(1 - \frac{z^2}{3^2}\right)\cdots\left(1 - \frac{z^2}{n^2}\right),$$

$$n = 1, 2, 3, \ldots$$

For what values of $z$ is $|P_n(z)|$ larger than $|P_0(z)|, |P_1(z)|, \ldots, |P_{n-1}(z)|, |P_{n+1}(z)|, \ldots$? ($P_n(z)$ is the $n$-th partial product in the product expansion of $\frac{\sin \pi z}{\pi}$.)

**14.** We assume that the real functions $f(t)$ and $\varphi(t)$ are defined on the interval $a \leqq t \leqq b$, that $f(t)$ is positive and continuous and $\varphi(t)$ properly integrable. Then

$$\left| \int_a^b f(t)\, e^{i\varphi(t)}\, dt \right| \leqq \int_a^b f(t)\, dt.$$

Equality holds if and only if the function $\varphi(t)$ assumes the same value mod $2\pi$ at all its points of continuity.

**15.** Let $\varphi(t)$ be defined for $t \geqq 0$ and be properly integrable over any finite interval. If

$$\int_0^\infty e^{-(t+i\varphi(t))}\, dt = P, \qquad \int_0^\infty e^{-2(t+i\varphi(t))}\, dt = Q,$$

then

$$|4P^2 - 2Q| \leqq 3.$$

Equality holds if and only if $\varphi(t)$ assumes the same value mod $2\pi$ at all its points of continuity.

## § 2. Location of the Roots of Algebraic Equations

We consider polynomials of degree $n$

$$P(z) = a_0 z^n + a_1 z^{n-1} + a_2 z^{n-2} + \cdots + a_{n-1} z + a_n$$

with arbitrary coefficients; $a_0 \neq 0$ is often assumed. The complex number $z_0$ is called a zero of this polynomial if

$$a_0 z_0^n + a_1 z_0^{n-1} + a_2 z_0^{n-2} + \cdots + a_{n-1} z_0 + a_n = 0.$$

($z_0$ is a root of the algebraic equation $P(z) = 0$.) If $z_1, z_2, \ldots, z_n$ are the $n$ zeros of the polynomial $P(z)$ we can write

$$P(z) = a_0(z - z_1)(z - z_2) \cdots (z - z_n)$$

as is proved in algebra.

**16.** A polynomial of the form

$$z^n - p_1 z^{n-1} - p_2 z^{n-2} - \cdots - p_{n-1} z - p_n,$$

where $p_1 \geqq 0, p_2 \geqq 0, \ldots, p_n \geqq 0, p_1 + p_2 + \cdots + p_n > 0$, has just one positive zero.

**17.** If $z_0$ is a zero of the polynomial

$$z^n + a_1 z^{n-1} + a_2 z^{n-2} + \cdots + a_n$$

then $|z_0|$ is not larger than the only positive zero $\zeta$ of the polynomial $z^n - |a_1| z^{n-1} - |a_2| z^{n-2} - \cdots - |a_n|$.

**18.** Assume $a_n \neq 0$. The absolute value of none of the zeros of the polynomial

$$P(z) = z^n + a_1 z^{n-1} + a_2 z^{n-2} + \cdots + a_n$$

is smaller than the only positive zero $\zeta$ of the polynomial

$$z^n + |a_1| z^{n-1} + |a_2| z^{n-2} + \cdots + |a_{n-1}| z - |a_n|.$$

**19.** All the zeros of the polynomial $z^n + c$ are on the circle centred at $z = 0$ with radius $|c|^{1/n}$.

***20.** Let $c_1, c_2, \ldots, c_n$ be positive numbers and

$$c_1 + c_2 + \cdots + c_n \leqq 1.$$

The absolute values of the zeros of the polynomial

$$z^n + a_1 z^{n-1} + a_2 z^{n-2} + \cdots + a_n$$

are not larger than

$$M = \max\left(\frac{|a_1|}{c_1}, \sqrt{\frac{|a_2|}{c_2}}, \ldots, \sqrt[n]{\frac{|a_n|}{c_n}}\right).$$

***21.** The absolute values of the roots of the equation

$$z^n + a_1 z^{n-1} + a_2 z^{n-2} + \cdots + a_n = 0$$

are not larger than the largest among the numbers

$$n\,|a_1|, \quad \sqrt{n\,|a_2|}, \quad \sqrt[3]{n\,|a_3|}, \quad \ldots, \quad \sqrt[n]{n\,|a_n|};$$

also they are not larger than the largest of the numbers

$$\sqrt[k]{\frac{2^n-1}{\binom{n}{k}}|a_k|}\,, \qquad k = 1, 2, 3, \ldots, n;$$

and they are certainly smaller than the largest of the numbers

$$2\sqrt[k]{|a_k|}\,, \qquad k = 1, 2, \ldots, n.$$

**22.** Assume

$$p_0 > p_1 > p_2 \cdots > p_n > 0.$$

The polynomial

$$p_0 + p_1 z + p_2 z^2 + \cdots + p_n z^n$$

cannot have a zero in the unit disc $|z| \leqq 1$.

**23.** Suppose that all the coefficients $p_1, p_2, \ldots, p_n$ of the polynomial

$$p_0 z^n + p_1 z^{n-1} + \cdots + p_{n-1} z + p_n$$

are positive. Then the zeros of this polynomial lie in the annulus $\alpha \leqq |z| \leqq \beta$, where $\alpha$ is the smallest, $\beta$ the largest among the values

$$\frac{p_1}{p_0}, \quad \frac{p_2}{p_1}, \quad \frac{p_3}{p_2}, \quad \ldots, \quad \frac{p_n}{p_{n-1}}.$$

**24.** Let $a_0, a_1, a_2, \ldots, a_n$ be digits (in the ordinary decimal notation, that is integers between 0 and 9 inclusively) $n \geqq 1$, $a_n \geqq 1$. Then the zeros of the polynomial

$$a_0 + a_1 z + a_2 z^2 + \cdots + a_n z^n$$

are either in the open left half-plane or in the open disk

$$|z| < \frac{1+\sqrt{37}}{2}.$$

The best upper bound that may replace the last number is between 3 and 4.

**25.** We assume that all the zeros of the polynomial

$$P(z) = a_0 z^n + a_1 z^{n-1} + \cdots + a_{n-1} z + a_n$$

are in the upper half-plane $\Im z > 0$. Let $a_\nu = \alpha_\nu + i\beta_\nu$, $\alpha_\nu$, $\beta_\nu$ real, $\nu = 0, 1, 2, \ldots, n$, and

$$U(z) = \alpha_0 z^n + \alpha_1 z^{n-1} + \cdots + \alpha_{n-1} z + \alpha_n,$$
$$V(z) = \beta_0 z^n + \beta_1 z^{n-1} + \cdots + \beta_{n-1} z + \beta_n.$$

Then the polynomials $U(z)$ and $V(z)$ have only real zeros.

**26.** Let $P(z) = 0$ stand for an algebraic equation of degree $n$ all the zeros of which are in the unit circle $|z| < 1$. Replacing each coefficient of $P(z)$ by its conjugate we obtain the polynomial $\overline{P}(z)$. We define $P^*(z) = z^n \overline{P}(z^{-1})$. The roots of the equation $P(z) + P^*(z) = 0$ are all on the unit circle $|z| = 1$.

**27.** Suppose that the polynomial $P(z)$ of degree $n$, $n \geqq 2$, assumes the values $\alpha$ and $\beta$ for $z = a$ and $z = b$, respectively, where $a \neq b$ and $\alpha \neq \beta$. Let $\mathfrak{C}$ denote the closed domain bounded by two arcs of circle the boundary whereof is the set of those points at which the line segment $a$, $b$ subtends the angle $\frac{\pi}{n}$. Show that to each point $\gamma$ on the line connecting $\alpha$ and $\beta$ there exists a point $z$ in $\mathfrak{C}$ such that $\gamma = P(z)$.

## § 3. Zeros of Polynomials, Continued. A Theorem of Gauss

**28.** If all the complex numbers $z_1, z_2, \ldots, z_n$ (considered as points in the complex plane) are on the same side of a straight line passing through the origin, then

$$z_1 + z_2 + \cdots + z_n \neq 0, \qquad \frac{1}{z_1} + \frac{1}{z_2} + \cdots + \frac{1}{z_n} \neq 0.$$

**29.** Suppose $z_1, z_2, \ldots, z_n$ are arbitrary complex numbers that add up to zero. Any straight line $l$ through the origin separates the numbers $z_1, z_2, \ldots, z_n$ so that there are some $z_\nu$'s on each side of $l$ unless all the $z_\nu$'s lie on $l$ itself.

**30.** Let $z_1, z_2, \ldots, z_n$ be arbitrary points of the complex plane, $m_1 > 0, m_2 > 0, \ldots, m_n > 0$, $m_1 + m_2 + \cdots + m_n = 1$ and

$$z = m_1 z_1 + m_2 z_2 + \cdots + m_n z_n.$$

Then there are points $z_\nu$ on both sides of any straight line through $z$ except when all the $z_\nu$'s lie on that straight line.

We can interpret the numbers $m_1, m_2, \ldots, m_n$ as masses fixed at the points $z_1, z_2, \ldots, z_n$. Then the point $z$ defined in **30** is the center of gravity of this mass distribution. If we consider all such mass distributions at the points $z_1, z_2, \ldots, z_n$ the corresponding centers of gravity cover the interior of a convex polygon, the smallest one containing the points $z_1, z_2, \ldots, z_n$. The only exception arises when all the points are on a straight line. Then the centers of gravity fill out the interior of the smallest line segment that contains all the points $z_1, z_2, \ldots, z_n$.

**31.** The derivative $P'(z)$ of $P(z)$ cannot have any zeros outside the smallest convex polygon that contains all the zeros of $P(z)$ (considered as points in the complex plane). Those zeros of $P'(z)$ that are not zeros of $P(z)$ lie in the interior of the smallest convex polygon (the smallest line segment) that contains the zeros of $P(z)$.

**32.** Let $z_1, z_2, \ldots, z_n$ be arbitrary complex numbers, $z_\mu \neq z_\nu$ for all $\mu \neq \nu$, $\mu, \nu = 1, 2, \ldots, n$. We consider all the polynomials $P(z)$ that vanish only at the points $z_1, z_2, \ldots, z_n$ (having there zeros of arbitrary order). The set of the zeros of the derivatives $P'(z)$ of all these polynomials is everywhere dense in the smallest convex polygon that contains $z_1, z_2, \ldots, z_n$.

**33.** Let $P(z)$ denote a polynomial. The zeros of $cP'(z) - P(z)$, $c \neq 0$ lie in the smallest (infinite) convex polygon that contains the rays parallel to the vector $c$ starting from the zeros of $P(z)$.

A zero of $cP'(z) - P(z)$ appears on the boundary of this region only in one of the following two cases: (a) the zero in question is also a zero of $P(z)$; (b) the region in question degenerates into a ray.

**34.** Let $\varrho_1, \varrho_2, \ldots, \varrho_p$ be positive numbers, $a_1, a_2, \ldots, a_p$ arbitrary complex numbers, and let the polynomials $A(z)$ and $B(z)$ of degree $p$ and $p - 1$ resp. be related by

$$\frac{B(z)}{A(z)} = \frac{\varrho_1}{z - a_1} + \frac{\varrho_2}{z - a_2} + \cdots + \frac{\varrho_p}{z - a_p}.$$

Suppose that the polynomial $P(z)$ is a divisor of $A(z)\,P''(z) + 2B(z)\,P'(z)$, i.e.

$$A(z)\,P''(z) + 2B(z)\,P'(z) = C(z)\,P(z),$$

where $C(z)$ denotes a polynomial. Then the zeros of $P(z)$ lie in the smallest convex polygon that contains the numbers $a_1, a_2, \ldots, a_p$.

**35.** If a polynomial $f(z)$ whose coefficients are all real has only real zeros then this is true also for its derivative $f'(z)$. If $f(z)$ has complex zeros then they appear in pairs, the two zeros forming a pair are mirror images to each other with respect to the real axis; they are complex conjugates. We draw all those disks the "vertical" diameters of which are the line segments connecting the conjugate zeros of such pairs. If $f'(z)$ has any complex zeros they lie in these disks. [Examine the imaginary part of $\frac{f'(z)}{f(z)}$.]

## § 4. Sequences of Complex Numbers

**36.** Assume that the numbers $z_1, z_2, \ldots, z_n, \ldots$ are all in the sector $-\alpha \leqq \arg z \leqq \alpha$, $\alpha < \frac{\pi}{2}$. Then the series

$$z_1 + z_2 + \cdots + z_n + \cdots \quad \text{and} \quad |z_1| + |z_2| + \cdots + |z_n| + \cdots$$

are either both convergent or both divergent.

**37.** Suppose that the numbers $z_1, z_2, \ldots, z_n, \ldots$ are all in the half-plane $\Re z \geqq 0$ and that the two series

$$z_1 + z_2 + \cdots + z_n + \cdots \quad \text{and} \quad z_1^2 + z_2^2 + \cdots + z_n^2 + \cdots$$

converge. Then $|z_1|^2 + |z_2|^2 + \cdots + |z_n|^2 + \cdots$ converges too.

**38.** There exist complex sequences $z_1, z_2, \ldots, z_n, \ldots$ for which all the series

$$z_1^k + z_2^k + \cdots + z_n^k + \cdots, \qquad k = 1, 2, 3, \ldots$$

converge and all the series

$$|z_1|^k + |z_2|^k + \cdots + |z_n|^k + \cdots, \qquad k = 1, 2, 3, \ldots$$

diverge.

**39.** Let $z_1, z_2, \ldots, z_n, \ldots$ be arbitrary complex numbers. If there exists a positive distance $\delta$ such that $|z_l - z_k| \geqq \delta$ for $l < k$, $l, k = 1, 2, 3, \ldots$ the convergence exponent of the sequence $|z_1|, |z_2|, |z_3|, \ldots$ is at most 2. [I **114**.]

**40.** The limit points of the complex numbers

$$\frac{1^{i\alpha} + 2^{i\alpha} + 3^{i\alpha} + \cdots + n^{i\alpha}}{n}, \ \alpha \text{ real}, \alpha \gtrless 0, \qquad n = 1, 2, 3, \ldots$$

fill out the entire circle with radius $(1 + \alpha^2)^{-1/2}$ and center at the origin. [The expression in question is closely related to a sum of rectangles.]

**41.** Find the locus of the limit points of the complex sequence $z_1, z_2, \ldots, z_n, \ldots$, where

$$z_n = \left(1 + \frac{i}{1}\right)\left(1 + \frac{i}{2}\right)\left(1 + \frac{i}{3}\right)\cdots\left(1 + \frac{i}{n}\right).$$

**42.** Put

$$\left(1 + \frac{i}{\sqrt{1}}\right)\left(1 + \frac{i}{\sqrt{2}}\right)\cdots\left(1 + \frac{i}{\sqrt{n}}\right) = z_n$$

and connect the points $z_{n-1}$ and $z_n$ by a straight line. The distance between these two points is always 1. The polygonal line connecting the successive points approaches with increasing $n$ an Archimedean spiral; if $z_n = r_n e^{i\varphi_n}$, $r_n > 0$, $0 < \varphi_n - \varphi_{n-1} < \frac{\pi}{2}$, then

$$\lim_{n\to\infty} \frac{r_n - r_{n-1}}{\varphi_n - \varphi_{n-1}} = \frac{1}{2}, \qquad \lim_{n\to\infty} \frac{r_n}{\varphi_n} = \frac{1}{2}.$$

**43.** Let $t$ be a fixed real number and put $z = 2ne^{\frac{it}{\sqrt{n}}}$. Then

$$\lim_{n\to\infty} \sqrt{\frac{n}{\pi}}\, \frac{2^z n!}{z(z-1)(z-2)\cdots(z-n)} = e^{-t^2}.$$

[II **59**; II **10** slightly modified.]

## § 5. Sequences of Complex Numbers, Continued: Transformation of Sequences

By means of the infinite triangular array

$$\begin{array}{lll} a_{00}, & & \\ a_{10}, & a_{11}, & \\ a_{20}, & a_{21}, & a_{22}, \\ \dots\dots & \dots\dots & \dots\dots \end{array}$$

we transform an arbitrary infinite sequence $z_0, z_1, \dots, z_n, \dots$ into a new sequence $w_0, w_1, w_2, \dots, w_n, \dots$

$$w_n = a_{n0}z_0 + a_{n1}z_1 + a_{n2}z_2 + \cdots + a_{nn}z_n\,, \qquad n = 0, 1, 2, \dots$$

The triangular array is called *convergence preserving* if it transforms *every* convergent sequence $z_0, z_1, z_2, \dots, z_n, \dots$ into a convergent sequence $w_0, w_1, w_2, \dots, w_n, \dots$ (Cf. I, Chap. 2.) The array is convergence preserving if it fulfills the following condition, consisting of two parts:

(1) $\lim\limits_{n\to\infty} a_{n\nu} = a_\nu$ exists for all fixed $\nu$;

(2) with the notation

$$\sum_{\nu=0}^{n} a_{n\nu} = \sigma_n, \qquad \sum_{\nu=0}^{n} |a_{n\nu}| = \zeta_n,$$

the sequence $\sigma_0, \sigma_1, \sigma_2, \dots, \sigma_n, \dots$ is convergent and the sequence $\zeta_0, \zeta_1, \zeta_2, \dots, \zeta_n, \dots$ is bounded. [O. Toeplitz: Prace mat.-fiz. Vol. 22, pp. 113–119 (1911); H. Steinhaus: Prace mat.-fiz. Vol. 22, pp. 121–134 (1911); T. Kojima: Tôhoku Math. J. Vol. 12, pp. 291–326 (1917); I. Schur: J. reine angew. Math. Vol. 151, pp. 79–111 (1921).]

**44.** Prove the easier part of the above mentioned proposition: If the conditions (1) and (2) are satisfied the array preserves convergence. [I **66**, I **80**.]

**45.** What conditions must the series $u_0 + u_1 + u_2 + \cdots + u_n + \cdots$ satisfy in order that its Cauchy product [I **34**, II **23**, VIII, Chap. 1, § 5] with any convergent series $v_0 + v_1 + v_2 + \cdots + v_n + \cdots$ results in a convergent series

$$u_0v_0 + (u_0v_1 + u_1v_0) + (u_0v_2 + u_1v_1 + u_2v_0) + \cdots \\ + (u_0v_n + u_1v_{n-1} + \cdots + u_{n-1}v_1 + u_nv_0) + \cdots ?$$

**46.** What condition must the series $u_1 + u_2 + \cdots + u_n + \cdots$ satisfy in order that its Dirichlet product [VIII, Chap. 1, § 5] with any convergent series $v_1 + v_2 + \cdots + v_n + \cdots$ results in a convergent series

$$u_1v_1 + (u_1v_2 + u_2v_1) + (u_1v_3 + u_3v_1) + \cdots + \sum_{t/n} u_t v_{\frac{n}{t}} + \cdots ?$$

**47.** The sequence of factors

$$\gamma_0, \quad \gamma_1, \quad \gamma_2, \quad \ldots, \quad \gamma_n, \quad \ldots$$

turns any convergent series $a_0 + a_1 + a_2 + \cdots + a_n + \cdots$ into a convergent series

$$\gamma_0 a_0 + \gamma_1 a_1 + \gamma_2 a_2 + \cdots + \gamma_n a_n + \cdots$$

if and only if the series

$$|\gamma_0 - \gamma_1| + |\gamma_1 - \gamma_2| + \cdots + |\gamma_n - \gamma_{n+1}| + \cdots$$

converges.

**48.** The existence of

$$\lim_{n\to\infty} (u_1 + u_2 + \cdots + u_{n-1} + cu_n) = \alpha$$

implies the existence of

$$\lim_{n\to\infty} (u_1 + u_2 + \cdots + u_{n-1} + u_n) = \alpha$$

in two cases only: if $c = 0$ or if $\Re c > \frac{1}{2}$, but not if $\Re c \leqq \frac{1}{2}$, $c \neq 0$.

**49.** Let $u_0, u_1, u_2, \ldots, u_n, \ldots$ be arbitrary complex numbers. For what values of $c$ does the existence of

$$\lim_{n\to\infty} \left( u_n + c\,\frac{u_0 + u_1 + \cdots + u_n}{n+1} \right)$$

imply the existence of $\lim\limits_{n\to\infty} u_n$?

**50.** If the Dirichlet series

$$a_1 1^{-s} + a_2 2^{-s} + a_3 3^{-s} + \cdots + a_n n^{-s} + \cdots$$

converges for $s = \sigma + i\tau$, $\sigma$, $\tau$ real, $\sigma > 0$, then

$$\lim_{t\to 1-0} (1-t)^{\sigma} (a_1 t + a_2 t^2 + a_3 t^3 + \cdots + a_n t^n + \cdots) = 0. \qquad \text{[I 92.]}$$

## § 6. Rearrangement of Infinite Series

**51.** If every subseries of a series with complex terms converges the series converges absolutely.

**52.** Assume that the series $|z_1| + |z_2| + \cdots + |z_n| + \cdots$ diverges. Then there exists a *direction of accumulation*, that is a real number $\alpha$

such that those terms of the series $z_1 + z_2 + \cdots + z_n + \cdots$ that are contained in the sector $\alpha - \varepsilon < \arg z < \alpha + \varepsilon$ constitute an absolutely divergent subseries for any $\varepsilon > 0$.

**53.** If $\lim_{n \to \infty} z_n = 0$ and if the positive real axis is a direction of accumulation of the conditionally convergent series $z_1 + z_2 + z_3 + \cdots$ then there exists a subseries $z_{r_1} + z_{r_2} + z_{r_3} + \cdots$ the real part of which diverges to $+\infty$ and the imaginary part converges to a finite number.

**54.** If the series $z_1 + z_2 + z_3 + \cdots$ is convergent, but not absolutely convergent, any value represented by a point of a certain straight line can be obtained as the sum of the series rearranged in a suitable order. [Consider two complementary subseries shifted relatively to each other; **52**, **53**, **I 133**, **I 134**.]

# Chapter 2

# Mappings and Vector Fields

If we associate each point $z$ of some domain $\mathfrak{D}$ of the $z$-plane with a certain complex value $w$ according to a given law then $w$ is called a function of $z$. Two geometrical interpretations of the functional relation are particularly useful. One uses one plane, the other two planes. The value $w$ belonging to the point $z$ (or, if more expedient, $\overline{w}$) can be thought of as a vector acting on the point $z$; in this way a *vector field* is defined in the domain $\mathfrak{D}$. In the other interpretation, the value $w$ associated with the point $z$ in the $z$-plane is conceived as a point in another complex plane ($w$-plane). In this way the domain $\mathfrak{D}$ is *mapped* onto a certain point set of the $w$-plane.

## § 1. The Cauchy-Riemann Differential Equations

Let $u(x, y)$ and $v(x, y)$ be two real functions of the two real variables $x$ and $y$. Then $w = u + iv$ is a function of the variable $z = x + iy$. The function $w = u + iv$ of $z = x + iy$ is called *analytic* in a certain open region if $u$ and $v$ are continuous as well as their first partial derivatives and satisfy the Cauchy-Riemann differential equations

$$\frac{\partial u}{\partial x} = \frac{\partial v}{\partial y}, \quad \frac{\partial u}{\partial y} = -\frac{\partial v}{\partial x}.$$

Observe the combination

$$\frac{\partial}{\partial x}(u + iv) = \frac{1}{i}\frac{\partial}{\partial y}(u + iv) = \frac{dw}{dz}.$$

**55.** Are the functions

$$z, \quad z^2, \quad |z|, \quad \bar{z}$$

analytic?

**55.1.** Assume that $f(z)$ is analytic, use the notation

$$w = u + iv = f(z) = f(x + iy)$$

as above and use subscripts to denote partial derivatives in the usual way. Verify that

$$u_x^2 + v_x^2 = u_y^2 + v_y^2 = u_x^2 + u_y^2 = v_x^2 + v_y^2 = u_x v_y - u_y v_x = \left|\frac{dw}{dz}\right|^2.$$

**55.2** (continued). Prove that

$$u_{xx} + u_{yy} = v_{xx} + v_{yy} = 0.$$

**55.3** (continued). Let $\varphi(x, y)$ and $\psi(x, y)$ denote functions of the two real variables $x$ and $y$ having continuous derivatives of first order; they can also be considered as functions of $u$ and $v$ where $f'(z) \neq 0$. Verify that

$$\varphi_x\psi_x + \varphi_y\psi_y = (\varphi_u\psi_u + \varphi_v\psi_v)\left|\frac{dw}{dz}\right|^2.$$

**55.4** (continued). If $\varphi(x, y)$ has continuous derivatives of the second order, also

$$\varphi_{xx} + \varphi_{yy} = (\varphi_{uu} + \varphi_{vv})\left|\frac{dw}{dz}\right|^2.$$

**55.5** (continued). Assume that $a$, $b$, $c$ and $d$ are real constants,

$$ad - bc = 1,$$

and consider

$$w = \frac{az + b}{cz + d}.$$

Then

$$\frac{\varphi_u^2 + \varphi_v^2}{\varphi_x^2 + \varphi_y^2} = \frac{\varphi_{uu} + \varphi_{vv}}{\varphi_{xx} + \varphi_{yy}} = \frac{y^2}{v^2}.$$

**56.** Find the analytic function of $z$ that vanishes for $z = 0$ and has the real part

$$u = \frac{x(1 + x^2 + y^2)}{1 + 2x^2 - 2y^2 + (x^2 + y^2)^2}.$$

**57.** We denote by $a$ and $b$, $a < b$, two fixed real numbers, by $z$ a variable point in the half-plane $\Im z > 0$ and by $\omega$ the variable angle under which the interval $[a, b]$ is seen from the point $z$. If possible find an analytic function the real part of which is $\omega$.

**58.** Show that for any analytic function $f(z) = f(x + iy)$

$$\left(\frac{\partial^2}{\partial x^2} + \frac{\partial^2}{\partial y^2}\right) |f(x + iy)|^2 = 4\,|f'(x + iy)|^2.$$

**59.** Show that for any analytic function of $z = x + iy$

$$\left(\frac{\partial^2}{\partial x^2} + \frac{\partial^2}{\partial y^2}\right) \log\left(1 + |f(x + iy)|^2\right) = \frac{4\,|f'(x + iy)|^2}{(1 + |f(x + iy)|^2)^2}.$$

## § 2. Some Particular Elementary Mappings

The Cauchy-Riemann differential equations express the fact that an analytic function brings about a *conformal* mapping of the $z$-plane into the $w$-plane. (Preservation of the angles including sense.)

The import of the Cauchy-Riemann differential equations for vector-fields will be discussed later. Cf. § 3.

**60.** We consider an orthogonal coordinate system $\xi$, $\eta$, $\zeta$ in three dimensional space. An arbitrary point $(\xi, \eta, \zeta)$ of the sphere $\xi^2 + \eta^2 + \zeta^2 = 1$ (unit sphere) is projected from the point $(0, 0, 1)$ (north-pole of the sphere) into the plane $\zeta = 0$ (equatorial plane). Let the projection be $(x, y, 0)$. Express $x + iy$ in terms of $\xi$, $\eta$, $\zeta$ and $\xi$, $\eta$, $\zeta$ in terms of $x$ and $y$. (Stereographic projection.)

**61** (continued). Let the point $P$ on the plane $\zeta = 0$ be the stereographic projection of the point $P'$ on the unit sphere. A rotation through the angle $\pi$ of the unit sphere around the $\xi$-axis moves the point $P'$ to the point $P''$. This point $P''$ is then projected stereographically into the point $P'''$ of the $\xi$, $\eta$-plane. Let the point $P$ have the coordinates $x, y, 0$ and the point $P'''$ the coordinates $u$, $v$, $0$. Express $u + iv$ in terms of $x + iy$.

We introduce on the unit sphere the geographic coordinates $\theta$ and $\varphi$ (longitude and latitude) whereby

$$-\pi < \theta \leqq \pi, \qquad -\frac{\pi}{2} \leqq \varphi \leqq \frac{\pi}{2}.$$

The sphere is described by

$$\xi = \cos\varphi\cos\theta, \quad \eta = \cos\varphi\sin\theta, \quad \zeta = \sin\varphi.$$

We consider now the circular cylinder tangent to the unit sphere $\xi^2 + \eta^2 + \zeta^2 = 1$ along the equator (great circle in the plane $\zeta = 0$). Imagine a system of coordinates $(u, v)$ on the cylinder that becomes a cartesian system when the cylinder is unrolled. Let the point $u = 0$,

$v = 0$ coincide with (1, 0, 0), the positive $u$-axis be a generatrix pointing upwards and let the $v$-axis on the cylinder coincide with the equator. The values of $v$ varie on the cylinder in the same sense as $\theta$ from $-\pi$ to $\pi$. The points obtained in this way fill out an infinite strip of width $2\pi$ with the $u$-axis as center-line when the cylinder is unrolled.

Mercator's projection establishes a conformal one to one correspondence between the unit sphere and the $u$, $v$-cylinder (unrolled into a strip). The point on the cylinder corresponding to the point $\varphi$, $\theta$ on the sphere has the coordinates

$$u = \log \tan\left(\frac{\varphi}{2} + \frac{\pi}{4}\right), \qquad v = \theta.$$

**62.** Into which lines does the Mercator projection transform the meridians and parallel circles? What are their images under the stereographic projection?

**63** (continued). The point $P$ of the unit sphere is stereographically projected onto the point $(x, y, 0)$ of the plane $z = 0$ and the image of $P$ under Mercator's projection is $u$, $v$ on the cylinder. Express $x + iy$ in terms of $u + iv$.

**64.** Suppose $z = e^w$. To what curves in the $z$-plane do the two families of lines $\Re w =$ const. and $\Im w =$ const. in the $w$-plane (which are orthogonal to each other) correspond?

**65.** Along which curves of the $z$-plane is the real part of $z^2$ constant? Along which curves is the imaginary part constant? The two families of curves form an orthogonal system; why?

**66.** Which curves in the $z$-plane are transformed by $w = \sqrt{z}$ into the lines $\Re w =$ const. in the $w$-plane? Same question for $\Im w =$ const.

**67.** The mapping $w = \cos z$ transforms the line $\Re z =$ const. of the $z$-plane into hyperbolas, the lines $\Im z =$ const. into ellipses of the $w$-plane.

**68.** Consider the function $z = w + e^w$, $z = x + iy$, $w = u + iv$. Find the equations of the curves in the $x$, $y$-plane that are mapped onto the lines $u =$ const. and $v =$ const. respectively. What corresponds to the lines $v = 0$, $v = \pi$?

**69.** Given the function $w = e^z$ compute the area of the image of the square $a - \varepsilon \leqq x \leqq a + \varepsilon$, $-\varepsilon \leqq y \leqq \varepsilon$, $0 < \varepsilon < \pi$, $z = x + iy$. Find the ratio of the two areas and its limit as $\varepsilon$ converges to 0.

We define the *linear enlargement* (enlargement ratio, stretching, change of scale) of the mapping $w = f(z)$ at a point $z$ where $f(z)$ is regular to be the ratio of the length of the line element at the point $w = f(z)$ (in the $w$-plane) to the length of the line element at the point $z$ (in the $z$-plane).

This ratio is equal to $|f'(z)|$. We define the *area enlargement* to be the analogous ratio of the area elements; it is equal to $|f'(z)|^2$. A curve $L$ in the $z$-plane is therefore transformed into a curve in the $w$-plane of length

$$\int_L |f'(z)|\,|dz|.$$

An area $A$ in the $z = x + iy$-plane is mapped onto an area in the $w$-plane of size

$$\iint_A |f'(z)|^2\,dx\,dy.$$

The change in direction of the line element under the mapping $w = f(z)$ is equal to $\arg f'(z)$. It is called the *rotation* at the point $z$ and it is determined up to a multiple of $2\pi$ at any point where $f'(z) \neq 0$. The branch for which $-\pi < \arg f'(z) \leqq \pi$ is usually adopted.

**70.** The function $w = \cos z$, $z = x + iy$, yields a one to one conformal mapping of the rectangle

$$0 < x_1 \leqq x \leqq x_2 < \frac{\pi}{2}, \qquad 0 < y_1 \leqq y \leqq y_2,$$

onto a domain bounded by parts of confocal ellipses and hyperbolas [**67**]. Compute the area of this domain.

**71.** Consider the function $w = z^2$. What is the locus of those points at which the linear enlargement equals some given constant? Analogous question for the rotation.

**72.** Let $a$ be an increasing positive parameter. Determine the region onto which the function $w = e^z$, $z = x + iy$, maps the variable square $-a < x < a$, $-a < y < a$. Up to what value of $a$ is this region covered only once? For what values of $a$ is the image covered exactly $n$ times?

**73.** We examine the image of the closed disk $|z| \leqq r$ under the function $w = e^z$. Suppose $r$ is continuously increasing. There is on the ray $\arg w = \alpha$ a point that is covered by the image of the disk growing with $r$ at least as often as any other point of that ray for all values of $r$. Where is this point?

The regular function $w = f(z)$ is called *schlicht* (or univalent) in the region $\mathfrak{R}$ if it does not assume in $\mathfrak{R}$ any value more than once. E.g. the function $f(z) = z^2$ is schlicht in the upper half-plane $\mathfrak{J}z > 0$; the function $\sqrt{z}$ is schlicht in the $z$-plane cut open along the positive real axis; the function $e^z$ is schlicht in the horizontal strip $-\pi < \mathfrak{J}z \leqq \pi$ but not in any wider horizontal strip [**72**] etc. A schlicht function $w = f(z)$ establishes a conformal one to one correspondence between the region $\mathfrak{R}$ and a region $\mathfrak{S}$ of the $w$-plane. Very often it is useful to consider the point at infinity as an ordinary point. In fact, stereographic projection onto the sphere maps the point at infinity onto a point that plays no special role on the sphere. If $f(z)$ is schlicht in the region $\mathfrak{R}$ the derivative $f'(z)$ does not vanish in $\mathfrak{R}$. The converse is not true [**72**].

**74.** The function $w = z^2 + 2z + 3$ is schlicht in the open disk $|z| < 1$.

**75.** The function $w = z^2$ is schlicht in the upper half-plane $\mathfrak{I}z > 0$ and maps it onto the $w$-plane cut along the non-negative real axis.

**76.** Let $\alpha$ be real and $|a| < 1$. The function

$$w = e^{i\alpha}\frac{z-a}{1-\bar{a}z}$$

maps the unit disk $|z| \leqq 1$ univalently onto itself. [**5**.] What is the locus of those points at which the linear enlargement equals some given constant?

**77.** Assume that $C$ is a circle inside the unit disk. Then there exists a transformation of the unit disk onto itself of the type

$$w = e^{i\alpha}\frac{z-a}{1-\bar{a}z}$$

that maps the circle $C$ onto a circle centred at the origin.

**78.** Find a function that transforms the upper half-plane $\mathfrak{I}z > 0$ onto the disk $|w| < 1$ so that $z = i$ is transformed into $w = 0$.

**79.** The function $w = \frac{1}{2}\left(z + \frac{1}{z}\right)$ is schlicht in the open unit disk and maps it onto the $w$-plane cut open along the real line segment $-1 \leqq w \leqq 1$. What curves correspond to the rays starting from the origin? What is the image of the unit circle?

**80.** Find a function that maps the annulus $0 < r_1 < |z| < r_2$ onto the area bounded by the two confocal ellipses

$$|w-2| + |w+2| = 4a_1, \quad |w-2| + |w+2| = 4a_2, \quad 1 < a_1 < a_2.$$

[The result of **79** can be used if the given constants satisfy the relation

$$\frac{a_1 - \sqrt{a_1^2 - 1}}{r_1} = \frac{a_2 - \sqrt{a_2^2 - 1}}{r_2}$$

(the roots are positive).]

**81.** Transform the upper half of the unit disk $|z| < 1$, $\mathfrak{I}z > 0$ onto the upper half-plane [**79**]. At which points of the $z$-plane is the linear enlargement equal to $\frac{1}{2}$? At which points is the rotation $\pm\frac{\pi}{2}$?

**82.** Map the upper half of the unit disk $|z| < 1$, $\mathfrak{I}z > 0$ onto the $w$-plane cut along the non-negative real axis in such a way that

$z = 0$ corresponds to $w = 0$, $z = 1$ to $w = 1$, $z = i$ to $w + \infty$. Which point corresponds to $z = -1$?

**83.** Let $0 \leqq \alpha < \beta < 2\pi$. The function

$$w = (e^{-i\alpha} z)^{\frac{2\pi}{\beta - \alpha}}$$

maps the sector $\alpha < \arg z < \beta$ onto the $w$-plane cut open along the non-negative real axis.

**84.** Let $0 \leqq \alpha < \beta < 2\pi$. Map the circular sector

$$\alpha < \arg z < \beta, \qquad |z| < 1$$

onto the unit disk $|w| < 1$.

## § 3. Vector Fields

We use a special notation (slightly different from the one used in other parts of the chapter) to discuss vector fields that are defined by analytic functions of a complex variable. The independent variable is denoted by

$$z = x + iy = re^{i\vartheta},$$

$x, y, r, \vartheta$ real, $r \geqq 0$. Let

$$f = f(z) = \varphi + i\psi = \varphi(x, y) + i\psi(x, y)$$

be an analytic function of $z$; $\varphi$, $\psi$ real. We write

$$\frac{df}{dz} = w = u - iv,$$

$u$, $v$ real; $w = u - iv$ is again an analytic function. Its Cauchy-Riemann equations, obtained by separating the real and imaginary parts of

$$\frac{\partial}{\partial x}(u - iv) = \frac{1}{i}\frac{\partial}{\partial y}(u - iv),$$

are

$$\frac{\partial u}{\partial y} = \frac{\partial v}{\partial x}, \qquad \frac{\partial u}{\partial x} = -\frac{\partial v}{\partial y}.$$

We assign the vector $\overline{w} = u + iv$ to the point $z = x + iy$. In this way we obtain a vector field in the horizontal $z$-plane. This, in turn, defines a *three-dimensional* vector field. It is obtained by assigning to a given point in space with vertical projection $z = x + iy$ the same vector $\overline{w}$ that is assigned to the point $z$. In this way all points along the same perpendicular to the $z$-plane play the same role.

This vector field is *irrotational*: the first Cauchy-Riemann differential equation

$$\frac{\partial v}{\partial x} - \frac{\partial u}{\partial y} = 0$$

expresses the fact that the rotation vanishes. (It is obvious that the other two components of the rotation vanish.) Furthermore the vector field is a *solenoidal* vector field: the second Cauchy-Riemann differential equation

$$\frac{\partial u}{\partial x} + \frac{\partial v}{\partial y} = 0$$

shows that the divergence vanishes. (The third term usually appearing in the expression for the divergence is obviously 0.)

**85.** Show that

$$u = \frac{\partial \varphi}{\partial x}, \qquad v = \frac{\partial \varphi}{\partial y}.$$

($\varphi(x, y)$ is the *potential* of the vector field; the curves $\varphi(x, y) = \text{const.}$ are the *level lines.*) Furthermore

$$u = \frac{\partial \psi}{\partial y}, \qquad v = -\frac{\partial \psi}{\partial x}.$$

($\psi(x, y)$ is the *conjugate potential* or *stream function* or *stream potential*; the lines $\psi(x, y) = \text{const.}$ are the *stream lines* or *the lines of force*, depending on the physical interpretation of the vector $\overline{w}$.)

**86.** The lines $\varphi(x, y) = \text{const.}$ and the lines $\psi(x, y) = \text{const.}$ are orthogonal to each other.

**87.** We have

$$\frac{\partial^2 \varphi}{\partial x^2} + \frac{\partial^2 \varphi}{\partial y^2} = 0 \qquad \text{(Laplace's equation).}$$

**88.** Connect two points $z_1 = x_1 + iy_1$ and $z_2 = x_2 + iy_2$ in the vector field by a curve $L$ whose line element forms the angle $\tau$ with the positive $x$-axis. Prove that

$$\int_L (u \cos \tau + v \sin \tau)\, ds = \varphi(x_2, y_2) - \varphi(x_1, y_1),$$

i.e. the line integral of the tangential component of $\overline{w}$ is equal to the potential difference (work).

**89.** Using the same notation as in **88** establish the relation

$$\int_L (u \sin \tau - v \cos \tau)\, ds = \psi(x_2, y_2) - \psi(x_1, y_1),$$

i.e. the line integral of the normal component of $\overline{w}$ is equal to the change of the stream function (flux of force). (The normal to the curve $L$ is pointing to the right as one moves from $z_1$ to $z_2$:

$$-i\,(\cos \tau + i \sin \tau) = \sin \tau - i \cos \tau.)$$

The vector field generated by an analytic function can be interpreted as an electrostatic, magnetostatic or gravitational field. The vector field may also be interpreted as the field of a steady flow of heat or electricity. In this case the vector $\overline{w}$ is the *gradient* and as such it is proportional to the *intensity of the flow*. Finally the vector field can be also thought of as the field of the irrotational steady flow of an incompressible fluid.

**90.** The steady flow of an incompressible fluid of constant density $\varrho$ and variable pressure $p$, subject to no forces, and moving parallel to the complex plane is described by the equations[1]

$$u\frac{\partial u}{\partial x} + v\frac{\partial u}{\partial y} + \frac{1}{\varrho}\frac{\partial p}{\partial x} = 0, \quad u\frac{\partial v}{\partial x} + v\frac{\partial v}{\partial y} + \frac{1}{\varrho}\frac{\partial p}{\partial y} = 0,$$

$$\frac{\partial u}{\partial x} + \frac{\partial v}{\partial y} = 0.$$

If $w = u - iv$ is an analytic function the components of the vector $\overline{w} = u + iv$ and

$$p = p_0 - \frac{\varrho}{2}(u^2 + v^2) \quad \text{(Bernoulli's equation)}$$

($p_0$ constant) satisfy these equations.

**91.** The function

$$w = \frac{1}{z}$$

determines a vector field. Find the direction and absolute value of $\overline{w}$ at the point $z = re^{i\vartheta}$, the potential, the conjugate potential, the level lines and the stream lines. (That part of the vector field that lies in the annulus $0 < r_1 < |z| < r_2$ can be considered as the electrostatic field between two condenser plates of a Leyden jar, or as the field of the heat flow in the chimney of a factory.)

**92.** Let $\varphi$ denote the potential and $\psi$ the conjugate potential of the vector field described in **91**. Suppose that $z_1$ and $z_2$ are arbitrary points on the circles $|z| = r_1$ and $|z| = r_2$, respectively, and that $\varphi_1$ and $\varphi_2$ are the values of the potential at $z_1$ and $z_2$. Find

$$\varphi_2 - \varphi_1$$

(difference of the potentials between the two condenser plates of the Leyden jar).

The conjugate potential $\psi$ in **91** turns out to be infinitely multivalued. Study the change in value along an arbitrary closed curve $L$ without double points that contains the circle $|z| = r_1$ and lies in the annulus $r_1 < |z| < r_2$. For a given point $z$ of $L$ let $\psi$ be the value of the stream

[1] Cf. e.g. A. Sommerfeld: Mechanics of Deformable Bodies. New York: Academic Press 1950, p. 86.

potential in $z$ before, and $\psi'$ be the value after, describing $L$ exactly once in the positive sense. Evaluate

$$\psi' - \psi.$$

(Flow of force passing from one condenser plate of the Leyden jar to the other.) Furthermore find

$$\frac{\frac{1}{4\pi}(\psi' - \psi)}{\varphi_2 - \varphi_1}.$$

(Capacity of the cylindrical condenser plate per unit of the generatrix.)

**93.** Discuss the same questions as in **91** for the vector field determined by

$$w = -\frac{i}{z}.$$

(Stationary magnetic field of force generated by an infinite straight conductor perpendicular to the $z$-plane.) Are the potential $\varphi$ and the conjugate potential $\psi$ uniquely determined in this field?

**94.** Two infinite straight conductors are perpendicular to the $z$-plane piercing it at the points $z = -1$ and $z = 1$. They carry currents of the same intensity but in opposite directions. Determine the stream lines and the level lines in the magnetic field so generated.

**95.** There are $n$ infinite straight conductors perpendicular to the $z$-plane piercing it at the points $z_1, z_2, \ldots, z_n$. They carry currents of the same direction. There exist at most $n - 1$ points in the $z$-plane where the generated magnetic force vanishes (points of equilibrium); these points lie in the smallest convex polygon that contains the points $z_1, z_2, \ldots, z_n$. [When all the vectors of the field are rotated through 90° the last statement is an evident consequence of the mechanical interpretation of the field.]

**96.** Two confocal ellipses are given with foci at $z = -2$, $z = 2$ and semi-axes $2a_1$, $2b_1$ and $2a_2$, $2b_2$

$$a_1^2 - b_1^2 = a_2^2 - b_2^2 = 1.$$

Find an irrotational and solenoidal vector field in the region bounded by the two ellipses such that these ellipses are level lines. (Electrostatic field in a condenser whose plates are confocal elliptic cylinders.) What do the stream lines and level lines look like? Find the capacity [**92**]. [Map the region between the ellipses onto the region bounded by two concentric circles. Cf. **80** and **91**.]

**97.** Suppose $0 < a < b$, $\alpha < \beta < \alpha + 2\pi$. In the domain defined by the inequalities

$$a \leqq |z| \leqq b, \qquad \alpha \leqq \arg z \leqq \beta$$

determine an irrotational solenoidal vector field for which the bounding circular arcs are stream lines and the bounding straight segments are level lines (electric current in a plate of constant thickness).

We call $\psi_1$ and $\psi_2$ the values of the conjugate potential on $|z| = a$ and $|z| = b$ resp., $\varphi_1$ and $\varphi_2$ the values of the potential on $\arg z = \alpha$ and $\arg z = \beta$ resp. Compute

$$\frac{\varphi_2 - \varphi_1}{\psi_2 - \psi_1}.$$

(Resistance, except for a factor depending on the thickness and the specific resistance of the plate.)

When the problem is more difficult it is advisable to consider at the same time three (possibly four) planes: the $z$-plane (stream plane); the $\overline{w}$-plane (velocity plane) and the $f$-plane (potential plane); and possibly the $w$-plane. The notation suggests the flow of a fluid. Since $f = \varphi + i\psi$ and $w = u - iv = \frac{df}{dz}$ are analytic functions of the complex variable $z = x + iy$ the $z$-, $w$- and $f$-planes are conformally mapped onto each other. These three planes are mapped onto the $\overline{w}$-plane with preservation of the angles but with reversed orientation. The $\overline{w}$-plane in particular is the mirror image of the $w$-plane with respect to the real axis. The stream lines and the level lines in the stream plane ($z$-plane) correspond to lines parallel to the axes in the potential plane ($f$-plane). The two undetermined real constants contained in $\varphi$ and $\psi$ correspond to a translation of the $f$-plane.

**98.** Find an irrotational solenoidal vector field outside the unit circle ($|z| \geqq 1$); $\overline{w}$ ought to be $= 1$ for $z = \infty$ and tangent to the unit circle. (A fluid passing a circular pillar; at a considerable distance off the pillar the flow is uniform.) [By reasons of symmetry the two segments of the real axis inside the vector field have to be stream lines. We may expect that the horizontal component of $\overline{w}$ everywhere points to the right. Find the mapping of the stream plane onto the potential plane.]

**99.** At which points of the vector field determined in **98** does the velocity vector vanish? (Stagnation points.) In how many points of the field does $\overline{w}$ assume the same value? Where is the pressure $p$ minimal, where is it maximal? [**90**.] What is the resultant pressure exerted on the pillar? Rotate all the vectors of the field through 90°: Give a physical interpretation of the vector field so obtained.

**100.** The figure represents the contours of a vector field determined by the following conditions: The irrotational and solenoidal field covers the entire upper, and part of the lower, half-plane; it is symmetric with respect to the imaginary axis. For $z = \infty$ the velocity $\overline{w}$ becomes $-i$, for $z = 0$ we have $\overline{w} = 0$. The following stream lines are known: the positive imaginary axis, the two pieces of real axis from $z = 0$ to $z = l$ and from $z = 0$ to $z = -l$ ($C$ to $A$, $A$ to $B$ and $A$ to $D$ in the figure);

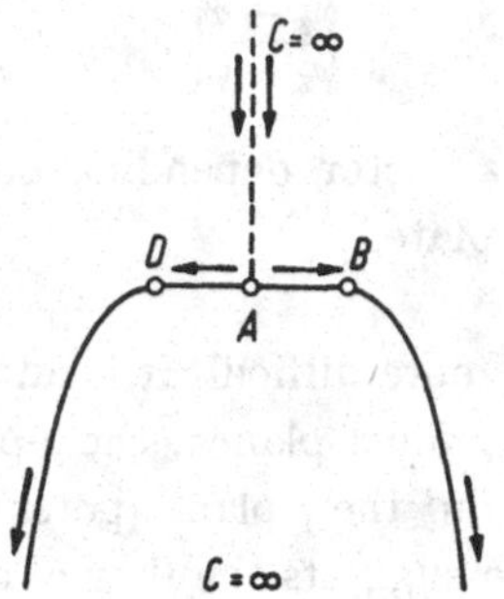


Stream plane ($z$-plane)

the direction of $\overline{w}$ is indicated by arrows. The two remaining curvilinear parts of the boundary, which lead from $z = l$ and $z = -l$ ($B$ and $D$ in the figure) to $-i\infty$ should be determined so that they are at the same time stream lines and lines of constant velocity, i.e. $\overline{w}$ is tangent to, and $|\overline{w}|$ constant on, both curves. Draw the contours of the images of the field in the $\overline{w}$-plane and in the $f$-plane. (Stagnant water, "dead water", or infinite wake of constant pressure, behind a planklike barrier perpendicular to the direction of flow.) The segment from $z = -l$ to $z = l$ represents the barrier, the stagnation point $z = 0$ is at its center. The undisturbed flow is assumed to be homogeneous with constant velocity $-i$. Throughout the wake the pressure is constant; this implies, according to Bernoulli's equation [**90**], that $|\overline{w}| =$ const. along the stream line that separates the stagnant from the flowing water.

**101** (continued). With the help of a conformal mapping express $w$ in terms of $f$ and then $z$ in terms of $f$ [**82**]. Determine the width of the wake at a great distance.

**102** (continued). Assuming that the density is $\varrho = 1$, compute the total pressure against the barrier.

Chapter 3

# Some Geometrical Aspects of Complex Variables

## § 1. Mappings of the Circle. Curvature and Support Function

**103.** Suppose that the point $z$ is moving with uniform angular velocity 1 on the circle $|z| = r$. Find the vector (its magnitude and direction) that represents the velocity of the image $w$ (in the $w$-plane) of the moving point $z$ under the mapping $w = f(z)$.

**104.** We consider the image of the circle $|z| = r$ in the $w$-plane under the mapping $w = f(z)$. What is the shortest distance of the tangent at the point $w = f(z)$ from the origin of the $w$-plane?

**105.** The point $z$ moves with uniform angular velocity 1 on the circle $|z| = r$. What is the angular velocity of the vector drawn from the origin to the point $w = f(z)$ in the $w$-plane?

**106.** The image of the circle $|z| = r$ under the mapping $w = f(z)$ has at the point $w = f(z)$ the curvature

$$\frac{1}{\varrho} = \frac{1 + \Re z \dfrac{f''(z)}{f'(z)}}{|zf'(z)|}.$$

[We are dealing with the angular velocity of the rotation of the tangent.]

**107** (continued). The sign of the curvature depends on the situation of a fixed point (e.g. the origin) not on the trajectory of the moving point $w = f(z)$: The fixed point may be to the right or to the left of this trajectory and on its concave or convex side. Explain this dependency. [Note the example $w = z^n + a$, $n$ real, $a$ complex.]

**108.** We assume that in the motion defined in **103** the point $w = f(z)$ describes in the positive direction a closed curve without double points. This curve is everywhere convex (seen from outside) if and only if for $|z| = r$

$$\Re z \frac{f''(z)}{f'(z)} > -1.$$

**109.** A closed curve without double points is called *starshaped* with respect to a point in its interior if any ray from this point intersects the curve at exactly one point (all the points of the curve can be "seen" from this point). Let the image of the circle $|z| = r$ under the mapping $w = f(z)$ be a closed curve without double points described in the positive sense [**103**]. The curve is star-shaped with respect to the origin $w = 0$ if and only if on $|z| = r$

$$\Re z \frac{f'(z)}{f(z)} > 0.$$

**110.** The image of the circle $|z| = r$ under $w = f(z)$ is convex if and only if the function $w = zf'(z)$ maps $|z| = r$ onto a curve that is star-shaped with respect to the origin.

**111.** The set of points with respect to which a closed curve is star-shaped forms a convex set. Prove this purely geometric proposition for analytic curves with the help of **109**.

Let $\mathfrak{K}$ denote a finite convex domain in the $w = u + iv$-plane. For a fixed angle $\varphi$ the expression

$$u \cos \varphi + v \sin \varphi = \Re w e^{\overline{i\varphi}}$$

assumes a certain maximum $h(\varphi)$ in $\mathfrak{K}$. The function $h(\varphi)$ is periodic with period $2\pi$ and is called the *support function* of $\mathfrak{K}$. The straight line

$$u \cos \varphi + v \sin \varphi - h(\varphi) = 0$$

is a *line of support*; its normal points away from $\mathfrak{K}$ and forms the angle $\varphi$ with the positive $u$-axis. If $\mathfrak{K}$ extends to infinity these definitions are modified insofar as a finite maximum exists only in a sector with an opening $\leqq \pi$. The two cases: the infinite strip and the half-plane are exceptions; then there exist only two lines of support, or only one, respectively.

**112.** Find the support function of the convex domain that consists of the point $a = |a| e^{i\alpha}$.

**113.** The function $w = f(z)$ establishes a one to one correspondence between the disk $|z| \leqq r$ and the convex domain $\mathfrak{K}$. Suppose that $f(z)$ is regular and $f'(z) \neq 0$ at a certain point $z$ on the boundary, $|z| = r$. In this case a definite line of support (tangent) passes through the boundary point $w = f(z)$ of $\mathfrak{K}$. Express the corresponding quantities $\varphi$ any $h(\varphi)$ in terms of $f(z)$.

**114.** The function $w = \log(1 + z)$ maps the disk $|z| = 1$ onto an infinite domain contained in the strip $-\frac{\pi}{2} < \Im w < \frac{\pi}{2}$. Its support function (defined only for $-\frac{\pi}{2} \leqq \varphi \leqq \frac{\pi}{2}$) is

$$h(\varphi) = \cos \varphi \cdot \log(2 \cos \varphi) + \varphi \sin \varphi.$$

**115.** The function $w = \frac{2}{i} \arcsin iz$ maps the disk $|z| \leqq 1$ onto a finite convex domain which has two corners and lies in the strip

$-\pi \leqq \Im w \leqq \pi$. Its support function is

$$h(\varphi) = \begin{cases} \cos\varphi \log\left(\sqrt{\cos 2\varphi} + \sqrt{2}\cos\varphi\right)^2 + 2\sin\varphi \arcsin\left(\sqrt{2}\sin\varphi\right) & \text{for} \quad 0 \leqq \varphi \leqq \frac{\pi}{4}; \\ \pi \sin\varphi & \text{for} \quad \frac{\pi}{4} \leqq \varphi \leqq \frac{\pi}{2}; \end{cases}$$

$$h(\varphi + \pi) = h(-\varphi) = h(\varphi).$$

**116.** The equation $we^{-w+1} = z$ defines $w = f(z)$. It maps the disk $|z| \leqq 1$ onto a finite convex domain which has one corner and lies in the half-plane $\Re w \leqq 1$. Its support function is

$$h(\varphi) = \cos\varphi \quad \text{for} \quad -\frac{\pi}{4} \leqq \varphi \leqq \frac{\pi}{4};$$

in the sector $\frac{\pi}{4} < \varphi < \frac{7\pi}{4}$ it is given in parametric form

$$h(\varphi)\, e^{i\varphi} = \frac{w}{1-w}\, \Re(1-w)$$

whereby $w$ describes the boundary, $|we^{-w+1}| = 1$.

## § 2. Mean Values Along a Circle

**117.** With $z = e^{i\vartheta}$ we find for $k, l = 0, 1, 2, 3, \ldots$

$$\frac{1}{2\pi}\int_0^{2\pi} z^k \bar{z}^l \, d\vartheta = \begin{cases} 0 & \text{for} \quad k \neq l, \\ 1 & \text{for} \quad k = l. \end{cases}$$

The functions $1, z, z^2, z^3, \ldots$ form an orthogonal system on the unit circle.

**118.** Let $f(z)$ denote a regular function on the disk $|z| \leqq r$. The arithmetic mean [II **48**] of $f(z)$ on the circle $|z| = r$ is defined by

$$\frac{1}{2\pi}\int_0^{2\pi} f(re^{i\vartheta})\, d\vartheta = \lim_{n\to\infty} \frac{f(r) + f(r\omega_n) + f(r\omega_n^2) + \cdots + f(r\omega_n^{n-1})}{n},$$

where $\omega_n = e^{\frac{2\pi i}{n}}$. Show that

$$\frac{1}{2\pi}\int_0^{2\pi} f(re^{i\vartheta})\, d\vartheta = f(0).$$

This means: If an analytic function is regular at every point of a closed disk its value at the center of the disk is equal to the arithmetic mean of its values on the bounding circle.

**119.** Let the function $f(z)$ be regular and different from zero for $|z| \leqq r$. Prove that the geometric mean of $|f(z)|$ on the circle $|z| = r$

$$e^{\frac{1}{2\pi}\int\limits_0^{2\pi} \log|f(re^{i\vartheta})|\,d\vartheta} = \lim_{n\to\infty} \sqrt[n]{|f(r)\, f(r\omega_n)\, f(r\omega_n^2) \cdots f(r\omega_n^{n-1})|} = |f(0)|.$$

[$\log f(z)$ is regular for $|z| \leqq r$.]

**120.** The function $f(z)$ is regular in the disk $|z| \leqq r$ and does not vanish for $z = 0$; the zeros of $f(z)$ in the disk are $z_1, z_2, \ldots, z_n$, where each multiple zero is represented with its multiplicity. Then the geometric mean of $|f(z)|$ on the circle $|z| = r$ is

$$e^{\frac{1}{2\pi}\int\limits_0^{2\pi} \log|f(re^{i\vartheta})|\,d\vartheta} = |f(0)| \frac{r^n}{|z_1 z_2 \cdots z_n|}.$$

[$f(z) = (z - z_1)(z - z_2) \cdots (z - z_n) f^*(z)$, $f^*(z)$ regular and different from 0 for $|z| \leqq r$.]

**121.** Under the hypothesis of **120** the geometric mean of $|f(z)|$ on the disk $|z| \leqq r$,

$$\mathfrak{g}(r) = e^{\frac{1}{\pi r^2}\int\limits_0^r\int\limits_0^{2\pi} \log|f(\varrho e^{i\vartheta})|\varrho\,d\varrho\,d\vartheta}$$

is always smaller than the geometric mean of $|f(z)|$ on the circle $|z| = r$,

$$\mathfrak{G}(r) = e^{\frac{1}{2\pi}\int\limits_0^{2\pi} \log|f(re^{i\vartheta})|\,d\vartheta}.$$

We have in fact

$$\frac{\mathfrak{g}(r)}{\mathfrak{G}(r)} = e^{-\frac{n}{2}\left(1 - \frac{|z_1|^2 + |z_2|^2 + \cdots + |z_n|^2}{nr^2}\right)}.$$

**122.** Let $f(z) = a_0 + a_1 z + a_2 z^2 + \cdots + a_n z^n + \cdots$ be a regular function for $|z| \leqq r$. The arithmetic mean of $|f(z)|^2$ on the circle $|z| = r$ is

$$\frac{1}{2\pi}\int\limits_0^{2\pi} |f(re^{i\vartheta})|^2\,d\vartheta = |a_0|^2 + |a_1|^2 r^2 + |a_2|^2 r^4 + \cdots + |a_n|^2 r^{2n} + \cdots.$$

**123.** Assume that $f(z) = a_0 + a_1 z + a_2 z^2 + \cdots + a_n z^n + \cdots$ is regular for $|z| \leqq 1$. The partial sums

$$s_n(z) = a_0 + a_1 z + a_2 z^2 + \cdots + a_n z^n, \qquad n = 0, 1, 2, \ldots$$

of the power series of $f(z)$ have the following minimum property: If $P(z)$ denotes any polynomial of degree $n$ the integral

$$\frac{1}{2\pi}\int\limits_0^{2\pi} |f(e^{i\vartheta}) - P(e^{i\vartheta})|^2\,d\vartheta$$

is a m inimum if and only if $P_n(z) = s_n(z)$. The minimum equals

$$|a_{n+1}|^2 + |a_{n+2}|^2 + |a_{n+3}|^2 + \cdots.$$

## § 3. Mappings of the Disk. Area

**124.** We assume that the function

$$f(z) = a_0 + a_1 z + a_2 z^2 + \cdots + a_n z^n + \cdots$$

is regular for $|z| \leqq r$. The mapping $w = f(z)$ transforms the disk $|z| \leqq r$ into a domain of the $w$-plane. The $w$-values to which several $z$-values, $|z| \leqq r$, correspond have to be counted with the proper multiplicity. The area of the image is

$$\pi(|a_1|^2 r^2 + 2|a_2|^2 r^4 + 3|a_3|^2 r^6 + \cdots + n|a_n|^2 r^{2n} + \cdots).$$

(The area is the additive combination of the areas of the images of the disk $|z| = r$ under the functions $w = a_n z^n$, $n = 0, 1, 2, \ldots$)

**125.** Let

$$w = f(z) = \sum_{n=-\infty}^{\infty} a_n z^n = \cdots + a_{-n} z^{-n} + a_{-n+1} z^{-n+1} + \cdots + a_{-1} z^{-1}$$
$$+ a_0 + a_1 z + \cdots + a_n z^n + \cdots$$

be regular in the annulus $r \leqq |z| \leqq R$. The area of the image (count covering with the proper multiplicity) is equal to

$$\pi \sum_{n=-\infty}^{\infty} n |a_n|^2 (R^{2n} - r^{2n}).$$

**126.** Assume that

$$\varphi(z) = cz + c_0 + \frac{c_1}{z} + \frac{c_2}{z^2} + \cdots + \frac{c_n}{z^n} + \cdots, \qquad c \neq 0,$$

is regular and schlicht for $|z| \geqq r$. The range of $w = \varphi(z)$ is a proper subset of the $w$-plane. The area of its complement is

$$\pi\left(|c|^2 r^2 - \frac{|c_1|^2}{r^2} - \frac{2|c_2|^2}{r^4} - \frac{3|c_3|^2}{r^6} - \cdots\right).$$

**127.** Suppose that the function

$$w = f(z) = \sum_{n=-\infty}^{\infty} a_n z^n = \cdots + a_{-n} z^{-n} + a_{-n+1} z^{-n+1} + \cdots + a_{-1} z^{-1}$$
$$+ a_0 + a_1 z + \cdots + a_n z^n + \cdots$$

is regular and establishes a one to one relationship between the points of the circle $|z| = r$ and its image $L$. The area of the domain bounded by

$L$ is

$$\pi \sum_{n=-\infty}^{\infty} n\,|a_n|^2\, r^{2n}.$$

The area is considered positive or negative depending on whether the moving image $w$ of the point $z$ describing the circle in positive sense leaves the area enclosed by $L$ on the left or on the right.

**128.** Let $f(z)$ be regular on the disk $|z| \leqq r$ and let $J(\varrho)$ denote the area of the image of the disk $|z| \leqq \varrho$, $0 \leqq \varrho \leqq r$, under the mapping $w = f(z)$. Then

$$4\int_0^r \frac{J(\varrho)}{\varrho}\,d\varrho = \int_0^{2\pi} |f(re^{i\vartheta})|^2\,d\vartheta - 2\pi\,|f(0)|^2.$$

**129.** Assume that the function

$$w = \varphi(z) = cz + c_0 + \frac{c_1}{z} + \frac{c_2}{z^2} + \cdots + \frac{c_n}{z^n} + \cdots$$

is regular outside the circle $|z| = r$ and that it maps the domain $|z| \geqq r$ univalently into the closed exterior of a curve $L$ in the $w$-plane. We assume a homogeneous mass distribution on the circle $|z| = r$ and on the curve $L$ in the $w$-plane, a distribution such that arcs which correspond to each other under the mapping $w = \varphi(z)$ carry the same mass. The mass distribution defined in this way on $L$ has a certain center of gravity $\xi$ (conformal center of gravity of $L$). We find

$$\xi = c_0.$$

## § 4. The Modular Graph. The Maximum Principle

Let the function $f(z) = u + iv$ be regular in a domain $\mathfrak{D}$ of the $z = x + iy$-plane which we conceive as horizontal. We assign to each point $z$ in $\mathfrak{D}$ the point over the $z$-plane with cartesian coordinates $x, y, \zeta$, where

$$\zeta = |f(z)|^2 = u^2 + v^2.$$

The surface obtained in this way appropriately represents the variation of the modulus of the function $f(z)$. We will call it the *modular graph*[1]. Jensen [Acta Math. Vol. 36, p. 195 (1912)] calls it an "analytic landscape".

**130.** We take the cylinder over the disk $|z| \leqq r$ and intersect it with the modular graph of the function

$$f(z) = a_0 + a_1 z + a_2 z^2 + \cdots + a_n z^n + \cdots.$$

[1] There is no connection with the modular group or with elliptic functions.

The volume of that part of the cylinder that is contained between the $z$-plane and the modular graph

$$= \pi r^2 \left( \frac{|a_0|^2}{1} + \frac{|a_1|^2 r^2}{2} + \frac{|a_2|^2 r^4}{3} + \cdots + \frac{|a_n|^2 r^{2n}}{n+1} + \cdots \right).$$

**131.** Let $\gamma(z)$ denote the angle between the $x$, $y$-plane and the tangential plane of the modular graph at $z$, $|f(z)|^2$. Then

$$\tan \gamma(z) = 2\,|f(z)|\,|f'(z)|.$$

**132.** A point of the modular graph with a horizontal tangent plane belongs to one of two types, it is a "pit" or a "saddle point": If the tangent plane is the $x$, $y$-plane, then there are only isolated points of the graph on it, pits. If the tangent plane is not the $x$, $y$-plane it intersects the modular graph along a curve (level line) $2n$ branches of which meet under equal angles, $\frac{2\pi}{2n}$, at the point of contact which is a saddle point. (For a mountain pass, a saddle point in an actual landscape, we expect $n = 2$.) The $2n$ regions of the modular graph determined locally in this way lie alternately above and below the tangent plane. (All the pits are in the $x$, $y$-plane, all the saddle points above, at diverse heights.)

**133.** The intersection of the modular graph of a polynomial with real zeros only and of a plane perpendicular to the $x$-axis is a convex curve the lowest point of which lies in the $x$, $\zeta$-plane.

**134.** Let $f(z)$ be a regular and single-valued function in the disk $|z - z_0| \leqq r$ and $M$ be the maximum of $|f(z)|$ for $z$ on the circle $|z - z_0| = r$. Then

$$|f(z_0)| \leqq M.$$

Equality holds if and only if $f(z)$ is a constant.

**135.** The function $f(z)$ is assumed to be regular and single-valued in the domain $\mathfrak{D}$. The maximum of $|f(z)|$ on the boundary of $\mathfrak{D}$ is denoted by $M$. Then

$$|f(z)| < M$$

in the interior of $\mathfrak{D}$ unless $f(z)$ is a constant. (The *Maximum Modulus Principle*, or briefly *Maximum Principle*; cf. Chap. 6.)

**136.** What does the maximum principle say about the modular graph?

**137.** The $n$ points $P_1, P_2, \ldots, P_n$ are given in a plane, $P$ is a variable point in this plane. In any domain $\mathfrak{D}$ the function of the point $P$

$$\overline{PP_1} \cdot \overline{PP_2} \cdots \overline{PP_n}$$

($\overline{PP_\nu}$ is the distance between the points $P$ and $P_\nu$) assumes its maximum on the boundary.

**138.** Let $f(z)$ be regular, single-valued and non-vanishing in the domain $\mathfrak{D}$. If $f(z)$ is not a constant, $|f(z)|$ can assume its minimum only at boundary points of $\mathfrak{D}$.

**139.** We assume that the given points $P_1, P_2, \ldots, P_n$ are all inside a circle of radius $R$ and $P$ is moving along this circle. Then

$$\sqrt[n]{\overline{PP_1} \cdot \overline{PP_2} \cdots \overline{PP_n}}$$

(the geometric mean of the $n$ distances $\overline{PP_\nu}$) attains a maximum $> R$ and a minimum $< R$ unless all the $P_\nu$'s coincide with the center of the circle.

**140** (continued). The same statement as in **139** is true for the maximum of the arithmetic mean

$$\frac{\overline{PP_1} + \overline{PP_2} + \cdots + \overline{PP_n}}{n}$$

of the $n$ distances $\overline{PP_\nu}$, but it is not true for the minimum.

**141** (continued). The same statement as in **139** holds for the minimum of the harmonic mean

$$\frac{n}{\dfrac{1}{\overline{PP_1}} + \dfrac{1}{\overline{PP_2}} + \cdots + \dfrac{1}{\overline{PP_n}}}$$

of the $n$ distances but not for the maximum.

**142.** Consider the domain bounded by a closed level line ($|f(z)|$ is constant along this curve) without selfintersections and lying inside the region where $f(z)$ is regular. It contains at least one zero of $f(z)$ unless $f(z)$ is a constant.

**143.** Given $n$ points $P_1, P_2, \ldots, P_n$ in a plane in which $P$ varies. The locus of the points $P$ for which the product of the distances

$$\overline{PP_1} \cdot \overline{PP_2} \cdots \overline{PP_n} = \text{const.}$$

is a "lemniscate with $n$ foci". (The standard lemniscate represents the special case $n = 2$, cf. **4**.) Show that a lemniscate with $n$ foci can never consist of more than $n$ separate closed branches.

**144.** Let the function $f(z)$ be regular on the disk $|z| \leqq r$ and $z_0$, $|z_0| = r$, be a point at which $|f(z)|$ assumes its maximum. Then $z_0 \frac{f'(z_0)}{f(z_0)}$ is real and positive. [**103**, **132**.]

# Chapter 4

## Cauchy's Theorem. The Argument Principle

### § 1. Cauchy's Formula

**145.** Put

$$\omega = e^{\frac{2\pi i}{n}}, \quad z_\nu = a\omega^\nu, \quad \zeta_\nu = \frac{z_{\nu-1} + z_\nu}{2},$$

$$\nu = 1, 2, \ldots, n; \quad z_0 = z_n, \quad a \text{ fixed}, \quad a \neq 0.$$

Compute the sum

$$\frac{z_1 - z_0}{\zeta_1} + \frac{z_2 - z_1}{\zeta_2} + \frac{z_3 - z_2}{\zeta_3} + \cdots + \frac{z_n - z_{n-1}}{\zeta_n},$$

which converges to the integral $\oint \frac{dz}{z}$ along $|z| = |a|$ as $n \to \infty$.

**146.** Let $k$ denote an integer different from $-1$ and $L$ a closed curve without double points and of finite length; if $k \leqq -2$, $L$ does not pass through $z = 0$; furthermore the points $z_1, z_2, \ldots, z_n$ are consecutive points on $L$. Show that the integral $\oint_L z^k\, dz$ vanishes; try to approximate it by a sum of the form

$$z_1^k(z_1 - z_0) + z_2^k(z_2 - z_1) + \cdots + z_n^k(z_n - z_{n-1}), \quad z_0 = z_n. \quad \text{[II 1, II 2.]}$$

**147.** Evaluate

$$\oint \frac{dz}{1 + z^4}$$

along the ellipse $z = x + iy$, $x^2 - xy + y^2 + x + y = 0$.

**148.** Show that

$$\int_0^{\frac{\pi}{2}} \frac{x\, d\vartheta}{x^2 + \sin^2 \vartheta} = \frac{\pi}{2\sqrt{1 + x^2}} \quad \text{when} \quad x > 0.$$

**149.** Prove the formula

$$\int_0^{2\pi} \frac{(1 + 2\cos\vartheta)^n \cos n\vartheta}{1 - r - 2r\cos\vartheta}\, d\vartheta = \frac{2\pi}{\sqrt{1 - 2r - 3r^2}} \left(\frac{1 - r - \sqrt{1 - 2r - 3r^2}}{2r^2}\right)^n,$$

$$-1 < r < \frac{1}{3}, \qquad n = 0, 1, 2, \ldots$$

**150.** Evaluate the curvilinear integral

$$\oint \frac{(1 - x^2 - y^2)\, y\, dx + (1 + x^2 + y^2)\, x\, dy}{1 + 2x^2 - 2y^2 + (x^2 + y^2)^2}$$

along an ellipse with the foci $(0, -1)$ and $(0, 1)$.

**151.** We have for $0 < \Re s < 1$

$$\int_0^\infty x^{s-1} e^{-ix}\, dx = \Gamma(s)\, e^{-\frac{i\pi s}{2}}.$$

**152.** The equation

$$\int_0^\infty \frac{\sin(x^n)}{x^n}\, dx = \frac{1}{n-1}\Gamma\left(\frac{1}{n}\right)\sin\left(\frac{n-1}{n}\frac{\pi}{2}\right)$$

holds for $n > 1$.

**153.** We find, assuming $\mu > 0$, $0 < \alpha < \frac{\pi}{2}$, $n = 0, 1, 2, \ldots$, that

$$\int_0^\infty e^{-x^\mu \cos\alpha} \sin(x^\mu \sin\alpha)\, x^n\, dx = \frac{1}{\mu}\Gamma\left(\frac{n+1}{\mu}\right)\sin\frac{(n+1)\alpha}{\mu}.$$

Note the special case $\alpha = \mu\pi$.

**154.** Assume $\mu > 0$, $x > 0$, $\mu$ fixed, $x$ variable. Show that

$$\lim_{x\to\infty} x^{\mu+1}\int_0^{+\infty} e^{-t^\mu}\cos xt\, dt = \Gamma(\mu+1)\sin\frac{\mu\pi}{2}.$$

**155.** Let $a > 0$. The integral

$$J(\alpha) = \frac{1}{2\pi i}\int_{a-i\infty}^{a+i\infty} \frac{e^{\alpha s}}{s^2}\, ds$$

along the straight line $s = a + it$, $-\infty < t < \infty$, parallel to the imaginary axis converges absolutely for all real values of $\alpha$. The integral turns out to be

$$J(\alpha) = \begin{cases} 0 & \text{if} \quad \alpha \leqq 0, \\ \alpha & \text{if} \quad \alpha \geqq 0. \end{cases}$$

**156.** We call $\mu(t)$ the largest term of the series

$$1 + \frac{t}{1!} + \frac{t^2}{2!} + \cdots + \frac{t^n}{n!} + \cdots$$

Assume $\lambda > 0$ and let $z$ be the only positive root of the equation

$$\lambda - z - e^{-z} = 0.$$

Then

$$\int_0^\infty \mu(t)\, e^{-\lambda t}\, dt = \frac{1}{z}.$$

$$\left[\frac{1}{\pi}\int_{-\infty}^{+\infty} \frac{\sin\frac{u}{2}\, e^{i(n+\frac{1}{2}-t)u}}{u}\, du = \begin{cases} 1 & \text{if} \quad n < t < n+1, \\ 0 & \text{if} \quad t < n \quad \text{or} \quad t > n+1. \end{cases}\right]$$

**157.** The Legendre polynomials can be defined as the coefficients of the expansion into a power series [VI **91**]

$$\frac{1}{\sqrt{1-2zx+z^2}} = \frac{P_0(x)}{z} + \frac{P_1(x)}{z^2} + \frac{P_2(x)}{z^3} + \cdots + \frac{P_n(x)}{z^{n+1}} + \cdots.$$

Deduce from this Laplace's formula (VI **86**)

$$P_n(x) = \frac{1}{\pi}\int_{-1}^{1}\left(x + \alpha\sqrt{x^2-1}\right)^n \frac{d\alpha}{\sqrt{1-\alpha^2}}$$

and the Dirichlet-Mehler formula

$$P_n(\cos\vartheta) = \frac{2}{\pi}\int_0^{\vartheta}\frac{\cos\left(n+\frac{1}{2}\right)t}{\sqrt{2(\cos t - \cos\vartheta)}}\,dt = \frac{2}{\pi}\int_{\vartheta}^{\pi}\frac{\sin\left(n+\frac{1}{2}\right)t}{\sqrt{2(\cos\vartheta - \cos t)}}\,dt,$$

$$0 < \vartheta < \pi.$$

(The square roots are positive.)

**158.** Let $\mathfrak{H}$ denote the half-strip

$$\Re z > 0, \qquad -\pi < \Im z < \pi$$

and $L$ the boundary of $\mathfrak{H}$ consisting of three straight pieces. We orient $L$ so that $\mathfrak{H}$ is on the right hand side of $L$. The integral

$$\frac{1}{2\pi i}\int_L \frac{e^{e^\zeta}}{\zeta - z}\,d\zeta = E(z)$$

defines a function $E(z)$ for points $z$ on the left hand side of $L$. Show that $E(z)$ is an entire function which assumes real values for real $z$.

**159** (continued). We find

$$\frac{1}{2\pi i}\int_L e^{e^\zeta}\,d\zeta = 1.$$

**160** (continued). The function

$$z^2\left(E(z) + \frac{1}{z}\right)$$

is bounded outside $\mathfrak{H}$. Inside $\mathfrak{H}$ the function

$$z^2\left(E(z) - e^{e^z} + \frac{1}{z}\right)$$

is bounded.

**161.** We define

$$\frac{2^z}{z(z-1)(z-2)\cdots(z-n)} = f_n(z).$$

Show that

$$\lim_{n\to\infty} \frac{\oint\limits_{|z|=2n} f_n(z)\,dz}{\oint\limits_{|z|=2n} |f_n(z)|\,|dz|} = i,$$

where the integrals are computed along the positively oriented circles. [II **217**.]

**162.** Assume that $f(z)$ is regular in the disk $|z| \leqq r$ and different from 0 on the circle $|z| = r$. The largest value of $\Re z \dfrac{f'(z)}{f(z)}$ on $|z| = r$ is at least equal to the number of zeros of $f(z)$ in $|z| < r$.

**163.** Let $z_1, z_2, \ldots, z_n$ be arbitrary but distinct complex numbers and $L$ be a closed continuous curve without double points, enclosing all the $z_\nu$'s. The function $f(z)$ is supposed to be regular inside of, and on, $L$. Then

$$P(z) = \frac{1}{2\pi i} \oint\limits_L \frac{f(\zeta)}{\omega(\zeta)} \frac{\omega(\zeta) - \omega(z)}{\zeta - z}\, d\zeta,$$

where $\omega(z) = (z - z_1)(z - z_2) \cdots (z - z_n)$, is the uniquely determined polynomial of degree $n - 1$ that coincides with $f(z)$ at the points $z_1, z_2, \ldots, z_n$.

**164.** The function $f(z)$ is analytic on the segment $a \leqq z \leqq b$ of the real axis and it assumes there real values. The closed curve $L$ is continuous without double points and encloses the segment $a \leqq z \leqq b$; $f(z)$ is regular inside $L$. Let $z_1, z_2, \ldots, z_n$ denote arbitrary points of the real segment $[a, b]$. Then there exists a point $z_0$ on $[a, b]$ such that

$$\oint\limits_L \frac{f(z)}{(z - z_1)(z - z_2)\cdots(z - z_n)}\, dz = \oint\limits_L \frac{f(z)}{(z - z_0)^n}\, dz.$$

**165.** Assume that the entire function $F(z)$ satisfies the inequality

$$|F(x + iy)| < Ce^{\varrho|y|}$$

in the entire $z$-plane, $z = x + iy$; $C$ and $\varrho$ are positive constants. Then

$$\frac{d}{dz}\left(\frac{F(z)}{\sin \varrho z}\right) = -\sum_{n=-\infty}^{+\infty} \frac{\varrho(-1)^n F\left(\frac{n\pi}{\varrho}\right)}{(\varrho z - n\pi)^2}.$$

Example: $F(z) = \cos \varrho z$.

**166.** The entire function $G(z)$ is supposed to satisfy the inequality

$$|G(x + iy)| < Ce^{\varrho|y|}$$

on the entire $z$-plane, $z = x + iy$; $C$, $\varrho$ positive constants; in addition assume that $G(z)$ is an odd function, $G(-z) = -G(z)$. Then

$$\frac{G(z)}{2\varrho z \cos \varrho z} = \sum_{n=0}^{\infty} \frac{(-1)^n G\left(\frac{(n+\frac{1}{2})\pi}{\varrho}\right)}{((n+\frac{1}{2})\pi)^2 - \varrho^2 z^2}.$$

Example: $G(z) = \sin \varrho z$.

**167.** We suppose that the function $f(z)$ is regular in the disk $|z| \leqq 1$. Then

$$\int_0^1 f(x)\,dx = \frac{1}{2\pi i} \oint_{|z|=1} f(z) \log z\,dz = \frac{1}{2\pi i} \oint_{|z|=1} f(z)(\log z - i\pi)\,dz,$$

$$\int_0^1 x^k f(x)\,dx = \frac{1}{e^{2\pi i k} - 1} \oint_{|z|=1} z^k f(z)\,dz, \qquad k > -1; \quad k \neq 0, 1, 2, \ldots$$

We integrate with respect to $x$ along a straight line from 0 to 1, with respect to $z$ along the positively oriented unit circle. We start at the point $z = 1$ with the branch of $\log z$ that is real, and the branch of $z^k$ that is positive, for positive $z$.

**168.** The function $f(z)$ is regular in the unit disk $|z| \leqq 1$ and satisfies the condition

$$\int_0^{2\pi} |f(e^{i\vartheta})|\,d\vartheta = 1.$$

Let $k > -1$. Then

$$\left|\int_0^1 x^k f(x)\,dx\right| \leqq \begin{cases} \dfrac{1}{2} & \text{when } k \text{ is an integer}, \\[2ex] \dfrac{1}{2\,|\sin k\pi|}, & \text{when } k \text{ is not an integer}. \end{cases}$$

**169.** Let $\alpha > -2$. The quadratic form of the infinitely many real variables $x_1, x_2, x_3, \ldots$

$$\sum_{\lambda=1}^{\infty} \sum_{\mu=1}^{\infty} \frac{x_\lambda x_\mu}{\lambda + \mu + \alpha}$$

is bounded, i.e. there exists a constant $M$ independent of $n$ such that

$$\left|\sum_{\lambda=1}^{n} \sum_{\mu=1}^{n} \frac{x_\lambda x_\mu}{\lambda + \mu + \alpha}\right| < M$$

whenever the variables $x_1, x_2, \ldots, x_n$ satisfy the condition $x_1^2 + x_2^2 + \cdots + x_n^2 = 1$; $n = 1, 2, 3, \ldots$ We may choose $M = \pi$ if $\alpha$ is an integer and $M = \dfrac{\pi}{|\sin \alpha\pi|}$ if $\alpha$ is not an integer.

**170.** We assume that the functions $f_1(z), f_2(z), \ldots, f_n(z), \ldots$ are regular in the open region $\mathfrak{R}$ and that they converge uniformly to the function

$f(z)$ in any closed domain inside $\mathfrak{R}$. Then the limit function $f(z)$ is regular in $\mathfrak{R}$.

**171.** The complex function $f(z) = u(x, y) + iv(x, y)$ of the real variables $x$ and $y$ is defined and continuous in a region $\mathfrak{R}$ of the $z = x + iy$ plane. Moreover we assume that the integral

$$\oint f(z)\, dz$$

along any circle inside $\mathfrak{R}$ vanishes. Then $f(z)$ is an analytic function of the complex variable $z$, regular in the entire region $\mathfrak{R}$. [Compute the variation of the area integral

$$F_r(z) = \iint\limits_{\xi^2+\eta^2 \leq r^2} f(z + \xi + i\eta)\, d\xi\, d\eta$$

when the real or the imaginary part of $z$ varies.]

## § 2. Poisson's and Jensen's Formulas

**172.** The function $f(z)$ is assumed to be analytic in the open disk $|z| < 1$, bounded on the closed disk $|z| \leq 1$ and continuous, possibly with the exception of a finite number of points. Then

$$\frac{1}{2\pi} \int_0^{2\pi} f(e^{i\vartheta})\, d\vartheta = f(0).$$

(More general than **118.**)

**173.** The function $f(z)$ is regular in the disk $|z| \leq R$; let $0 < r < R$. Then we have Poisson's formula

$$f(re^{i\vartheta}) = \frac{1}{2\pi} \int_0^{2\pi} f(Re^{i\Theta}) \frac{R^2 - r^2}{R^2 - 2Rr\cos(\Theta - \vartheta) + r^2}\, d\Theta.$$

**174.** Suppose that the function $f(z)$ is regular and bounded in the half-plane $\mathfrak{R}z \geq 0$. Then, for $x > 0$,

$$f(x + iy) = \frac{1}{\pi} \int_{-\infty}^{+\infty} f(i\eta)\, d\arctan\frac{\eta - y}{x}.$$

**175.** We assume that the function $f(z)$ is meromorphic in the disk $|z| \leq 1$, regular and non-zero on the boundary and at the origin. The zeros of $f(z)$ in $|z| \leq 1$ are $a_1, a_2, \ldots, a_m$ and the poles $b_1, b_2, \ldots, b_n$ (multiple zeros and poles are listed with correct multiplicity). Then we have Jensen's formula

$$\log|f(0)| + \log\frac{1}{|a_1|} + \log\frac{1}{|a_2|} + \cdots + \log\frac{1}{|a_m|}$$

$$- \log\frac{1}{|b_1|} - \log\frac{1}{|b_2|} - \cdots - \log\frac{1}{|b_n|} = \frac{1}{2\pi} \int_0^{2\pi} \log|f(e^{i\vartheta})|\, d\vartheta.$$

[Draw circles of radius $\varepsilon$ around the zeros and poles in the open disk $|z| < 1$ so that these circles do not have points in common with each other nor with the circle $|z| = 1$. Connect the $\varepsilon$-circles with the unit circle by paths that do not intersect each other (e.g. by radii of the unit circle if all the poles and zeros have different argument, cf. diagram).

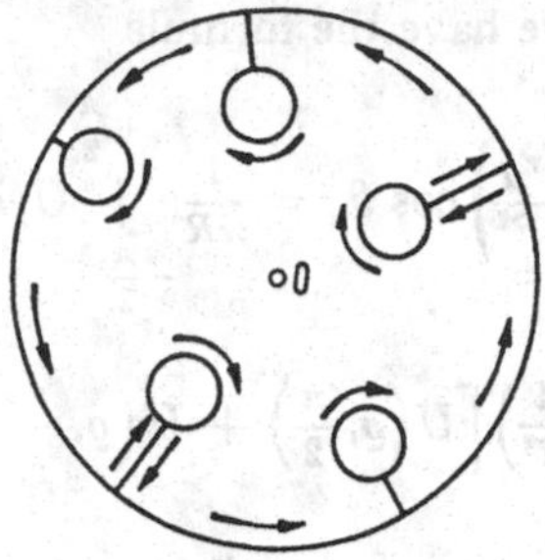

By excluding the $\varepsilon$-disks and the connecting paths we reduce the disk $|z| < 1$ to a simply connected region $\mathfrak{R}_\varepsilon$. Compute the integral $\int \frac{\log f(z)}{z}\, dz$ along the positively oriented boundary of $\mathfrak{R}_\varepsilon$.]

**176.** The function $f(z)$ is meromorphic in the disk $|z| \leqq R$, regular and different from 0 on the circle $|z| = R$ and it has inside the circle the zeros $a_1, a_2, \ldots, a_m$ and the poles $b_1, b_2, \ldots, b_n$, counted with correct multiplicities. If the point $z = re^{i\vartheta}$, $r < R$, is neither a zero nor a pole we have

$$\log |f(z)| + \sum_{\mu=1}^{m} \log \left| \frac{R^2 - \bar{a}_\mu z}{(a_\mu - z) R} \right| - \sum_{\nu=1}^{n} \log \left| \frac{R^2 - \bar{b}_\nu z}{(b_\nu - z) R} \right|$$

$$= \frac{1}{2\pi} \int_0^{2\pi} \log |f(Re^{i\Theta})| \frac{R^2 - r^2}{R^2 - 2Rr\cos(\Theta - \vartheta) + r^2}\, d\Theta.$$

**177.** The function $f(z)$ is meromorphic in the half-plane $\Re z \geqq 0$, regular and non-zero on its boundary and it has inside the half-plane the zeros $a_1, a_2, \ldots, a_m$ and the poles $b_1, b_2, \ldots, b_n$, counted with correct multiplicity. If $f(z)$ is regular at infinity (but also under weaker conditions on its behaviour at infinity) and if $f(z)$ is regular and non-zero at $z = x + iy$, $x > 0$, we have

$$\log |f(z)| + \sum_{\mu=1}^{m} \log \left| \frac{z + \bar{a}_\mu}{z - a_\mu} \right| - \sum_{\nu=1}^{n} \log \left| \frac{z + \bar{b}_\nu}{z - b_\nu} \right|$$

$$= \frac{1}{\pi} \int_{-\infty}^{+\infty} \log |f(i\eta)|\, d \arctan \frac{\eta - y}{x}.$$

**178.** The function $f(z)$ is regular in the domain

$$(\mathfrak{D}) \qquad r \leqq |z| \leqq R, \quad -\frac{\pi}{2} \leqq \arg z \leqq \frac{\pi}{2},$$

different from zero on the boundary of $\mathfrak{D}$ and it has in $\mathfrak{D}$ the zeros $a_1, a_2, \ldots, a_m$, $a_\mu = r_\mu e^{i\vartheta_\mu}$, $\mu = 1, 2, \ldots, m$. Using the definition $\log |f(\varrho e^{i\vartheta})| = U(\varrho, \vartheta)$ we have the formula

$$\sum_{r<r_\mu<R} \left(\frac{1}{r_\mu} - \frac{r_\mu}{R^2}\right) \cos\vartheta_\mu = \frac{1}{\pi R} \int_{-\frac{\pi}{2}}^{+\frac{\pi}{2}} U(R, \vartheta) \cos\vartheta \, d\vartheta$$

$$+ \frac{1}{2\pi} \int_r^R \left(\frac{1}{\varrho^2} - \frac{1}{R^2}\right)\left[U\left(\varrho, \frac{\pi}{2}\right) + U\left(\varrho, -\frac{\pi}{2}\right)\right] d\varrho + \chi(R),$$

where $\chi(R)$ is bounded as $R \to \infty$; $r$, $f(z)$ are fixed.

[$\oint \log f(z) \left(\frac{1}{z^2} + \frac{1}{R^2}\right) \frac{dz}{i}$ has to be computed along a path analogous to the one in **175**.]

## § 3. The Argument Principle

**179.** Prove **25** by examining the change of arctan $\frac{V(x)}{U(x)}$ as $x$ increases from $-\infty$ to $+\infty$ along the real axis.

We consider a closed, continuous, oriented curve in the $z$-plane that avoids the origin. If, starting from an arbitrary point, $z$ describes the entire curve in the given direction (returning to its starting point) the argument of $z$ changes continuously and its total variation is a multiple, $2\pi n$, of $2\pi$. The integer $n$ is called the *winding number* of the curve.

**180.** Every ray from the origin intersects the curve in question at least $|n|$ times.

In the sequel (**181**—**194**) $L$ denotes a closed continuous curve without double points and $\mathfrak{D}$ the closed interior of $L$. The function $f(z)$ is assumed to be regular in $\mathfrak{D}$, except possibly at finitely many poles, finite and non-zero on $L$. As $z$ moves along $L$ in the positive sense the point $w = f(z)$ describes a certain closed continuous curve the winding number of which is equal to the number of zeros inside $L$ minus the number of poles inside $L$. [The *Argument Principle.* Cf. Hille, Vol. I, p. 253.] The proposition remains true also when $f(z)$ is only continuous and non-zero on $L$.

**181.** The functions $\varphi(z)$ and $\psi(z)$ are regular in $\mathfrak{D}$, except possibly at finitely many poles, finite and different from 0 on the boundary $L$ of $\mathfrak{D}$. We define $f(z) = \varphi(z)\,\psi(z)$. The winding number of the image under $w = f(z)$ of $L$ is equal to the sum of the winding numbers of the images of $L$ under $w = \varphi(z)$ and $w = \psi(z)$ respectively.

**182.** Prove the argument principle for a polynomial.

**183.** The argument principle implies: If $\varphi(z)$ is regular and non-zero in the domain $\mathfrak{D}$ the winding number of the image of $L$ (the boundary curve of $\mathfrak{D}$) generated by $w = \varphi(z)$ vanishes; i.e. the argument of $\varphi(z)$ is a single-valued function on $L$. Deduce the general statement of the principle from this particular case.

**184.** The real trigonometric polynomial

$$a_m \cos m\vartheta + b_m \sin m\vartheta + a_{m+1} \cos (m+1)\vartheta + b_{m+1} \sin (m+1)\vartheta + \cdots + \\ + a_n \cos n\vartheta + b_n \sin n\vartheta$$

has at least $2m$ and at most $2n$ zeros in the interval $0 \leqq \vartheta < 2\pi$. [Examine

$$P(z) = (a_m - ib_m)\,z^m + (a_{m+1} - ib_{m+1})\,z^{m+1} + \cdots + (a_n - ib_n)\,z^n.]$$

**185.** Let $0 < a_1 < a_2 < \cdots < a_n$. Then the trigonometric polynomial

$$a_0 + a_1 \cos \vartheta + a_2 \cos 2\vartheta + \cdots + a_n \cos n\vartheta$$

has $2n$ distinct zeros in the interval $0 \leqq \vartheta < 2\pi$ $2n$. [**22.**] [Consequently it has real zeros only, **VI 14.**]

**186.** The function $f(z)$ is meromorphic in the interior of the curve $L$ and regular on the curve $L$. If $|a|$ is larger than the maximum of $|f(z)|$ on $L$ then $f(z)$ assumes the value $a$ inside $L$ just as often as it has poles there.

**187.** The function

$$w = e^{\pi z} - e^{-\pi z}$$

assumes any value $w$ with positive real part once and only once in the half-strip $\Re z > 0,\ -\frac{1}{2} < \Im z < \frac{1}{2}$.

**188.** Suppose that $f(z)$ is regular in the closed disk $|z| \leqq r$ and univalent on the circle $|z| = r$, that is it assumes there no value more than once. Then the image of $|z| = r$ under $w = f(z)$ has the same orientation as $|z| = r$ and the function $f(z)$ is univalent (schlicht) also in the disk $|z| < r$.

**189.** The zeros of the function $\int\limits_0^z e^{-\frac{x^2}{2}}\,dx$ lie with the exception of

$z = 0$ inside the region $\Re z^2 < 0$. [Cornu spiral, cf. e.g. A. Sommerfeld: Optics. New York: Academic Press 1954, pp. 243—244].

**190.** The function $f(z)$ is single-valued, regular and does not assume a certain value $a$ in the annulus $r < |z| < R$. All the closed continuous curves without double points that enclose the circle $|z| = r$ and that lie inside the annulus are mapped by the function $w = f(z) - a$ onto curves in the $w$-plane with the same winding number.

**191.** The function $f(z)$ is regular in the domain $\mathfrak{D}$ and its absolute value on the boundary curve $L$ of $\mathfrak{D}$ is constant. As $z$ moves on the curve $L$ the argument of $f(z)$ changes monotonically. (Whence a new proof of **142**.)

**192.** The function $f(z)$ has one zero more than $f'(z)$ inside $L$ under the hypothesis of **191**. (More informative than **142**.) This means geometrically: Inside a closed level line without double points the modular graph has more pits than saddle points, namely precisely one more.

**193.** If $f(z)$ is regular in the domain $\mathfrak{D}$ and $f'(z)$ does not vanish in $\mathfrak{D}$ the mapping $w = f(z)$ of $\mathfrak{D}$ is not necessarily schlicht [**72**]. If however $|f(z)|$ is constant on the boundary of $\mathfrak{D}$ the mapping in question has to be schlicht.

## § 4. Rouché's Theorem

**194.** We suppose that $f(z)$ and $\varphi(z)$ are two functions that are regular in the interior of $\mathfrak{D}$, continuous in the closed domain $\mathfrak{D}$ and that furthermore $|f(z)| > |\varphi(z)|$ on the boundary $L$ of $\mathfrak{D}$. Then the function $f(z) + \varphi(z)$ has exactly the same number of zeros inside $\mathfrak{D}$ as $f(z)$.

**195.** Let $\lambda$ be real, $\lambda > 1$. The equation

$$ze^{\lambda - z} = 1$$

has exactly one root in the disk $|z| = 1$. This root is real and positive.

**196.** Let $\lambda$ be real, $\lambda > 1$. The equation

$$\lambda - z - e^{-z} = 0$$

has only one root in the half-plane $\Re z \geqq 0$ which, consequently, is real.

**197.** A function (not necessarily schlicht) that maps the closed unit disk onto a domain contained in the open unit disk has exactly one fixed point. I.e. if $f(z)$ is regular in the disk $|z| \leqq 1$ and if $|f(z)| < 1$ in $|z| \leqq 1$ then the equation $f(z) - z = 0$ has exactly one root in $|z| \leqq 1$.

**198.** The entire function $\frac{1}{\Gamma(z)}$ assumes each value infinitely often in the half-strip

$$-d < \Im z < d, \qquad \Re z < 0 \qquad (d \text{ arbitrary}).$$

**199.** Let $f(t)$ be a real-valued twice continuously differentiable function on the interval $0 \leqq t \leqq 1$. If $|f(1)| > |f(0)|$ the entire function

$$F(z) = \int_0^1 f(t) \sin zt \, dt$$

has infinitely many real zeros and only a finite number of complex zeros; if $0 < |f(1)| < |f(0)|$ it has only a finite number of real, and infinitely many complex, zeros. [The zeros of $F(z)$ behave with respect to reality like the zeros of $f(0) - f(1) \cos z$.]

**200.** Let $a$ be a constant, $|a| > 2.5$. The power series

$$1 + \frac{z}{a} + \frac{z^2}{a^4} + \frac{z^3}{a^9} + \cdots + \frac{z^n}{a^{n^2}} + \cdots = F(z)$$

defines an entire function which does not vanish on the boundary of the annulus

$$|a|^{2n-2} < |z| < |a|^{2n}$$

and has exactly one zero inside the annulus, $n = 1, 2, \ldots$ [Examine the maximum term on the circle $|z| = |a|^{2n}$, I **117**.]

**201** (continuation of **170**). Let $\mathfrak{S}$ denote the set of all zeros of all the functions $f_n(z)$, $n = 1, 2, 3, \ldots$ in $\mathfrak{R}$. If the limit function $f(z)$ does not vanish identically its zeros in $\mathfrak{R}$ are identical with the limit points of $\mathfrak{S}$ in $\mathfrak{R}$. (The term "limit point" is used here to mean a point an arbitrary fixed neighborhood of which contains at least one zero of $f_n(z)$ for all sufficiently large $n$.)

**202.** The functions

$$f_1(z), \quad f_2(z), \quad \ldots, \quad f_n(z), \quad \ldots$$

are schlicht in the unit disk $|z| < 1$ and converge in any smaller disk $|z| \leqq r < 1$ uniformly to a not everywhere constant limit function $f(z)$. Then the function $f(z)$ is schlicht in the unit disk $|z| < 1$.

**203.** Let $g_1(z), g_2(z), \ldots, g_n(z), \ldots$ be entire functions which have real zeros only. If

$$\lim_{n \to \infty} g_n(z) = g(z)$$

uniformly in any finite domain, the entire function $g(z)$ can have only real zeros.

**204.** Assume

$$0 < a_0 \leqq a_1 \leqq \cdots \leqq a_n; \qquad a \geqq 0, \qquad d > 0.$$

The entire function

$$\sum_{\nu=0}^{n} a_\nu \cos (a + \nu d) z$$

has only real zeros.

**205.** Suppose that $f(t)$ is a positive valued, never decreasing function, defined on the interval $0 \leqq t < 1$ and that its integral $\int_0^1 f(t)\, dt$ is finite. The entire function

$$\int_0^1 f(t) \cos zt\, dt$$

has real zeros only. [**185.**]

**206.** The domain $\mathfrak{D}$ contains the segment $a \leqq z \leqq b$ of the real axis. The functions $f_1(z), f_2(z), \ldots, f_n(z), \ldots$ are regular in $\mathfrak{D}$, they assume real values for real $z$ and they have no zeros on $[a, b]$. If these functions converge in $\mathfrak{D}$ uniformly to a not identically vanishing limit function $f(z)$ then $f(z)$ has no zero on the segment $a \leqq z \leqq b$.—This statement is *false*.

**206.1.** The analytic functions $f_1(z), f_2(z), \ldots, f_n(z)$ are regular and single-valued in the connected closed domain $\mathfrak{D}$; let $c_1, c_2, \ldots, c_n$ denote constants. If the function

$$c_1 f_1(z) + c_2 f_2(z) + \cdots + c_n f_n(z)$$

does not vanish identically the number of its zeros in $\mathfrak{D}$ cannot exceed a certain upper bound which depends on $f_1(z), f_2(z), \ldots, f_n(z)$ and $\mathfrak{D}$ but does not depend on $c_1, c_2, \ldots, c_n$. (**206.2** is less general but more precise.)

**206.2.** Let $\lambda_1, \lambda_2, \ldots, \lambda_l$ denote real numbers

$$\lambda_1 < \lambda_2 < \cdots < \lambda_l$$

and $m_1, m_2, \ldots, m_l$ positive integers,

$$m_1 \geqq 1, \ldots, m_l \geqq 1, \qquad m_1 + m_2 + \cdots + m_l = n.$$

Let $f_1(z), f_2(z), \ldots, f_n(z)$ stand for the functions

$$e^{\lambda_1 z}, ze^{\lambda_1 z}, \ldots, z^{m_1-1}e^{\lambda_1 z}$$

$$e^{\lambda_2 z}, ze^{\lambda_2 z}, \ldots, z^{m_2-1}e^{\lambda_2 z}$$

$$\cdots\cdots\cdots\cdots\cdots\cdots$$

$$e^{\lambda_l}, ze^{\lambda_l z}, \ldots, z^{m_l-1}e^{\lambda_l z}$$

taken in this order, and $N$ for the number of those zeros of the function

$$c_1 f_1(z) + c_2 f_2(z) + \cdots + c_n f_n(z)$$

that are contained in the horizontal strip

$$\alpha \leqq \Im z \leqq \beta.$$

Assuming that

$$|c_1| + |c_2| + \cdots + |c_{m_1}| > 0, \qquad |c_{n-m_l+1}| + \cdots + |c_{n-1}| + |c_n| > 0$$

show that

$$\frac{(\lambda_l - \lambda_1)(\beta - \alpha)}{2\pi} - n + 1 \leqq N \leqq \frac{(\lambda_l - \lambda_1)(\beta - \alpha)}{2\pi} + n - 1.$$

(**206.1** is more general but less precise.)

# Chapter 5

## Sequences of Analytic Functions

### § 1. Lagrange's Series. Applications

The power series

$$a_1 z + a_2 z^2 + \cdots + a_n z^n + \cdots = w$$

which converges not only for $z = 0$ and for which $a_1 \neq 0$ establishes a conformal one to one mapping of a certain neighbourhood of $z = 0$ onto a certain neighbourhood of $w = 0$. Consequently the relationship between $z$ and $w$ can also be represented by the expansion

$$b_1 w + b_2 w^2 + \cdots + b_n w^n + \cdots = z,$$

$a_1 b_1 = 1$. To compute the second series from the first we set

$$\frac{1}{a_1 + a_2 z + a_3 z^2 + \cdots + a_n z^{n-1} + \cdots} = \varphi(z).$$

The equation

$$w = \frac{z}{\varphi(z)},$$

where $\varphi(z)$ is regular in a neighbourhood of $z = 0$, $\varphi(0) \neq 0$, implies

$$z = \sum_{n=1}^{\infty} \frac{w^n}{n!} \left[\frac{d^{n-1}[\varphi(x)]^n}{dx^{n-1}}\right]_{x=0}.$$

More generally, if $f(z)$ is regular in a neighbourhood of $z = 0$, then

$$(L) \qquad f(z) = f(0) + \sum_{n=1}^{\infty} \frac{w^n}{n!} \left[\frac{d^{n-1} f'(x)[\varphi(x)]^n}{dx^{n-1}}\right]_{x=0}.$$

(Bürmann-Lagrange series, cf. Hurwitz-Courant, p. 137; Whittaker and Watson, p. 129.)

**207.** We use the same notation and hypothesis as before and expand

$$\frac{f(z)}{1 - w\varphi'(z)} = \sum_{n=0}^{\infty} \frac{w^n}{n!} \left[\frac{d^n f(x) [\varphi(x)]^n}{dx^n}\right]_{x=0}.$$

Derive this formula from Lagrange's formula or Lagrange's formula from this one by correctly using the generality of both formulas. [The existence of one formula for a certain $f(z)$ implies immediately the other formula for another $f(z)$.]

**208.** Prove the formula in **207** directly by expressing the coefficient of $w^n$ as a Cauchy integral.

**209.** Expand in ascending powers of $w$ the solution of the transcendental equation

$$ze^{-z} = w$$

that vanishes for $w = 0$.

**210** (continued). Expand $e^{\alpha z}$ in powers of $w = ze^{-z}$ where $\alpha$ is an arbitrary constant.

**211.** Expand in ascending powers of $w$ the solution $x$ of the trinomial equation

$$1 - x + wx^\beta = 0$$

that becomes 1 for $w = 0$.

**212** (continued). Expand $x^\alpha$ in powers of $w$, where $\alpha$ denotes an arbitrary constant. ($x^\alpha = y$ is the solution of the trinomial equation

$$1 - y^{1/\alpha} + wy^{\beta/\alpha} = 0.)$$

**213** (continued). Note the cases $\beta = 0, 1, 2, -1, \frac{1}{2}$ and derive **209**, **210** by taking the limit in **211**, **212**.

**214.** Evaluate the sum of the power series

$$1 + \sum_{n=1}^{\infty} \frac{(n + \alpha)^n w^n}{n!}.$$

What is its radius of convergence?

**215.** Prove **156** with the help of the results of **214**.

**216.** Let $\alpha$ and $\beta$ denote rational numbers. Then the series

$$1 + \binom{\alpha + \beta}{1} w + \binom{\alpha + 2\beta}{2} w^2 + \cdots + \binom{\alpha + n\beta}{n} w^n + \cdots$$

represents an algebraic function of $w$.

**217.** Arrange the successive powers of the trinomial $1 + w + w^2$ in a regular triangular array

$$\begin{array}{c} \mathbf{1} \\ 1 + \mathbf{w} + w^2 \\ 1 + 2w + \mathbf{3w^2} + 2w^3 + w^4 \\ 1 + 3w + 6w^2 + \mathbf{7w^3} + 6w^4 + 3w^5 + w^6 \\ \cdots\cdots\cdots\cdots\cdots\cdots\cdots\cdots\cdots\cdots \end{array}$$

The sum of the middle terms (in boldface) is

$$1 + w + 3w^2 + 7w^3 + \cdots = \frac{1}{\sqrt{1 - 2w - 3w^2}}.$$

**218.** Arrange the successive powers of the binomial $1 + w$ in a regular triangular array (Pascal's triangle)

$$\begin{array}{c} \mathbf{1} \\ 1 + w \\ 1 + \mathbf{2w} + w^2 \\ 1 + 3w + 3w^2 + w^3 \\ 1 + 4w + \mathbf{6w^2} + 4w^3 + w^4 \\ \cdots\cdots\cdots\cdots\cdots\cdots\cdots \end{array}$$

Find the sum of the middle terms (in boldface) and, more generally, the sum of any column.

**219.** Find the generating functions of the polynomials $P_n(x)$, $P_n^{(\alpha,\beta)}(x)$, $L_n^{(\alpha)}(x)$, defined by the formulas

(1) $$P_n(x) = \frac{1}{2^n n!} \frac{d^n}{dx^n} (x^2 - 1)^n \quad \text{(Legendre's polynomials)};$$

(2) $$(1 - x)^\alpha (1 + x)^\beta P_n^{(\alpha,\beta)}(x) = \frac{(-1)^n}{2^n\, n!} \frac{d^n}{dx^n} (1 - x)^{n+\alpha} (1 + x)^{n+\beta}$$

$$\alpha > -1, \quad \beta > -1 \quad \text{(Jacobi's polynomials)};$$

(3) $$e^{-x} x^\alpha L_n^{(\alpha)}(x) = \frac{1}{n!} \frac{d^n}{dx^n} e^{-x} x^{n+\alpha}, \quad \alpha > -1$$

(generalized Laguerre's polynomials).

(Cf. VI **84**, VI **98**, VI **99**. The generating function of the Legendre polynomials is the series

$$P_0(x) + P_1(x)\, w + P_2(x)\, w^2 + \cdots + P_n(x)\, w^n + \cdots$$
$$= 1 + xw + \frac{3x^2 - 1}{2} w^2 + \cdots,$$

the sum of which has to be found as a function of $x$ and $w$; similarly in the other two cases.)

We define as usual

$$\Delta F(z) = F(z+1) - F(z),$$

$$\Delta^2 F(z) = \Delta[\Delta F(z)] = F(z+2) - 2F(z+1) + F(z),$$

$$\dots$$

$$\Delta^n F(z) = F(z+n) - \binom{n}{1} F(z+n-1) + \binom{n}{2} F(z+n-2) - \cdots + (-1)^n F(z)$$

$$\dots$$

**220.** Let $s$ be a constant of sufficiently small modulus. Then the following formulas are valid for $F(z) = e^{sz}$:

$$(1)\quad F(z) = F(0) + \frac{z}{1!}\Delta F(0) + \frac{z(z-1)}{2!}\Delta^2 F(0) + \cdots + \frac{z(z-1)\cdots(z-n+1)}{n!}\Delta^n F(0) + \cdots;$$

$$(2)\quad F'(z) = \Delta F(z) - \frac{1}{2}\Delta^2 F(z) + \frac{1}{3}\Delta^3 F(z) - \cdots + (-1)^{n-1}\frac{1}{n}\Delta^n F(z) + \cdots;$$

$$(3)\quad F(z) = F(0) + \frac{z}{1!}F'(1) + \frac{z(z-2)}{2!}F''(2) + \cdots + \frac{z(z-n)^{n-1}}{n!}F^{(n)}(n) + \cdots$$ **[210]**;

$$(4)\quad F(z) = F(0) + \sum_{n=1}^{\infty} \frac{z^2(z^2-1^2)(z^2-2^2)\cdots[z^2-(n-1)^2]}{(2n)!}\Delta^{2n}F(-n) + \sum_{n=1}^{\infty} \frac{z(z^2-1^2)(z^2-2^2)\cdots[z^2-(n-1)^2]}{(2n-1)!}\,\frac{\Delta^{2n-2}[F(-n+2)-F(-n)]}{2}$$

**[212, 216]**.

**221.** The four formulas mentioned in **220** hold for any polynomial $F(z)$. (In this case the series are obviously finite.)

**222.** The formulas (1), (2) given in **220** are also valid for any rational function $F(z)$ if the real part of $z$ is larger than the real part of any of the finite poles of $F(z)$; formula (1) requires the additional condition that $F(z)$ be regular for $z = 0, 1, 2, 3, \ldots$—Do the formulas (3) and (4) hold for rational but not entire functions?

In the sequel (**223**–**226**) we use the notation

$$\Delta^n a_k = a_{k+n} - \binom{n}{1} a_{k+n-1} + \binom{n}{2} a_{k+n-2} - \cdots + (-1)^n a_k.$$

**223.**

$$(1-z)^n \sum_{k=-\infty}^{\infty} a_k z^k = \sum_{k=-\infty}^{\infty} \Delta^n a_k z^{n+k}$$

where $a_k$, $k = 0, \pm 1, \pm 2, \ldots$, denote arbitrary constants.

**224.** Define

$$F(z) = a_0 + a_1 z + a_2 z^2 + \cdots + a_n z^n + \cdots$$

and establish the relation

$$\frac{1}{1+t} F\left(\frac{t}{1+t}\right) = a_0 + \Delta a_0 t + \Delta^2 a_0 t^2 + \cdots + \Delta^n a_0 t^n + \cdots.$$

**225.** Define

$$F(z) = a_0 + 2a_1 z + 2a_2 z^2 + \cdots + 2a_n z^n + \cdots \qquad a_{-n} = a_n$$

and show that

$$\frac{1}{\sqrt{1+4t}} F\left(\frac{1+2t-\sqrt{1+4t}}{2t}\right) = a_0 + \Delta^2 a_{-1} t + \Delta^4 a_{-2} t^2 + \cdots +$$

$$+ \Delta^{2n} a_{-n} t^n + \cdots.$$

**226.** Define

$$F(z) = 2a_1 z + 2a_2 z^2 + 2a_3 z^3 + \cdots + 2a_n z^n + \cdots, \qquad a_{-n} = -a_n$$

and show that

$$\frac{1}{t} F\left(\frac{1+2t-\sqrt{1+4t}}{2t}\right) = a_1 - a_{-1} + (\Delta^2 a_0 - \Delta^2 a_{-2})\, t$$

$$+ (\Delta^4 a_{-1} - \Delta^4 a_{-3})\, t^2 + \cdots + (\Delta^{2n} a_{-n+1} - \Delta^{2n} a_{-n-1})\, t^n + \cdots.$$

**227.**

$$\prod_{n=1}^{\infty} \left(1 + \frac{z(1-z)}{n\,(n+1)}\right) = \frac{\sin \pi z}{\pi z(1-z)}.$$

**228.** $\sin \pi z$ is a single-valued function of $w = z(1-z)$. The expansion of $\sin \pi z$ in powers of $w$ contains only positive coefficients (except the constant term) [**227**].

**229.** Prove that

$$\left[\frac{d^n (\pi - x)^{-n-1} \cos x}{dx^n}\right]_{x=0} > 0, \qquad n = 0, 1, 2, \ldots$$

## § 2. The Real Part of a Power Series

**230.** We assume that the function
$f(z) = a_0 + a_1 z + a_2 z^2 + \cdots + a_n z^n + \cdots$ is regular in the disk $|z| < R$. Express the coefficients $a_1, a_2, \ldots, a_n, \ldots$ in terms of the real and the imaginary part resp. of $f(z)$ on the circle $|z| = r$, $0 < r < R$.

**231** (continued). We set $\Re f(re^{i\vartheta}) = U(r, \vartheta)$ and assume that $f(0)$ is real and $|z| < r$. Then

$$f(z) = \frac{1}{2\pi} \int_0^{2\pi} U(r, \vartheta) \frac{r + ze^{-i\vartheta}}{r - ze^{-i\vartheta}} \, d\vartheta .$$

**232** (continued). Suppose $f(z)$ does not vanish on $|z| = r$ and that its zeros in the disk $|z| < r$ are $c_1, c_2, \ldots, c_m$. Then we have for $|z| < r$

$$\log f(z) = i\gamma + \sum_{\mu=1}^{m} \log \frac{(z - c_\mu)\, r}{r^2 - \bar{c}_\mu z} + \frac{1}{2\pi} \int_0^{2\pi} \log |f(re^{i\vartheta})| \frac{r + ze^{-i\vartheta}}{r - ze^{-i\vartheta}} \, d\vartheta ,$$

where $\gamma$ is a real constant. [Consequence of **231** as **120** is a consequence of **119**.]

**233.** The function $f(z)$ is regular, its real part is positive in the open disk $|z| < R$ and continuous in the closed disk $|z| \leqq R$. If the real part becomes identically zero on an arc of the circle the imaginary part of $f(z)$ changes on this arc always in the same sense: it decreases as $\arg z$ increases.

**234.** Let $f(z) = a_0 + a_1 z + a_2 z^2 + \cdots + a_n z^n + \cdots$ be regular in the disk $|z| < R$, $f(re^{i\vartheta}) = U(r, \vartheta) + iV(r, \vartheta)$, $U(r, \vartheta)$, $V(r, \vartheta)$ are real. Then the equation

$$\int_0^{2\pi} [U(r, \vartheta)]^2 \, d\vartheta = \int_0^{2\pi} [V(r, \vartheta)]^2 \, d\vartheta$$

holds for $0 < r < R$ provided that it holds for $r = 0$.

**235.** The function

$$f(z) = \frac{1}{2} + a_1 z + \cdots + a_n z^n + \cdots$$

is assumed to be regular and to have positive real part in the open disk $|z| < 1$. Then $|a_n| \leqq 1$, $n = 1, 2, \ldots$ In none of these inequalities can we replace 1 by a smaller number.

**236.** Suppose that the function
$f(z) = a_0 + a_1 z + a_2 z^2 + \cdots + a_n z^n + \cdots$ is regular and that $\Re f(z) < A$ on the disk $|z| < R$. Then the inequality

$$|a_0| + |a_1| r + |a_2| r^2 + \cdots + |a_n| r^n + \cdots \leqq |a_0| + \frac{2r}{R - r} (A - \Re a_0)$$

holds for $0 < r < R$. Example: $f(z) = \dfrac{z+1}{z-1}$, $R = 1$, $A = 0$.

**237.** We assume that the Laurent series

$$\psi(z) = \sum_{n=-\infty}^{\infty} a_n z^n$$

converges on the annulus $0 < |z| < \infty$ (sphere from which two points have been removed) and that $z = 0$ and $z = \infty$ are essential isolated singular points. The maximum of the real part of $\psi(z)$ on the circle $|z| = r$ is denoted by $A(r)$. $A(r)$ increases to $\infty$ faster than any power of $r$ as $r \to \infty$ and faster than any power of $\frac{1}{r}$ as $r \to 0$. Precisely

$$\lim_{r\to\infty} \frac{\log A(r)}{\log r} = +\infty, \quad \lim_{r\to 0} \frac{\log A(r)}{\log \frac{1}{r}} = +\infty.$$

**238.** We assume that the function $f(z) = a_0 + a_1 z + \cdots + a_n z^n + \cdots$ is regular in the disk $|z| < R$ and we denote by $\Delta(f)$ the largest oscillation of the real part of $f(z)$ in $|z| < R$, i.e. $\Delta(f)$ is the least upper bound of $|\Re f(z_1) - \Re f(z_2)|$ for $|z_1| < R$, $|z_2| < R$. Then we have

$$|a_1| R \leqq \frac{2}{\pi} \Delta(f).$$

The factor $\frac{2}{\pi}$ cannot be replaced by a smaller one.

Interpret this proposition geometrically.

**239.** Let the function $f(z) = a_0 + a_1 z + \cdots + a_n z^n + \cdots$ be regular in the disk $|z| < R$ and denote by $D(f)$ the largest oscillation of $f(z)$ for $|z| < R$, i.e. $D(f)$ is the least upper bound of $|f(z_1) - f(z_2)|$ for $|z_1| < R$, $|z_2| < R$. Then we have

$$|a_1| R \leqq \frac{1}{2} D(f).$$

The factor $\frac{1}{2}$ can not be replaced by a smaller one. Interpret this proposition geometrically.

**240.** Let the function $f(z)$ satisfy the following conditions:

(1) $f(z)$ is regular, $|f(z)| \leqq M$ in the disk $|z - s| \leqq r$;

(2) $f(z)$ does not vanish in the closed half-disk $|z - s| \leqq r$, $\Re(z - s) \geqq 0$;

(3) $f(z)$ has in the disk $|z - s| \leqq \frac{2}{3} r$ the zeros $c_1, c_2, \ldots, c_l$. Then

$$-\Re \frac{f'(s)}{f(s)} \leqq \frac{2}{r} \log \frac{M}{|f(s)|} - \sum_{\lambda=1}^{l} \Re \frac{1}{s - c_\lambda}.$$

[We may assume $s = 0$; **232**, **120**.]

## § 3. Poles on the Circle of Convergence

**241.** Suppose that the unit circle is the circle of convergence of a certain power series and that there are only poles of first order on this circle (no other singularities). Then the sequence of coefficients is bounded.

**242.** If there is only one singular point $z_0$ on the circle of convergence of $\sum_{n=0}^{\infty} a_n z^n$ and if $z_0$ is a pole then

$$\lim_{n\to\infty} \frac{a_n}{a_{n+1}} = z_0.$$

**243.** Let $\sum_{n=0}^{\infty} a_n z^n$ be the expansion into a power series of a rational function whose denominator (relative prime to the numerator) has degree $q$. The radius of convergence is called $\varrho$ and $A_n$ denotes the largest among the $q$ numbers $|a_n|, |a_{n-1}|, \ldots, |a_{n-q+1}|$. Then

$$\lim_{n\to\infty} \sqrt[n]{A_n} = \frac{1}{\varrho}.$$

(lim not lim sup!)

**244.** Let $\nu_n$ be the number of non-zero coefficients among the $n$ coefficients $a_0, a_1, \ldots, a_{n-1}$. If there are only poles (and no other singularities) on the circle of convergence of the power series $a_0 + a_1 z + a_2 z^2 + \cdots + a_n z^n + \cdots$ the number of such poles is not smaller than

$$\limsup_{n\to\infty} \frac{n}{\nu_n}.$$

Example: $1 + z^k + z^{2k} + z^{3k} + \cdots = \frac{1}{1 - z^k}$.

**245.** We assume that the coefficients $a_0, a_1, \ldots, a_n, \ldots$ of the power series $a_0 + a_1 z + \cdots + a_n z^n + \cdots$ are real and that $\varrho e^{i\alpha}$ and $\varrho e^{-i\alpha}$ are poles and the only singularities on the circle of convergence, $0 < \alpha < \pi$. We call $V_n$ the number of changes of sign in the sequence $a_0, a_1, \ldots, a_{n-1}, a_n$. Then

$$\lim_{n\to\infty} \frac{V_n}{n} = \frac{\alpha}{\pi}. \qquad \text{[VIII 14.]}$$

**246.** If there is also a pole among the singularities on the circle of convergence the power series converges at no point of the circle of convergence.

**247.** Suppose that the point $z = 1$ is a regular point of the power series

$$f(z) = a_1 z + a_2 z^2 + \cdots + a_n z^n + \cdots$$

that converges inside the unit disk. Then the Dirichlet series, assumed convergent for certain values of $s$,

$$D(s) = a_1 1^{-s} + a_2 2^{-s} + \cdots + a_n n^{-s} + \cdots$$

defines an entire function. In the case where the point $z = 1$ is a pole of order $h$ of the function $f(z)$, $D(s)$ is a meromorphic function which can have poles only at the points $s = 1, 2, \ldots, h$ (only the last one must be a pole); the poles are simple. [$D(s)\,\Gamma(s) = \int_0^\infty x^{s-1} f(e^{-x})\,dx$, cf. **II 117**; $\int_\varrho^\infty x^{s-1} f(e^{-x})\,dx$ is an entire function of $s$ if $\varrho > 0$.]

## § 4. Identically Vanishing Power Series

**248.** The sequence $a_0, a_1, a_2, \ldots, a_n, \ldots$ is assumed to satisfy the condition

$$\limsup_{n\to\infty} \frac{\log |a_n|}{\sqrt{n}} = -h, \qquad h > 0.$$

Then the series

$$\Phi(s) = 2a_0 + a_1(e^s + e^{-s}) + a_2\left(e^{\sqrt{2}s} + e^{-\sqrt{2}s}\right) + \cdots + a_n\left(e^{\sqrt{n}s} + e^{-\sqrt{n}s}\right) + \cdots$$

converges in the infinite strip

$$-h < \Re s < h$$

of the $s$-plane. The convergence is absolute and uniform in any closed interior strip $-h + \varepsilon \leqq \Re s \leqq h - \varepsilon$, $\varepsilon > 0$; there the series defines an analytic function $\Phi(s)$. The function $\Phi(s)$ vanishes identically only if all the coefficients $a_0, a_1, a_2, \ldots, a_n, \ldots$ vanish. [Compute

$$F(u) = \frac{1}{2\pi i} \int_{a-i\infty}^{a+i\infty} \Phi(s) \frac{e^{-us}}{s^2}\,ds, \qquad 0 < a < h, \qquad u > 0; \textbf{155}.]$$

**249.** We assume that the power series

$$f(z) = a_0 + a_1 z + a_2 z^2 + \cdots + a_n z^n + \cdots$$

converges inside the unit disk and that $f(z)$ and all its derivatives tend to 0 as $z$ tends to 1 on the real axis, i.e. $\lim_{z=x\to 1} f^{(n)}(z) = 0$, $n = 0, 1, 2, \ldots$ Then there are two possibilities:

(1) $f(z)$ vanishes identically;

(2) $z = 1$ is a singular point for $f(z)$.

If the power series converges in a disk larger than the unit disk, i.e. if

$$\limsup_{n\to\infty} \frac{\log|a_n|}{n} < 0,$$

we are necessarily dealing with the first case. The weaker condition

$$\limsup_{n\to\infty} \frac{\log|a_n|}{\sqrt{n}} < 0$$

admits the same conclusion. [Construct the function $\Phi(s)$ of **248**.]

**250.** The proposition **249** is not valid if the condition on the coefficients $a_0, a_1, a_2, \ldots, a_n, \ldots$ of $f(z)$, namely $\limsup\limits_{n\to\infty} \frac{\log|a_n|}{\sqrt{n}} < 0$, is replaced by

$$\limsup_{n\to\infty} \frac{\log|a_n|}{n^\mu} < 0, \qquad 0 < \mu < \tfrac{1}{2}.$$

[Put

$$f(z) = \int_0^\infty e^{-x^\mu \cos\mu\pi} \sin(x^\mu \sin\mu\pi)\, e^{-x(1-z)}\, dx; \qquad \textbf{153}, \text{II } \textbf{222}.]$$

## § 5. Propagation of Convergence

The following examples show that the convergence of sequences of analytic functions is often "contagious".

**251.** If the series

$$g(z) + g'(z) + g''(z) + \cdots + g^{(n)}(z) + \cdots$$

converges at one single point at which $g(z)$ is regular then $g(z)$ is an entire function and the series converges at every point. The convergence is uniform in any finite domain of the $z$-plane.

**252.** If the sequence

$$|g'(z)|, \quad \sqrt{|g''(z)|}, \quad \ldots, \quad \sqrt[n]{|g^{(n)}(z)|}, \quad \ldots$$

is bounded at a single point of the $z$-plane then $g(z)$ is an entire function and the sequence stays bounded at all the points of the $z$-plane. It even has the same limit superior at all the points.

**253.** Let

$$a_0, \quad a_1, \quad a_2, \quad \ldots, \quad a_n, \quad \ldots$$

$$c_0, \quad c_1, \quad c_2, \quad \ldots, \quad c_n, \quad \ldots$$

be two infinite sequences, the second one being arbitrary, the first one such that $a_n \neq 0$, $a_m \neq a_n$ when $m \neq n$, $m, n = 0, 1, 2, \ldots$ and that

$$\frac{1}{a_0} + \frac{1}{a_1} + \frac{1}{a_2} + \cdots + \frac{1}{a_n} + \cdots$$

converges absolutely. The equations

$$Q_n(a_0) = c_0, \quad Q_n(a_1) = c_1, \quad \ldots, \quad Q_n(a_n) = c_n$$

define a unique polynomial $Q_n(z)$ of degree not larger than $n$. If the sequence

$$Q_0(z), \quad Q_1(z), \quad Q_2(z), \quad \ldots, \quad Q_n(z), \quad \ldots$$

converges at a single point $z$ different from $a_0, a_1, a_2, \ldots$ it converges at every point $z$, even uniformly in any finite domain of the $z$-plane.

**254.** Let

$$\ldots, \quad c_{-n}, \quad c_{-n+1}, \quad \ldots, \quad c_0, \quad \ldots, \quad c_{n-1}, \quad c_n, \quad \ldots$$

denote a sequence that is infinite in two directions and $Q_{2n}(z)$ be the polynomial of degree $2n$ that satisfies the equations

$$Q_{2n}(-n) = c_{-n}, \quad Q_{2n}(-n+1) = c_{-n+1}, \quad \ldots, \quad Q_{2n}(0) = c_0, \quad \ldots,$$
$$Q_{2n}(n-1) = c_{n-1}, \quad Q_{2n}(n) = c_n.$$

If the sequence of polynomials

$$Q_0(z), \quad Q_2(z), \quad Q_4(z), \quad \ldots, \quad Q_{2n}(z), \quad \ldots$$

converges for two different non-integral values of $z$ it converges at every point $z$, even uniformly in any finite domain of the $z$-plane.

**255.** The sequence $c_0, c_1, c_2, \ldots, c_n, \ldots$ is given. A polynomial $Q_n(z)$ of degree not exceding $n$ can be found [VI **76**] for which

$$Q_n(0) = c_0, \quad Q_n'(1) = c_1, \quad Q_n''(2) = c_2, \quad \ldots, \quad Q_n^{(n)}(n) = c_n.$$

These $n+1$ conditions define $Q_n(z)$ uniquely [VI **75**]. If the sequence

$$Q_0(z), \quad Q_1(z), \quad Q_2(z), \quad \ldots, \quad Q_n(z), \quad \ldots$$

converges at a single point $z$, $z \neq 0$, it converges at every point $z$, in fact uniformly in every finite domain of the $z$-plane.

**256.** We assume that in the unit disk, $|z| < 1$, the functions $f_0(z), f_1(z), f_2(z), \ldots, f_n(z), \ldots$ are regular and different from zero and that their absolute values are smaller than 1. If $\lim_{n\to\infty} f_n(0) = 0$, then $\lim_{n\to\infty} f_n(z) = 0$ in the entire open disk $|z| < 1$; the convergence is actually uniform in every smaller disk $|z| \leq r < 1$.

**257.** The harmonic functions

$$u_0(x, y), \quad u_1(x, y), \quad u_2(x, y), \quad \ldots, \quad u_n(x, y), \quad \ldots$$

are assumed to be regular and positive in a certain open region $\mathfrak{R}$ of the $x, y$-plane. If the infinite series

$$u_0(x, y) + u_1(x, y) + u_2(x, y) + \cdots + u_n(x, y) + \cdots$$

converges at a single point of $\mathfrak{R}$ it converges everywhere in $\mathfrak{R}$; in fact, it converges uniformly in any closed subdomain of $\mathfrak{R}$.

**258.** We suppose that the functions of the sequence $f_0(z), f_1(z), f_2(z), \ldots, f_n(z), \ldots$ are analytic in the open region $\mathfrak{R}$ and that the sequence of their real parts converges uniformly in every closed subdomain of $\mathfrak{R}$. Then the sequence of their imaginary parts either diverges at all points or it converges uniformly in any closed subdomain of $\mathfrak{R}$.

## § 6. Convergence in Separated Regions

**259.** The series

$$\frac{z}{1+z} + \frac{z^2}{(1+z)(1+z^2)} + \frac{z^4}{(1+z)(1+z^2)(1+z^4)} + \frac{z^8}{(1+z)(1+z^2)(1+z^4)(1+z^8)} + \cdots$$

converges uniformly in any domain that lies either entirely inside or entirely outside the unit circle and its sum is $z$ or 1 according as $|z| < 1$ or $|z| > 1$ [I **14**].

**260.** Let $\alpha$ denote an arbitrary constant, $\alpha \neq 0$. The series

$$1 + \sum_{n=1}^{\infty} \frac{\alpha(\alpha+n)^{n-1} x^n e^{-nx}}{n!}$$

converges uniformly for all positive values of $x$; it represents the function $e^{\alpha x}$ for $0 \leqq x \leqq 1$ and a *different* analytic function for $1 < x < \infty$.

**261.** The sequence of functions $f_1(z), f_2(z), \ldots, f_n(z), \ldots$

$$f_n(z) = \frac{\left[\frac{n}{1}\right]1^z + \left[\frac{n}{2}\right]2^z + \cdots + \left[\frac{n}{n}\right]n^z}{n(1^{z-1} + 2^{z-1} + \cdots + n^{z-1})}, \qquad n = 1, 2, 3, \ldots,$$

converges uniformly in any finite domain that does not contain the imaginary axis.

**262.** Let

$$\alpha > 0, \quad \beta > 0, \quad \alpha + \beta = 1,$$

and put

$$\varphi(z) = \alpha z + \beta \frac{1}{z}.$$

The sequence of iterated functions

$$\varphi(z), \quad \varphi[\varphi(z)], \quad \varphi\{\varphi[\varphi(z)]\}, \quad \ldots$$

converges to $+1$, when $\mathfrak{R}z > 0$, to $-1$ when $\mathfrak{R}z < 0$ and diverges when $\mathfrak{R}z = 0$.

## § 7. The Order of Growth of Certain Sequences of Polynomials

**263.** Let $h(\varphi)$ denote the support function of the infinite convex domain considered in **114**. The sequence $f_1(z), f_2(z), \ldots, f_n(z), \ldots$

$$f_n(z) = \frac{\sqrt{z}\,(z-1)\,(z-2)\cdots(z-n+1)\,(z-n)}{n!}\,e^{-rh(\varphi)}, \qquad n = 1, 2, 3, \ldots,$$

where $z = re^{i\varphi}$, is bounded in the entire half-plane $\Re z \geqq 0$ [**12**, II **220**]; $h(\varphi)$ is the smallest function of the angle $\varphi$ that keeps the sequence bounded.

**264.** Let $h(\varphi)$ denote the support function of the convex domain considered in **115**. The sequence $f_1(z), f_2(z), \ldots, f_n(z), \ldots$

$$f_n(z) = z\left(1 - \frac{z^2}{1^2}\right)\left(1 - \frac{z^2}{2^2}\right)\cdots\left(1 - \frac{z^2}{n^2}\right)e^{-rh(\varphi)}, \quad n = 1, 2, 3, \ldots,$$

where $z = re^{i\varphi}$, is bounded in the entire $z$-plane [**13**, II **221**]; $h(\varphi)$ is the smallest function of the angle $\varphi$ that keeps the sequence bounded.

**265.** Let $h(\varphi)$ denote the support function of the convex domain considered in **116**. Then

$$\left|1 + \frac{z}{n}\right|^n e^{-rh(\varphi)} \leqq 1, \qquad n = 1, 2, 3, \ldots,$$

$z = re^{i\varphi}$, in the entire plane.

# Chapter 6

# The Maximum Principle

## § 1. The Maximum Principle of Analytic Functions

The values that an analytic function assumes in the different parts of its domain of existence are related to each other: they are connected by analytic continuation and it is impossible to modify the values in one part without inducing a change throughout. Therefore an analytic function can be compared to an organism the main characteristic of which is exactly this: Action on any part calls forth a reaction of the entire system. E.g. the propagation of convergence [**251**–**258**] can be compared to the spreading of an infection. Mr. Borel advanced ingenious reflections upon similar comparisons[1]. We shall examine in what manner

[1] E. Borel: Méthodes et problèmes de théorie des fonctions. Paris: Gauthier-Villars 1922. Introduction.

the moduli of the values are related that the function assumes in different parts of its domain of existence.

Let the function $f(z)$ be regular in the circle $|z| < R$; the maximum of its absolute value on the circle $|z| = r$, $r < R$, is denoted by $M(r)$.

**266.** The maximum of $|f(z)|$ on the disk $|z| \leqq r$ is $M(r)$.

**267.** The maximum $M(r)$ is monotone increasing with $r$ unless $f(z)$ is a constant.

**268.** We assume that the function $f(z)$ is regular in the simply connected region $|z| > R$. $M(r)$ denotes the maximum of $|f(z)|$ on the circle $|z| = r$, $r > R$. Then $M(r)$ is also the maximum of $|f(z)|$ for $|z| \geqq r$ and $M(r)$ is monotone decreasing unless $f(z)$ is a constant.

**269.** Let $f(z)$ denote a polynomial of degree $n$; then

$$\frac{M(r_1)}{r_1^n} \geqq \frac{M(r_2)}{r_2^n}, \qquad 0 < r_1 < r_2.$$

Equality is attained only if the polynomial is of the form $cz^n$.

**270.** Suppose that $f(z)$ is a polynomial of degree $n$ and that

$$|f(z)| \leqq M$$

on the real interval $-1 \leqq z \leqq 1$. Then we have

$$|f(z)| \leqq M(a+b)^n$$

for any $z$ outside this interval; $a$ and $b$ are the semi-axes of the ellipse through $z$ and with foci $-1$ and $1$. What does the proposition imply for $z \to \infty$?

**271.** We assume that $f(z)$ is a polynomial of degree $n$, that $E_1$ and $E_2$ denote two homofocal ellipses with semi-axes $a_1$, $b_1$ and $a_2$, $b_2$, $a_1 < a_2$, $b_1 < b_2$. The maximum of $|f(z)|$ on $E_1$ and $E_2$ is denoted by $M_1$ and $M_2$ resp.; then

$$\frac{M_1}{(a_1+b_1)^n} \geqq \frac{M_2}{(a_2+b_2)^n}.$$

Derive **269** and **270** from this proposition.

**272.** If an analytic function is regular in a closed disk and not a constant its absolute value at the center of the disk is smaller than the arithmetic mean of its absolute value on the boundary of the disk.

**273.** If the absolute value of an analytic function $f(z)$ is constant in an open set of the $z$-plane (e.g. in a disk) $f(z)$ must be a constant.

**274.** We suppose that the functions $\varphi(z)$ and $\psi(z)$ are regular in the closed disk $|z| \leqq 1$ and non-zero in the open disk $|z| < 1$ and that, besides, $\varphi(0)$ and $\psi(0)$ are real and positive. If $\varphi(z)$ and $\psi(z)$ have the same modulus on the circle $|z| = 1$ then $\varphi(z) = \psi(z)$ identically.

**275.** The function $f(z)$ is regular and single-valued in the interior of the closed domain $\mathfrak{D}$ and continuous in $\mathfrak{D}$, boundary included; the maximum of $|f(z)|$ on the boundary of $\mathfrak{D}$ is called $M$. Under these conditions we have the inequality

$$|f(z)| < M$$

in the interior of $\mathfrak{D}$ unless $f(z)$ is a constant. [This statement is stronger than **135**.]

**276.** Let $\mathfrak{D}$ denote a domain, $\zeta$ be an inner point of $\mathfrak{D}$ and $\mathfrak{B}$ be the set of those boundary points of $\mathfrak{D}$ whose distance to $\zeta$ does not exceed $\varrho$. The circle of radius $\varrho$ and center $\zeta$ is assumed to have an arc that does not belong to $\mathfrak{D}$ and the length of which is not smaller than $\frac{2\pi\varrho}{n}$, $n$ integer.

We suppose that the function $f(z)$ is regular and single-valued in the interior of $\mathfrak{D}$ and continuous on the boundary and that, in particular, $|f(z)| \leqq a$ at the points of $\mathfrak{B}$ and $|f(z)| \leqq A$ at the remaining boundary points of $\mathfrak{D}$; $a < A$. Then

$$|f(\zeta)| \leqq a^{\frac{1}{n}} A^{1-\frac{1}{n}}.$$

[Examine the product $\prod_{\nu=0}^{n-1} f[\zeta + (z-\zeta)\,\omega^{-\nu}]$ in a suitable domain, $\omega = e^{2\pi i/n}$.]

**277.** We assume that the function $f(z)$ is regular and bounded in the sector $0 < \arg z < \alpha$, continuous on the real axis and that $\lim_{x\to\infty} f(x) = 0$, $x$ real, $x > 0$. Then the limit relation

$$\lim_{|z|\to\infty} f(z) = 0$$

holds uniformly in any sector $0 \leqq \arg z \leqq \alpha - \varepsilon < \alpha$.

**278.** We denote by $M$ a positive constant and by $\mathfrak{R}$ a connected open region. The analytic function $f(z)$ is assumed to have the following properties:

(1) $f(z)$ is regular at every point of $\mathfrak{R}$;

(2) $f(z)$ is single-valued in $\mathfrak{R}$;

(3) each boundary point of $\mathfrak{R}$ has for every positive number $\varepsilon$ a neighbourhood such that at every point $z$ of $\mathfrak{R}$ in this neighbourhood the inequality

$$|f(z)| < M + \varepsilon$$

is satisfied.

These conditions imply

$$|f(z)| \leqq M \quad \text{for } z \text{ in } \mathfrak{R}$$

and even $|f(z)| < M$ unless $f(z)$ is a constant. [Stronger than **275**.]

**279.** Let $f(z)$ be regular and bounded on the disk $|z| < 1$ and let

$$\lim_{r \to 1} f(re^{i\vartheta}) = 0$$

hold uniformly in a sector $\alpha \leqq \vartheta \leqq \beta$, $\alpha < \beta$. Then $f(z)$ vanishes identically.

## § 2. Schwarz's Lemma

**280.** The function $f(z)$ is assumed to be regular and $|f(z)| < 1$ in the disk $|z| < 1$. If $f(0) = 0$ either the stricter inequality $|f(z)| < |z|$ holds for $z \neq 0$ or $f(z) = e^{i\alpha} z$, $\alpha = $ real.

**281.** We denote by $z = \varphi(\zeta)$ and $w = \psi(\zeta)$ two schlicht maps of the unit disk $|\zeta| < 1$ into the regions $\mathfrak{R}$ and $\mathfrak{S}$ of the $z$- and $w$-plane resp.; the images of the origin $\zeta = 0$ are the points $z = z_0$ and $w = w_0$ resp. In addition let $0 < \varrho < 1$ and $\mathfrak{r}$ and $\mathfrak{s}$ be the images of the disk $|\zeta| \leqq \varrho$ under the above mentioned maps. We assume that $w = f(z)$ is a regular analytic function in $\mathfrak{R}$ the range of which belongs to $\mathfrak{S}$ and that $f(z_0) = w_0$. Then $f(z)$ assumes in the subdomain $\mathfrak{r}$ of $\mathfrak{R}$ values that belong to the subdomain $\mathfrak{s}$ of $\mathfrak{S}$. These values are in the interior of $\mathfrak{s}$ unless $f(z)$ is a schlicht function that maps the region $\mathfrak{R}$ onto the region $\mathfrak{S}$.

**282.** Let $f(z)$ be regular and $|f(z)| < 1$ for $|z| < 1$. Then

$$|f(z) - f(0)| \leqq |z| \frac{1 - |f(0)|^2}{1 - |f(0)|\,|z|}, \qquad 0 < |z| < 1.$$

The relation will be an equality only for the linear function

$$f(z) = \frac{e^{i\alpha} z + w_0}{1 + \overline{w}_0 e^{i\alpha} z}, \quad \alpha \text{ real}.$$

**283.** Suppose that $f(z)$ is regular for $|z| < R$ and that $A(r)$ denotes the maximum of the real part of $f(z)$ for $|z| \leqq r$, $0 \leqq r < R$. Then we have the inequality

$$A(r) \leqq \frac{R - r}{R + r} A(0) + \frac{2r}{R + r} A(R), \qquad 0 < r < R,$$

where $\lim\limits_{r \to R - 0} A(r) = A(R)$ [$A(r)$ increases monotonically with $r$, **313**]. There is equality only for the linear function

$$f(z) = \frac{R w_0 + [\overline{w}_0 - 2A(R)]\, e^{i\alpha} z}{R - e^{i\alpha} z}, \qquad \alpha \text{ real}.$$

**284** (continued). The maximum $M(r)$ of the absolute value of $f(z)$ in the disk $|z| \leqq r$ is restricted by

$$M(r) \leqq M(0) + \frac{2r}{R-r}\,[A(R) - A(0)] \leqq \frac{R+r}{R-r}\,M(0) + \frac{2r}{R-r}\,A(R).$$

**285.** We assume that $f(z)$ is regular, non-zero and bounded for $|z| < R$. Then

$$M(r) \leqq M(0)^{\frac{R-r}{R+r}}\,M(R)^{\frac{2r}{R+r}}, \qquad 0 < r < R,$$

where $\lim\limits_{r \to R-0} M(r) = M(R)$.

**286.** The functions $f_1(z), f_2(z), f_3(z), \ldots, f_n(z), \ldots$ are supposed to be regular, non-zero, and of absolute value smaller than 1, for $|z| < 1$. If the series

$$f_1(0) + f_2(0) + f_3(0) + \cdots + f_n(0) + \cdots$$

is absolutely convergent, the series

$$[f_1(z)]^2 + [f_2(z)]^2 + [f_3(z)]^2 + \cdots + [f_n(z)]^2 + \cdots$$

is absolutely convergent for $|z| \leqq \frac{1}{3}$.

**287.** We assume that the function $f(z)$ is regular, has positive real part for $|z| < 1$ and that $f(0)$ is real. Then we have

$$f(0)\,\frac{1-|z|}{1+|z|} \leqq \Re f(z) \leqq f(0)\,\frac{1+|z|}{1-|z|}, \qquad |\Im f(z)| \leqq f(0)\,\frac{2\,|z|}{1-|z|^2},$$

$$f(0)\,\frac{1-|z|}{1+|z|} \leqq |f(z)| \leqq f(0)\,\frac{1+|z|}{1-|z|}, \qquad 0 < |z| < 1.$$

There is equality only if

$$f(z) = w_0\,\frac{1+e^{i\alpha}z}{1-e^{i\alpha}z}, \qquad w_0,\ \alpha \text{ real},\ w_0 > 0.$$

**288.** Let $f(z)$ be a regular function with $|\Re f(z)| < 1$ for $|z| < 1$. If $f(0) = 0$ the stronger inequality

$$|\Re f(z)| \leqq \frac{4}{\pi} \arctan |z|, \qquad 0 < |z| < 1,$$

holds; in addition

$$|\Im f(z)| \leqq \frac{2}{\pi} \log \frac{1+|z|}{1-|z|}, \qquad 0 < |z| < 1.$$

Equality occurs if and only if

$$f(z) = \frac{2}{\pi i} \log \frac{1+e^{i\alpha}z}{1-e^{i\alpha}z}, \qquad \alpha \text{ real}.$$

**289.** Suppose that the function $f(z)$ is regular for $|z| < R$ and that $\Delta$ denotes the oscillation of its real part in the disk, i.e

$$|\Re f(z_1) - \Re f(z_2)| < \Delta$$

for $|z_1| < R$, $|z_2| < R$. Then the oscillation of its real part in the smaller disk $|z| \leqq r$, $r < R$, is

$$|\Re f(z_1) - \Re f(z_2)| \leqq \frac{4\Delta}{\pi} \arctan \frac{r}{R}, \qquad |z_1| \leqq r, \qquad |z_2| \leqq r.$$

The oscillation of the imaginary part is restricted by the inequality

$$|\Im f(z_1) - \Im f(z_2)| \leqq \frac{2\Delta}{\pi} \log \frac{R+r}{R-r}, \qquad |z_1| \leqq r, \qquad |z_2| \leqq r.$$

**290.** We denote by $\mathfrak{T}$ the infinite region of the $z$-plane that is symmetric with respect to the $x$-axis and whose points are characterized by the inequalities

$$x > 0, \qquad -k(x) < y < k(x),$$

where $k(x)$ is a positive and continuous function defined for $x \geqq 0$. Then there exists a positive function $h(x)$ that depends only on $\mathfrak{T}$ and has the following properties: If $F(z)$ is regular in $\mathfrak{T}$ and bounded from below, $|F(z)| > c$, then

$$\frac{\log |F(x)|}{h(x)}$$

is bounded from above as $x$ increases to infinity. (The theorem is particularly interesting when $k(x)$ converges monotonically to 0; i.e. when the region $\mathfrak{T}$ is "tapering"; whereas an analytic function can increase arbitrarily fast along a ray (IV **180**) the rate of increase is limited if the function grows to infinity in some properly chosen neighbourhood of the ray. The result can be formulated with reference to the points of the modular graph somewhat vaguely so: If none should fall below a certain minimum standard none may rise arbitrarily high.)

**291.** Let $f(z)$ be a regular function and $|f(z)| < 1$ in the unit disk $|z| < 1$; in addition let $f(z)$ be regular at $z = 1$ and $f(0) = 0$, $f(1) = 1$. Then $f'(1)$ must be real and $f'(1) \geqq 1$.

**292.** We assume that the function $f(z)$ is regular and $|f(z)| < 1$ in the unit disk $|z| < 1$, and that, besides, $f(z)$ is regular for $z = 1$ and $f(1) = 1$. Then the derivative $f'(1)$ is real and

$$f'(1) \frac{1 - |f(z)|^2}{|1 - f(z)|^2} \geqq \frac{1 - |z|^2}{|1 - z|^2}, \qquad |z| < 1.$$

**293.** Suppose that $f(z)$ is regular and $\Im f(z) > 0$ in the upper half-plane $\Im z > 0$ and that, besides, $f(z)$ is regular at the point $z = a$ of the real axis and $f(a) = b$, $b$ real. Then $f'(a)$ is real and positive and we have the inequality

$$\Im \frac{1}{b - f(z)} \geqq \Im \frac{1}{(a - z) f'(a)} \qquad \text{for } \Im z > 0.$$

**294.** Let the function $f(z)$ be regular, have the zeros $z_1, z_2, \ldots, z_n$ and let $f(z)$ be bounded, $|f(z)| \leqq M$, in the unit disk $|z| < 1$. Then the stronger inequality

$$|f(z)| \leqq \left| \frac{z - z_1}{1 - \bar{z}_1 z} \cdot \frac{z - z_2}{1 - \bar{z}_2 z} \cdots \frac{z - z_n}{1 - \bar{z}_n z} \right| M$$

holds for $|z| < 1$. We have equality either at every point or at no point of the open disk $|z| < 1$. (Proposition **280** is a special case. $n = 1$, $z_1 = 0$.)

**295.** Let the function $f(z)$ be regular, have the zeros $z_1, z_2, \ldots, z_n$ and let $f(z)$ be bounded, $|f(z)| \leqq M$, in the open half-plane $\Re z > 0$. Then the stronger inequality

$$|f(z)| \leqq \left| \frac{z_1 - z}{\bar{z}_1 + z} \cdot \frac{z_2 - z}{\bar{z}_2 + z} \cdots \frac{z_n - z}{\bar{z}_n + z} \right| M$$

holds for all $z$ with $\Re z > 0$. We have equality either at every point or at no point of the right half-plane $\Re z > 0$.

**296.** A function that is meromorphic in a closed disk and of constant absolute value on the boundary circle is a rational function; in fact it is, up to a constant factor, the product of linear fractional functions that map the disk in question either onto the interior or the exterior of the unit circle.

**297.** We assume that the function $f(z)$ is regular and bounded in the disk $|z| < 1$ and vanishes at the points $z_1, z_2, z_3, \ldots$ Then

$$(1 - |z_1|) + (1 - |z_2|) + (1 - |z_3|) + \cdots$$

(the sum of the distances of the zeros from the unit circle) is finite or else $f(z) \equiv 0$.

**298.** We assume that the function $f(z)$ is regular and bounded in the half-plane $\Re z > 0$ and vanishes at the points $z_1, z_2, z_3, \ldots$ outside the unit disk in the half-plane, i.e. $|z_n| > 1$, $\Re z_n > 0$, $n = 1, 2, 3, \ldots$ Then the sum of the series

$$\Re \frac{1}{z_1} + \Re \frac{1}{z_2} + \Re \frac{1}{z_3} + \cdots$$

is finite or else $f(z) \equiv 0$.

## § 3. Hadamard's Three Circle Theorem

**299.** The sum of the absolute values of several analytic functions attains its maximum on the boundary. Here is a more detailed statement: The functions $f_1(z), f_2(z), f_3(z), \ldots, f_n(z)$ are supposed to be regular and single-valued in the domain $\mathfrak{D}$. Then the function

$$\varphi(z) = |f_1(z)| + |f_2(z)| + \cdots + |f_n(z)|,$$

which is continuous in $\mathfrak{D}$, assumes its maximum on the boundary of $\mathfrak{D}$.

**300** (continued). The function $\varphi(z)$ assumes its maximum only on the boundary of $\mathfrak{D}$ unless all the functions $f_1(z), f_2(z), \ldots, f_n(z)$ are constants.

**301.** In three dimensional space, the $n$ points $P_1, P_2, \ldots, P_n$ are given and $P$ denotes a variable point. The function

$$\varphi(P) = \overline{PP_1} \cdot \overline{PP_2} \cdots \overline{PP_n}$$

($\overline{PP_\nu}$ is the distance between $P$ and $P_\nu$) of the point $P$ assumes its maximum in any domain on the boundary. (Generalization of **137**.)

**302.** We assume that the functions $f_1(z), f_2(z), \ldots, f_n(z)$ are regular and single-valued in the domain $\mathfrak{D}$. Let $p_1, p_2, \ldots, p_n$ denote positive numbers. The function

$$\varphi(z) = |f_1(z)|^{p_1} + |f_2(z)|^{p_2} + \cdots + |f_n(z)|^{p_n}$$

is continuous in $\mathfrak{D}$. It reaches its maximum only on the boundary of $\mathfrak{D}$ unless all the functions $f_1(z), f_2(z), \ldots, f_n(z)$ are constants.

**303.** The function $f(z)$ is supposed to be regular in the multiply connected closed domain $\mathfrak{D}$ and $|f(z)|$ single-valued in $\mathfrak{D}$. [$f(z)$ is not necessarily single-valued.] The absolute value $|f(z)|$ attains its maximum at a boundary point of $\mathfrak{D}$. The maximum cannot be attained at an inner point of $\mathfrak{D}$ unless $f(z)$ is a constant.

**304.** Let the function $f(z)$ be regular in the disk $|z| < R$. Suppose

$$0 < r_1 < r_2 < r_3 < R.$$

Then

$$\log M(r_2) \leqq \frac{\log r_2 - \log r_1}{\log r_3 - \log r_1} \log M(r_3) + \frac{\log r_3 - \log r_2}{\log r_3 - \log r_1} \log M(r_1).$$

This means that in an orthogonal system of coordinates the graph of $\log M(r)$ as a function of $\log r$ appears as a convex curve. (Hadamard's three circle theorem.) [Examine $z^\alpha f(z)$ with suitably chosen $\alpha$.]

**305** (continued). The function $\log M(r)$ is a strictly convex function of $\log r$ unless $f(z)$ is of the form $az^\alpha$, $a$, $\alpha$ constants, $\alpha$ real: this is the only type of function for which the inequality in **304** becomes an equality.

**306.** Suppose that the function $f(z)$ is regular for $|z| < R$ but not of the form $cz^n$, $c$ constant, $n$ integer; let

$$I_2(r) = \frac{1}{2\pi} \int_0^{2\pi} |f(re^{i\vartheta})|^2 \, d\vartheta$$

denote the arithmetic mean of $|f(z)|^2$ on the circle $|z| = r$, $r < R$. The function $I_2(r)$ is monotone increasing with $r$ and $\log I_2(r)$ is a strictly convex function of $\log r$.

**307.** Let $f(z)$ be regular for $|z| < R$;

$$\mathfrak{G}(r) = e^{\frac{1}{2\pi} \int_0^{2\pi} \log|f(re^{i\vartheta})| d\vartheta}$$

denotes the geometric mean of $|f(z)|$ on the circle $|z| = r$, $r < R$. The function $\mathfrak{G}(r)$ is monotone increasing with $r$ and a convex function of $\log r$ (in the wide sense).

**308.** The function $f(z)$ is supposed to be regular for $|z| < R$ and not a constant. We put

$$I(r) = \frac{1}{2\pi} \int_0^{2\pi} |f(re^{i\vartheta})| \, d\vartheta, \qquad r < R.$$

The function $I(r)$ is monotone increasing with $r$ and $\log I(r)$ is a convex function of $\log r$. [**299**, **304**.]

**309.** Let $f(z)$ be regular and not a constant in the disk $|z| < R$. The function $w = f(z)$ maps the circle $|z| = r$, $r < R$, in the $z$-plane onto a curve in the $w$-plane with length $l(r)$. The ratio $\frac{l(r)}{2\pi r}$ is monotone increasing with $r$.

**310.** We suppose that $f(z)$ is regular and not constant for $|z| < R$ and that $p$ is a positive number. We define

$$I_p(r) = \frac{1}{2\pi} \int_0^{2\pi} |f(re^{i\vartheta})|^p \, d\vartheta, \qquad r < R.$$

The function $I_p(r)$ is monotone increasing with $r$ and $\log I_p(r)$ is a convex function of $\log r$. (Cf. **306**, **308**, **307**, **267** and **304**, for an analogous case see IV **19**.)

## § 4. Harmonic Functions

**311.** An analytic function that is regular in the closed disk $\mathfrak{K}$ cannot assume real values at all the boundary points of $\mathfrak{K}$ except if it is a real constant.

**312.** If a harmonic function is regular in a closed disk its absolute value at the center does not exceed the arithmetic mean of its absolute values on the boundary. Under what conditions is there equality?

**313.** A harmonic function is supposed to be regular and single-valued in a domain $\mathfrak{D}$. Then it attains its maximum and its minimum on the boundary and only on the boundary unless it is a constant.

**314.** A harmonic function that is regular in the domain $\mathfrak{D}$ and vanishes at all the boundary points of $\mathfrak{D}$ is identically zero.

**315.** The equilibrium described in solution **31** is not stable.

**316.** A harmonic function is assumed to be single-valued in the domain $\mathfrak{D}$ and regular with the exception of finitely many points at which it becomes $-\infty$ (i.e. the function converges to $-\infty$ as $z$ approaches such a point). Then it assumes its maximum on the boundary of $\mathfrak{D}$.

**317.** Let the function $f(z)$ be regular in the disk $|z| \leqq R$. The function $w = f(z)$ maps the disk onto a certain domain of the $w$-plane that is assumed to be star-shaped with respect to $w = 0$. Let $f(0) = 0$. Then the images of the concentric circles $|z| = r$, $r < R$, are also star-shaped with respect to the origin.

**318.** Suppose that the function $f(z)$ is regular in the disk $|z| \leqq R$ and that it maps the disk onto a convex domain of the $w$-plane. Then the image of an arbitrary circle in the open disk $|z| < R$ is also convex.

**319.** Let $u_1(x, y), u_2(x, y), \ldots, u_n(x, y)$, $z = x + iy$, be regular harmonic functions in a domain $\mathfrak{D}$. The continuous function

$$|u_1(x, y)| + |u_2(x, y)| + \cdots + |u_n(x, y)|$$

assumes its maximum on the boundary of $\mathfrak{D}$.

**320.** Consider a regular harmonic function in the disk $|z| < R$. We denote by $A(r)$ its maximum on the circle $|z| = r$, $r < R$. When $0 < r_1 < r_2 < r_3 < R$ we have

$$A(r_2) \leqq \frac{\log r_2 - \log r_1}{\log r_3 - \log r_1} A(r_3) + \frac{\log r_3 - \log r_2}{\log r_3 - \log r_1} A(r_1),$$

i.e. $A(r)$ is a convex function of $\log r$.

**321.** Deduce Hadamard's three circle theorem **304** from **320** and vice versa **320** from Hadamard's three circle theorem.

## § 5. The Phragmén-Lindelöf Method

**322.** Let $\alpha$ be given, $0 < \alpha < \frac{\pi}{2}$. The function $f(z)$ is assumed to be regular in the sector $-\alpha \leqq \vartheta \leqq \alpha$ $(z = re^{i\vartheta})$. In addition $f(z)$ has the following properties:

(1) there exist two positive constants $A$ and $B$ such that

$$|f(z)| < Ae^{B|z|} \quad \text{for} \quad -\alpha \leqq \vartheta \leqq \alpha, \quad z = re^{i\vartheta};$$

(2) $$|f(z)| \leqq 1 \quad \text{for} \quad \vartheta = -\alpha \text{ and } \vartheta = \alpha$$

(i.e. on the boundary of the sector considered). Then the inequality $|f(z)| \leqq 1$ holds in the entire sector.

**Proof:** Let $\lambda$ be a fixed number, $1 < \lambda < \frac{\pi}{2\alpha}$. Compare $f(z)$ with the function $e^{z^\lambda}$ where we choose the branch of $z^\lambda$ that maps the positive real axis onto itself. The "comparison function" $e^{z^\lambda}$ is regular in the entire closed sector except at the origin $z = 0$ where $e^{z^\lambda}$ is continuous. We have

$$|e^{z^\lambda}| = e^{r^\lambda \cos \lambda\alpha} \geqq 1$$

on the two bounding rays $\vartheta = -\alpha$ and $\vartheta = \alpha$ because $0 < \lambda\alpha < \frac{\pi}{2}$. On the circular arc $|z| = r$, $-\alpha \leqq \vartheta \leqq \alpha$, we have

$$|e^{z^\lambda}| = e^{r^\lambda \cos \lambda\vartheta} \geqq e^{r^\lambda \cos \lambda\alpha}.$$

(The comparison function almost satisfies the conditions of the theorem but the conclusion is not valid. Imagine that $\lambda = 1$ or that $\lambda = \frac{\pi}{2\alpha}$ — which values are, in fact, excluded; in the first forbidden limit case condition (1) is satisfied, in the other condition (2), but in no case are both conditions fulfilled and in no case is $e^{z^\lambda}$ bounded inside the sector in question.)

Examine now the function $f(z)\, e^{-\varepsilon z^\lambda}$, where $\varepsilon > 0$, at a certain inner point $z_0$ of the sector. Enclose the point $z_0$ in the finite (circular) sector bounded by the rays $\vartheta = -\alpha$, $\vartheta = \alpha$ and the arc of the circle $|z| = r$ between these rays; $r$ is subject to the conditions

$$r > |z_0|, \quad r > \left(\frac{2B}{\varepsilon \cos \lambda\alpha}\right)^{\frac{1}{\lambda - 1}}, \quad r > \frac{\log A}{B}.$$

By virtue of condition (2) we have on the rectilinear boundary of the sector

$$|f(z)\, e^{-\varepsilon z^\lambda}| \leqq 1 \cdot e^{-\varepsilon r^\lambda \cos \lambda\alpha} \leqq 1.$$

By virtue of condition (1) and the inequalities $\varepsilon r^\lambda \cos \lambda\alpha > 2Br$, $e^{Br} > A$, we have on the bounding arc $|z| = r$.

$$|f(z) e^{-\varepsilon z^\lambda}| < Ae^{Br} e^{-\varepsilon r^\lambda \cos \lambda\alpha} < Ae^{-Br} < 1.$$

The maximum principle together with these inequalities implies that at the inner point $z_0$

$$|f(z_0)\, e^{-\varepsilon z_0^\lambda}| \leqq 1.$$

This inequality holds for any $\varepsilon$, however small, which concludes the proof.

**323.** Condition (1) of **322** can be generalized insofar as the inequality

$$|f(z)| < Ae^{B|z|}$$

need not be required to hold in the entire sector but only on the arcs of the circles $|z| = r_1$, $|z| = r_2, \ldots$, $|z| = r_n, \ldots$ intercepted by the rays $\vartheta = -\alpha$, $\vartheta = \alpha$, $\lim_{n\to\infty} r_n = \infty$. The conclusion, namely $|f(z)| \leqq 1$ in the entire sector, remains the same. What more general curves can replace the circular arcs?

**324.** We modify condition (2) of **322** in the following way: There exist in the sector $-\alpha \leqq \vartheta \leqq \alpha$ two curves $\Gamma_1$ and $\Gamma_2$ connecting the points $z = 0$ and $z = \infty$ that do not intersect and along which $|f(z)| \leqq 1$. This modified condition together with condition (1) as stated in **322** implies the inequality $|f(z)| \leqq 1$ in the domain bounded by $\Gamma_1$ and $\Gamma_2$.

**325.** The function $f(z)$ is assumed to be regular in the half-plane $\Re z \geqq 0$ and to satisfy the following conditions:

(1) there exist two constants $A$ and $B$, $A > 0$, $B > 0$ such that in the entire half-plane

$$|f(z)| \leqq Ae^{B|z|};$$

(2) we have for $r \geqq 0$

$$|f(ir)| \leqq 1, \quad |f(-ir)| \leqq 1;$$

(3)
$$\limsup_{r\to+\infty} \frac{\log|f(r)|}{r} \leqq 0.$$

Then $f(z)$ is bounded by 1 in the entire half-plane:

$$|f(z)| \leqq 1 \quad \text{for} \quad \Re z \geqq 0.$$

Proof: In treating **322** we extracted the desired conclusion from the maximum principle by introducing a variable parameter (the number $\varepsilon$); now we introduce two parameters. Assume $\eta > 0$. By virtue of condition (3) the function $|f(r)\, e^{-\eta r}|$ of the variable $r$ converges to 0 as $r \to \infty$; it reaches its maximum $F_\eta$ at a certain point $r_0$, $r_0 \geqq 0$. If $r_0 = 0$ the maximum is $F_\eta \leqq 1$ because of (2). We choose a fixed number $\lambda$, $1 < \lambda < 2$ $\left(\text{e.g. } \lambda = \frac{3}{2}\right)$ and study the analytic function

$$f(z)\, e^{-\eta z} e^{-\varepsilon e^{-i\lambda\pi/4} z^\lambda}$$

in the sector $0 \leqq \vartheta \leqq \frac{\pi}{2}$, $\varepsilon > 0$; take the branch of $z^\lambda$ that maps the positive real axis onto itself. We see that in the above mentioned sector

$$\left|f(z)\, e^{-\eta z} e^{-\varepsilon e^{-\frac{i\lambda\pi}{4}} z^\lambda}\right| = |f(z)\, e^{-\eta z}|\, e^{-\varepsilon r^\lambda \cos\left(\lambda\vartheta - \frac{\lambda\pi}{4}\right)}$$

$$\cos\left(\lambda\vartheta - \frac{\lambda\pi}{4}\right) \geqq \cos\left(\pm\frac{\lambda\pi}{4}\right) > 0.$$

In the same way as in **322** we derive from (1), (2) and (3) that for $0 \leqq \vartheta \leqq \frac{\pi}{2}$ the absolute value $|f(z)\, e^{-\eta z}|$ cannot be larger than 1 or $F_\eta$ whichever is larger. The same can be said for the sector $-\frac{\pi}{2} \leqq \vartheta \leqq 0$. We claim that $F_\eta \leqq 1$: If $F_\eta$ were larger than 1 we would have $|f(z)\, e^{-\eta z}| \leqq F_\eta$ on the entire half-plane and at the point $t = r_0$ the maximum would be reached, $|f(r_0)\, e^{-\eta r_0}| = F_\eta$ (see above). This is impossible because the maximum cannot be attained at an inner point $z = r_0$. Hence $F_\eta \leqq 1$, consequently

$$|f(z)\, e^{-\eta z}| \leqq 1, \qquad \Re z \geqq 0.$$

This is valid for any $\eta$ which concludes the proof. Notice that in condition (3) the positive real axis may be replaced by any ray from $z = 0$ that goes to infinity in the half-plane $\Re z \geqq 0$. Such a ray cuts the half-plane into two sectors, both with an angle smaller than $\pi$: Only this fact is essential [**322**, also **330**].

**326.** Let the function $f(z)$ be regular in the half-plane $\Re z \geqq 0$ and satisfy the following conditions:

(1) there exist two constants $A$ and $B$, $A > 0$, $B > 0$, such that in the entire half-plane

$$|f(z)| < A e^{B|z|};$$

(2) $f(z)$ is bounded on the imaginary axis,

$$|f(ir)| \leqq 1, \qquad |f(-ir)| \leqq 1, \qquad r \geqq 0;$$

(3) there exists an angle $\alpha$, $-\frac{\pi}{2} < \alpha < \frac{\pi}{2}$ such that

$$\lim_{r\to+\infty} \frac{\log|f(re^{i\alpha})|}{r} = -\infty.$$

Such a function must vanish identically. [Examine the function $e^{\omega z} f(z)$, $\omega > 0$.]

**327.** The function $f(z)$ is supposed to be regular in the half-plane $\Re z \geqq 0$ and to satisfy the conditions:

(1) there exist two constants $A$ and $B$, $A > 0$, $B > 0$, such that on the entire half-plane

$$|f(z)| < Ae^{B|z|};$$

(2) there exist two constants $C$ and $\gamma$, $C > 0$, $\gamma > 0$ such that for $r \geqq 0$

$$|f(\pm ir)| \leqq Ce^{-\gamma r}.$$

A function satisfying these conditions must vanish identically. [Examine the function $f(z)\, e^{-\beta z \log(z+1)}$.]

**328.** The function $\sin \pi z$ is the smallest function that is analytic for $\Re z \geqq 0$ and that vanishes at the points $z = 0, 1, 2, 3, \ldots$ More precisely, the following proposition holds:

We assume that the function $f(z)$ is analytic in the half-plane $\Re z \geqq 0$ and that it satisfies the conditions:

(1) there exist two constants $A$, $B$, $A > 0$, $B > 0$, such that for $\Re z \geqq 0$

$$|f(z)| < Ae^{B|z|};$$

(2) there exist two constants $C$ and $\gamma$, $C > 0$, $\gamma > 0$ such that for $r \geqq 0$

$$|f(\pm ir)| \leqq Ce^{(\pi-\gamma)r};$$

(3) $f(z)$ has the zeros $0, 1, 2, \ldots, n, \ldots$

Such a function vanishes identically.

**329.** Let $\omega(x)$ be a positive function of the positive variable $x$ that increases with $x$ and tends to $+\infty$ as $x$ increases to $+\infty$. A function $f(z)$, regular in the half-plane $\Re z \geqq 0$, that satisfies the inequality

$$|f(z)| \geqq e^{\omega(|z|)|z|} \quad \text{for} \quad \Re z \geqq 0$$

*does not exist.*

**330.** Suppose that the function $f(z)$ is regular at any finite point of the sector $\alpha \leqq \vartheta \leqq \beta$, bounded by 1, $|f(z)| \leqq 1$, on the two rays $\vartheta = \alpha$ and $\vartheta = \beta$ and that, furthermore, there exists a positive constant $\delta$ such that

$$|f(z)| \exp\left(-|z|^{\frac{\pi}{\beta-\alpha}-\delta}\right)$$

is bounded for $\alpha \leqq \vartheta \leqq \beta$. Then $|f(z)| \leqq 1$ at every inner point of the sector $\alpha \leqq \vartheta \leqq \beta$. [Comparison function $\exp\left(z^{\frac{\pi}{\beta-\alpha}-\sigma}\right)$.]

**331.** Let $f(z)$ be regular in the sector $\alpha \leqq \vartheta \leqq \beta$. If $|f(z)| \leqq 1$ on the bounding rays $\vartheta = \alpha$ and $\vartheta = \beta$ and if there exists for every $\varepsilon > 0$

an $r_0$ such that

$$|f(re^{i\vartheta})| < e^{\varepsilon r^{\frac{\pi}{\beta-\alpha}}} \quad \text{for} \quad r > r_0$$

in the above mentioned sector then the stronger inequality

$$|f(z)| \leqq 1$$

holds in the entire sector. [Method of **325**.]

**332.** The function $g(z)$ is assumed to be an entire function, $M(r)$ be the maximum of $|g(z)|$ on the circle $|z| = r$. If

$$\lim_{r\to\infty} \frac{\log M(r)}{\sqrt{r}} = 0$$

then $g(z)$ cannot be bounded along any ray. [E.g. $g(z)$ is not bounded along the negative real axis.]

**333.** Suppose that the function $f(z)$ is not a constant and that it is regular in the half-strip $\mathfrak{S}$ defined by the inequalities

$$x \geqq 0, \qquad -\frac{\pi}{2} \leqq y \leqq \frac{\pi}{2}, \qquad z = x + iy.$$

If there exist two constants $A$ and $a$, $A > 0$, $0 < a < 1$, such that in $\mathfrak{S}$

$$|f(x + iy)| < e^{Ae^{ax}},$$

and if

$$|f(z)| \leqq 1$$

on the boundary of $\mathfrak{S}$ (i.e. for $x = 0$, $-\frac{\pi}{2} \leqq y \leqq \frac{\pi}{2}$ and for $x \geqq 0$, $y = \pm\frac{\pi}{2}$) then $f(z)$ satisfies the strict inequality

$$|f(z)| < 1$$

in the interior of $\mathfrak{S}$. [The comparison function is of the type $e^{e^{bz}}$.]

**334.** Let $\omega(x)$ have the same properties as in **329**. Every function $f(z)$ that is regular in the half-strip

$$x \geqq 0, \qquad -\frac{\pi}{2} \leqq y \leqq \frac{\pi}{2}, \qquad z = x + iy$$

must satisfy the inequality

$$|f(x + iy)| < e^{\omega(x)e^x}$$

at least at one point $z = x + iy$ of the half-strip.

**335.** The assumptions of **278** are weakened insofar as (3) is satisfied in all but possibly finitely many boundary points $z_1, z_2, \ldots, z_n$ of $\mathfrak{R}$. An other assumption, however, is added, namely that there exists a positive number $M'$ for which the inequality

$$|f(z)| < M'$$

holds everywhere in $\mathfrak{R}$. (Only the case $M' > M$ is interesting.) This modification of the hypothesis does not change the conclusion of **278** that under those conditions $|f(z)| \leqq M$, $|f(z)| < M$ respectively. [In the case where the point at infinity belongs to $\mathfrak{R}$ and all the boundary points of $\mathfrak{R}$ lie in the disk $|z| < r$ we examine the comparison function $(2r)^n \prod_{\nu=1}^{n} (z - z_\nu)^{-1}$.]

**336.** The domain $\mathfrak{D}$ is supposed to lie in the half-plane $\Im z \geqq 0$; the boundary contains a finite number of segments of the real axis; $\Omega$ denotes the sum of the angles under which the segments are seen from an inner point $\zeta$ of $\mathfrak{D}$. Assume that $f(z)$ is regular and single-valued in the interior of $\mathfrak{D}$ and continuous on the boundary of $\mathfrak{D}$, that $|f(z)| \leqq A$ at the real boundary points, $|f(z)| \leqq a$ on the remaining boundary of $\mathfrak{D}$, $0 < a < A$. Then

$$|f(\zeta)| \leqq A^{\frac{\Omega}{\pi}} a^{1-\frac{\Omega}{\pi}}. \qquad [57.]$$

**337.** The function $f(z)$ is supposed to be regular on that piece of the Riemann surface of $\log z$ that covers the annulus $0 < |z| \leqq 1$. If $f(z)$ is bounded and in particular if $|f(z)| \leqq 1$ for $|z| = 1$ then $|f(z)| \leqq 1$ in the entire domain.

**338.** Let $g(z)$ denote an entire function but not a constant; $\mathfrak{R}$ be a connected region on the boundary of which (more exactly: at whose boundary points different from $z = \infty$) $|g(z)| = k$ and in the interior of which $|g(z)| > k$, $k > 0$. Then the point $z = \infty$ is necessarily a boundary point of $\mathfrak{R}$ and $g(z)$ is not bounded in $\mathfrak{R}$.

**339.** Let $\Gamma_1$ and $\Gamma_2$ be two continuous curves that have a common starting point, extend to $\infty$ and enclose together with $z = \infty$ a certain region $\mathfrak{R}$ (e.g. two rays enclosing a sector). We assume that no point of the negative real axis belongs to $\mathfrak{R}$.

The function $f(z)$ is supposed to be regular on $\Gamma_1$, $\Gamma_2$ and in the enclosed region; in addition $\lim f(z) = 0$ as $z$ tends to $\infty$ along $\Gamma_1$ and $\Gamma_2$. If $f(z)$ is bounded in $\mathfrak{R}$ then $\lim f(z) = 0$ as $z$ goes to $\infty$ along any path in $\mathfrak{R}$. [Examine $\dfrac{\log z}{A + \varepsilon \log z} f(z)$.]

**340.** Let the curves $\Gamma_1$ and $\Gamma_2$ have the properties described in **339**. Let $f(z)$ be bounded and regular in the region between $\Gamma_1$ and $\Gamma_2$ and assume, in addition, that $\lim f(z) = a$ as $z$ tends to $\infty$ along $\Gamma_1$ and $\lim f(z) = b$ as $z$ tends to $\infty$ along $\Gamma_2$. Then we have $a = b$. [Consider $\left(f(z) - \frac{a+b}{2}\right)^2 - \left(\frac{a-b}{2}\right)^2$.]

# Solutions

## Part One

## Infinite Series and Infinite Sequences

***1.** [Cf. HSI, pp. 238, 252—253, ex. 20.] $A_{100} = 292$ [**2**].

***2.** [For an intuitive solution see G. Pólya: Amer. Math. Monthly Vol. 63, pp. 689—697 (1956). Cf. MD, Vol. 1, p. 97, ex. **3.84.**]

$$\begin{aligned}\sum_{n=0}^{\infty} A_n \zeta^n = {} & (1 + \zeta + \zeta^2 + \zeta^3 + \cdots + \zeta^x + \cdots)\\ & (1 + \zeta^5 + \zeta^{10} + \zeta^{15} + \cdots + \zeta^{5y} + \cdots)\\ & (1 + \zeta^{10} + \zeta^{20} + \zeta^{30} + \cdots + \zeta^{10z} + \cdots)\\ & (1 + \zeta^{25} + \zeta^{50} + \zeta^{75} + \cdots + \zeta^{25u} + \cdots)\\ & (1 + \zeta^{50} + \zeta^{100} + \zeta^{150} + \cdots + \zeta^{50v} + \cdots)\\ = {} & \frac{1}{(1-\zeta)(1-\zeta^5)(1-\zeta^{10})(1-\zeta^{25})(1-\zeta^{50})}.\end{aligned}$$

For the numerical computation of the coefficients $A_n$ expand successively the functions

$$(1-\zeta)^{-1},$$

$$(1-\zeta)^{-1}(1-\zeta^5)^{-1},$$

$$(1-\zeta)^{-1}(1-\zeta^5)^{-1}(1-\zeta^{10})^{-1},$$

$$(1-\zeta)^{-1}(1-\zeta^5)^{-1}(1-\zeta^{10})^{-1}(1-\zeta^{25})^{-1},$$

$$(1-\zeta)^{-1}(1-\zeta^5)^{-1}(1-\zeta^{10})^{-1}(1-\zeta^{25})^{-1}(1-\zeta^{50})^{-1}.$$

It is convenient to dispose the coefficients needed for the computation of $A_{100}$ in a rectangular array.

***3.** $B_5 = 15$ [**4**].

**4.** The coefficient of $\zeta^n$ in the expansion of

$$(\zeta + \zeta^2 + \zeta^3 + \zeta^4)^s$$

is equal to the number of sums of value $n$ with $s$ terms of value 1, 2, 3, 4, where the order of the terms is taken into account. Therefore we have

$$1 + \sum_{n=1}^{\infty} B_n \zeta^n = 1 + (\zeta + \zeta^2 + \zeta^3 + \zeta^4) + (\zeta + \zeta^2 + \zeta^3 + \zeta^4)^2 + \cdots$$
$$= \frac{1}{1 - \zeta - \zeta^2 - \zeta^3 - \zeta^4}.$$

For the numerical computation use the relation

$$B_n = B_{n-1} + B_{n-2} + B_{n-3} + B_{n-4},$$

which follows from the definition of $B_n$ or from the above equation.

**5.** $4 = C_{78}$ [**7**].

**6.** $20 = D_{78}$ [**8**].

**7.** $\sum_{n=0}^{99} C_n \zeta^n = (1 + \zeta)^2 (1 + \zeta^2) (1 + \zeta^5) (1 + \zeta^{10})^2 (1 + \zeta^{20}) (1 + \zeta^{50}).$

**8.** $\sum_{n=-99}^{99} D_n \zeta^n = (\zeta^{-1} + 1 + \zeta)^2 (\zeta^{-2} + 1 + \zeta^2) (\zeta^{-5} + 1 + \zeta^5)$

$$(\zeta^{-10} + 1 + \zeta^{10})^2 (\zeta^{-20} + 1 + \zeta^{20}) (\zeta^{-50} + 1 + \zeta^{50}).$$

**9.** [Cf. Euler: Introductio in Analysin infinitorum, Chap. 16, De Partitione Numerorum; Opera Omnia, Ser. 1, Vol. 8, pp. 313—338. Leipzig and Berlin: B. G. Teubner 1922; also e.g. W. Ahrens: Mathematische Unterhaltungen und Spiele, 2nd Ed., Vol. 1, pp. 88—98, Vol. 2, p. 329. Leipzig: B. G. Teubner 1910, 1918.] The "change problem" [**2**]:

$$\frac{1}{(1 - \zeta^{a_1})(1 - \zeta^{a_2}) \cdots (1 - \zeta^{a_l})} = \sum_{n=0}^{\infty} A_n \zeta^n.$$

$A_n$ denotes the number of non-negative integral solutions of the Diophantine equation

$$a_1 x_1 + a_2 x_2 + \cdots + a_l x_l = n.$$

The "postage stamp problem" [**4**]:

$$\frac{1}{1 - \zeta^{a_1} - \zeta^{a_2} - \cdots - \zeta^{a_l}} = \sum_{n=0}^{\infty} B_n \zeta^n.$$

The "first weighing problem" [**5**] (all the weights on one pan):

$$(1 + \zeta^{a_1})(1 + \zeta^{a_2}) \cdots (1 + \zeta^{a_l}) = \sum_{n=0}^{\infty} C_n \zeta^n.$$

The "second weighing problem" [**6**] (weights may be placed on both pans):

$$(\zeta^{-a_1}+1+\zeta^{a_1})(\zeta^{-a_2}+1+\zeta^{a_2})\cdots(\zeta^{-a_l}+1+\zeta^{a_l})=\sum_{n=-\infty}^{\infty}D_n\zeta^n.$$

**10.** This problem is equivalent to the following: We have to weigh on one pan of the scales an object of $n$ units with $p$ different weights of one unit. According to **9** the number $C_n$ of the different possibilities is the coefficient of $\zeta^n$ in the expansion of

$$(1+\zeta)^p=1+\binom{p}{1}\zeta+\binom{p}{2}\zeta^2+\cdots+\binom{p}{n}\zeta^n+\cdots+\zeta^p,$$

therefore:

$$C_n=\binom{p}{n}=\frac{p!}{n!\,(n-p)!}.$$

In abstract terms: The number of different subsets of $n$ elements contained in a set of $p$ elements is $\binom{p}{n}$.

**11.** This problem is equivalent to the following: Someone owns quarters minted in $p$ different years. In how many different ways can he pay out $n$ quarters? According to **9** the number of different ways is $A_n$, the coefficient of $\zeta^n$ in the expansion of

$$\frac{1}{(1-\zeta)^p}=1+\binom{-p}{1}(-\zeta)+\cdots+\binom{-p}{n}(-\zeta)^n+\cdots,$$

thus

$$A_n=\frac{p(p+1)\cdots(p+n-1)}{1\cdot 2\cdots n}=\binom{p+n-1}{p-1}.$$

**12.** According to **11** this number is

$$\binom{p+(n-p)-1}{p-1}=\binom{n-1}{p-1}.$$

The result follows also directly from the expression $(\zeta+\zeta^2+\cdots)^p$.

***13.** Identical with **11**. Another solution: Consider the $p$-fold series

$$\sum_{\nu_1,\nu_2,\ldots,\nu_p=0,1,2,3,\ldots} x_1^{\nu_1}x_2^{\nu_2}\cdots x_p^{\nu_p}=(1-x_1)^{-1}(1-x_2)^{-1}\cdots(1-x_p)^{-1}$$

and identify the $x_i$'s with $\zeta$.

Intuitive solution: Consider $n+p-1$ places in a row. At the left hand end there are a certain number of places filled with $x_1$, then a place filled with a multiplication point, then a certain number of places filled with $x_2$, then a multiplication point and so on, as shown:

$$x_1x_1\cdots x_1\cdot x_2x_2\cdots x_2\cdot x_3\cdots x_{p-1}\cdot x_px_p\cdots x_p.$$

We have to choose $p-1$ among the $n+p-1$ places for the $p-1$ multiplication points which can be done in $\binom{n+p-1}{p-1}$ different ways, by **10**. ("Combination with repetition of $p$ different elements taken $n$ at a time" is the traditional term. In some of the cases that must be admitted the multiplication points are placed in an unorthodox way.)

**14.** According to the first weighing problem [**9** extended to infinitely many weights we have to consider

$$(1+\zeta)(1+\zeta^2)(1+\zeta^4)(1+\zeta^8)\cdots = \frac{1-\zeta^2}{1-\zeta}\cdot\frac{1-\zeta^4}{1-\zeta^2}\cdot\frac{1-\zeta^8}{1-\zeta^4}\cdot\frac{1-\zeta^{16}}{1-\zeta^8}\cdots$$
$$= \frac{1}{1-\zeta} = 1+\zeta+\zeta^2+\zeta^3+\cdots.$$

Cf. **16**, **17**.

**15.**

$$(\zeta^{-1}+1+\zeta)(\zeta^{-3}+1+\zeta^3)\cdots(\zeta^{-3^n}+1+\zeta^{3^n})$$
$$=\zeta^{-1}\frac{\zeta^3-1}{\zeta-1}\zeta^{-3}\frac{\zeta^9-1}{\zeta^3-1}\cdots\zeta^{-3^n}\frac{\zeta^{3^{n+1}}-1}{\zeta^{3^n}-1}=\zeta^{-N}\frac{\zeta^{3^{n+1}}-1}{\zeta-1}$$
$$=\zeta^{-N}+\zeta^{-N+1}+\cdots+\zeta^{N-1}+\zeta^N,\quad N=\frac{3^{n+1}-1}{2}.$$

**16.** $a_n = a^{F_n}$, where $F_n$ is the number of digits 1 in the binary representation of $n$ (its expansion in powers of 2).

**17.** [E. Catalan, Problem: Nouv. Corresp. Math. Vol. 6, p. 143 (1880). Solved by E. Cesàro: Nouv. Corresp. Math. Vol. 6, p. 276 (1880).] The series in question results from the expansion into a power series of

$$(1-a\zeta)(1-b\zeta^2)(1-c\zeta^4)(1-d\zeta^8)\ldots$$

on setting $\zeta=1$. To determine the sign of a coefficient it is sufficient to examine the case where $a=b=\cdots=1$. According to **16** the sign is given by $(-1)^{F_n}$ where $F_n$ denotes the number of ones in the binary expansion of $n$.

**18.**

$$\frac{1-\zeta^{10}}{1-\zeta}\cdot\frac{1-\zeta^{100}}{1-\zeta^{10}}\cdot\frac{1-\zeta^{1000}}{1-\zeta^{100}}\cdots=\frac{1}{1-\zeta}.$$

This problem is not contained in **9**. The result, however, is well known: Any positive integer admits a unique representation in the decimal notation. Cf. **14**.

**18.1.** There are $l$ kinds of coins and we have a limited number of each kind, $p_1$ coins of the first kind, each worth $a_1$ cents, etc. In how many ways can we pay $n$ cents?

Given the positive integers

$$a_1, a_2, \ldots, a_l,$$
$$p_1, p_2, \ldots, p_l,$$

find $E_n$, the number of solutions of the equation

$$a_1x_1 + a_2x_2 + \cdots + a_lx_l = n$$

in integers $x_1, x_2, \ldots, x_n$ subject to the condition

$$0 \leqq x_1 \leqq p_1, \quad 0 \leqq x_2 \leqq p_2, \quad \ldots, \quad 0 \leqq x_l \leqq p_l.$$

By the method of solution **2**

$$\sum_{n=0}^{\infty} E_n x^n = \frac{1 - x^{a_1(p_1+1)}}{1 - x^{a_1}} \frac{1 - x^{a_2(p_2+1)}}{1 - x^{a_2}} \cdots \frac{1 - x^{a_l(p_l+1)}}{1 - x^{a_l}}.$$

Particular cases:

$$E_n = C_n \text{ when } p_1 = p_2 = \cdots = p_l = 1,$$
$$E_n = A_n \text{ when } p_1 = p_2 = \cdots = p_l = \infty,$$
$$E_{a+n} = D_n \text{ when } p_1 = p_2 = \cdots = 2 \text{ and } a_1 + a_2 + \cdots + a_l = a.$$

To encompass **18** we admit $l = \infty$ (properly interpreted). We have $p_1 = p_2 = \cdots = 9$, $a_1 = 1$, $a_2 = 10$, $a_3 = 100, \ldots$ in the case **18**.

**18.2.** Particular case of **18.1**; $l = 3$, $a_1 = a_2 = a_3 = 1$, $p_1 = p_2 = p_3 = n$,

$$E_{2n+1} = \binom{2n+3}{2} - 3\binom{n+2}{2} = \binom{n+1}{2}.$$

***19.** [Euler, l.c. **9.**] First solution: According to solution **14** we have

$$(1+\zeta)(1+\zeta^2)(1+\zeta^4)(1+\zeta^8)\cdots = \frac{1}{1-\zeta},$$
$$(1+\zeta^3)(1+\zeta^6)(1+\zeta^{12})(1+\zeta^{24})\cdots = \frac{1}{1-\zeta^3},$$
$$(1+\zeta^5)(1+\zeta^{10})(1+\zeta^{20})(1+\zeta^{40})\cdots = \frac{1}{1-\zeta^5}, \quad \text{etc.}$$

Second solution:

$$K(\zeta) = \prod_{n=1}^{\infty} (1+\zeta^n)(1-\zeta^{2n-1})$$

is invariant under the substitution of $\zeta^2$ for $\zeta$ because

$$1 - \zeta^{4n-2} = (1+\zeta^{2n-1})(1-\zeta^{2n-1}),$$

i.e. $[|\zeta| < 1]$

$$K(\zeta) = K(\zeta^2) = K(\zeta^4) = K(\zeta^8) = \cdots = K(0) = 1.$$

Third solution:

$$\prod_{n=1}^{\infty}(1+\zeta^n)=\prod_{n=1}^{\infty}\frac{(1-\zeta^{2n})}{(1-\zeta^n)}=\prod_{n=1}^{\infty}\frac{1}{(1-\zeta^{2n-1})}.$$

**20.** [Euler, l.c. **9**.] Interpret the coefficients in the expansion of the functions in **19**. The result states that the first weighing problem with all the integers as weights admits as many solutions as the change problem with the odd integers as coins.

**21.** If we omit the restriction that the terms in the sum have to be smaller than $n$, i.e. if we admit also the representation $n = n$, then we are dealing with the postage stamp problem [**9**]. The number of different sums is equal to the coefficient of $\zeta^n$ in

$$\frac{1}{1-\zeta-\zeta^2-\cdots-\zeta^n},$$

or in

$$\frac{1}{1-\zeta-\zeta^2-\zeta^3-\cdots}=\frac{1}{1-\dfrac{\zeta}{1-\zeta}}=\frac{1-\zeta}{1-2\zeta}=(1-\zeta)(1+2\zeta+4\zeta^2+\cdots)$$

$$=1+\zeta+2\zeta^2+4\zeta^3+\cdots+2^{n-1}\zeta^n+\cdots.$$

Intuitive solution: The interval $0 \leqq x \leqq n$ appears as a sum of subintervals of integral lengths if some of the $n-1$ points $1, 2, 3, \ldots, n-1$ are chosen as points of division. For each of these points there are two possibilities, to be chosen or not, independently of the other $n-2$ points, and so the total number of possible choices is $2^{n-1}$. [MD, Vol. 2, p. 189, problem 3.40.1.]

**22.** [E. Catalan, Problem: Mathésis (Gand) Vol. 2, p. 158 (1882). Solved by E. Cesàro: Mathésis (Gand) Vol. 3, p. 87 (1883).] The total number of solutions is equal to the coefficient of $\zeta^n$ in the expansion of

$$\frac{1}{(1-\zeta)(1-\zeta^2)}+\frac{\zeta}{(1-\zeta^2)(1-\zeta^3)}+\frac{\zeta^2}{(1-\zeta^3)(1-\zeta^4)}+\cdots+$$

$$+\frac{\zeta^\nu}{(1-\zeta^{\nu+1})(1-\zeta^{\nu+2})}+\cdots=\frac{1}{\zeta(1-\zeta)}\sum_{\nu=0}^{\infty}\left(\frac{1}{1-\zeta^{\nu+1}}-\frac{1}{1-\zeta^{\nu+2}}\right)$$

$$=\frac{1}{\zeta(1-\zeta)}\left(\frac{1}{1-\zeta}-1\right)=\sum_{n=0}^{\infty}(n+1)\zeta^n.$$

**23.** [E. Catalan: Nouv. Annls Math. Ser. 3, Vol. 1, p. 528 (1882). Solved by E. Cesàro: Nouv. Annls Math. Ser. 3, Vol. 2, p. 380 (1883).] The number of solutions is equal to the coefficient of $\zeta^{n-1}$ in the expansion of

$$\frac{1}{(1-\zeta)(1-\zeta^2)}+\frac{\zeta^2}{(1-\zeta^2)(1-\zeta^3)}+\frac{\zeta^4}{(1-\zeta^3)(1-\zeta^4)}+\cdots+$$

$$+\frac{\zeta^{2\nu}}{(1-\zeta^{\nu+1})(1-\zeta^{\nu+2})}+\cdots=\frac{1}{1-\zeta}\sum_{\nu=0}^{\infty}\zeta^{\nu-1}\left(\frac{1}{1-\zeta^{\nu+1}}-\frac{1}{1-\zeta^{\nu+2}}\right)$$

$$=\frac{1}{\zeta^2(1-\zeta)^2}-\frac{1}{\zeta^3}\sum_{\nu=1}^{\infty}\frac{\zeta^\nu}{1-\zeta^\nu}.$$

We have

$$\sum_{\nu=1}^{\infty}\frac{\zeta^\nu}{1-\zeta^\nu}=\tau(1)\,\zeta+\tau(2)\,\zeta+\cdots+\tau(\nu)\,\zeta^\nu+\cdots,$$

where $\tau(\nu)$ denotes the number of divisors of $\nu$ [VIII **74**].

**24.** [E. Cesàro, Problem: Mathésis (Gand) Vol. 2, p. 208 (1882).] The number of solutions is equal to the absolute term ($\zeta$-free) of the infinite sum

$$\sum_{\nu=1}^{\infty}\frac{\zeta^{\nu^2-(2\nu+1)n}}{(1-\zeta^{\nu^2})(1-\zeta^{(\nu+1)^2})}=\sum_{\nu=1}^{\infty}(\zeta^{-(2\nu+1)n}+\zeta^{-(2\nu+1)(n-1)}+\zeta^{-(2\nu+1)(n-2)}+\cdots$$

$$+1+\zeta^{2\nu+1}+\zeta^{2(2\nu+1)}+\cdots)\left(\frac{1}{1-\zeta^{\nu^2}}-\frac{1}{1-\zeta^{(\nu+1)^2}}\right).$$

It is therefore sufficient to show that the absolute term of the following sum is equal to 1 for $k\geqq 1$:

$$\sum_{\nu=1}^{\infty}\zeta^{-(2\nu+1)k}\left(\frac{1}{1-\zeta^{\nu^2}}-\frac{1}{1-\zeta^{(\nu+1)^2}}\right)=\sum_{\nu=1}^{\infty}\frac{\zeta^{-(2\nu+1)k}-\zeta^{-(2\nu-1)k}}{1-\zeta^{\nu^2}}+\frac{\zeta^{-k}}{1-\zeta}.$$

A multiple of $\nu^2$ is, however, equal to $(2\nu+1)\,k$ or to $(2\nu-1)\,k$ if and only if $\nu^2$ is a factor of $k$. Therefore the absolute term of the sum on the right hand side vanishes, which concludes the proof.

**25.** [Cf. G. H. Hardy: Some Famous Problems of the Theory of Numbers and in Particular Waring's Problem. Oxford 1920, pp. 9, 10.] "Change problem" **9**; put $\omega=e^{2\pi i/3}$:

$$\frac{1}{(1-\zeta)(1-\zeta^2)(1-\zeta^3)}$$

$$=\frac{1}{6(1-\zeta)^3}+\frac{1}{4(1-\zeta)^2}+\frac{17}{72(1-\zeta)}+\frac{1}{8(1+\zeta)}+\frac{1}{9(1-\omega\zeta)}+\frac{1}{9(1-\omega^2\zeta)}$$

$$=\sum_{n=0}^{\infty}\left(\frac{(n+3)^2}{12}-\frac{7}{72}+\frac{(-1)^n}{8}+\frac{2}{9}\cos\frac{2n\pi}{3}\right)\zeta^n.$$

It is

$$\left|-\frac{7}{72}+\frac{(-1)^n}{8}+\frac{2}{9}\cos\frac{2n\pi}{3}\right|\leqq\frac{32}{72}<\frac{1}{2}.$$

***26.** [Proposition by P. Paoli; cf. Interméd. math. Vol. 1, p. 247—248 (1894). Ch. Hermite, Problem: Nouv. Annls Math. Ser. 1, Vol. 17, p. 32 (1858). Solved by L. Rassicod: Nouv. Annls Math. Ser. 1, Vol. 17, p. 126—130 (1858).] Cf. the more general **27.1** and the more detailed **27.2**.

**27.** [Cf. Laguerre: Oeuvres, Vol. 1. Paris: Gauthier-Villars 1898, pp. 218—220.] A decomposition into partial fractions similar to the one in **25** furnishes for $\sum_{n=0}^{\infty} A_n\zeta^n = (1-\zeta^{a_1})^{-1}(1-\zeta^{a_2})^{-1}\cdots(1-\zeta^{a_l})^{-1}$ the "principal term"

$$\frac{1}{a_1a_2\cdots a_l}\frac{1}{(1-\zeta)^l}.$$

Since $a_1, a_2, \ldots, a_l$ do not have a common factor the denominator of the other terms is of degree $l-1$ at most. The statement follows from this [solution III **242**]. Cf. **28.** Notice also that the $l$-dimensional volume characterized by the inequalities

$$x_1 \geqq 0,\ x_2 \geqq 0,\ x_3 \geqq 0, \ldots, x_l \geqq 0,\ a_1x_1 + a_2x_2 + a_3x_3 + \cdots + a_lx_l \leqq n$$

in $l$-dimensional space is

$$\frac{1}{l!}\cdot\frac{n}{a_1}\cdot\frac{n}{a_2}\cdots\frac{n}{a_l} \quad \text{and} \quad A_n \sim \frac{d}{dn}\left(\frac{1}{l!}\cdot\frac{n}{a_1}\cdot\frac{n}{a_2}\cdots\frac{n}{a_l}\right).$$

**27.1.** [Cf. E. Netto: Lehrbuch der Combinatorik, 2nd Ed. Leipzig and Berlin: B. G. Teubner 1927, pp. 319—320.] We assert that

$$(*) \qquad \frac{1}{(1-\xi^{a_1})(1-\xi^{a_2})\cdots(1-\xi^{a_l})} = R\left(\frac{1}{1-\xi}\right) + \frac{S(\xi)}{1-\xi^{a_1a_2\cdots a_l}}$$

where $R(x)$ and $S(x)$ are polynomials, of degree $l$ and smaller than $a_1a_2\cdots a_l$, respectively.

By our assumption concerning the $a_i$'s the denominator of the left hand side of (*) is the product of $(1-\xi)^l$ by a polynomial which has no multiple roots and is, therefore, a divisor of $1-\xi^{a_1a_2\cdots a_l}$. Based on this fact the decomposition in partial fractions yields (*). The expansions of the two terms on the right hand side of (*) yield the two terms into which $A_n$ is decomposed, cf. III **242** and VIII **158**, respectively.

**27.2.** We consider various values of $n$ one after the other.

(1) We consider the case $n < ab$. Let $x', y', x'', y''$ be non-negative integers such that

$$ax' + by' = n, \quad ax'' + by'' = n.$$

Then

$$0 \leqq x' < b, \quad 0 \leqq x'' < b,$$

$$a(x'-x'') = -b(y'-y'')$$

and so $x' - x''$ is divisible by $b$. Yet

$$-b < x' - x'' < b$$

and so

$$x' - x'' = 0.$$

That is, the number of solutions is $A_n < 2$.

(2) By a slight modification of the foregoing $A_{ab} \leqq 2$. Yet

$$a \cdot b + b \cdot 0 = a \cdot 0 + b \cdot a = ab$$

and so $A_{ab} = 2$.

(3) By the argument of the solution **27.1**

$$\frac{(1-\xi^{ab})(1-\xi)}{(1-\xi^a)(1-\xi^b)} = T(\xi)$$

is a polynomial

$$T(\xi) = \xi^{ab-a-b+1} + \cdots + 1,$$

$$T(1) = \frac{ab}{ab} = 1.$$

Hence

$$\frac{T(\xi) - T(1)}{1-\xi} = -\xi^{ab-a-b} + \cdots$$

is also a polynomial. Now

$$\begin{aligned} \sum_{n=0}^{\infty} (A_n - A_{n-ab})\,\xi^n &= (1-\xi^{ab}) \sum_{n=0}^{\infty} A_n \xi^n \\ &= \frac{T(\xi)}{1-\xi} \\ &= \frac{T(\xi) - T(1)}{1-\xi} + \frac{1}{1-\xi} \\ &= \cdots - \xi^{ab-a-b} + \sum_{n=0}^{\infty} \xi^n \end{aligned}$$

and so

$$A_{ab-a-b} = 0,$$

$$A_n = A_{n-ab} + 1 \geqq 1 \quad \text{when} \quad n > ab - a - b.$$

(4) By the last result

$$A_n = A_{n - \left[\frac{n}{ab}\right] ab} + \left[\frac{n}{ab}\right].$$

Yet

$$n - \left[\frac{n}{ab}\right] ab < n - \left(\frac{n}{ab} - 1\right) ab = ab$$

and so, by (1),

$$A_{n - \left[\frac{n}{ab}\right] ab} = 0 \text{ or } 1, \text{ which proves } \mathbf{26}.$$

**28.** Put $p = 3$ in **11** and **12**, respectively.

**29.** Let $k$ be a non-negative integer. The number of solutions of $|x_1| + |x_2| + \cdots + |x_p| = k$ is equal to the coefficient $a_k$ of $\zeta^k$ in the expansion of

$$(1 + 2\zeta + 2\zeta^2 + 2\zeta^3 + \cdots)^p = \left(\frac{1+\zeta}{1-\zeta}\right)^p = \sum_{k=0}^{\infty} a_k\zeta^k.$$

The number of lattice points in the octahedron is therefore equal to

$$a_0 + a_1 + a_2 + \cdots + a_n,$$

the coefficient of $\zeta^n$ in the expansion of

$$\frac{(1+\zeta)^p}{(1-\zeta)^{p+1}} = \frac{(2\zeta + 1 - \zeta)^p}{(1-\zeta)^{p+1}} =$$

$$= (2\zeta)^p(1-\zeta)^{-p-1} + \binom{p}{1}(2\zeta)^{p-1}(1-\zeta)^{-p} + \binom{p}{2}(2\zeta)^{p-2}(1-\zeta)^{-p+1} + \cdots,$$

i.e. equal to

$$2^p\binom{n}{p} + 2^{p-1}\binom{p}{1}\binom{n}{p-1} + 2^{p-2}\binom{p}{2}\binom{n}{p-2} + \cdots + 1.$$

**30.** [G. Pólya: Math. Ann. Vol. 74, p. 204 (1913).] The number of lattice points is equal to the sum of the coefficients of $\zeta^{-s}, \zeta^{-s+1}, \ldots, 1, \ldots, \zeta^{s-1}, \zeta^s$ in the expansion of

$$(\zeta^{-n} + \zeta^{-n+1} + \cdots + \zeta^{-1} + 1 + \zeta + \cdots + \zeta^{n-1} + \zeta^n)^3.$$

The following relation between the series $\sum_{\nu=-k}^{k} a_\nu\zeta^\nu$ and its coefficients $a_\nu$ holds in general:

$$\frac{1}{2\pi}\int_0^{2\pi}\left(\sum_{\nu=-k}^{k} a_\nu\zeta^\nu\right)\zeta^{-r}\,dt = a_r, \qquad \zeta = e^{it},\ -k \leqq r \leqq k$$

and

$$\sum_{\nu=-m}^{m} \zeta^\nu = \frac{\zeta^{-\frac{2m+1}{2}} - \zeta^{\frac{2m+1}{2}}}{\zeta^{-\frac{1}{2}} - \zeta^{\frac{1}{2}}} = \frac{\sin\frac{2m+1}{2}t}{\sin\frac{t}{2}}, \qquad \zeta = e^{it}.$$

**31.** [Cf. Ch. Hermite, Problem: Nouv. Annls Math. Ser. 2, Vol. 7, p. 335 (1868). Solved by V. Schlegel: Nouv. Annls Math. Ser. 2, Vol. 8, p. 91 (1869).] Since $z = n - x - y$ we have $x + y < n$, $x > 0$, $y > 0$; also $x \leqq n - x$, $y - x \leqq n - x - y \leqq x + y$.

Consequently the number of solutions in question is equal to the number of solutions of the inequalities

$$1 \leqq x \leqq \frac{n}{2}, \qquad \frac{n}{2} - x \leqq y \leqq \frac{n}{2}, \qquad y > 0, \qquad x + y < n.$$

**31.1.** (1) You are one of $n$ persons forming an assembly which elects a committee consisting of $r$ of its members. There are $\binom{n}{r}$ possible committees [**10**], namely $\binom{n-1}{r-1}$ to which you belong and $\binom{n-1}{r}$ to which you do not belong.

(2) $(1+x)^n = (1+x)^{n-1}(1+x)$. Use the binomial theorem and then consider the coefficients of $x^r$ on both sides.

(3) Straightforward verification if you use the usual expression of the binomial coefficients.

**31.2.** Pass from $r-1$ to $r$ by mathematical induction; use **31.1**.

**32.** Compare the coefficients of $z^n$ in the identity

$$(1+z)^n (1+z)^n = (1+z)^{2n}.$$

**33.** Compare the coefficients of $z^{2n}$ in the identity

$$(1+z)^{2n}(1-z)^{2n} = (1-z^2)^{2n}.$$

**34.** Evident.

**34.1.** From

$$e^x \sum \frac{(-1)^n a_n x^n}{n!} = \sum \frac{b_n x^n}{n!}$$

there follows

$$e^x \sum \frac{(-1)^n b_n x^n}{n!} = \sum \frac{a_n x^n}{n!}$$

[**34**]. See also VII **54.1**.

**35.** We have

$$x^{n|h} = h^n n! \binom{\frac{x}{h}}{n}$$

therefore

$$\sum_{n=0}^{\infty} \frac{x^{n|h}}{n!} z^n = (1+hz)^{\frac{x}{h}}.$$

Apply **34**.

**36.** Cf. **35**.

**37.** Differentiate the identity

$$1-(1-x)^n = \binom{n}{1}x - \binom{n}{2}x^2 + \cdots + (-1)^{n-1}\binom{n}{n}x^n$$

and put $x=1$.

**38.** [Problem from Ed. Times. Cf. Mathésis (Gand) Ser. 2, Vol. 1, p. 104 (1891). Solved by Greenstreet et al.: Mathésis (Gand) Ser. 2,

Vol. 1, p. 236 (1891).] The sum in question is equal to

$$\int_0^1 \frac{1-(1-x)^n}{x}\,dx = \int_0^1 \frac{1-x^n}{1-x}\,dx = \int_0^1 (1+x+x^2+\cdots+x^{n-1})\,dx$$
$$= 1+\frac{1}{2}+\frac{1}{3}+\cdots+\frac{1}{n}.$$

**39.** The general term of the left hand side sum is the coefficient of $x^{2n+1}$ in $\frac{1}{2}(1+2x)^{n+k+1}(-x^2)^{n-k}$. Therefore we have to consider the coefficient of $x^{2n+1}$ in

$$\sum_{k=0}^{n} \tfrac{1}{2}(1+2x)^{n+k+1}\,(-x^2)^{n-k} = \tfrac{1}{2}(1+2x)^{n+1}\,\frac{(1+2x)^{n+1}-(-x^2)^{n+1}}{(1+x)^2}$$

or in the power series of $\frac{1}{2}(1+2x)^{2n+2}\,(1+x)^{-2}$. Division leads to a polynomial of degree $2n$, the remainder is $\frac{1}{2}[1-(2n+2)\,(2x+2)]$. The coefficient of $x^{2n+1}$ in

$$\frac{1}{2}\,\frac{1}{(1+x)^2} - \frac{2n+2}{1+x}$$

is equal to $-\frac{1}{2}(2n+2)+2n+2 = n+1$.

**40.** Decompose the sum into three terms according to

$$(\nu-n\alpha)^2 = n^2\alpha^2 - (2n\alpha-1)\,\nu + \nu(\nu-1).$$

The formula

$$\sum_{\nu=0}^{n}\binom{n}{\nu}p^\nu q^{n-\nu} = (p+q)^n$$

and its first and second derivatives with respect to $p$ furnish the three terms if $p$ is replaced by $x$ and $q$ by $1-x$. Cf. II **144**.

**41.** It is sufficient to consider $n \geqq p$ and

$$\varphi(x) = \frac{x(x-1)\cdots(x-p+1)}{p!} = \binom{x}{p}, \qquad \psi(x) = \frac{1}{2^p}\binom{x}{p}.$$

We have now

$$\sum_{\nu=p}^{n}\binom{n}{\nu}\binom{\nu}{p} = \frac{n!}{p!\,(n-p)!}\sum_{\nu=p}^{n}\binom{n-p}{\nu-p} = \binom{n}{p}2^{n-p} = 2^n\psi(n),$$

$$\sum_{\nu=p}^{n}(-1)^\nu\binom{n}{\nu}\binom{\nu}{p} = \frac{n!}{p!\,(n-p)!}\sum_{\nu=p}^{n}(-1)^\nu\binom{n-p}{\nu-p} = \begin{cases} 0 \text{ for } n>p, \\ (-1)^n \text{ for } n=p. \end{cases}$$

**42.** This is a special case of **40** with $x=\alpha=1/2$. It is also a special case of **41** with

$$\varphi(x) = (2x-n)^2 = n^2-4xn+4x^2 = n^2-4(n-1)\,x+4x\,(x-1)$$
$$\psi(x) = n^2-2(n-1)\,x+x(x-1).$$

We have $\psi(n)=n$.

**43.** Special case of **41** for $\varphi(x) = (2x - n)^2$ [**42**]. We obtain $a_n = 0$ for $n \neq 2$, $a_n = 4$ for $n = 2$.

**43.1.** (1) First multiply both sides by $e^x$, then use **34** and **38**.

(2) Let $y$ denote either the left hand, or the right hand, side of the desired identity and verify that, in both cases, $y = 0$ for $x = 0$ and

$$\frac{dy}{dx} = \frac{1 - e^{-x}}{x}.$$

**44.** Write $f(x) = c_n(x - x_1)(x - x_2) \cdots (x - x_n)$. We have

$$\left(z\frac{d}{dz} - x_\nu\right)z^k = (k - x_\nu)\, z^k.$$

**45.** [G. Darboux, Problem: Nouv. Annls Math. Ser. 2, Vol. 7, p. 138 (1868).] We deduce from **44**

$$\sum_{k=0}^{\infty} \frac{f(k)}{k!} z^k = f\left(z\frac{d}{dz}\right) e^z = e^z g(z),$$

where $g(z)$ is a polynomial with integral coefficients.

**46.** [E. Cesàro: Elementares Lehrbuch der algebraischen Analysis und der Infinitesimalrechnung. Leipzig and Berlin: B. G. Teubner 1904, p. 872.] The functions $f_{n+1}$ and $f_n$ are connected by the recursion formula

$$f_{n+1}(z) = z[f_n'(z)(1 - z) + (n + 1) f_n(z)].$$

Therefore the coefficients of $f_n(z) = a_1^{(n)}z + a_2^{(n)}z^2 + \cdots + a_n^{(n)}z^n$ are linked by the relations

$$a_\nu^{(n+1)} = \nu a_\nu^{(n)} + (n - \nu + 2)\, a_{\nu-1}^{(n)}, \quad \nu = 1, 2, \ldots, n + 1; \quad a_0^{(n)} = a_{n+1}^{(n)} = 0.$$

This together with $f_1(z) = z$ concludes the proof. The value of $f_n(1)$ is determined by

$$f_{n+1}(1) = (n + 1)\, f_n(1).$$

**47.** [Cf. N. H. Abel: Oeuvres, Vol. 2, Nouvelle édition. Christiania: Grøndahl & Son 1881, p. 14.] If $g(x)$ is constant then the proposition is a consequence of **44**. In assuming that the proposition holds for polynomials of a degree smaller than the degree of $g(x)$ we write

$$g(x) = (x - x_1)\, g_1(x).$$

$$g\left(z\frac{d}{dz}\right) y = g_1\left(z\frac{d}{dz}\right)\left[\left(z\frac{d}{dz} - x_1\right) y\right]$$

$$= g_1\left(z\frac{d}{dz}\right)\left(\frac{f(0)}{g_1(0)} + \frac{f(1)}{g_1(1)} z + \frac{f(2)}{g_1(2)} z^2 + \cdots\right) = f\left(z\frac{d}{dz}\right)\frac{1}{1 - z}.$$

The given differential equation is soluble by successive quadratures because this is so for the equation

$$\left(z\frac{d}{dz} - x_0\right)y = zy' - x_0 y = \varphi(z).$$

**48.** According to **44** both sides are equal to

$$f(1)\,z + \frac{f(1)\,f(2)}{g(1)}\,z^2 + \frac{f(1)\,f(2)\,f(3)}{g(1)\,g(2)}\,z^3 + \cdots + \frac{f(1)\,f(2)\cdots f(n-1)\,f(n)}{g(1)\,g(2)\cdots g(n-1)}\,z^n + \cdots.$$

**49.** Apply **48** putting $f(x) = (x - \frac{1}{2})^2$, $g(x) = x^2$.

**50.** Comparing the coefficients of $z^n$ on both sides of the functional equation we get

$$A_n(q^n - 1) = A_{n-1}q^n, \qquad n = 1\ 2, 3, \ldots; \quad A_0 = 1,$$

hence

$$A_n = \frac{q^{\frac{n(n+1)}{2}}}{(q-1)(q^2-1)\cdots(q^n-1)}, \qquad n = 1, 2, 3, \ldots$$

**51.** According to the functional equation [**50**]

$$B_n(1 - q^n) = B_{n-1}q, \qquad n = 1, 2, 3, \ldots; \quad B_0 = 1,$$

therefore

$$B_n = \frac{q^n}{(1-q)(1-q^2)\cdots(1-q^n)}, \qquad n = 1, 2, 3, \ldots$$

**52.** [Cf. R. Appell and E. Lacour: Principes de la théorie des fonctions elliptiques et applications. Paris: Gauthier-Villars 1897, p. 398. For closely similar preceding work of Gauss see l.c. solution **55**.] Call $\varphi_n(z)$ the expression in question:

$$\varphi_n(q^2 z) = \varphi_n(z)\,\frac{1 + q^{2n+1}z}{qz + q^{2n}}.$$

From this identity follows

$$C_\nu q^{2\nu+1}(1 - q^{2n-2\nu}) = C_{\nu+1}(1 - q^{2n+2\nu+2}), \qquad \nu = 0, 1, \ldots, n-1,$$

$$C_n = q^{n^2},$$

i.e.

$$C_\nu = \frac{(1 - q^{2n+2\nu+2})(1 - q^{2n+2\nu+4})\cdots(1 - q^{4n})}{(1-q^2)(1-q^4)\cdots(1 - q^{2n-2\nu})}\,q^{\nu^2}, \quad \nu = 0, 1, \ldots, n-1.$$

**53.** [Jacobi: Fundamenta nova theoriae functionum ellipticarum, § 64,; Werke, Vol. 1. Berlin: G. Reimer 1881, p. 234.] Take the limit $n \to \infty$ in **52**. [**181**.]

**54.** [Euler: Commentationes arithmeticae, Vol. 1; Opera Omnia, Ser. 1, Vol. 2. Leipzig and Berlin: B. G. Teubner 1915, pp. 249–250.] Special case of **53**: $q \mid q^{3/2}$, $z = -q^{1/2}$.

**55.** [Gauss: Summatio quarundam serierum singularium, Opera, Vol. 2. Göttingen: Ges. d. Wiss. 1863, pp. 9–45.] Special case of **53**: replace $q$ by $q^{1/2}$ and $z$ by $q^{1/2}$, and apply **19** or the procedure of the third solution of **19**.

**56.** [Jacobi, l.c. **53**, § 66; Werke, Vol. 1, p. 237.] Put $z = -1$ in **53** and use **19**.

**57.** Setting $-q^n z = a_n$ we obtain

$$\begin{aligned} 1+G(z)-G(qz) &= 1+\sum_{n=1}^{\infty}\frac{q^n z}{1-q^n}(1-qz)(1-q^2z)\cdots(1-q^{n-1}z)(q^n-1) \\ &= 1+a_1+a_2(1+a_1)+a_3(1+a_1)(1+a_2) \\ &\quad + a_4(1+a_1)(1+a_2)(1+a_3)+\cdots \\ &= (1+a_1)(1+a_2)+a_3(1+a_1)(1+a_2) \\ &\quad + a_4(1+a_1)(1+a_2)(1+a_3)+\cdots \\ &= (1+a_1)(1+a_2)(1+a_3) \\ &\quad + a_4(1+a_1)(1+a_2)(1+a_3)+\cdots, \text{ etc.} \end{aligned}$$

**58.**

$$D_0 = G(0) = \frac{q}{1-q}+\frac{q^2}{1-q^2}+\frac{q^3}{1-q^3}+\cdots+\frac{q^n}{1-q^n}+\cdots;$$

applying **50** and **57** we find

$$G(z)-G(qz)=\sum_{n=1}^{\infty}A_n z^n, \qquad G(qz)-G(q^2z)=\sum_{n=1}^{\infty}A_n q^n z^n,$$

$$G(q^2z)-G(q^3z)=\sum_{n=1}^{\infty}A_n q^{2n}z^n, \ldots,$$

hence addition of the first $m$ equations and taking the limit $m\to\infty$ yield

$$G(z)-G(0)=\sum_{n=1}^{\infty}\frac{A_n}{1-q^n}z^n.$$

**59.** Obviously $G(1)=0$. The functional equation in **57** implies with the help of complete induction that

$$G(q^{-n})=\sum_{k=1}^{n}\frac{q^k}{1-q^k}\left(1-\frac{1}{q^n}\right)\left(1-\frac{q}{q^n}\right)\cdots\left(1-\frac{q^{k-1}}{q^n}\right)=-n, \quad q^{-1}=a.$$

Introducing $(1-q)\,n = y$, $1-\frac{y}{n} = q$ we get

$$\sum_{k=1}^{n} \frac{q^k}{1+q+q^2+\cdots+q^{k-1}} \left[1-\left(1-\frac{y}{n}\right)^{-n}\right]\left[1-\left(1-\frac{y}{n}\right)^{-n+1}\right]\cdots$$
$$\left[1-\left(1-\frac{y}{n}\right)^{-n+k-1}\right] = -(1-q)\,n.$$

Let $n\to\infty$ for $y$ fixed, and so $q\to 1$:

$$\sum_{k=1}^{\infty} \frac{1}{k}(1-e^y)^k = -y. \qquad [\mathbf{181}.]$$

**60.** If we regard as known the sum of the series (cf. **59**) we have

$$f(z) = \frac{1}{2z}\log\frac{1+z}{1-z}$$

and the solution is straightforward. Without supposing such knowledge we equate the coefficients of $z^{2n}$ on both sides and so obtain

$$\sum_{k=0}^{n} (-1)^{n-k} \frac{2^{2k}}{2k+1}\binom{n+k}{n-k} = \frac{1}{2n+1}.$$

To verify this observe that

$$\frac{2n+1}{2k+1}\binom{n+k}{n-k} = \binom{n+k+1}{2k+1} + \binom{n+k}{2k+1}$$

and apply **39**.

**60.1.** If $l$ and $m$ are positive integers

$$\frac{1-q^l}{1-q^m} = \frac{1+q+\cdots+q^{l-1}}{1+q+\cdots+q^{m-1}}.$$

**60.2.** Obvious in the Gaussian as in the ordinary case.

**60.3.** Define $f(x)$ as the left hand side of the desired identity. Then

$$(1+x)\,f(qx) = f(x)\,(1+q^n x).$$

Proceed hence as in **50**, **51**, and **52**.

**60.4.** Multiply the identity **60.3** by $1+q^n x$ and compare coefficients. Or verify directly from the initial definition.

**60.5.** Obvious for $k=0$ and $k=n$. For $0<k<n$ from **60.4** by mathematical induction except the last clause about symmetry for which use the initial definition and substitute $q^{-1}$ for $q$.

**60.6.** Start from **60.3**, substitute first

$$2n \text{ for } n, \quad q^2 \text{ for } q,$$

then

$$zq^{-2n+1} \text{ for } x$$

and multiply finally by

$$q^{n^2} z^{-n} = \frac{q}{z} \cdot \frac{q^3}{z} \cdots \frac{q^{2n-1}}{z}.$$

The result is, in fact, identical with **52**.

**60.7.** According as $n = 2m$ or $n = 2m + 1$, the left hand side in **60.3** reduces for $q = -1$ to

$$(1 - x^2)^m \qquad \text{or} \qquad (1 - x^2)^m (1 + x).$$

**60.8.** By virtue of **60.4**

$$\begin{aligned} F(q, n) &= \sum_{k=0}^{n} (-1)^k \left( \begin{bmatrix} n-1 \\ k \end{bmatrix} + \begin{bmatrix} n-1 \\ k-1 \end{bmatrix} q^{n-k} \right) \\ &= \sum_{k=0}^{n-1} (-1)^k \begin{bmatrix} n-1 \\ k \end{bmatrix} + \sum_{k=1}^{n} (-1)^k \begin{bmatrix} n-1 \\ k-1 \end{bmatrix} q^{n-k} \\ &= \sum_{k=0}^{n-1} (-1)^k (1 - q^{n-k-1}) \begin{bmatrix} n-1 \\ k \end{bmatrix} \\ &= \sum_{k=0}^{n-2} (-1)^k (1 - q^{n-1}) \begin{bmatrix} n-2 \\ k \end{bmatrix} \qquad \text{Q.E.D.} \end{aligned}$$

**60.9.** [See e.g. MD, Vol. 1, pp. 68–78.] Obvious when $r = 0$ or $s = 0$. Use this remark, the relation

$$\binom{n+1}{r} = \binom{n}{r} + \binom{n}{r-1}$$

and mathematical induction to pass from $n$ to $n + 1$.

**60.10.** [G. Pólya: J. Combinatorial Theory, Vol. 6, p. 105 (1969).] Fairly obvious when $\alpha = 0$ or $\alpha = r(n - r)$. Use these cases, **60.4**, and mathematical induction.

**60.11.** Cut the area under the zig-zag path into vertical strips of width 1 by equidistant parallels to the $y$-axis. The top of such a strip is an $x$-segment (unit segment parallel to the $x$-axis) of the path and the area of the strip is the number of $y$-segments of the path preceding (forming an inversion with) that top $x$-segment.

The "Gaussian" analogues of polynomial coefficients can be used to determine the number of certain more general letter sequences having a given number of inversions. [Cf. G. Pólya: Proceedings of the Second Chapel Hill Conference on Combinatorial Mathematics and its Applications: Chapel Hill N. C. 1970, pp. 381–384.]

**61.** Follows from the definition; for the product use **34**.

**62.** If $a_1, a_2, \ldots, a_n$ are positive we have

$$1 + a_\nu z \ll e^{a_\nu z}, \qquad \nu = 1, 2, \ldots, n.$$

According to **61**

$$(1 + a_1 z)(1 + a_2 z) \cdots (1 + a_n z) \ll e^{(a_1 + a_2 + \cdots + a_n)z}.$$

From this the particular case $a_1 = a_2 = \cdots = a_n = \frac{1}{n}$ follows.

**63.** $A(z) \ll P(z)$ implies

$$\int_0^z A(z)\,dz \ll \int_0^z P(z)\,dz \quad \text{and} \quad e^{A(z)} \ll e^{P(z)}.$$

Therefore

$$\frac{f'(z)}{f(z)} - \frac{1}{z} \ll \frac{2}{1-z}$$

leads to

$$\log \frac{f(z)}{z} \ll \log \frac{1}{(1-z)^2}, \qquad \frac{f(z)}{z} \ll \frac{1}{(1-z)^2} = \sum_{n=1}^{\infty} n z^{n-1}.$$

**64.** a) It is obvious from the combinatorial meaning of $A_n$, $B_n$, $C_n$, defined in **9**, that

$$0 \leqq C_n \leqq A_n \leqq B_n.$$

b) The first part of the statement follows immediately from

$$1 + z^a \ll \frac{1}{1 - z^a} = 1 + z^a + z^{2a} + \cdots,$$

where we put $a = a_1, a_2, \ldots, a_l$ and multiply [**61**]. The second part follows from

$$\frac{(1 - z^{a_1} - z^{a_2} - \cdots - z^{a_{m-1}})(1 - z^{a_m})}{1 - z^{a_1} - z^{a_2} - \cdots - z^{a_{m-1}} - z^{a_m}}$$

$$= 1 + \frac{(z^{a_1} + z^{a_2} + \cdots + z^{a_{m-1}})\, z^{a_m}}{1 - z^{a_1} - z^{a_2} - \cdots - z^{a_{m-1}} - z^{a_m}} \gg 1$$

for $m = 2, 3, \ldots, l$; multiplication yields

$$\frac{(1 - z^{a_1})(1 - z^{a_2}) \cdots (1 - z^{a_l})}{1 - z^{a_1} - z^{a_2} - \cdots - z^{a_l}} \gg 1.$$

Multiply both sides by $[(1 - z^{a_1})(1 - z^{a_2}) \cdots (1 - z^{a_l})]^{-1}$.

**64.1.** We shall use the notation

$$\left\{\frac{1}{z^m}\right\} = \frac{c_m}{z^m} + \frac{c_{m+1}}{z^{m+1}} + \frac{c_{m+2}}{z^{m+2}} + \cdots$$

for any expansion in negative powers in which the coefficient of $z^{-k}$ vanishes for $k = 0, 1, 2, \ldots, m-1$. We shall also extend the concept of a majorant series and the use of the symbol $\ll$ to expansions in positive and negative powers.

$$z^n + a_1 z^{n-1} + a_2 z^{n-2} + \cdots + a_n = (z - z_1)(z - z_2) \cdots (z - z_n)$$

$$= z^n \exp\left(\log\left(1 - \frac{z_1}{z}\right) + \log\left(1 - \frac{z_2}{z}\right) + \cdots + \log\left(1 - \frac{z_n}{z}\right)\right)$$

$$= z^n \exp\left(-\frac{s_1}{z} - \frac{s_2}{2z^2} - \frac{s_3}{3z^3} - \cdots\right)$$

$$\ll z^n \exp\left(\frac{1}{z} + \frac{1}{2z^2} + \cdots + \frac{1}{nz^n} + \frac{|s_{n+1}|}{(n+1)\, z^{n+1}} + \cdots\right)$$

$$= \frac{z^n}{1 - \frac{1}{z}} \exp\left(1 + \left\{\frac{1}{z^{n+1}}\right\}\right) = z^n + z^{n-1} + z^{n-2} + \cdots + 1 + \left\{\frac{1}{z}\right\}$$

and so $|a_k| \leqq 1$ for $k = 1, 2, \ldots, n$; apply III **21**.

**64.2.** For the polynomial

$$\frac{z^{n+1} - 1}{z - 1} = z^n + z^{n-1} + \cdots + 1$$

$a_k = 1$ and $s_k = -1$ for $k = 1, 2, \ldots, n$.

**65.** Obvious.

**66.** Assume $s_n = 0$ for $n \neq \nu$, $s_\nu = 1$, therefore $t_n = p_{n\nu}$ for $n \geqq \nu$. If $\lim_{n\to\infty} t_n = \lim_{n\to\infty} s_n$ holds for this special sequence then $\lim_{n\to\infty} p_{n\nu} = 0$, i.e. the condition is necessary. Suppose the condition is satisfied. Let $\varepsilon$ be any positive number. Find $N$ such that $|s_n - s| < \frac{\varepsilon}{2}$ for all $n > N$; in addition $n$ has to be such that $p_{n0}, p_{n1}, \ldots, p_{nN}$ are all smaller than $\frac{\varepsilon}{4(N+1)\, M}$, where $M$ denotes the maximum of $|s_\nu|$. From

$$t_n - s = p_{n0}(s_0 - s) + p_{n1}(s_1 - s) + \cdots + p_{nn}(s_n - s)$$

we now deduce

$$|t_n - s| < (N+1)\, 2M \frac{\varepsilon}{4(N+1)\, M}$$

$$+ \frac{\varepsilon}{2}(p_{n,N+1} + p_{n,N+2} + \cdots + p_{nn}) \leqq \frac{\varepsilon}{2} + \frac{\varepsilon}{2}.$$

**67.** Special case of **66**: $p_{n\nu} = \frac{1}{n+1}$.

**68.** Equivalent to **67**: $\log p_n = s_n$.

**68.1.** Equivalent to **68**, $p_n = \frac{a_n}{a_{n-1}}$. This reformulation is useful in dealing with power series.

**69.** Special case of **68**: put

$$p_0 = 1, \quad p_1 = \left(\frac{2}{1}\right)^1, \quad p_2 = \left(\frac{3}{2}\right)^2, \quad \ldots, \quad p_n = \left(\frac{n+1}{n}\right)^n,$$

then we have

$$p_0 p_1 p_2 \cdots p_n = \frac{(n+1)^{n+1}}{(n+1)!}.$$

**70.** Special case of **66**: put

$$s_n = \frac{a_n}{b_n}, \quad p_{n\nu} = \frac{b_\nu}{b_0 + b_1 + b_2 + \cdots + b_n}, \quad t_n = \frac{a_0 + a_1 + a_2 + \cdots + a_n}{b_0 + b_1 + b_2 + \cdots + b_n};$$

observe

$$\lim_{n\to\infty} p_{n\nu} = 0.$$

**71.** Special case of **70**: $a_n = (n+1)^{\alpha-1}$, $b_n = (n+1)^\alpha - n^\alpha$. Since

$$\lim_{n\to\infty} \frac{(n+1)^\alpha - n^\alpha}{n^{\alpha-1}} = \lim_{n\to\infty} \frac{(1+1/n)^\alpha - 1^\alpha}{1/n} = \left(\frac{d}{dx} x^\alpha\right)_{x=1} = \alpha,$$

we obtain

$$\lim_{n\to\infty} \frac{1^{\alpha-1} + 2^{\alpha-1} + \cdots + (n+1)^{\alpha-1}}{(n+1)^\alpha} = \lim_{n\to\infty} \frac{(n+1)^{\alpha-1}}{(n+1)^\alpha - n^\alpha} = \frac{1}{\alpha}.$$

**72.** [For **72**–**74** cf. N. E. Nörlund: Lunds Universitets Arsskrift, N. F. Avd. 2, Vol. 16, No. 3 (1919).] Special case of **66**:

$$p_{n\nu} = \frac{p_{n-\nu}}{p_0 + p_1 + p_2 + \cdots + p_n} \leqq \frac{p_{n-\nu}}{p_0 + p_1 + p_2 + \cdots + p_{n-\nu}}.$$

**73.** Special case of **66**. Set

$$p_0 + p_1 + p_2 + \cdots + p_n = P_n, \quad q_0 + q_1 + q_2 + \cdots + q_n = Q_n,$$
$$r_0 + r_1 + r_2 + \cdots + r_n = R_n,$$

and

$$p_{n\nu} = \frac{p_{n-\nu} Q_\nu}{p_0 Q_n + p_1 Q_{n-1} + \cdots + p_n Q_0} \leqq \frac{p_{n-\nu}}{p_0 + p_1 + \cdots + p_{n-\nu}}.$$

We obtain [cf. **74**]

$$\frac{r_n}{R_n} = \frac{p_0 q_n + p_1 q_{n-1} + \cdots + p_n q_0}{p_0 Q_n + p_1 Q_{n-1} + \cdots + p_n Q_0} = p_{n0} \frac{q_0}{Q_0} + p_{n1} \frac{q_1}{Q_1} + \cdots + p_{nn} \frac{q_n}{Q_n}.$$

**74.** We write

$$\mathfrak{p}_n = \frac{s_0 p_n + s_1 p_{n-1} + \cdots + s_n p_0}{p_0 + p_1 + \cdots + p_n}, \quad \mathfrak{q}_n = \frac{s_0 q_n + s_1 q_{n-1} + \cdots + s_n q_0}{q_0 + q_1 + \cdots + q_n},$$

$$\mathfrak{r}_n = \frac{s_0 r_n + s_1 r_{n-1} + \cdots + s_n r_0}{r_0 + r_1 + \cdots + r_n}$$

($r_n$ as in **73**). Then

$$\mathfrak{r}_n = \frac{p_n\mathfrak{Q}_0\mathfrak{q}_0 + p_{n-1}\mathfrak{Q}_1\mathfrak{q}_1 + \cdots + p_0\mathfrak{Q}_n\mathfrak{q}_n}{p_n\mathfrak{Q}_0 + p_{n-1}\mathfrak{Q}_1 + \cdots + p_0\mathfrak{Q}_n}$$

$$= \frac{q_nP_0\mathfrak{p}_0 + q_{n-1}P_1\mathfrak{p}_1 + \cdots + q_0P_n\mathfrak{p}_n}{q_nP_0 + q_{n-1}P_1 + \cdots + q_0P_n}$$

consequently [**66**, **73**] $\lim_{n\to\infty} \mathfrak{p}_n = \lim_{n\to\infty} \mathfrak{q}_n = \lim_{n\to\infty} \mathfrak{r}_n$. To establish the above identities we use the power series whose coefficients are the sequences in question [**34**]:

$$\sum_{k=0}^{\infty} z^k \sum_{l=0}^{\infty} p_l z^l = \sum_{n=0}^{\infty} P_n z^n, \qquad \sum_{k=0}^{\infty} z^k \sum_{l=0}^{\infty} q_l z^l = \sum_{n=0}^{\infty} Q_n z^n,$$

$$\sum_{k=0}^{\infty} z^k \sum_{l=0}^{\infty} r_l z^l = \sum_{n=0}^{\infty} R_n z^n = \sum_{k=0}^{\infty} P_k z^k \sum_{l=0}^{\infty} q_l z^l = \sum_{k=0}^{\infty} p_k z^k \sum_{l=0}^{\infty} Q_l z^l,$$

$$\sum_{n=0}^{\infty} P_n\mathfrak{p}_n z^n = \sum_{k=0}^{\infty} p_k z^k \sum_{l=0}^{\infty} s_l z^l, \qquad \sum_{n=0}^{\infty} Q_n\mathfrak{q}_n z^n = \sum_{k=0}^{\infty} q_k z^k \sum_{l=0}^{\infty} s_l z^l,$$

$$\sum_{n=0}^{\infty} R_n\mathfrak{r}_n z^n = \sum_{k=0}^{\infty} p_k z^k \sum_{l=0}^{\infty} q_l z^l \sum_{m=0}^{\infty} s_m z^m$$

$$= \sum_{k=0}^{\infty} p_k z^k \sum_{l=0}^{\infty} Q_l\mathfrak{q}_l z^l = \sum_{k=0}^{\infty} q_k z^k \sum_{l=0}^{\infty} P_l\mathfrak{p}_l z^l.$$

**75.** Put

$$t_n = (a_1 + a_2 + \cdots + a_n)\, n^{-\sigma}, \qquad s_n = a_1 1^{-\sigma} + a_2 2^{-\sigma} + \cdots + a_n n^{-\sigma},$$

and $\lim_{n\to\infty} s_n = s$. The expression

$$t_n - n^{-\sigma}(n+1)^{\sigma}(s_n - s) = n^{-\sigma} \sum_{\nu=1}^{n} (s_\nu - s)\,[\nu^{\sigma} - (\nu+1)^{\sigma}] + n^{-\sigma}s$$

converges to 0 [**66**].

**76.** [E. Cesàro: Nouv. Annls Math. Ser. 3, Vol. 9, pp. 353—367 (1890).] According to **70** the limit is equal to

$$\lim_{n\to\infty} \frac{p_nP_n^{-1}}{\log P_n - \log P_{n-1}} = \lim_{n\to\infty} \frac{p_nP_n^{-1}}{-\log\left(1 - p_nP_n^{-1}\right)}.$$

**77.** [I. Schur.] Set $\sum_{\nu=1}^{n} p_\nu = P_n$, $\sum_{\nu=1}^{n} q_\nu = Q_n$ and assume $\beta > 0$. We have $\lim_{n\to\infty} Q_n = \lim_{n\to\infty} nq_n = \infty$, because otherwise we would have $q_n \sim \frac{q}{n}$, $q > 0$; i.e. [**76**] $Q_n \sim q \log n \to \infty$: contradiction. Also $\lim_{n\to\infty} P_n = \lim_{n\to\infty} np_n = \infty$. If $\alpha > 0$ the same argument as for $Q_n$ can be used. For $\alpha = 0$ $\lim_{n\to\infty} np_n P_n^{-1} = \infty$, all the more $\lim_{n\to\infty} np_n = \infty$, consequently $\sum_{\nu=1}^{\infty} p_\nu$ diverges. Therefore the series $\sum_{n=1}^{\infty} np_n q_n$ is divergent. If $\alpha = 0$ we conclude $nq_n < KQ_n$, $K$ independent of $n$,

$$\sum_{\nu=1}^{n} \nu p_\nu q_\nu < K \sum_{\nu=1}^{n} p_\nu Q_\nu \leqq KQ_n \sum_{\nu=1}^{n} p_\nu = KP_n Q_n,$$

i.e.

$$\lim_{n\to\infty} \frac{\sum_{\nu=1}^{n} \nu p_\nu q_\nu}{n^2 p_n q_n} = \lim_{n\to\infty} \frac{P_n}{np_n} \cdot \lim_{n\to\infty} \frac{Q_n}{nq_n} = 0.$$

In the case $\alpha > 0$ replace the proposition by

$$\lim_{n\to\infty} \frac{P_n Q_n}{\sum_{\nu=1}^{n} \nu p_\nu q_\nu} = \alpha + \beta.$$

Apply **70**:

$$a_n = P_n Q_n - P_{n-1} Q_{n-1}, \qquad b_n = np_n q_n,$$

$$\frac{a_n}{b_n} = \frac{P_n}{np_n} + \frac{Q_n}{nq_n} - \frac{1}{n} \to \alpha + \beta = s.$$

**78.** Example: $a_1 = a_2 = a_3 = \cdots = 1$. Now assume that $a_n \geqq a_{n+1}$, $a_n \to 0$ and

$$a_1 + a_2 + \cdots + a_n - na_n \leqq K \text{ for } n = 1, 2, 3, \ldots$$

For a given $m$ find $n$ such that $a_n \leqq \frac{1}{2} a_m$. From

$$K \geqq a_1 + a_2 + \cdots + a_m - ma_n + (a_{m+1} + \cdots + a_n) - (n - m)\, a_n \geqq m(a_m - a_n) \geqq \tfrac{1}{2} ma_m$$

follows that $a_1 + a_2 + \cdots + a_m \leqq K + ma_m \leqq 3K$ for $m = 1, 2, 3, \ldots$ We are dealing here with a transformation under special conditions; **66** in itself is no help.

**79.** Contains **65** as a particular case; the proof is the same. If $p_{kl} = 0$ for $l > k$ the matrix is called triangular (cf. **65**, **66**) or more specifically *lower triangular*. If $p_{kl} = 0$ for $l < k$ the matrix is termed *upper triangular*.

**80.** Contains **66** as a special case. Proof analogous.

**81.** With $s_n = nc_n + (n + 1)\, c_{n+1} + \cdots$ we can write

$$t_n = \frac{1}{n} s_n + \left(\frac{2}{n+1} - \frac{1}{n}\right) s_{n+1} + \left(\frac{3}{n+2} - \frac{1}{n+1}\right) s_{n+2} + \cdots.$$

Obviously $\lim_{n\to\infty} s_n = 0$, which implies $\lim_{n\to\infty} t_n = 0$ [**80**].

**82.** [G. H. Hardy and J. E. Littlewood: Rend. Circ. Mat. Palermo Vol. 41, pp. 50–51 (1916); cf. also T. Carleman: Ark. Mat. Astr. Fys. Vol. 15, No. 11 (1920).] Put

$$a_0 + a_1 + a_2 + \cdots + a_n = s_n, \qquad \frac{f^{(n)}(\alpha)}{n!} = b_n,$$

$$b_0 + b_1(1-\alpha) + \cdots + b_n(1-\alpha)^n = t_n.$$

It is known from analysis [Hurwitz-Courant, pp. 32–33; Hille, Vol. I, p. 128.] that for $|y| < 1 - \alpha$

$$b_0 + b_1 y + \cdots + b_n y^n + \cdots = a_0 + a_1(\alpha + y) + \cdots + a_n(\alpha + y)^n + \cdots$$

holds identically. Consequently

$$\sum_{k=0}^{\infty} (1-\alpha)^{n-k} y^k \sum_{l=0}^{\infty} b_l y^l = \frac{(1-\alpha)^{n+1}}{1-(\alpha+y)} \sum_{l=0}^{\infty} a_l(\alpha+y)^l$$

$$= (1-\alpha)^{n+1} \sum_{l=0}^{\infty} s_l(\alpha+y)^l.$$

The coefficient of $y^n$ on the left hand side is $t_n$ and on the right hand side

$$(1-\alpha)^{n+1}\left[s_n + \binom{n+1}{1}\alpha s_{n+1} + \binom{n+2}{2}\alpha^2 s_{n+2} + \binom{n+3}{3}\alpha^3 s_{n+3} + \cdots\right] = t_n.$$

$\sum_{\nu=0}^{\infty} p_{n\nu} = 1$ (binomial formula). The present transformation has an "upper triangular matrix" whereas the matrix considered in **65** should be termed "lower triangular".

**83.** Cf. the analogous propositions of **65** and **79**.

**84.** Cf. the analogous propositions of **66** and **80**.

**85.** In **84** put

$$s_n = \frac{a_n}{b_n}, \qquad \varphi_n(t) = \frac{b_n t^n}{b_0 + b_1 t + b_2 t^2 + \cdots + b_n t^n + \cdots}.$$

For given $\nu$ and $\varepsilon$, $\varepsilon > 0$, choose $n$ so that

$$b_0 + b_1 + b_2 + \cdots + b_n > \frac{b_\nu}{\varepsilon}.$$

We have

$$\varphi_\nu(t) < \frac{b_\nu t^\nu}{b_0 + b_1 t + b_2 t^2 + \cdots + b_n t^n}.$$

The right hand side converges to a value smaller than $\varepsilon$ as $t \to 1$. The proposition holds also if the radius of convergence is $\varrho$ instead of 1, $\varrho > 0$.

**86.** [N. H. Abel, l.c. **47**, Vol. 1, p. 223.] According to **85**

$$a_0 + a_1 t + a_2 t^2 + \cdots + a_n t^n + \cdots = \frac{\sum_{n=0}^{\infty} (a_0 + a_1 + \cdots + a_n)\, t^n}{\sum_{n=0}^{\infty} t^n}$$

$$\to \lim_{n\to\infty} \frac{a_0 + a_1 + \cdots + a_n}{1} = s.$$

**87.** [G. Frobenius: J. reine angew. Math. Vol. 89, pp. 262–264 (1880).] It follows from the hypothesis that $n^{-1}a_n$ is bounded, therefore $\sum_{n=0}^{\infty} a_n t^n$ converges for $|t| < 1$. According to **85** we have

$$a_0 + a_1 t + a_2 t^2 + \cdots + a_n t^n + \cdots$$

$$= \frac{\sum_{n=0}^{\infty} (a_0 + a_1 + \cdots + a_n)\, t^n}{\sum_{n=0}^{\infty} t^n} = \frac{\sum_{n=0}^{\infty} (s_0 + s_1 + \cdots + s_n)\, t^n}{\sum_{n=0}^{\infty} (n+1)\, t^n}$$

$$\to \lim_{n\to\infty} \frac{s_0 + s_1 + \cdots + s_n}{n+1} = s.$$

**88.** Multiplying numerator and denominator with the geometrical series we obtain

$$\frac{a_0 + a_1 t + a_2 t^2 + \cdots + a_n t^n + \cdots}{b_0 + b_1 t + b_2 t^2 + \cdots + b_n t^n + \cdots} = \frac{\sum_{n=0}^{\infty} (a_0 + a_1 + \cdots + a_n)\, t^n}{\sum_{n=0}^{\infty} (b_0 + b_1 + \cdots + b_n)\, t^n}$$

$$\to \lim_{n\to\infty} \frac{a_0 + a_1 + \cdots + a_n}{b_0 + b_1 + \cdots + b_n} = s.$$

**89.** Set $\varphi(z) = \log\left(1 + \frac{\alpha}{z}\right) - \alpha \log\left(1 + \frac{1}{z}\right)$. According to **156**

$$\sum_{\nu=1}^{\infty} \left[\log\left(1 + \frac{\alpha}{\nu}\right) - \alpha \log\left(1 + \frac{1}{\nu}\right)\right]$$

converges, i.e. there exists a finite limit

$$\lim_{n\to\infty} \frac{n^{\alpha-1}\, n!}{\alpha(\alpha+1)\cdots(\alpha+n-1)} > 0.$$

In **85** put $a_n = n^{\alpha-1}$ and $b_n = \frac{\alpha(\alpha+1)\cdots(\alpha+n-1)}{n!}$.

**90.** The integral in question expanded in powers of $k^2$ is

$$\frac{\pi}{2} \sum_{n=1}^{\infty} \left(\frac{1\cdot 3 \cdots (2n-1)}{2\cdot 4 \cdots 2n}\right)^2 k^{2n}.$$

Special case of **85**: $a_n = \frac{\pi}{2}\left(\frac{1\cdot 3 \cdots (2n-1)}{2\cdot 4 \cdots 2n}\right)^2$, $b_n = \frac{1}{n}$, $s = \frac{1}{2}$, $k^2 = t$.

**91.** [Cf. O. Perron: Die Lehre von den Kettenbrüchen. Leipzig: B. G. Teubner 1913, p. 353, formula (24); R. G. Archibald: An Introduction to the Theory of Numbers. Columbus/Ohio: Charles E. Merrill Publishing 1970, p. 176.] Recursion formulas for $A_n$ and $B_n$:

$$A_{n+2} = (2n+1)\,A_{n+1} + aA_n, \qquad B_{n+2} = (2n+1)\,B_{n+1} + aB_n,$$

$$n = 0, 1, 2, \ldots; \qquad A_0 = B_1 = 1, \qquad A_1 = B_0 = 0.$$

This leads to the differential equation

$$y'' = 2xy'' + y' + ay,$$

where $y$ stands for $F(x)$ or $G(x)$. Substituting $v^2$ for $a(1-2x)$ we find $\frac{d^2y}{dv^2} - y = 0$, $y = c_1e^v + c_2e^{-v}$, $c_1$ and $c_2$ constant, i.e.

$$F(x) = \frac{e^{v-\sqrt{a}} + e^{-v+\sqrt{a}}}{2}, \qquad G(x) = \frac{-e^{v-\sqrt{a}} + e^{-v+\sqrt{a}}}{2\sqrt{a}}.$$

Put $2x = t$. Then **85** may be applied in the following manner:

$$\lim_{n\to\infty} \frac{A_n}{B_n} = \lim_{n\to\infty} \frac{n\dfrac{A_n}{n!}\,\dfrac{1}{2^{n-1}}}{n\dfrac{B_n}{n!}\,\dfrac{1}{2^{n-1}}} = \lim_{t\to 1-0} \frac{F'\left(\dfrac{t}{2}\right)}{G'\left(\dfrac{t}{2}\right)} = \sqrt{a}\,\frac{e^{\sqrt{a}} - e^{-\sqrt{a}}}{e^{\sqrt{a}} + e^{-\sqrt{a}}}.$$

The power series for $G'\left(\frac{t}{2}\right)$ is divergent for $t = 1$ because all its coefficients are non-negative and $\lim_{t\to 1-0} G'\left(\frac{t}{2}\right) = \infty$.

**92.** Special case of **88**.

$$(1-t)^{-\sigma} = \sum_{n=0}^{\infty} b_n t^n = \sum_{n=0}^{\infty} \binom{\sigma+n-1}{n} t^n.$$

Since $(1-t)^{-\sigma-1} = \sum_{n=0}^{\infty} (b_0 + b_1 + \cdots + b_n)\,t^n$ we have

$$b_0 + b_1 + b_2 + \cdots + b_n = \binom{\sigma+n}{n} = \frac{(\sigma+1)(\sigma+2)\cdots(\sigma+n)}{n!} \sim bn^\sigma,$$

$$b > 0.$$

[Solution **89**.] According to **75** we obtain

$$\lim_{n\to\infty} \frac{a_0 + a_1 + a_2 + \cdots + a_n}{b_0 + b_1 + b_2 + \cdots + b_n} = 0.$$

**93.** According to solution **89** we have

$$\lim_{t\to 1-0}(1-t)^{3/2}\sum_{n=1}^{\infty}\left([\sqrt{n}]-2\left[\sqrt{\frac{n}{2}}\right]\right)t^n=\lim_{n\to\infty}\frac{[\sqrt{n}]-2\left[\sqrt{\frac{n}{2}}\right]}{\frac{3}{2}\cdot\frac{5}{4}\cdot\frac{7}{6}\cdots\frac{2n+1}{2n}}<0.$$

The limit is $-\frac{(\sqrt{2}-1)\sqrt{\pi}}{2}$, as $\frac{3}{2}\cdot\frac{5}{4}\cdot\frac{7}{6}\cdots\frac{2n+1}{2n}\sim 2\sqrt{\frac{n}{\pi}}$. [II **202**].

**94.** The statement **85** is true not only for $t\to 1$ but also for $t\to\infty$. Cf. **84**. The sum of the series

$$b_0+b_1t+b_2t^2+\cdots+b_nt^n+\cdots$$

increases to infinity as $t\to\infty$ because $b_n>0$.

**95.** Application of **94**: $a_n=\frac{s_n}{n!}$, $b_n=\frac{1}{n!}$. [Borel's summation, Knopp, p. 471.]

**96.** We write $s_n=a_0+a_1+\cdots+a_n$, $s_{-1}=0$. Then using partial integration for the subtrahend we get [**95**]

$$\int_0^t e^{-x}g(x)\,dx=\sum_{\nu=0}^{\infty}\frac{s_\nu-s_{\nu-1}}{\nu!}\int_0^t e^{-x}x^\nu\,dx=\sum_{\nu=0}^{\infty}s_\nu\int_0^t\left(\frac{x^\nu}{\nu!}-\frac{x^{\nu+1}}{(\nu+1)!}\right)e^{-x}\,dx$$

$$=\sum_{\nu=0}^{\infty}s_\nu\frac{t^{\nu+1}}{(\nu+1)!}e^{-t}.$$

**97.** In **96** put $a_n=0$ for $n$ odd

$$a_n=(-1)^m\frac{1}{2}\cdot\frac{3}{4}\cdot\frac{5}{6}\cdots\frac{2m-1}{2m}\qquad\text{for}\qquad n=2m.$$

We have $s=\left(\frac{1}{\sqrt{1-z}}\right)_{z=-1}=\frac{1}{\sqrt{2}}$. Similarly we obtain for $-1\leqq x\leqq 1$

$$\int_0^{\infty}e^{-t}J_0(xt)\,dt=\frac{1}{\sqrt{1+x^2}}.$$

**98.** [For a special case see M. Fekete: Math. Z. Vol. 17, p. 233 (1939).] It is sufficient to consider the case where the lower bound $\alpha$ is finite. Assume $\varepsilon>0$ and $\frac{a_m}{m}<\alpha+\varepsilon$. Any number $n$ can be written in the form $n=qm+r$ where $r$ is an integer, $0\leqq r\leqq m-1$. We define $a_0=0$. Then we have

$$a_n=a_{qm+r}\leqq a_m+a_m+\cdots+a_m+a_r=qa_m+a_r,$$

$$\frac{a_n}{n}=\frac{a_{qm+r}}{qm+r}\leqq\frac{qa_m+a_r}{qm+r}=\frac{a_m}{m}\,\frac{qm}{qm+r}+\frac{a_r}{n},$$

$$\alpha\leqq\frac{a_n}{n}<(\alpha+\varepsilon)\frac{qm}{qm+r}+\frac{a_r}{n}.$$

***99.** Since $2a_m - 1 < a_{2m} < 2a_m + 1$ we have

(*) $$\left|\frac{a_{2m}}{2m} - \frac{a_m}{m}\right| < \frac{1}{2m}.$$

The series

$$\frac{a_1}{1} + \left(\frac{a_2}{2} - \frac{a_1}{1}\right) + \left(\frac{a_4}{4} - \frac{a_2}{2}\right) + \left(\frac{a_8}{8} - \frac{a_4}{4}\right) + \cdots = \lim_{n\to\infty} \frac{a_{2^n}}{2^n} = \omega$$

is convergent because

$$|a_1| + 2^{-1} + 2^{-2} + 2^{-3} + \cdots$$

is, on account of (*), a majorant series. Write the integer $n$ in the binary system, i.e.

$$n = 2^m + \varepsilon_1 2^{m-1} + \cdots + \varepsilon_m,$$

where $\varepsilon_1, \varepsilon_2, \ldots, \varepsilon_m$ are 0 or 1; according to the hypothesis

$$a_{2^m} + \varepsilon_1 a_{2^{m-1}} + \cdots + \varepsilon_m a_1 - (\varepsilon_1 + \varepsilon_2 + \cdots + \varepsilon_m)$$
$$\leqq a_n \leqq a_{2^m} + \varepsilon_1 a_{2^{m-1}} + \cdots + \varepsilon_m a_1 + (\varepsilon_1 + \varepsilon_2 + \cdots + \varepsilon_m),$$
$$\left|\frac{a_n}{n} - \frac{2^m}{n}\frac{a_{2^m}}{2^m} - \frac{\varepsilon_1 2^{m-1}}{n}\frac{a_{2^{m-1}}}{2^{m-1}} - \cdots - \frac{\varepsilon_m}{n}\frac{a_1}{1}\right| \leqq \frac{m}{n} \leqq \frac{\log n}{n \log 2}.$$

Applying **66** with

$$s_0 = 0, \quad s_1 = \frac{a_1}{1}, \ldots, \quad s_{m-1} = \frac{a_{2^{m-1}}}{2^{m-1}}, \quad s_m = \frac{a_{2^m}}{2^m}, \ldots,$$

$$p_{n0} = 0,\ p_{n1} = \frac{\varepsilon_m}{n}, \ldots,\ p_{n,m-1} = \frac{\varepsilon_1 2^{m-1}}{n},\ p_{nm} = \frac{2^m}{n},\ p_{n,m+1} = 0, \ldots,$$

we conclude $\lim_{n\to\infty} n^{-1} a_n = \omega$. By virtue of (*) we finally obtain

$$\left|\omega - \frac{a_m}{m}\right| \leqq \left|\frac{a_{2m}}{2m} - \frac{a_m}{m}\right| + \left|\frac{a_{4m}}{4m} - \frac{a_{2m}}{2m}\right| + \cdots < \frac{1}{2m} + \frac{1}{4m} + \cdots = \frac{1}{m}.$$

Another proof: From

$$a_{m+n} + 1 < (a_m + 1) + (a_n + 1), \qquad 1 - a_{m+n} < (1 - a_m) + (1 - a_n)$$

we conclude by **98** that

$$\lim_{n\to\infty} \frac{a_n + 1}{n} = \omega, \quad \lim_{n\to\infty} \frac{1 - a_n}{n} = -\omega,$$

where $-\omega$, being a lower bound, cannot be $+\infty$, and so $\omega$ is not $-\infty$. We conclude further that

$$\frac{a_n + 1}{n} \geqq \omega, \quad \frac{1 - a_n}{n} \geqq -\omega.$$

**100.** [L. Fejér: C. R. Acad. Sci. (Paris) Sér. A—B, Vol. 142, pp. 501—503 (1906).] The proof will show that it is sufficient to discuss the case of bounded partial sums $s_1, s_2, s_3, \ldots, s_n, \ldots$ Put $\liminf_{n\to\infty} s_n = m$ and $\limsup_{n\to\infty} s_n = M$, $l$ is a positive integer, $l > 2$ and $\frac{M-m}{l} = \delta$. Divide the number line into $l$ intervals by the points

$$-\infty, \quad m+\delta, \quad m+2\delta, \quad \ldots, \quad M-2\delta, \quad M-\delta, \quad \infty.$$

Choose $N$ such that $|s_n - s_{n+1}| < \delta$ for $n > N$. Let, furthermore, $s_{n_1}$, $n_1 > N$, be in the first (infinite) interval, $s_{n_2}$, $n_2 > n_1$, in the last (infinite) interval. In each of the $l-2$ intervals of length $\delta$ there will be at least one point $s_{n_1+k}$ $(0 < k < n_2 - n_1)$. A similar argument holds if the sequence does not "slowly increase" but is "slowly decreasing".

**101.** [Cf. G. Szegö, Problem: Arch. Math. Phys. Ser. 3, Vol. 23, p. 361 (1914). Solved by P. Veress: Arch. Math. Phys. Ser. 3, Vol. 25, p. 88 (1917).] The interval (0, 1). Cf. **102.**

**102.** [G. Pólya: Rend. Circ. Mat. Palermo Vol. 34, pp. 108—109 (1912).] There are subsequences with arbitrarily high subscripts $t_{n_1}, t_{n_1+1}, \ldots, t_{n_2}$, that descend arbitrarily slowly from the lim sup to the lim inf of the sequence. The detailed proof follows the lines of solution **100.**

**103.**

$$\frac{\nu_n}{n+\nu_n} - \frac{\nu_{n+1}}{n+1+\nu_{n+1}}$$

$$= \frac{n(\nu_n - \nu_{n+1}) + \nu_n}{(n+\nu_n)(n+1+\nu_{n+1})} \leqq \frac{\nu_n}{(n+\nu_n)(n+1+\nu_{n+1})} < \frac{1}{n}. \quad [\mathbf{102.}]$$

**104.** Let $s_1, s_2, s_3, \ldots, s_n, \ldots, \lim_{n\to\infty} s_n = s$, be the sequence in question. Choose $s_{\nu_1}$ anywhere in the interval $s - \frac{1}{2}, s + \frac{1}{2}$, and more generally $s_{\nu_n}$ in the interval $s - \frac{1}{2^n}, s + \frac{1}{2^n}$; $\nu_1 < \nu_2 < \nu_3 < \cdots$. The terms of the series $s_{\nu_1} + (s_{\nu_2} - s_{\nu_1}) + (s_{\nu_3} - s_{\nu_2}) + \cdots$ are not larger than the terms of $|s_{\nu_1}| + \left(\frac{1}{2} + \frac{1}{4}\right) + \left(\frac{1}{4} + \frac{1}{8}\right) + \cdots$.

**105.** Only finitely many terms of the sequence are below a certain fixed number. Among finitely many numbers there is a smallest one.

**106.** If the Weierstrass least upper and greatest lower bound of the sequence coincide there is nothing to prove. If they are different then at least one of them is different from the limit of the sequence. This bound is equal to the largest or smallest term of the sequence.

**107.** The smallest among the numbers $l_1, l_2, l_3, \ldots, l_m$ ($m$ given) is called $\eta$, $\eta > 0$. According to the hypothesis there are terms of the sequence that are smaller than $\eta$. Let $n$ be the smallest index for which $l_n < \eta$. Then we have

$$n > m; \qquad l_n < l_1, l_n < l_2, \ldots, l_n < l_{n-1}.$$

**108.** Apply **105** to the sequence $l_k^{-1}, l_{k+1}^{-1}, l_{k+2}^{-1}, \ldots$

**109.** [G. Pólya: Math. Ann. Vol. 88, pp. 170-171 (1923).] We call $l_m$ an "outstanding term" of the sequence if $l_m$ is larger than all the following terms. According to the hypothesis and to **108** there are infinitely many outstanding terms:

$$l_{n_1}, l_{n_2}, l_{n_3}, \ldots, \qquad l_{n_1} > l_{n_2} > l_{n_3} > \cdots.$$

If $l_\nu$ is not oustanding it lies between two consecutive outstanding terms (for $\nu > n_1$), i.e. $n_{r-1} < \nu < n_r$. We find successively $l_{n_r-1} \leqq l_{n_r}$, $l_{n_r-2} \leqq l_{n_r}, \ldots, l_\nu \leqq l_{n_r}$, consequently

$$(*) \qquad l_\nu s_\nu < l_{n_r} s_{n_r}.$$

From this we conclude

$$\limsup_{r\to\infty} l_{n_r} s_{n_r} = +\infty.$$

Otherwise $l_{n_r} s_{n_r}$ and consequently, according to (*), the sequence $l_1 s_1, l_2 s_2, \ldots$ would be bounded, contrary to the hypothesis. Apply now **107** to the sequence

$$l_{n_1}^{-1} s_{n_1}^{-1}, \quad l_{n_2}^{-1} s_{n_2}^{-1}, \quad l_{n_3}^{-1} s_{n_3}^{-1}, \ldots$$

**110.** [Concerning **110–112** cf. A. Wiman: Acta Math. Vol. 37, pp. 305–326 (1914); G. Pólya: Acta Math. Vol. 40, pp. 311–319 (1916); G. Valiron: Ann. Sci. Ecole Norm. sup. (Paris) Ser. 3, Vol. 37, pp. 221–225 (1920); W. Saxer: Math. Z. Vol. 17, pp. 206–227 (1923).]

Analytic proof: We have $\lim_{m\to\infty} (L_m - mA) = +\infty$. Let $L_n - nA$ [**105**] be the minimum of the sequence

$$L_0 - 0, \quad L_1 - A, \quad L_2 - 2A, \quad L_3 - 3A, \quad \ldots$$

Obviously

$$L_{n-\mu} - (n-\mu)A \geqq L_n - nA, \qquad L_{n+\nu} - (n+\nu)A \geqq L_n - nA$$

for $\mu = 1, 2, \ldots, n$; $\nu = 1, 2, 3, \ldots$; $n = 0$ is excluded by the assumption on $A$.

Geometric proof: From the given points draw vertical rays upwards, construct the smallest convex figure (infinite polygon) containing these rays and determine the line of support[1] of slope $A$. We call $(n, L_n)$ the corner (or one of the corners) through which this line of support passes. The lines connecting $(n, L_n)$ with the points to the left (in the convex domain) have a slope smaller than $A$, and with the points to the right (in the convex domain) a slope larger than $A$.

**111.** [Cf. G. Pólya, Problem: Arch. Math. Phys. Ser. 3, Vol. 24, p. 282 (1916).] Write

$$l_1 + l_2 + \cdots + l_m = L_m, \qquad m = 1, 2, 3, \ldots, \qquad L_0 = 0 \quad [\mathbf{110}].$$

Since $L_1 - A < 0$ the difference $L_0 - 0$ cannot be the minimum mentioned in the solution of **110**. Then $l_{n+1} \geqq A$, therefore $l_{n+1}$, and consequently $n$, increases to infinity simultaneously with $A$.

**112.** Put $l_1 + l_2 + \cdots + l_m = L_m$, $m = 1, 2, 3, \ldots$, $L_0 = 0$. We find

$$\lim_{m\to\infty} \frac{L_m - mA}{m} = -A \quad [\mathbf{67}].$$

The sequence

$$L_0 - 0, \quad L_1 - A, \quad L_2 - 2A, \quad \ldots, \quad L_m - mA, \ldots$$

tends to $-\infty$. Let the maximum be $L_n - nA$. The inequalities in question are satisfied for this subscript $n$. There are in the sequence $L_0, L_1, \ldots, L_m, \ldots$ infinitely many terms larger than all preceding terms [**107**]. Let $L_s$ be one of them.

$$\frac{l_s}{1}, \quad \frac{l_{s-1} + l_s}{2}, \ldots, \quad \frac{l_1 + l_2 + \cdots + l_s}{s}$$

are all positive. If $A$ is smaller than their minimum the subscript $n$ belonging to $A$ is $\geqq s$. — The points $(n, L_n)$ are to be enclosed in an infinite polygon convex from above.

**113.** Set $\limsup_{m\to\infty} \frac{\log m}{\log r_m} = S$. Then we have a) $S \geqq \lambda$. This is obvious for $S = \infty$. If $S$ is finite we find for $\varepsilon > 0$ and for large enough $m$ $\log m < (S + \varepsilon) \log r_m$. Therefore

$$r_m^{-S-\varepsilon} < m^{-1}, \qquad r^{-S-2\varepsilon} < m^{-\frac{S+2\varepsilon}{S+\varepsilon}}, \qquad \sum_{m=1}^{\infty} r_m^{-S-2\varepsilon} \text{ converges},$$

[1] By line of support of a closed set $\mathfrak{M}$ we mean a line that contains at least one point of $\mathfrak{M}$ but such that one of the open half-planes determined by this line does not contain any point of $\mathfrak{M}$. The intersection of the closed half-planes defined by all possible lines of support containing $\mathfrak{M}$ is a convex domain $\mathfrak{K}$, the smallest that contains $\mathfrak{M}$ (the convex hull of $\mathfrak{M}$). Every line of support of $\mathfrak{M}$ is a line of support of $\mathfrak{K}$ and vice versa. These concepts can be easily adapted to the case where the ideal point (point at infinity) belongs to the set $\mathfrak{M}$, as is the case here. Cf. III, Chap. 3, §1.

i.e. $S + 2\varepsilon \geqq \lambda$, $S \geqq \lambda$. Furthermore b) $S \leqq \lambda$. For $\lambda = \infty$ this is evident. If $\lambda$ is finite $\sum_{m=1}^{\infty} r_m^{-\lambda-\varepsilon}$ converges for $\varepsilon > 0$. Therefore, by the well known particular case of **139**, where $\varepsilon_n = 1$, $mr_m^{-\lambda-\varepsilon} \to 0$, i.e. $\frac{\log m}{\log r_m} < \lambda + \varepsilon$ for large enough $m$, $S \leqq \lambda + \varepsilon$, $S \leqq \lambda$.

**114.** Since the assumption on the $x_m$'s is independent of the numbering we may assume that $|x_m| = r_m$, $0 < r_1 \leqq r_2 \leqq r_3 \leqq \cdots$. We enclose each number $x_\nu$, $\nu = 1, 2, 3, \ldots, m$ in an interval with center $x_\nu$ and length $\delta$. These intervals do not have inner points in common and lie completely in the interval $\left[-r_m - \frac{\delta}{2}, r_m + \frac{\delta}{2}\right]$. Therefore

$$m\delta < 2r_m + \delta, \quad \text{i.e.} \quad \limsup_{m\to\infty} \frac{\log m}{\log r_m} \leqq 1.$$

**115.** According to **113** $\lim_{m\to\infty} mr_m^{-\beta} = 0$. Apply **107** with $l_m = mr_m^{-\beta}$.

**116.** We have $\limsup_{m\to\infty} mr_m^{-\alpha} = +\infty$ because otherwise we could find a constant $K$ independent of $m$ such that $mr_m^{-\alpha} < K$, thus

$$\frac{1}{r_m^{\alpha(1+\varepsilon)}} < \frac{K^{1+\varepsilon}}{m^{1+\varepsilon}} \quad \text{for} \quad \alpha < \alpha(1+\varepsilon) < \lambda,$$

which is contradictory to the hypothesis that $\lambda$ is the convergence exponent of the sequence $r_1, r_2, \ldots, r_m, \ldots$ Furthermore $mr_m^{-\beta} \to 0$. Apply **109** with

$$l_m = \frac{m^{\frac{1}{\beta}}}{r_m}, \qquad s_m = m^{\frac{1}{\alpha} - \frac{1}{\beta}}, \quad l_m s_m = \frac{m^{\frac{1}{\alpha}}}{r_m}.$$

**117.** The maximum term's index is $m = 0$ if $0 \leqq x < r_1$; its index is $m$ if $r_m < x < r_{m+1}$ and $m \geqq 1$. The terms increase at first until the $m$-th term is reached (maximum term) and they decrease afterwards.

**118.** In **111** put $l_m = \log r_m - \log s_m$, $k = n - \mu$ and $k = n + \nu$ respectively. For given $A$ determine $n$ according to **111** and then draw $r$ from the relation $A = \log r - \log s_n$. It is obvious [**117**] that for $y = s_n$ the $n$-th term of the second power series becomes the maximum term.

**119.** Let $p_n x^n$ be an arbitrary term and choose $m$ such that $m > n$, $p_m > 0$. Then we have

$$p_m x^m > p_n x^n \quad \text{if} \quad x > \sqrt[m-n]{\frac{p_n}{p_m}}.$$

**120.** If for a certain $x$ a term $p_n x^n$ is larger than all preceding terms, i.e. if for a certain value of $x$ all the inequalities

$$x^\nu(p_n x^{n-\nu} - p_\nu) \geqq 0, \; \nu = 0, 1, 2, \ldots, n-1,$$

hold, then this remains true also for any larger value of $x$.

**121.** Let $m$ be arbitrary and choose $x$ so that $p_m x^m$ is the maximum term. Then

$$p_m \varrho^m \geqq p_m x^m \geqq p_0, \quad \frac{1}{p_m} \leqq \frac{\varrho^m}{p_0}.$$

On the other hand $p_m(\theta\varrho)^m$ is bounded for $m \to \infty$, $0 < \theta < 1$. Consequently

$$\lim_{m\to\infty} \sqrt[m]{p_m} = \frac{1}{\varrho}.$$

**122.** [l.c. **110.**] Suppose that for a certain positive value $\bar{z}$ the central subscript of the series $\sum_{m=0}^{\infty} \frac{a_m}{b_m} z^m$ is $n$ [**121**] and that $\bar{y}$ is a value for which the same $n$ is the central subscript of $\sum_{m=0}^{\infty} b_m y^m$. Define $\bar{x}$ by $\bar{z} = \frac{\bar{x}}{\bar{y}}$. Then we have

$$\frac{a_k}{b_k}\frac{\bar{x}^k}{\bar{y}^k} \leqq \frac{a_n}{b_n}\frac{\bar{x}^n}{\bar{y}^n}, \quad b_k\bar{y}^k \leqq b_n\bar{y}^n, \; k = 0, 1, 2, \ldots$$

**123.** [l.c. **110.**] Let

$$n_1, n_2, \ldots, n_k, \ldots$$

be the successive values of the central subscript of the series $\sum_{m=0}^{\infty} \frac{a_m}{b_m} z^m$. Assume that the term with the specific subscript $k$ is the maximum term in the interval $(\zeta_{k-1}, \zeta_k)$ and that $y_k$ is the value for which $b_k y^k$ becomes the maximum term of the series $\sum_{m=0}^{\infty} b_m y^m$. The method used for the solution **122** associates these values $y_1, y_2, \ldots$ with values of $x$ that belong to the intervals

$$(0, y_1\zeta_1), \quad (y_2\zeta_1, y_2\zeta_2), \quad (y_3\zeta_2, y_3\zeta_3), \ldots, \quad (y_k\zeta_{k-1}, y_k\zeta_k), \ldots$$

The exceptional values $x^*$ with which no $y$ is associated must lie in the intervals

$$(y_1\zeta_1, y_2\zeta_1), \quad (y_2\zeta_2, y_3\zeta_2), \ldots, \quad (y_{k-1}\zeta_{k-1}, y_k\zeta_{k-1}), \ldots$$

Thus the values $\log x^*$ have to fall into intervals of length

$$\log\frac{y_2}{y_1}+\log\frac{y_3}{y_2}+\cdots+\log\frac{y_k}{y_{k-1}}+\cdots=\lim_{k\to\infty}\log\frac{y_k}{y_1}=\log\frac{\varrho}{y_1},$$

where $\varrho$ denotes the radius of convergence of $\sum_{n=0}^{\infty} b_n y^n$.

**124.** [A. J. Kempner: Amer. Math. Monthly Vol. 21, pp. 48–50.] All the non-negative integers between $0 = 00\cdots000$ and $10^m - 1 = 99\cdots999$ that are written with the nine digits 0, 1, 2, ..., 8 are obtained by lining up these nine digits in all possible ways $m$ at a time. Thus we get a total of $9^m$ numbers. Let $r_n$ be the $n$-th non-negative integer that is written without the figure 9. If $10^{m-1} - 1 < r_n < 10^m - 1$ then $n \leqq 9^m$. Therefore

$$\limsup_{n\to\infty}\frac{\log n}{\log r_n}\leqq\frac{\log 9}{\log 10}<1 \qquad \textbf{[113]}.$$

More directly: The number of terms of the subseries in question with values between $10^{m-1}-1$ and $10^m - 1$ is $9^m - 9^{m-1}$. Consequently the sum of the subseries is smaller than

$$\frac{9-1}{1}+\frac{9^2-9}{10}+\frac{9^3-9^2}{100}+\cdots=8\left(1+\frac{9}{10}+\frac{9^2}{100}+\cdots\right)=80.$$

**125.** Consider the two subseries which contain all positive and all negative terms, respectively.

**126.** [K. Knopp: J. reine angew. Math. Vol. 142, pp. 292–293 (1913).] No. Example: Let $b_1 + b_2 + b_3 + \cdots$ be convergent and $|b_1| + |b_2| + |b_3| + \cdots$ be divergent. Put

$$a_1 = b_1,\ a_2 = a_3 = \frac{b_2}{2!},\ a_4 = a_5 = \cdots = a_9 = \frac{b_3}{3!},\ a_{10} = \cdots = a_{33} = \frac{b_4}{4!},\ \ldots$$

Noticing that $n!$ is divisible by $l$ if $n \geqq l$ and collecting all the terms which belong to the same $b_m$ we transform the subseries $a_k + a_{k+l} + a_{k+2l} + \cdots$ into the series $\frac{1}{l}b_1 + \frac{1}{l}b_2 + \frac{1}{l}b_3 + \cdots$ except for a finite number of terms.

**127.** No [**128**].

**128.** No. – Use $b_n$ of **126**. Suppose that the functions $\varphi(x)$ and $\Phi(x)$ assume only positive integral values and are strictly increasing: $0 < \varphi(1) < \varphi(2) < \cdots$, $0 < \Phi(1) < \Phi(2) < \Phi(3) < \cdots$; $\varphi(n)$, $\Phi(n)$ integers. Define a new series $a_1 + a_2 + a_3 + \cdots$ with the general term

$$a_\nu = \frac{b_m}{\Phi(m)-\Phi(m-1)} \text{ for } \Phi(m-1) < \nu \leqq \Phi(m);\ a_1 = a_2 = \cdots = a_{\Phi(1)} = \frac{b_1}{\Phi(1)}.$$

The inequality $\varphi(t_m) \leqq \Phi(m) < \varphi(t_m + 1)$ determines the integer $t_m$ completely. Collecting the terms that belong to the same $b_m$ we transform the series $a_{\varphi(1)} + a_{\varphi(2)} + \cdots$ into the series $\sum_{m=1}^{\infty} \frac{t_m - t_{m-1}}{\Phi(m) - \Phi(m-1)} b_m$. If we set $\Phi(x) = 2^{x^2}$ we obtain a series $a_1 + a_2 + a_3 + \cdots$ which furnishes a counter example for the problems **126**—**128**: If $\varphi(x)$ is a polynomial of degree $\geqq 2$ or if $\varphi(x) = kl^x$ the sequence $\frac{t_m - t_{m-1}}{\Phi(m) - \Phi(m-1)}$ is, beyond a certain $m$, monotone decreasing [Knopp, p. 314].

If $\varphi(x) = k + lx$ we transform the contracted series by adding an absolutely convergent series into the series $l^{-1}(b_1 + b_2 + \cdots + b_n + \cdots)$ [**126**].

**129.** [A. Haar.] Since the series $s_l = a_l + a_{2l} + a_{3l} + \cdots$ is of the same type as $s_1$ it is sufficient to show $a_1 = 0$. We denote the first $m$ prime numbers by $p_1, p_2, \ldots, p_m$. The series

$$\begin{aligned} s_1 &- (s_{p_1} + s_{p_2} + \cdots + s_{p_m}) \\ &+ (s_{p_1 p_2} + s_{p_1 p_3} + \cdots) \\ &- \cdots \\ &+ (-1)^m s_{p_1 p_2 \cdots p_m} \end{aligned}$$

contains only $a_1$ and the $a_n$'s whose subscripts $n$ are not multiples of one of the prime numbers $p_1, p_2, \ldots, p_m$. Each of these $a_n$'s appears exactly once [VIII **26**]. I.e.

$$a_1 \leqq \sum_{n=p_m+1}^{\infty} |a_n|, \qquad a_1 = 0.$$

The condition "absolutely convergent" is essential as can easily be shown by the example $\sum_{n=1}^{\infty} \frac{\lambda(n)}{n}$ [VIII, Chap. 1, § 5].

**130.** [G. Cantor, cf. E. Hewitt and K. Stromberg: Real and Abstract Analysis. Springer: New York 1965, pp. 70—71.] We obtain the set of points in question by removing from the closed interval [0, 1] the open middle third of the interval, then apply the same process to the remaining two intervals and so on indefinitely. (This set is often called the Cantor discontinuum or the Cantor ternary set.)

**131.** [Cf. S. Kakeya: Proc. Tokyo math.-phys. Soc. Ser. 2, Vol. 7, p. 250 (1914); Tôhoku Sc. Rep. Vol. 3, p. 159 (1915).] Write

$$p_n + p_{n+1} + \cdots + p_{n+\nu} = P_{n,\nu}, \quad \lim_{\nu\to\infty} P_{n,\nu} = P_n, \; n = 1, 2, 3, \ldots, \nu = 0, 1, 2, \ldots$$

Assume that $p_{n_1}$ is the first term for which $p_{n_1} < \sigma$. Either there exists a $\nu_1$ such that $P_{n_1,\nu_1} < \sigma$, $P_{n_1,\nu_1+1} \geqq \sigma$, $\nu_1 \geqq 0$ or $P_{n_1} \leqq \sigma$. In the second case we have $P_{n_1} = \sigma$ because $P_{n_1} \geqq p_{n_1-1} \geqq \sigma$ (for $n_1 = 1$ this means $P_1 = s \geqq \sigma$), i.e. $\sigma$ may be represented as an infinite subseries. In the first case we determine the first term $p_{n_2}$ with $n_2 > n_1 + \nu_1$, $P_{n_1,\nu_1} + p_{n_2} < \sigma$. Either there exists a $\nu_2$ with $P_{n_1,\nu_1} + P_{n_2,\nu_2} < \sigma$, $P_{n_1,\nu_1} + P_{n_2,\nu_2+1} \geqq \sigma$, $\nu_2 \geqq 0$ or $P_{n_1,\nu_1} + P_{n_2} \leqq \sigma$. In the second case we have $P_{n_1,\nu_1} + P_{n_2} = \sigma$, because $P_{n_1,\nu_1} + P_{n_2} \geqq P_{n_1,\nu_1} + p_{n_2-1} \geqq \sigma$ ($n_2 > n_1 + \nu_1 + 1$, because $P_{n_1,\nu_1} + p_{n_1+\nu_1+1} = P_{n_1,\nu_1+1} \geqq \sigma$) i.e. $\sigma$ may again be written as an infinite subseries. If this procedure never terminates (if the first case occurs at every step) then $\sigma = P_{n_1,\nu_1} + P_{n_2,\nu_2} + P_{n_3,\nu_3} + \cdots$.

**132.** From the relations

$$p_n = p_{n+1} + p_{n+2} + p_{n+3} + \cdots$$

$$p_{n+1} = \qquad\quad p_{n+2} + p_{n+3} + \cdots$$

we gather $p_n = 2p_{n+1}$, thus $p_1 = \frac{1}{2}$, $p_2 = \frac{1}{4}, \ldots, p_n = \frac{1}{2^n}, \ldots$ The representation by *infinite* binary fractions is unique.

**132.1.** [G. Pólya, Problem: Amer. Math. Monthly Vol. 51, pp. 533—534 (1944). Solution Amer. Math. Monthly Vol. 53, pp. 279—282 (1946).] Define

$$R_n = \frac{3}{2}\,\frac{5}{4}\,\frac{7}{6}\cdots\frac{(2n+1)}{2n}.$$

Then

$$\left(1+\frac{1}{2}\right)\left(1+\frac{1}{4}\right)\cdots\left(1+\frac{1}{2n}\right) = R_n,$$

$$\left(1-\frac{1}{3}\right)\left(1-\frac{1}{5}\right)\cdots\left(1-\frac{1}{2n+1}\right) = \frac{1}{R_n},$$

$$R_n \sim n^{\frac{1}{2}}\, 2\pi^{-\frac{1}{2}} \quad \text{by} \quad \text{II } \mathbf{202}$$

and so the product of the first $m(p+q)$ factors of $P_{p,q}$ is

$$\frac{R_{mp}}{R_{mq}} \sim \left(\frac{p}{q}\right)^{\frac{1}{2}}.$$

**132.2.** See **132.1**. From $\log(1+x) = x - \frac{x^2}{2} + \frac{x^3}{3} - \cdots$ follows

$$\log P_{p,q} = S_{p,q} + A$$

where the infinite series $A$ is absolutely convergent and so its sum is independent of $p$ and $q$ (of the rearrangement of its terms). Another proof can be derived from the expression of Euler's constant given in the solution of II **18**.

**133.** Insertion of appropriate vanishing terms into the two complementary subseries reduces the proposition to the termwise addition of two convergent series.

***134.** We assume that all the terms of the divergent subseries $a_{r_1} + a_{r_2} + a_{r_3} + \cdots$ are non-negative. Then the complementary subseries $a_{s_1} + a_{s_2} + a_{s_3} + \cdots$ will be such that to each positive $\varepsilon$ there corresponds an integer $N$ so that

$$a_{s_m} + a_{s_{m+1}} + \cdots + a_{s_n} < \varepsilon$$

provided that $n > m > N$. After this remark the proof essentially coincides with the well known usual proof for Riemann's theorem on the rearrangement of the terms of conditionally convergent series. [Knopp, pp. 318; cf. also W. Threlfall: Math. Z. Vol. 24, pp. 212—214 (1926).]

**135.** From $p_1 \geqq p_2 \geqq p_3 \geqq \cdots$, $0 < m_1 < m_2 < m_3 < \cdots$ follows that

$$m_1 \geqq 1,\ m_2 \geqq 2,\ m_3 \geqq 3, \ldots,\ m_n \geqq n,$$

$$p_1 + p_2 + \cdots + p_n \geqq p_{m_1} + p_{m_2} + p_{m_3} + \cdots + p_{m_n}.$$

**136.** Determine the "red" subseries $p_{r_1} + p_{r_2} + p_{r_3} + \cdots$ so that

$$p_{r_n} < \min(2^{-n}, Q_n - Q_{n-1}), \qquad n = 1, 2, 3, \ldots, Q_0 = 0.$$

Then $p_{r_1} + p_{r_2} + \cdots + p_{r_n} \leqq Q_n$, the complete "red" subseries converges and $Q_n - (p_{r_1} + p_{r_2} + \cdots + p_{r_n})$ increases beyond all bounds. The terms of the complementary subseries are successively accommodated where the relations $\sum_{i=1}^{n} p_{r_i} < Q_n$ permit it. This construction only shifts the two complementary subseries relatively to each other.

**137.** [W. Sierpinski: Bull. int. Acad. pol. Sci. Lett. Cracovie 1911, p. 149.] To obtain $s'$ the divergence of the positive subseries is slowed down by the method used in **136**.

**138.** [E. Cesàro: Atti Accad. Naz. Lincei Rend. Cl. Sci. Fis. Mat. Natur Ser. 4, Vol. 4, 2nd Sem. p. 133 (1888); J. Bagnera: Darboux Bull. Ser. 2, Vol. 12, p. 227 (1888). Cf. also G. H. Hardy: Mess. Math. Ser. 2, Vol. 41, p. 17 (1911); H. Rademacher: Math. Z. Vol. 11, pp. 276—288 (1921).] Put $E_n = \varepsilon_1 + \varepsilon_2 + \cdots + \varepsilon_n$, $E_0 = 0$. We have now

$$\varepsilon_1 p_1 + \varepsilon_2 p_2 + \cdots + \varepsilon_n p_n = \sum_{\nu=1}^{n} (E_\nu - E_{\nu-1}) p_\nu = \sum_{\nu=1}^{n-1} E_\nu (p_\nu - p_{\nu+1}) + E_n p_n.$$

Suppose that $E_n > \alpha n$ for $n > N$, $\alpha > 0$. Then we have

$$\varepsilon_1 p_1 + \varepsilon_2 p_2 + \cdots + \varepsilon_n p_n > \sum_{\nu=1}^{N} E_\nu (p_\nu - p_{\nu+1}) + \alpha \sum_{\nu=N+1}^{n-1} \nu (p_\nu - p_{\nu+1}) + \alpha n p_n$$

$$= K + \alpha \sum_{\nu=N+1}^{n} p_\nu$$

where $K$ is independent of $n$; therefore the right hand side tends to $\infty$ in contradiction to the hypothesis.

**139.** [E. Lasker.] Set $E_n = \varepsilon_1 + \varepsilon_2 + \cdots + \varepsilon_n$ as in **138**. The sequence

$$(E) \qquad E_1, E_2, E_3, \ldots, E_n, \ldots$$

has the property that between two terms with different signs there must be a vanishing term. We distinguish two cases: (1) Infinitely many terms of the sequence $(E)$ vanish. (2) All but finitely many terms of the sequence $(E)$ have the same sign. Suppose they are positive. In the first case choose the subscript $M$ so that $E_M = 0$ and that we have for $M \leqq m < n$

$$\left| \sum_{\nu=m+1}^{n} \varepsilon_\nu p_\nu \right| = \left| \sum_{\nu=m+1}^{n} [(E_\nu - E_m) - (E_{\nu-1} - E_m)] \, p_\nu \right|$$

$$(*) \qquad = \left| \sum_{\nu=m+1}^{n-1} (E_\nu - E_m)(p_\nu - p_{\nu+1}) + (E_n - E_m) p_n \right| < \varepsilon.$$

Let $E_m$ denote the last vanishing term ahead of $E_n$ in the sequence $(E)$ i.e. $E_{m+1}, E_{m+2}, \ldots, E_n$ have the same sign. The inequality (*) implies $|(E_n - E_m) p_n| = |E_n p_n| < \varepsilon$. In the second case choose $M$ such that the inequality (*) holds for $M \leqq m < n$ and that $E_M, E_{M+1}, E_{M+2}, \ldots$ are all positive. Let $E_m$ be their minimum. Since in this case $E_\nu - E_m \geqq 0$ for $\nu > m$ the estimate (*) yields $(E_n - E_m) p_n < \varepsilon$, consequently

$$E_n p_n < \varepsilon + E_m p_n.$$

Since $m$ is fixed and $p_n$ converges to 0 we find for $n$ sufficiently large $E_n p_n < \varepsilon$.

**140.** The proposition is a consequence of the well-known representation of the remainder

$$f(x) - \left( f(0) + \frac{f'(0)}{1!} x + \frac{f''(0)}{2!} x^2 + \cdots + \frac{f^{(n)}(0)}{n!} x^n \right) = \frac{f^{(n+1)}(\theta x)}{(n+1)!} x^{n+1},$$

$$0 < \theta < 1,$$

because $f^{(n+1)}(\theta x) = \theta_n f^{(n+1)}(0)$ with $0 < \theta_n < 1$.

**141.** Consequence of **140**.

**142.** Integration of $\cos x \leqq 1$ (equality only for $x = 0, \pm 2\pi, \pm 4\pi, \pm 6\pi, \ldots$) implies for positive $x$

$$\sin x < x;\ 1 - \cos x < \frac{x^2}{2},\ \text{i.e.}\ \cos x > 1 - \frac{x^2}{2},$$

$$x - \sin x < \frac{x^3}{3!},\ \text{i.e.}\ \sin x > x - \frac{x^3}{3!},$$

$$\frac{x^2}{2} + \cos x - 1 < \frac{x^4}{4!},\ \text{i.e.}\ \cos x < 1 - \frac{x^2}{2!} + \frac{x^4}{4!},\ \text{etc.}$$

**143.** $\arctan x - \left(x - \frac{x^3}{3} + \frac{x^5}{5} - \cdots + (-1)^n \frac{x^{2n+1}}{2n+1}\right)$

$$= \int_0^x \frac{(-x^2)^{n+1}}{1+x^2}\,dx,\quad J_0(x) = \frac{2}{\pi}\int_0^1 \frac{\cos xt}{\sqrt{1-t^2}}\,dt.$$

**144.** Assume $a_0 > 0$, hence $a_1 < 0$, $a_2 > 0$, $a_3 < 0, \ldots$ Then $A - a_0 < 0$, $A - a_0 - a_1 > 0$, $A - a_0 - a_1 - a_2 < 0, \ldots$, i.e.

$$a_1 < A - a_0 < 0,$$

$$0 < A - a_0 - a_1 < a_2,$$

$$a_3 < A - a_0 - a_1 - a_2 < 0, \ldots,$$

whence the proposition follows. The proof runs similarly if $a_0 < 0$.

**145.** Suppose e.g. that $a_0 \leqq A$; then $a_1$ cannot be negative because in this case we would get $A - (a_0 + a_1) = |A - (a_0 + a_1)| \geqq |a_1| > |a_2|$ in contradiction to the hypothesis. Hence $a_1 > 0$. Since $0 \leqq A - a_0 < |a_1|$ we have furthermore $A - (a_0 + a_1) = A - a_0 - a_1 < 0$. By similar arguments we find $a_2 < 0$ and $a_0 + a_1 + a_2 < A$, $a_3 > 0$ and $a_0 + a_1 + a_2 + a_3 > A$ etc. In general, the terms of an enveloping series need not have alternating signs [**148**], yet the terms of a strictly enveloping series must have alternating signs.

**146.** If, in **145**, we assume only $|a_1| > |a_2| > \cdots > |a_n|$ we show in the same way that $a_1, a_2, \ldots, a_{n-1}$ have alternating signs and that the first partial series (with subscripts $\leqq n - 1$, $n = 2, 3, 4, \ldots$) are strictly enveloping. For $x$ sufficiently large and $n$ fixed we have in the present case

$$\left|\frac{a_1}{x}\right| > \left|\frac{a_2}{x^2}\right| > \cdots > \left|\frac{a_n}{x^n}\right|.$$

Thus the proposition is proved for the first $n - 1$ terms, consequently in general.

**147.** The hypothesis implies that the derivatives $f^{(n)}(t)$ are alternately positive monotone decreasing and negative monotone increasing. Let e.g. $f(t), f''(t), f^{IV}(t), \ldots$ be positive and monotone decreasing, $f'(t)$, $f'''(t), f^{V}(t), \ldots$ be negative and monotone increasing.

$$R_n = \int_0^\infty f(t)\cos xt\,dt - \left(-\frac{f'(0)}{x^2} + \frac{f'''(0)}{x^4} - \cdots + (-1)^n \frac{f^{(2n-1)}(0)}{x^{2n}}\right)$$

$$= \frac{(-1)^{n+1}}{x^{2n+1}} \int_0^\infty f^{(2n+1)}(t) \sin xt\,dt$$

$$= \frac{(-1)^{n+1}}{x^{2n+1}} \int_0^{\frac{\pi}{x}} \left[f^{(2n+1)}(t) - f^{(2n+1)}\left(t + \frac{\pi}{x}\right) + f^{(2n+1)}\left(t + \frac{2\pi}{x}\right) - \cdots\right] \sin xt\,dt.$$

Obviously $R_n$ has the same sign as $(-1)^{n+1} \dfrac{f^{(2n+1)}(0)}{x^{2n+2}}$. Apply **144**.

**148.** The sum of the first $2n - 2$ terms is

$$\frac{2}{3} + \frac{(-1)^n}{3 \cdot 2^{n-2}}, \qquad \frac{1}{3 \cdot 2^{n-2}} < \frac{3}{2^{n+1}}.$$

The sum of the first $2n - 1$ terms is

$$\frac{2}{3} + \frac{(-1)^{n+1}}{3 \cdot 2^{n+1}}, \qquad \frac{1}{3 \cdot 2^{n+1}} < \frac{1}{2^{n+1}}.$$

The signs of the terms do not alternate.

**149.** The graph consists of line segments forming a kind of spiral. Its shape justifies to some extent the expression "enveloping". We have

$$e^i = \frac{389}{720} + \frac{101}{120} i + \delta, \qquad |\delta| < \frac{1}{7!} < 2 \cdot 10^{-4}.$$

[Graph or **151**.] The result is accurate up to three decimals.

**150.** [H. Weyl.] a) $z$ lies on $\mathfrak{H}$, $|z| > 0$. The absolute value of the remainder

$$f(z) - \left(f(0) + \frac{f'(0)}{1!} z + \frac{f''(0)}{2!} z^2 + \cdots + \frac{f^{(n)}(0)}{n!} z^n\right) = \int_0^z \frac{(z-t)^n}{n!} f^{(n+1)}(t)\,dt$$

is smaller than

$$|f^{(n+1)}(0)| \int_0^{|z|} \frac{r^n}{n!}\,dr = \left|\frac{f^{(n+1)}(0)}{(n+1)!} z^{n+1}\right|.$$

b) Assume, more generally, that $f(z)$ is enveloped by the series $a_0 + a_1 z + a_2 z^2 + \cdots$ for $z$ on $\mathfrak{H}$. Suppose now that $z$ lies on $\overline{\mathfrak{H}}$, $|z| > 0$

and that $t$ is real and positive. Then $tz^{-1}$ lies on $\mathfrak{H}$. Therefore

$$\left|f\left(\frac{t}{z}\right) - a_0\right| < \left|\frac{a_1 t}{z}\right|;$$

thus the integral defining $F(z)$ is convergent. Furthermore

$$\left|F(z) - a_0 - \frac{1!\,a_1}{z} - \frac{2!\,a_2}{z^2} - \cdots - \frac{n!\,a_n}{z^n}\right| =$$

$$= \left|\int_0^\infty e^{-t}\left[f\left(\frac{t}{z}\right) - a_0 - \frac{a_1 t}{z} - \frac{a_2 t^2}{z^2} - \cdots - \frac{a_n t^n}{z^n}\right] dt\right| \leqq$$

$$\leqq \int_0^\infty e^{-t}\left|\frac{a_{n+1} t^{n+1}}{z^{n+1}}\right| dt = \left|\frac{(n+1)!\,a_{n+1}}{z^{n+1}}\right|.$$

**151.** [For $e^{-z}$ cf. E. Landau: Arch. Math. Phys. Ser. 3, Vol. 24, p. 104 (1915).] Cf. **150.** (For $\Re z = 0$, $z \neq 0$ the absolute values of the derivatives of $e^{-z}$ are constant, $e^{-z}$ however is not constant. Therefore the remarks of **150** are still valid.) (**141**.)

**152.** $e^{\frac{z^2}{2}} \int\limits_z^\infty e^{-\frac{t^2}{2}} dt = \frac{1}{z} \int\limits_0^\infty e^{-\frac{t^2}{2z^2}} e^{-t} dt.$

Since $\Re \frac{1}{z^2} \geqq 0$ we can apply **151** (**141** resp.).

**153.** $|a_n + b_n| = |a_n| + |b_n|$. Cf. definition.

**154.** [Cauchy: C. R. Acad. Sci. (Paris) Sér. A—B, Vol. 17, pp. 370—376 (1843); cf. also E. T. Whittaker and G. N. Watson, p. 136, ex. 7.]

$$z \coth z = 1 + \sum_{n=1}^{\infty} \frac{2z^2}{z^2 + n^2\pi^2} \qquad [\mathbf{151}, \mathbf{153}].$$

**155.** [Cf. Cauchy, l.c. **154**.] Its Maclaurin series envelops arc tan $z$ for $\Re z^2 \geqq 0$, $z \neq 0$. This has been proved in **143** for real $z$; we use the same formula for the complex $z$ in question.

**156.** It is sufficient to consider $\varphi(x) = a_0 + \frac{a_1}{x}$. We get

$$\varphi(1) + \varphi(2) + \cdots + \varphi(n) = a_0 n + a_1 \log n + O(1).$$

**157.** A necessary condition for convergence is that $\lim\limits_{n\to\infty} \varphi(n) = a_0 = 1$. Apply **156** to the function

$$\log \varphi(x) = \log\left(1 + \frac{a_1}{x} + \frac{a_2}{x^2} + \cdots\right) = \frac{a_1}{x} + \frac{a_2 - \frac{1}{2}a_1^2}{x^2} + \cdots.$$

**158.** The series in question is certainly convergent if $\varphi(n) = 0$ for a positive integer $n$. Therefore we assume that $\varphi(n) \neq 0$, $n = 1, 2, 3, \ldots$

We have [68]

$$\lim_{n\to\infty} \sqrt[n]{|\varphi(1)\,\varphi(2)\,\varphi(3)\cdots\varphi(n)|} = |a_0|.$$

Hence the series converges for $|a_0| < 1$ and diverges for $|a_0| > 1$. Assume $a_0 = 1$ and (for the sake of simplicity) $\varphi(n) > 0$, $n = 1, 2, 3, \ldots$ Then we get [157]

$$\log\varphi(1)\,\varphi(2)\cdots\varphi(n) = a_1\log n + b + \varepsilon_n,\quad \text{i.e.}\quad \varphi(1)\,\varphi(2)\cdots\varphi(n) = e^{b+\varepsilon_n}n^{a_1},$$

$\lim_{n\to\infty}\varepsilon_n = 0$, $b$ is an $n$-free constant. Consequently we have convergence for $a_1 < -1$, divergence for $a_1 \geqq -1$.—If $a_0 = -1$ we put

$$\varphi(n) = -\psi(n),\quad \varphi(1)\,\varphi(2)\cdots\varphi(n) = (-1)^n\,\psi(1)\,\psi(2)\cdots\psi(n).$$

The remainder of the series $\Sigma n^{-2}$ is $O(n^{-1})$. Hence, cf. also solution II 18,

$$\psi(1)\,\psi(2)\cdots\psi(n) = e^c n^{-a_1} + O(n^{-a_1-1}),$$

$c$ is independent of $n$. Therefore this series converges if and only if $\sum_{n=1}^{\infty}(-1)^n n^{-a_1}$ converges, i.e. for $a_1 > 0$. Summing up: the given series converges if and only if at least one of the following conditions is satisfied: a) $\varphi(n) = 0$ for some positive integer $n$; b) $|a_0| < 1$; c) $a_0 = 1$, $a_1 < -1$; d) $a_0 = -1$, $a_1 > 0$.

**159.** Special case of **158**:

$$\varphi(x) = 2 - e^{\frac{\alpha}{x}} = 1 - \frac{\alpha}{x} - \frac{\alpha^2}{2!}\frac{1}{x^2} + \cdots$$

convergent for $\alpha > 1$, and $\alpha = \log 2$; $\log 4, \log 8, \ldots$ are $> 1$.

**160.** $\int_0^1 e^{x\log\frac{1}{x}}\,dx = \sum_{n=0}^{\infty}\frac{1}{n!}\int_0^1 x^n\left(\log\frac{1}{x}\right)^n dx.$

Substitute $x^{n+1} = e^{-y}$.

**161.** Set $\sqrt{1+\sqrt{1+\cdots+\sqrt{1}}} = t_n$, then $t_n^2 = 1 + t_{n-1}$, $t_1 = 1$, $t_{n-1} < t_n$, $n = 2, 3, 4, \ldots$ For positive $x$ we have $x^2 < 1 + x$ if and only if $x < \frac{1}{2}(1+\sqrt{5})$, i.e. if $x$ is smaller than the positive root of the equation $x^2 - x - 1 = 0$. Hence $t_n^2 = 1 + t_{n-1} < 1 + t_n$, $t_{n-1} < t_n < \frac{1}{2}(1+\sqrt{5})$, $n = 2, 3, 4, \ldots$ and $\lim_{n\to\infty} t_n = t$ exists, $0 < t \leqq \frac{1}{2}(1+\sqrt{5})$, $t^2 = 1 + t$, i.e. $t = \frac{1}{2}(1+\sqrt{5})$. We proceed similarly in the case of the continued fraction where the recursion formula is

$$u_n = 1 + u_{n-1}^{-1},\quad u_1 = 1,\qquad n = 2, 3, 4, \ldots$$

**162.** [G. Pólya, Problem: Arch. Math. Phys. Ser. 3, Vol. 24, p. 84 (1916). Solved by G. Szegö: Arch. Math. Phys. Ser. 3, Vol. 25, pp. 88–89 (1917).] If (for the sake of simplicity for $\nu \geqq 1$) $\log\log a_\nu < \nu \log 2$, $a_\nu < e^{2^\nu}$, then $t_n < \sqrt{e^2 + \sqrt{e^4 + \cdots + \sqrt{e^{2^n}}}} < e^{\frac{1+\sqrt{5}}{2}}$ [**161**]. If, however, $a_n > e^{\beta^n}$, $\beta > 2$, then $t_n > e^{\left(\frac{\beta}{2}\right)^n}$. – If $a_n \leqq 1$, then, of course, $\frac{\log\log a_n}{n}$ must be interpreted as $-\infty$.

**163.** We prove

$$t_{n+1} - t_n < \frac{a_{n+1}}{2^n \sqrt{a_1 a_2 \cdots a_n a_{n+1}}}$$

by complete induction. Suppose that the corresponding relation is proved for the $n$ quantities $a_2, a_3, \ldots, a_{n+1}$, i.e. that

$$\sqrt{a_2 + \sqrt{a_3 + \cdots + \sqrt{a_n + \sqrt{a_{n+1}}}}} < t + s,$$

where

$$\sqrt{a_2 + \sqrt{a_3 + \cdots + \sqrt{a_n}}} = t, \qquad \frac{a_{n+1}}{2^{n-1}\sqrt{a_2 a_3 \cdots a_{n+1}}} = s.$$

Hence

$$t_{n+1}^2 < a_1 + t + s < \left(\sqrt{a_1 + t} + \frac{s}{2\sqrt{a_1 + t}}\right)^2 < \left(t_n + \frac{s}{2\sqrt{a_1}}\right)^2.$$

**164.** [Jacobi, l.c. **53**, § 52, Corollarium: Werke, Vol. 1, pp. 200–201.] Write $1 - q = a_0$, $1 + q^m = a_m$, $m = 1, 2, 4, 8, 16, \ldots$, then the $n+1$-th partial product is

$$\frac{a_0}{a_1}\left(\frac{a_0 a_1}{a_2}\right)^{\frac{1}{2}}\left(\frac{a_0 a_1 a_2}{a_4}\right)^{\frac{1}{4}}\left(\frac{a_0 a_1 a_2 a_4}{a_8}\right)^{\frac{1}{8}} \cdots \left(\frac{a_0 a_1 a_2 \cdots a_{2^{n-1}}}{a_{2^n}}\right)^{\frac{1}{2^n}}$$

$$= \frac{a_0^{2-2^{-n}}}{(a_1 a_2 a_4 a_8 \cdots a_{2^n})^{2^{-n}}}.$$

The product $a_1 a_2 a_4 a_8 \ldots$ converges. Cf. also VIII **78**.

**165.** Calling the sum in question $F(x)$ we find

$$F'(x) = F(x), \qquad F(x) = \text{const.} \cdot e^x.$$

**166.** $\varphi'(x) = \varphi(x)$, $\varphi(0) = 1$, $\varphi(x) = e^x = 1 + \frac{x}{1!} + \frac{x^2}{2!} + \cdots + \frac{x^n}{n!} + \cdots$

$\Delta\psi(x) = \psi(x)$, $\psi(0) = 1$, $\psi(x) = 2^x = 1 + \binom{x}{1} + \binom{x}{2} + \cdots + \binom{x}{n} + \cdots$

for $x > -1$. We have

$$\varphi_n(x) = \frac{x^n}{n!}, \psi(x) = \binom{x}{n} = \frac{x(x-1)(x-2)\cdots(x-n+1)}{n!}, \quad n = 1, 2, 3, \ldots$$

**167.**

$$\log \frac{x_n}{x_{n+1}} = \log \frac{y_n}{y_{n+1}} - \frac{1}{12n(n+1)}, \log \frac{y_n}{y_{n+1}} = \left(n + \frac{1}{2}\right) \log\left(1 + \frac{1}{n}\right) - 1;$$

$$\log \frac{1+x}{1-x} = 2\left(\frac{x}{1} + \frac{x^3}{3} + \frac{x^5}{5} + \cdots\right) \text{ yields for } x = \frac{1}{2n+1},$$

$$1 < \left(n + \frac{1}{2}\right) \log\left(1 + \frac{1}{n}\right) = 1 + \frac{1}{3(2n+1)^2} + \frac{1}{5(2n+1)^4} + \cdots$$

$$< 1 + \frac{1}{3[(2n+1)^2 - 1]} = 1 + \frac{1}{12n(n+1)},$$

therefore $x_n < x_{n+1}$, $y_n > y_{n+1}$. Part of **155** resp. II **205**. On the other hand, **167** together with II **202** implies II **205** for integral $n$.

**168.** [I. Schur.] The fact that $a_n$ is decreasing for $p \geqq \frac{1}{2}$ is obvious from the expansion

$$\log a_n = \frac{2(n+p)}{2n+1}\left(1 + \frac{1}{3(2n+1)^2} + \frac{1}{5(2n+1)^4} + \cdots\right)$$

$$= \left(1 + \frac{p - \frac{1}{2}}{n + \frac{1}{2}}\right)\left(1 + \frac{1}{3(2n+1)^2} + \frac{1}{5(2n+1)^4} + \cdots\right)$$

[solution **167**]. This leads to

$$\log a_n = 1 - \frac{\frac{1}{2} - p}{n + \frac{1}{2}} + \frac{1}{12n^2} + O\left(\frac{1}{n^3}\right),$$

thus

$$\log a_{n+1} - \log a_n = \frac{\frac{1}{2} - p}{(n + \frac{1}{2})(n + \frac{3}{2})} + O\left(\frac{1}{n^3}\right),$$

hence $a_n$ increases for $n$ larger than a certain subscript $N$ if $p < \frac{1}{2}$. If $p \leqq 0$ this is true already for $n \geqq 1$ as can easily be verified by expanding $\left(1 + \frac{1}{n}\right)^n$ with help of the binomial formula.

**169.** We write $a_n = \left(1 + \frac{1}{n}\right)^{n+\frac{1}{2}} \dfrac{1 + \frac{x}{n}}{\left(1 + \frac{1}{n}\right)^{\frac{1}{2}}}$; the first factor decreases [**168**]; the square of the second factor is $1 + \dfrac{2x - 1}{n+1} + \dfrac{x^2}{n(n+1)}$. The condition $x \geqq \frac{1}{2}$ is therefore sufficient. Now expand

$$\log a_n = 2n\left(\frac{1}{2n+1} + \frac{1}{3(2n+1)^3} + \frac{1}{5(2n+1)^5} + \cdots\right)$$

$$+ 2\left[\frac{x}{2n+x} + \frac{1}{3}\left(\frac{x}{2n+x}\right)^3 + \frac{1}{5}\left(\frac{x}{2n+x}\right)^5 + \cdots\right]$$

$$= \frac{2n}{2n+1} + \frac{2x}{2n+x} + \frac{1}{12n^2} + O\left(\frac{1}{n^3}\right).$$

Since $\log a_n - \log a_{n+1} = \frac{4x-2}{4n^2} + O\left(\frac{1}{n^3}\right)$ the condition $x \geqq \frac{1}{2}$ is also necessary.

**170.** [Cf. problem No. 1098, Nouv. Annls Math. Ser. 2, Vol. 11, p. 480 (1872). Solved by C. Moreau; Nouv. Annls Math. Ser. 2, Vol. 13, p. 61 (1874).] The first inequality means

$$\left(1+\frac{1}{n}\right)^{n+1} < e\left(1+\frac{1}{2n}\right)$$

and this is a consequence of the following inequality

$$f(x) = x + x\log\left(1+\frac{x}{2}\right) - (1+x)\log(1+x) > 0, \quad 0 < x \leqq \frac{1}{n}.$$

$$\left[f'(x) = \frac{x}{x+2} - \log\frac{1+x}{1+\frac{x}{2}} > \frac{x}{x+2} - \frac{1+x}{1+\frac{x}{2}} + 1 = 0, \quad f(0) = 0.\right]$$

The second inequality is equivalent to

$$e < \left(1+\frac{1}{n}\right)^n\left(1+\frac{1}{2n}\right) \qquad \textbf{[169]}.$$

**171.** [I. Schur.] The number $e$ lies in the second quarter of the interval because

$$\left(1+\frac{1}{n}\right)^n\left(1+\frac{1}{4n}\right) < e < \left(1+\frac{1}{n}\right)^n\left(1+\frac{1}{2n}\right), \qquad n = 1, 2, 3, \ldots$$

The first inequality follows from **170** because

$$1+\frac{1}{4n} < \left(1+\frac{1}{n}\right)\left(1+\frac{1}{2n}\right)^{-1},$$

the second inequality is contained in **169**.

**172.** [I. Schur.] We infer that $a_n$ is decreasing for $0 < x \leqq 2$ from the equation

$$\log a_n = (n+1)\log\frac{1+\frac{x}{2n+x}}{1-\frac{x}{2n+x}} = (2n+2)\sum_{\nu=1}^{\infty}\frac{1}{2\nu-1}\left(\frac{x}{2n+x}\right)^{2\nu-1}$$

$$= \sum_{\nu=1}^{\infty}\frac{x^{2\nu-1}}{2\nu-1}\,\frac{1}{(2n+x)^{2\nu-2}} + (2-x)\cdot\sum_{\nu=1}^{\infty}\frac{x^{2\nu-1}}{2\nu-1}\,\frac{1}{(2n+x)^{2\nu-1}}.$$

Furthermore

$$\log a_n = x + \frac{x^3}{3}\,\frac{1}{(2n+x)^2} + \frac{x(2-x)}{2n+x} + O\left(\frac{1}{n^3}\right),$$

$$\log a_n - \log a_{n+1} = \frac{2x(2-x)}{(2n+x)(2n+x+2)} + O\left(\frac{1}{n^3}\right),$$

i.e. $\log a_n - \log a_{n+1} < 0$ for $n$ sufficiently large, if $x < 0$ or $x > 2$. For $x = 0$ we have $a_n = 1$, $n = 1, 2, 3, \ldots$

**173.** [Proof based on a communication of E. Jacobsthal.] Cf. **174** for $\lim_{n\to\infty} \sin_n x$. We have

$$x - \frac{x^3}{6} < \sin x < x - \frac{x^3}{6} + \frac{x^5}{120}, \qquad x > 0 \qquad \textbf{[142]}.$$

From the binomial expansion we derive that

$$\frac{c}{\sqrt{n}} - \frac{1}{6}\left(\frac{c}{\sqrt{n}}\right)^3 > \frac{c}{\sqrt{n+1}} \text{ or } \frac{c}{\sqrt{n}} - \frac{1}{6}\left(\frac{c}{\sqrt{n}}\right)^3 + \frac{1}{120}\left(\frac{c}{\sqrt{n}}\right)^5 < \frac{c}{\sqrt{n+1}}$$

for constant $c$ and sufficiently large $n$, $n > N(c)$, according as $c < \sqrt{3}$ or $c > \sqrt{3}$. Let $c < \sqrt{3}$ and $\alpha > 0$, $\alpha$ fixed and such that $\sin_N x > \frac{c}{\sqrt{N+\alpha}}$. Then

$$\sin_{N+1} x > \sin \frac{c}{\sqrt{N+\alpha}} > \frac{c}{\sqrt{N+\alpha}} - \frac{1}{6}\left(\frac{c}{\sqrt{N+\alpha}}\right)^3 > \frac{c}{\sqrt{N+\alpha+1}},$$

thus $\sin_n x > \frac{c}{\sqrt{n+\alpha}}$, $n \geqq N$. Consequently for all $c < \sqrt{3}$, $\liminf_{n\to\infty} \sqrt{n}\sin_n x \geqq c$ i.e. $\geqq \sqrt{3}$. If $c > \sqrt{3}$ choose $m$ so large that $\sin_m x < \frac{c}{\sqrt{N+1}}$. In a similar way as in the first case we conclude $\sin_{m+1} x < \frac{c}{\sqrt{N+2}}$, $\sin_{m+2} < \frac{c}{\sqrt{N+3}}$, etc.

**174.** The sequence $v_n$ is decreasing, $v_n > 0$, therefore $\lim_{n\to\infty} v_n = v$ exists; $v = f(v)$ implies $v = 0$. Consequently it is sufficient to prove the proposition for small $x$. Let $b'$ be fixed, $b' > b$. For sufficiently small $x$ we have

$$x - ax^k < f(x) < x - ax^k + b'x^l$$

and when $n$ is sufficiently large, $n > N(c)$,

$$cn^{-\frac{1}{k-1}} - a\left(cn^{-\frac{1}{k-1}}\right)^k > c(n+1)^{-\frac{1}{k-1}}$$

or

$$cn^{-\frac{1}{k-1}} - a\left(cn^{-\frac{1}{k-1}}\right)^k + b'\left(cn^{-\frac{1}{k-1}}\right)^l < c(n+1)^{-\frac{1}{k-1}}$$

depending on whether

$$c < [(k-1)a]^{\frac{-1}{k-1}} \qquad \text{or} \qquad c > [(k-1)a]^{\frac{-1}{k-1}}.$$

Cf. **173**. The assumption on the sign of $b$ is not essential.

**175.** [J. Ouspensky, Problem: Arch. Math. Phys. Ser. 3, Vol. 20, p. 83 (1913).] Convergence for $s > 2$, divergence for $s \leqq 2$ [**173**].

**176.** [Cf. E. Cesàro, Problem: Nouv. Annls Math. Ser. 3, Vol. 7, p. 400 (1888). Solved by Audibert: Nouv. Annls Math. Ser. 3, Vol. 11, p. 35*

(1892).] The inequalities

$$x > \log \frac{e^x - 1}{x} > 0, \qquad x > 0; \qquad x < \log \frac{e^x - 1}{x} < 0, \qquad x < 0$$

imply that the sequence $u_n$ is steadily decreasing in the first case, $u_n > 0$, and increasing in the second case, $u_n < 0$. We have $\lim_{n \to \infty} u_n = u = 0$ because

$$u \gtrless \log \frac{e^u - 1}{u} \quad \text{for} \quad u \gtrless 0.$$

The recursion formula $e^{u_n} - 1 = u_n e^{u_{n+1}}$, $n = 1, 2, 3, \ldots$ yields

$$e^{u_1} = 1 + u_1 + u_1 u_2 + \cdots + u_1 u_2 \cdots u_{n-1} + u_1 u_2 \cdots u_n e^{u_{n+1}}$$

and $\lim_{n \to \infty} u_1 u_2 \cdots u_n e^{u_{n+1}} = 0$.

**177.** [C. A. Laisant, Problem: Nouv. Annls Math. Ser. 2, Vol. 9, p. 144 (1870). Solved by H. Rumpen: Nouv. Annls Math. Ser. 2, Vol. 11, p. 232 (1872).] $s = \frac{3}{4} \cos \varphi$. Notice that $4 \cos^3 \varphi = 3 \cos \varphi + \cos 3\varphi$.

**178.** [I. Schur, Problem: Arch. Math. Phys. Ser. 3, Vol. 27, p. 162 (1918). Cf. O. Szász: Sber. Berlin Math. Ges. Vol. 21, pp. 25–29 (1922).] If $\varepsilon > 0$ is so small that $|q| + \varepsilon < r$ then there exists a constant $A$ independent of $n$ and $\nu$ such that

$$\left| \frac{b_{n-\nu}}{b_n} \right| < A\,(|q| + \varepsilon)^\nu, \qquad \nu = 0, 1, \ldots, n; \qquad n = 0, 1, 2, \ldots$$

For $n > m$ we obtain

$$\frac{c_n}{b_n} - f(q) = \sum_{\nu=0}^{m} a_\nu \left( \frac{b_{n-\nu}}{b_n} - q^\nu \right) + \sum_{\nu=m+1}^{n} a_\nu \frac{b_{n-\nu}}{b_n} - \sum_{\nu=m+1}^{n} a_\nu q^\nu.$$

The sum of the last two terms is absolutely smaller than

$$A \sum_{\nu=m+1}^{\infty} |a_\nu| \, (|q| + \varepsilon)^\nu + \sum_{\nu=m+1}^{\infty} |a_\nu| \, |q|^\nu,$$

i.e. arbitrarily small with $m^{-1}$. Choose $m$ so large that these two terms are smaller than $\varepsilon$. For fixed $m$ choose $n$ so that the first term becomes absolutely smaller than $\varepsilon$.

**179.** [Special case of an important proposition in function theory by Vitali. Cf. E. Lindelöf: Bull. Soc. Math. France Vol. 41, p. 171 (1913).] We show only that $\lim_{n \to \infty} a_{n1} = 0$. (Then form $x^{-1} f_n(x) - a_{n1}$, etc.) Assume $\varepsilon > 0$ and $x$ so small that $0 < A \frac{x}{1-x} < \varepsilon$. Then we have

$$|a_{n1}| < x^{-1} |f_n(x)| + A \frac{x}{1-x} < x^{-1} |f_n(x)| + \varepsilon.$$

For fixed $x$ choose $n$ so large that $|f_n(x)| < \varepsilon x$.

**180.** We have $|a_k| \leqq A_k$, $k = 0, 1, 2, \ldots$, therefore $\sum_{k=0}^{\infty} a_k$ is convergent. Moreover

$$|s_n - s| \leqq |a_{n0} - a_0| + |a_{n1} - a_1| + \cdots + |a_{nm} - a_m| + 2 \sum_{k=m+1}^{\infty} A_k.$$

Assume $\varepsilon > 0$. Choose $m$ large enough to render the last term smaller than $\varepsilon$. Having fixed $m$ we select $n$ so large that $|a_{nk} - a_k| < \frac{\varepsilon}{m+1}$, $k = 0, 1, \ldots, m$. Then we get

$$|s_n - s| < 2\varepsilon.$$

**181.** a) Since the infinite product $\prod_{k=1}^{\infty} (1 - q^{2k})$ converges for $|q| < 1$ all its partial products lie between two positive numbers $a$ and $b$, $a < b$. Therefore $C_\nu$ as defined in **52** is bounded:

$$|C_\nu| < ba^{-2} q^{\nu^2}.$$

Apply **180**.

b) Let $y < 0$ in **59**, i.e. $q > 1$. Then we have

$$\frac{q^k}{1 + q + q^2 + \cdots + q^{k-1}} < q,$$

furthermore $\left(1 - \frac{y}{n}\right)^{-n+\nu} > e^y$, $\nu = 0, 1, 2, \ldots$, thus

$$\left[1 - \left(1 - \frac{y}{n}\right)^{-n}\right]\left[1 - \left(1 - \frac{y}{n}\right)^{-n+1}\right] \cdots \left[1 - \left(1 - \frac{y}{n}\right)^{-n+k-1}\right] < (1 - e^y)^k.$$

Apply **180**.

**181.1.** We derive from the definition of $U_i$ that

$$|s_i| \leqq U_i$$

and then from our assumption concerning $\Sigma\, U_i$ that

$$s_1 + s_2 + s_3 + \cdots + s_i + \cdots = S$$

is absolutely convergent. Define, for $m = 0, 1, 2, \ldots$,

$$r_{i,m} = a_{i,m+1} + a_{i,m+2} + \cdots,$$

(thus $r_{i,0} = s_i$) and define $S_M^*$ as the sum of the first $M$ terms of the series (*). These definitions are illustrated by the relation

$$S - S^*_{m(m+1)/2} = \sum_{i=1}^{m} r_{i,m+1-i} + \sum_{i=m+1}^{\infty} r_{i,0}.$$

Generally, for arbitrary $M$,

$$S - S_M^* = \sum_{i=1}^{\infty} r_{i,m_i}$$

where $m_i$ depends on $M$ and as $M \to \infty$

$$m_i \to \infty, \quad r_{i,m_i} \to 0$$

for any fixed $i$. Observe that

$$|r_{i,m}| \leqq 2U_i.$$

There follows, by virtue of **180**, that

$$S - S_M^* \to 0.$$

(The foregoing argument, without essential modification, shows that various "geometrically defined" rearrangements of the terms of the series (*) leave its convergence and its sum unchanged- square, rectangle, quarter of a circle instead of a triangle.)

**182.** We are dealing with the limit of the series

$$\sum_{k=1-n}^{\infty}{}' n^{\alpha-1}(n+k)^{\alpha-1}\,[(n+k)^\alpha - n^\alpha]^{-2} = \sum_{k=1-n}^{\infty}{}' \frac{1}{k^2}\varphi\left(\frac{k}{n}\right)$$

for $n \to \infty$; the term with subscript $k = 0$ must be omitted. The function $\varphi(x)$ is defined by

$$\varphi(x) = (1+x)^{\alpha-1}\left(\frac{x}{(1+x)^\alpha - 1}\right)^2;$$

$\varphi(x)$ is continuous for $-1 < x < \infty$ if we set $\varphi(0) = \alpha^{-2}$; $\varphi(x) \equiv 1$ if $\alpha = 1$; otherwise $\varphi(x) \sim x^{1-\alpha}$ for $x \to \infty$, $\varphi(x) \sim (1+x)^{\alpha-1}$ for $x \to -1$. If $\alpha = 1$ the value of the limit follows immediately from

$$1 + 2^{-2} + 3^{-2} + 4^{-2} + \cdots = \frac{\pi^2}{6}.$$

If $\alpha \neq 1$ the general term tends, for $k$ fixed and $n \to \infty$, to $k^{-2}\varphi(0)$.

If $\alpha > 1$ $\varphi(x)$ is bounded for $-1 < x < \infty$. If the maximum of $\varphi(x)$ is denoted by $M$ the series has as a majorant the series $\Sigma' M k^{-2}$ for $n = 1, 2, 3, \ldots$ [**180**.]

If $0 < \alpha < 1$ there exists a positive number $M$ such that

$$\varphi(x) < M \quad \text{for} \quad -\frac{1}{2} \leqq x \leqq 2,$$

$$\varphi(x) \leqq M(1+x)^{\alpha-1} \quad \text{for} \quad -1 < x \leqq -\frac{1}{2},$$

$$\varphi(x) \leqq M x^{1-\alpha} \quad \text{for} \quad x \geqq 2.$$

Consequently

$$\sum_{k=1-n}^{-\frac{1}{2}n} \frac{1}{k^2}\varphi\left(\frac{k}{n}\right) \leqq \sum_{k=1-n}^{-\frac{1}{2}n} \frac{M}{k^2}\left(\frac{n}{n+k}\right)^{1-\alpha} \leqq Mn^{1-\alpha}\sum_{k=\frac{1}{2}n}^{n-1} k^{-2} \to 0,$$

$$\varphi\left(\frac{k}{n}\right) \leqq Mk^{1-\alpha} \quad \text{for} \quad n = 1, 2, 3, \ldots, \quad k = 2n, 2n+1, \ldots$$

Hence the remaining part of the series (from $-\frac{1}{2}n$ to $\infty$) has $\Sigma' M\,|k|^{-1-\alpha}$ as a majorant. [**180**.]

**183.** We have

$$\sqrt{2+\sqrt{2+\sqrt{2+\cdots+\sqrt{2}}}} < \sqrt{2+\sqrt{2+\sqrt{2+\cdots}}} = 2,$$

therefore

$$a_n = \varepsilon_0\sqrt{2+\varepsilon_1\sqrt{2+\varepsilon_2\sqrt{2+\cdots+\varepsilon_n\sqrt{2}}}}$$

always makes sense. To prove

$$a_n = 2\sin\left(\frac{\pi}{4}\sum_{\nu=0}^{n}\frac{\varepsilon_0\varepsilon_1\cdots\varepsilon_\nu}{2^\nu}\right)$$

use mathematical induction. We have

$$\operatorname{sgn} a_n = \operatorname{sgn} 2\sin\left(\frac{\pi}{4}\sum_{\nu=0}^{n}\frac{\varepsilon_0\varepsilon_1\cdots\varepsilon_\nu}{2^\nu}\right) = \varepsilon_0,$$

and for $\varepsilon_0 \neq 0$

$$a_n^2 - 2 = \varepsilon_1\sqrt{2+\varepsilon_2\sqrt{2+\cdots+\varepsilon_n\sqrt{2}}},$$

$$4\sin^2\left(\frac{\pi}{4}\sum_{\nu=0}^{n}\frac{\varepsilon_0\varepsilon_1\cdots\varepsilon_\nu}{2^\nu}\right) - 2 = -2\cos\left(\frac{\pi}{2}\sum_{\nu=0}^{n}\frac{\varepsilon_0\varepsilon_1\cdots\varepsilon_\nu}{2^\nu}\right)$$

$$= -2\cos\left(\frac{\pi}{2}+\frac{\pi}{2}\sum_{\nu=1}^{n}\frac{\varepsilon_1\varepsilon_2\cdots\varepsilon_\nu}{2^\nu}\right)$$

$$= 2\sin\left(\frac{\pi}{4}\sum_{\nu=1}^{n}\frac{\varepsilon_1\varepsilon_2\cdots\varepsilon_\nu}{2^{\nu-1}}\right).$$

Take the limit $n \to \infty$.

**184.** [Cf. S. Pincherle: Atti Accad. Sci. Torino, Cl. Sci. Fis. Mat. Natur. Vol. 53, pp. 745–763 (1917–1918); Atti Accad. Naz. Lincei Rend. Cl. Sci. Fis. Mat. Natur. Ser. 5, Vol. 27, 2nd Sem. pp. 177–183 (1918).] Put $x = 2\cos\varphi$, $0 \leqq \varphi \leqq \pi$. Then the binary expansion is unique

$$\frac{2\varphi}{\pi} = g_0 + \frac{g_1}{2} + \frac{g_2}{2^2} + \cdots + \frac{g_n}{2^n} + \cdots, \qquad g_n = 0 \text{ or } 1,$$

except in the case of $\varphi = \frac{p}{2^q}\pi$, $p$, $q$, integers, $0 < p < 2^q$. In this case there are two representations possible. The equation

$$2\cos\varphi = \varepsilon_0\sqrt{2 + \varepsilon_1\sqrt{2 + \varepsilon_2\sqrt{2 + \cdots}}}$$

is [**183**] equivalent to

$$2\sin\frac{\pi}{4}\left(2 - \frac{4\varphi}{\pi}\right) = 2\sin\frac{\pi}{4}\left(\sum_{n=0}^{\infty}\frac{\varepsilon_0\varepsilon_1\cdots\varepsilon_n}{2^n}\right),$$

or, because both arguments are in the interval $\left(-\frac{\pi}{2}, \frac{\pi}{2}\right)$, equivalent to

$$2 - \frac{4\varphi}{\pi} = \sum_{n=0}^{\infty}\frac{\varepsilon_0\varepsilon_1\cdots\varepsilon_n}{2^n}, \qquad \frac{2\varphi}{\pi} = \sum_{n=0}^{\infty}\frac{\frac{1 - \varepsilon_0\varepsilon_1\cdots\varepsilon_n}{2}}{2^n};$$

therefore

$$g_n = \frac{1 - \varepsilon_0\varepsilon_1\cdots\varepsilon_n}{2}, \qquad n = 0, 1, 2, \ldots,$$

provided that the above mentioned exception is excluded. These equations determine the $\varepsilon_n$ from the $g_n$ uniquely (and vice versa). In the exceptional case $\varphi = \frac{p}{2^q}\pi$, $p$, $q$ integers, $0 < p < 2^q$, $q \geqq 2$, there are two representations of $\frac{2\varphi}{\pi}$ possible:

$$\frac{2\varphi}{\pi} = g_0 + \frac{g_1}{2} + \cdots + \frac{g_{q-2}}{2^{q-2}} + \frac{1}{2^{q-1}} + \frac{0}{2^q} + \frac{0}{2^{q+1}} + \cdots$$

$$= g_0 + \frac{g_1}{2} + \cdots + \frac{g_{q-2}}{2^{q-2}} + \frac{0}{2^{q-1}} + \frac{1}{2^q} + \frac{1}{2^{q+1}} + \cdots.$$

In this case $\varepsilon_0, \varepsilon_1, \ldots, \varepsilon_{q-2}$ are, as before, uniquely determined, $\varepsilon_q = -1$, $\varepsilon_{q+1} = \varepsilon_{q+2} = \cdots = 1$, $\varepsilon_{q-1}$ may be chosen $-1$ or $+1$. We have therefore

$$x = 2\cos\varphi = \varepsilon_0\sqrt{2 + \varepsilon_1\sqrt{2 + \varepsilon_2\sqrt{2 + \cdots + \varepsilon_{q-2}\sqrt{2}}}}.$$

According to **183** any number written in the above way has to be of the type $2\cos\frac{p}{2^q}\pi$, $p$, $q$ integers, $0 < p < 2^q$. If $q = 1$ we have to modify slightly: the numbers $g_0, \ldots, g_{q-2}, \varepsilon_0, \ldots, \varepsilon_{q-2}$ do not exist, $\varphi = \frac{\pi}{2}$, $x = 0$.

**185.** The sequence $g_n$ is periodic beyond a certain term if and only if this is true of the sequence $\varepsilon_n$.

**185.1.** [For precursory heuristic considerations see MD, Vol. 2, pp. 49—50 and 171.] We seek a solution of the system of 2 equations

$$u + 2v + 3w = 0$$

$$u + 8v + 27w = 0$$

with integral values for the three unknowns. We find

$$u = 5, \quad v = -4, \quad w = 1.$$

We note that

$$u + 32v + 243w = 120.$$

We define, for $m = 1, 2, 3, \ldots$,

$$a_{10m-9} = a_{10m-8} = a_{10m-7} = a_{10m-6} = a_{10m-5} = m^{-1/5},$$

$$a_{10m-4} = a_{10m-3} = a_{10m-2} = a_{10m-1} = -2m^{-1/5},$$

$$a_{10m} = 3m^{-1/5}$$

and

$$a_1^l + a_2^l + \cdots + a_n^l = s_n^{(l)}.$$

Then

$$s_{10m}^{(1)} = s_{10m}^{(3)} = 0,$$

$$s_{10m}^{(5)} = 120\left(1 + \frac{1}{2} + \frac{1}{3} + \cdots + \frac{1}{m}\right)$$

and so the series considered converges to 0 conditionally for $l = 1$ and $l = 3$, and diverges to $+\infty$ for $l = 5$. It is absolutely convergent for $l > 5$.

**185.2.** [G. Pólya, Problem: Amer. Math. Monthly Vol. 51, p. 593 (1944). Solved by N. J. Fine: Amer. Math. Monthly Vol. 53, pp. 283—284 (1946).] (1) The case where $D$ consists of just one odd number can be settled by an easily visible extension of solution **185.1**. (Why is $u + 32v + 243w \neq 0$? Give a reason avoiding numerical computation.)

(2) If $D$ consists of a finite number, say $h$, of different odd numbers, construct by (1) the corresponding sequences

$$a_{11}, a_{12}, \ldots, a_{1n}, \ldots,$$

$$a_{21}, a_{22}, \ldots, a_{2n}, \ldots,$$

$$\cdots\cdots\cdots\cdots\cdots\cdots$$

$$a_{h1}, a_{h2}, \ldots, a_{hn}, \ldots,$$

each of which yields a divergent series just for one required exponent and form the sequence

$$a_{11}, a_{21}, \ldots, a_{k1}, a_{12}, \ldots, a_{k2}, a_{13}, \ldots$$

which has the desired property.

(3) If $D$ has an infinity of members use the construction (1) infinitely often and form the desired sequence "diagonally" as suggested by **181.1**. You can fulfill the condition on $\Sigma U_n$ by considering, if necessary, instead of the sequence

$$a_{i1}, a_{i2}, \ldots, a_{in}, \ldots$$

obtained at first by the construction (1) the sequence

$$p_i a_{i1}, p_i a_{i2}, \ldots, p_i a_{in}, \ldots$$

where the positive numbers $p_1, p_2, p_3, \ldots$, decrease sufficiently rapidly.

***186.** List concretely the different possibilities for small $n$, use **187** for all $n \geqq 3$.

| $n\backslash k$ | 1 | 2 | 3 | 4 | 5 | 6 | 7 | 8 |
|---|---|---|---|---|---|---|---|---|
| 1 | 1 | | | | | | | |
| 2 | 1 | 1 | | | | | | |
| 3 | 1 | 3 | 1 | | | | | |
| 4 | 1 | 7 | 6 | 1 | | | | |
| 5 | 1 | 15 | 25 | 10 | 1 | | | |
| 6 | 1 | 31 | 90 | 65 | 15 | 1 | | |
| 7 | 1 | 63 | 301 | 350 | 140 | 21 | 1 | |
| 8 | 1 | 127 | 966 | 1701 | 1050 | 266 | 28 | 1 |

***187.** You belong to a set of $n + 1$ persons. In a partition of this set into $k$ subsets you may stand alone and form a subset by yourself; there are $S_{k-1}^n$ partitions of this kind. Or you may join one of the $k$ subsets already formed by the others; there are $kS_k^n$ partitions of this latter kind.

***188.** We set, without contradicting our original definition,

$$S_k^n = 0 \quad \text{if} \quad 0 \leqq n < k$$

and let $\Sigma$ stand for $\sum\limits_{n=0}^{\infty}$. By **187**

$$(z-k)\sum \frac{S_k^{n+1}}{z^{n+2}} = \sum \frac{S_k^{n+1} - kS_k^n}{z^{n+1}} = \sum \frac{S_{k-1}^n}{z^{n+1}}$$

and this proves our assertion for $k$ if it was assumed for $k - 1$. The case $k = 1$ is easy.

***189.** The right-hand side in **188**, decomposed into partial fractions, equals

$$\frac{1}{k!}\frac{1}{z-k}-\frac{1}{1!\,(k-1)!}\frac{1}{z-k+1}+\cdots+\frac{(-1)^{k-1}}{(k-1)!\,1!}\frac{1}{z-1}+\frac{(-1)^k}{k!\,z}.$$

Expand in powers of $z^{-1}$ and consider the coefficient of $z^{-n-1}$. For a combinatorial proof see VIII **22.1**.

***190.** From **189**, since

$$(e^z-1)^k=e^{kz}-\binom{k}{1}e^{(k-1)z}+\binom{k}{2}e^{(k-2)z}-\cdots+(-1)^k$$

$$=\sum_{n=0}^{\infty}\left[k^n-\binom{k}{1}(k-1)^n+\binom{k}{2}(k-2)^n-\cdots+(-1)^k\,0^n\right]\frac{z^n}{n!}.$$

***191.** Using the notation explained in the introduction to III **220**, we can present **189** in the form

$$S_k^n=\frac{\Delta^k 0^n}{k!}.$$

Apply III **220** (1) to $F(z)=z^n$, see III **221**; observe that $\Delta^k z^n=0$ for $k>n$ (for $z=0$ this follows from our argument).

***192.** We have to paint $n$ distinct objects ($n$ houses of a settlement, $n$ faces of a polyhedron) with $x$ different colors, by using just one color for each object. There are obviously $x^n$ different possibilities. If exactly $k$ among the $x$ colors are used, the objects of the same color form a subset, and the $k$ subsets so formed constitute a partition of the total set of $n$ objects. Hence the number of ways of using just $k$ different colors among the eligible $x$ is

$$S_k^n x(x-1)(x-2)\cdots(x-k+1).$$

As $k$ can be 1, 2, 3, ..., or $n$, this proves the assertion for any positive integer $x$, and hence for indeterminate $x$. In using the same facts as in solution **191**, but in the reverse order, we end up by proving **189**.

***193.** Use the definition of $T_n$ at the beginning and **190** at the end:

$$\sum_{n=0}^{\infty}\frac{T_n z^n}{n!}=1+\sum_{n=1}^{\infty}\sum_{k=1}^{n}\frac{S_k^n z^n}{n!}=1+\sum_{k=1}^{\infty}\sum_{n=k}^{\infty}\frac{S_k^n z^n}{n!}=\sum_{k=0}^{\infty}\frac{(e^z-1)^k}{k!}.$$

***194.** You belong to a set of $n+1$ persons and, in a partition of this set, to a subset of $k+1$ persons. This can happen in $\binom{n}{k}T_{n-k}$ different ways. In fact, the other people in your subset may be chosen in $\binom{n}{k}$ different ways [**10**] and, once they are chosen, the people remaining outside your subset can be partitioned in $T_{n-k}$ ways. Now, $k = 0, 1, 2, \ldots, n$ exhaust all possibilities.

***195.** Set

$$\sum_{n=0}^{\infty} \frac{T_n z^n}{n!} = y.$$

Then, by **34**, the relation **194** is equivalent to

$$\frac{dy}{dz} = e^z y \quad \text{or} \quad \frac{dy}{y} = e^z\, dz.$$

Integrate this differential equation and use the initial condition

$$y = T_0 = 1 \quad \text{for} \quad z = 0.$$

***196.** After differentiating $n$ times

$$e^{e^z - 1} = \frac{1}{e}\left(1 + \frac{e^z}{1!} + \frac{e^{2z}}{2!} + \frac{e^{3z}}{3!} + \cdots\right)$$

set $z = 0$ [**193**]. The result can be used to prove **45**.

***197.** List concretely the different possibilities for small $n$, use **198** for all $n \geqq 3$.

| $n \backslash k$ | 1 | 2 | 3 | 4 | 5 | 6 | 7 | 8 |
|---|---|---|---|---|---|---|---|---|
| 1 | 1 | | | | | | | |
| 2 | 1 | 1 | | | | | | |
| 3 | 2 | 3 | 1 | | | | | |
| 4 | 6 | 11 | 6 | 1 | | | | |
| 5 | 24 | 50 | 35 | 10 | 1 | | | |
| 6 | 120 | 274 | 225 | 85 | 15 | 1 | | |
| 7 | 720 | 1764 | 1624 | 735 | 175 | 21 | 1 | |
| 8 | 5040 | 13068 | 13132 | 6769 | 1960 | 322 | 28 | 1 |

***198.** In a permutation of $n + 1$ objects which is a product of $k$ distinct cycles a given object may form a cycle by itself; there are $s_{k-1}^n$ such permutations. Or the given object may enter at some place a cycle already formed by the $n$ other objects; there are $ns_k^n$ permutations of this latter kind.

***199.** Set $s_0^n = s_k^n = 0$ if $1 \leqq n < k$. By **198**

$$(x + n) \sum_{k=1}^{n} s_k^n x^k = \sum_{k=1}^{n+1} (s_{k-1}^n + ns_k^n)\, x^k = \sum_{k=1}^{n+1} s_k^{n+1} x^k.$$

Use mathematical induction.

***200.**

$$(1-t)^{-x} = \sum_{n=0}^{\infty} \frac{x(x+1)(x+2)\cdots(x+n-1)}{n!} t^n$$

$$= 1 + \sum_{n=1}^{\infty} \sum_{k=1}^{n} \frac{s_k^n x^k t^n}{n!}$$

$$= 1 + \sum_{k=1}^{\infty} x^k \sum_{n=k}^{\infty} \frac{s_k^n t^n}{n!}$$

$$= e^{-x\log(1-t)}$$

$$= \sum_{k=0}^{\infty} \frac{x^k}{k!} \left(\log \frac{1}{1-t}\right)^k.$$

We have first used the binomial expansion, then **199**. Consider the coefficient of $x^k$.

***201.** List concretely the different possibilities for small $n$, use **202** for all $n \geqq 4$.

| $n\backslash k$ | 1 | 2 | 3 | 4 |
|---|---|---|---|---|
| 1 | 0 | | | |
| 2 | 1 | 0 | | |
| 3 | 1 | 0 | | |
| 4 | 1 | 3 | 0 | |
| 5 | 1 | 10 | 0 | |
| 6 | 1 | 25 | 15 | 0 |
| 7 | 1 | 56 | 105 | 0 |
| 8 | 1 | 119 | 490 | 105 |

***202.** You belong to a set of $n+1$ persons. In a partition of this set into $k$ subsets of the desired kind you may be paired with another person to form a subset of 2; there are $n\tilde{S}_{k-1}^{n-1}$ partitions of this kind. Or you may belong to a subset of more than 2 persons; there are $k\tilde{S}_k^n$ partitions of this kind.

***203.** By mathematical induction on the basis of **202**, or by combinatorial considerations:

$$\tilde{S}_2^n = S_2^n - n = \frac{2^n - 2}{2} - n.$$

For $\tilde{S}_n^{2n}$ see MPR, Vol. 1, p. 118, ex. 11.

***204.** In a partition enumerated by $S^n_{n-a}$ there may be

$$n-a-1, \quad n-a-2, \ldots, \quad n-2a+1 \quad \text{or} \quad n-2a$$

subsets containing just one element each. Therefore

$$S^n_{n-a} = \binom{n}{a+1}\tilde{S}^{a+1}_1 + \binom{n}{a+2}\tilde{S}^{a+2}_2 + \cdots + \binom{n}{2a}\tilde{S}^{2a}_a.$$

Use numerical data from **201** for $a = 1, 2, 3$.

***205.**

| $n$ | 0 | 1 | 2 | 3 | 4 | 5 | 6 | 7 | 8 |
|---|---|---|---|---|---|---|---|---|---|
| $T_n$ | 1 | 1 | 2 | 5 | 15 | 52 | 203 | 877 | 4140 |
| $\tilde{T}_n$ | 1 | 0 | 1 | 1 | 4 | 11 | 41 | 162 | 715 |

***206.** Follow the line of solution **194**; observe that $\tilde{T}_0 = 1$, $\tilde{T}_1 = 0$.

***207.** By following the line of solution **195** you arrive at the differential equation

$$\frac{dy}{dz} = (e^z - 1)\,y.$$

***208.** From **193** and **207**

$$\sum \frac{\tilde{T}_n z^n}{n!} = e^{-z} \sum \frac{T_n z^n}{n!} \qquad [34].$$

***209.** Obvious for $n = 1$. Use **187** and mathematical induction.

***210.** (1) immediate consequence of **10**.

(2) from **190**. Or take $F(t) = e^{(t-1)w}$ and compute the $n$-th derivative of $F(e^x)$ at the point $x = 0$ by using **209**.

(3) See solution **200**. A proof independent of **186**—**209** is based on a fact quoted in solution VII **46**:

$$\sum_{k_1=0}^{\infty}\sum_{k_2=0}^{\infty}\sum_{k_3=0}^{\infty}\cdots Z_{k_1k_2k_3\ldots}\frac{z^{k_1+2k_2+3k_3+\cdots}w^{k_1+k_2+k_3+\cdots}}{n!}$$

$$= \sum_{k_1=0}^{\infty}\frac{1}{k_1!}\left(\frac{zw}{1}\right)^{k_1}\sum_{k_2=0}^{\infty}\frac{1}{k_2!}\left(\frac{z^2w}{2}\right)^{k_2}\sum_{k_3=0}^{\infty}\frac{1}{k_3!}\left(\frac{z^3w}{3}\right)^{k_3}\cdots$$

$$= e^{\frac{zw}{1}+\frac{z^2w}{2}+\frac{z^3w}{3}+\cdots} = e^{-w\log(1-z)}.$$

We can give an analogous proof for (2) by using the well known fact that the number of partitions of a set containing $n$ elements into $k_1$ subsets each containing 1 element, $k_2$ subsets each containg 2 elements,

$k_3$ subsets each containing 3 elements, etc. is

$$\frac{n!}{k_1!\,(1!)^{k_1}\cdot k_2!\,(2!)^{k_2}\cdot k_3!\,(3!)^{k_3}\cdots}$$

where, of course,

$$1k_1 + 2k_2 + 3k_3 + \cdots = n.$$

To rework **186–210** starting from **210** is left to the reader as a research project. Just one hint: If we let $f$ denote one of the expressions (1), (2) and (3), then

$$\frac{\partial f}{\partial z} - wf = f, \quad w\frac{\partial f}{\partial w}, \quad z\frac{\partial f}{\partial z}$$

which involves

$$\binom{n+1}{k} - \binom{n}{k-1} = \binom{n}{k}, \quad S_k^{n+1} - S_{k-1}^n = kS_k^n, \quad s_k^{n+1} - s_{k-1}^n = ns_k^n$$

respectively.

For additional material on the subject treated in the section just concluded see V **62.1,** VII **54.2,** VIII **22.1, 22.2, 22.3, 58.3, 247.1,** AI **191.1, 191.2.**

Part Two

# Integration

**1.**

$$\frac{r-1}{x_\nu^r} < \frac{1}{x_\nu^{r-1}\,x_{\nu-1}} + \frac{1}{x_\nu^{r-2}\,x_{\nu-1}^2} + \cdots + \frac{1}{x_\nu x_{\nu-1}^{r-1}} < \frac{r-1}{x_{\nu-1}^r}.$$

**2.**

$$(r+1)\,x_{\nu-1}^r < \frac{x_\nu^{r+1} - x_{\nu-1}^{r+1}}{x_\nu - x_{\nu-1}} < (r+1)\,x_\nu^r.$$

**3.**

$$\sum_{\nu=1}^{n} he^{a+(\nu-1)h}, \qquad \sum_{\nu=1}^{n} he^{a+\nu h},$$

where $h = x_\nu - x_{\nu-1} = \frac{b-a}{n}$.

$$\lim_{n\to\infty} he^a \frac{1-e^{nh}}{1-e^h} = \lim_{n\to\infty} h\frac{e^b - e^a}{e^h - 1} = e^b - e^a,$$

because

$$\lim_{h\to 0} \frac{e^h - 1}{h} = \left(\frac{de^x}{dx}\right)_{x=0} = 1.$$

**4.**

$$\sum_{\nu=1}^{n} \frac{aq^{\nu-1}(q-1)}{aq^\nu}, \qquad \sum_{\nu=1}^{n} \frac{aq^{\nu-1}(q-1)}{aq^{\nu-1}},$$

with $q = \frac{x_\nu}{x_{\nu-1}} = \sqrt[n]{\frac{b}{a}}$. We have [**3**]

$$\lim_{n\to\infty} n\left(\sqrt[n]{\frac{b}{a}} - 1\right) = \lim_{n\to\infty} \frac{e^{\frac{\log b - \log a}{n}} - 1}{\frac{\log b - \log a}{n}}(\log b - \log a) = \log b - \log a.$$

***5.** Set $1+\frac{1}{2}+\frac{1}{3}+\cdots+\frac{1}{n}=H_n$. Then

$$L_n=\frac{1}{n}\sum_{\nu=1}^{n}\frac{1}{1+\frac{\nu}{n}}=\frac{1}{n+1}+\frac{1}{n+2}+\cdots+\frac{1}{n+n}$$

$$=H_{2n}-H_n=H_{2n}-2(\tfrac{1}{2}H_n)$$

$$=1+\frac{1}{2}+\frac{1}{3}+\frac{1}{4}+\cdots+\frac{1}{2n-1}+\frac{1}{2n}$$

$$-\frac{2}{2}\qquad-\frac{2}{4}-\cdots\qquad-\frac{2}{2n}$$

$$=1-\frac{1}{2}+\frac{1}{3}-\frac{1}{4}+\cdots+\frac{1}{2n-1}-\frac{1}{2n}.$$

Moreover

$$U_n=\frac{1}{n}+\frac{1}{n+1}+\cdots+\frac{1}{2n-1}$$

$$=L_n+\frac{1}{n}-\frac{1}{2n}=L_n+\frac{1}{2n}.$$

**6.**

$$\lim_{n\to\infty}\frac{\pi}{n+1}\left(\frac{\sin\frac{\pi}{n+1}}{\frac{\pi}{n+1}}+\frac{\sin 2\frac{\pi}{n+1}}{2\frac{\pi}{n+1}}+\cdots+\frac{\sin n\frac{\pi}{n+1}}{n\frac{\pi}{n+1}}\right)=\int_0^\pi\frac{\sin x}{x}\,dx>0.$$

(VI **25**.)

**7.** The mean value theorem implies

$$F(b)-F(a)=\sum_{\nu=1}^{n}\left[F(x_\nu)-F(x_{\nu-1})\right]$$

$$=\sum_{\nu=1}^{n}f(\xi_\nu)\,(x_\nu-x_{\nu-1}),\qquad x_{\nu-1}<\xi_\nu<x_\nu.$$

Yet, according to the *definitions* we use *here*,

$$F(b)-F(a)=\int_a^b f(x)\,dx$$

need not be true. [Cf. V. Volterra: Giorn. Mat. Battaglini Vol. 19, p. 335 (1881).]

**8.** Let $\frac{k-1}{n}\leqq\xi<\frac{k}{n}$. Then

$$\Delta_n\leqq\sum_{\nu=k}^{n}\int_{\frac{\nu-1}{n}}^{\frac{\nu}{n}}\left[f(x)-f\left(\frac{\nu}{n}\right)\right]dx\leqq\frac{M-f\left(\frac{k}{n}\right)}{n}+\sum_{\nu=k+1}^{n}\frac{f\left(\frac{\nu-1}{n}\right)-f\left(\frac{\nu}{n}\right)}{n}$$

and

$$-\Delta_n \leqq \sum_{\nu=1}^{k} \int_{\frac{\nu-1}{n}}^{\frac{\nu}{n}} \left[f\left(\frac{\nu}{n}\right) - f(x)\right] dx \leqq \frac{f\left(\frac{k}{n}\right) - \min\left[f\left(\frac{k-1}{n}\right), f\left(\frac{k}{n}\right)\right]}{n}$$

$$+ \sum_{\nu=1}^{k-1} \frac{f\left(\frac{\nu}{n}\right) - f\left(\frac{\nu-1}{n}\right)}{n} = \frac{\max\left[f\left(\frac{k-1}{n}\right), f\left(\frac{k}{n}\right)\right] - f(0)}{n}.$$

**9.** [Cf. G. Pólya: Arch. Math. Phys. Ser. 3, Vol. 26, p. 198 (1917).] The least upper bound of the expression

$$|f(x_1) - f(x_0)| + |f(x_2) - f(x_1)| + \cdots + |f(x_n) - f(x_{n-1})|$$

for all possible subdivisions of the interval $[a, b]$ is called the *total variation* of the function $f(x)$ on $[a, b]$ (same notation as on p. 46). Functions of finite total variation are also called functions of "bounded variation".

$$|\Delta_n| \leqq \int_0^{\frac{1}{n}} \left[\sum_{\nu=1}^{n} \left|f\left(x + \frac{\nu-1}{n}\right) - f\left(\frac{\nu}{n}\right)\right|\right] dx \leqq \int_0^{\frac{1}{n}} V\, dx.$$

**10.**

$$-\Delta_n = \sum_{\nu=1}^{n} \int_{a+(\nu-1)\frac{b-a}{n}}^{a+\nu\frac{b-a}{n}} \left(a + \nu\frac{b-a}{n} - x\right) f'(\xi_\nu)\, dx,$$

where $a + (\nu - 1)\frac{b-a}{n} < \xi_\nu < a + \nu\frac{b-a}{n}$, thus

$$\frac{1}{2}\left(\frac{b-a}{n}\right)^2 \sum_{\nu=1}^{n} m_\nu \leqq -\Delta_n \leqq \frac{1}{2}\left(\frac{b-a}{n}\right)^2 \sum_{\nu=1}^{n} M_\nu;$$

$M_\nu$ and $m_\nu$ denote the least upper and the greatest lower bound of $f'(x)$ in the $\nu$-th subinterval. We obtain

$$\lim_{n\to\infty} n\,\Delta_n = \frac{b-a}{2}\,[f(a) - f(b)].$$

**11.**

$$\lim_{n\to\infty} n^2 \Delta_n' = \frac{(b-a)^2}{24}\,[f'(b) - f'(a)],$$

because

$$f(x) - f\left(a + (2\nu - 1)\frac{b-a}{2n}\right) = \left(x - a - (2\nu - 1)\frac{b-a}{2n}\right)$$

$$\times f'\left(a + (2\nu - 1)\frac{b-a}{2n}\right) + \frac{1}{2}\left(x - a - (2\nu - 1)\frac{b-a}{2n}\right)^2 f''(\xi_\nu).$$

The intergal of the linear term over the interval $\left[a+(\nu-1)\frac{b-a}{n}, a+\nu\frac{b-a}{n}\right]$ vanishes. Cf. **10**.

**12.** We have [**11**]

$$\Delta_n'' = \int\limits_a^{a+\frac{b-a}{2n+1}} [f(x)-f(a)]\,dx + \sum_{\nu=1}^{n} \int\limits_{a+(2\nu-1)\frac{b-a}{2n+1}}^{a+(2\nu+1)\frac{b-a}{2n+1}} \left[f(x)-f\left(a+2\nu\frac{b-a}{2n+1}\right)\right] dx$$

$$= \int\limits_a^{a+\frac{b-a}{2n+1}} (x-a)\,f'(\xi_0)\,dx + \sum_{\nu=1}^{n}\frac{1}{2} \int\limits_{a+(2\nu-1)\frac{b-a}{2n+1}}^{a+(2\nu+1)\frac{b-a}{2n+1}} \left(x-a-2\nu\frac{b-a}{2n+1}\right)^2 f''(\xi_\nu)\,dx.$$

The limit in question is

$$\frac{(b-a)^2}{24}[f'(b)+2f'(a)].$$

In the case of $f'(a)=0$ this result is a consequence of **11**: extend the function $f(x)$ considered in **11** to the left of $a$ by reflection and substitute $2n+1$ for $n$.

**13.** In **10** and **11** set $f(x)=\frac{1}{1+x}$, $a=0$, $b=1$. Cf. also **5**.

**14.** [For more details cf. G. N. Watson: Lond. Edin. Dubl. Phil. Mag. Ser. 3, Vol. 31, pp. 111–118 (1916).]

$$\sum_{\nu=1}^{n-1}\frac{1}{\sin\frac{\nu\pi}{n}}$$

$$=\frac{2}{\pi}\left(\frac{n}{1}+\frac{n}{2}+\cdots+\frac{n}{n-1}\right)+n\sum_{\nu=1}^{n-1}\left(\frac{1}{\sin\frac{\nu\pi}{n}}-\frac{1}{\frac{\nu\pi}{n}}-\frac{1}{\frac{(n-\nu)\pi}{n}}\right)\frac{1}{n}$$

$$=\frac{2n}{\pi}\left[\log n+C+O\left(\frac{1}{n}\right)\right]+n\left[\int_0^1\left(\frac{1}{\sin\pi x}-\frac{1}{\pi x}-\frac{1}{\pi(1-x)}\right)dx+O\left(\frac{1}{n}\right)\right]$$

$$=\frac{2n}{\pi}(\log n+C)-\frac{2n}{\pi}\log\frac{\pi}{2}+O(1),$$

with the help of **9**.

**15.** [E. Cesàro, Problem: Nouv. Annls Math. Ser. 3, Vol. 17, p. 112 (1888). Solved by G. Pólya: Nouv. Annls Math. Ser. 4, Vol. 11, pp. 373–381 (1911).] Put $f(x) = x \log x$, $a = 0$, $b = 1$ in **10**. Although the hypothesis of **10** is not completely satisfied the conclusion of **10** remains valid.

**16.** Write $P(x) = \frac{1}{2x} + \sum_{\nu=1}^{n} \frac{1}{x-\nu}$, $\beta = (1 - e^{-\alpha})^{-1}$. The equation $P(x) = \alpha$ has degree $n + 1$ and possesses one root in each of the intervals $(0, 1), (1, 2), (2, 3), \ldots, (n - 1, n), (n, \infty)$. The largest root is $x_n$. In **12** put $f(x) = \frac{1}{\beta - x}$, $a = 0$, $b = 1$. We find

$$\int_0^1 \frac{dx}{\beta - x} = \alpha = P(x_n) = P\big((n + \tfrac{1}{2})\,\beta\big) + \Delta_n'', \qquad \Delta_n'' > 0.$$

Since $P(x)$ is decreasing for $x > n$ we have $x_n < (n + \frac{1}{2})\,\beta$. The mean value theorem implies

$$0 < (n + \tfrac{1}{2})\beta - x_n = -\frac{\Delta_n''}{P'(\xi)} < -\frac{\Delta_n''}{P'((n + \frac{1}{2})\,\beta)}, \qquad x_n < \xi < (n + \tfrac{1}{2})\,\beta.$$

The quotient on the right hand side converges to 0 [**12**] because

$$n\,P'\big((n + \tfrac{1}{2})\,\beta\big) \to -\int_0^1 \frac{dx}{(\beta - x)^2}.$$

**17.** The equation $\alpha = \int_0^1 \frac{2\beta}{\beta^2 - x^2}\,dx$ leads to $\beta = \frac{1 + e^{-\alpha}}{1 - e^{-\alpha}}$. The proof is similar to the one given in **16**.

**18.** [J. Franel. Related to Euler's summation formula, cf. Knopp, p. 523.] We write $F(x) = \int_1^x f(\xi)\,d\xi$. Summation of the equations

$$\left.\begin{aligned} F(\nu + \tfrac{1}{2}) - F(\nu) &= \tfrac{1}{2}f(\nu) + \tfrac{1}{8}f'(\xi_\nu), \\ -F(\nu + \tfrac{1}{2}) + F(\nu + 1) &= \tfrac{1}{2}f(\nu + 1) - \tfrac{1}{8}f'(\eta_\nu) \end{aligned}\right\} \begin{aligned} &\nu < \xi_\nu < \nu + \tfrac{1}{2} < \eta_\nu < \nu + 1; \\ &\nu = 1, 2, 3, \ldots, n - 1, \end{aligned}$$

yields

$$\begin{aligned} &\tfrac{1}{2}f(1) + f(2) + f(3) + \cdots + f(n - 1) + \tfrac{1}{2}f(n) - F(n) \\ &= \tfrac{1}{8}[f'(\eta_1) - f'(\xi_1) + f'(\eta_2) - f'(\xi_2) + \cdots + f'(\eta_{n-1}) - f'(\xi_{n-1})]. \end{aligned}$$

The series

$$-f'(\xi_1) + f'(\eta_1) - f'(\xi_2) + f'(\eta_2) - f'(\xi_3) + f'(\eta_3) - \cdots = 8s$$

is convergent because the terms have alternating signs and their absolute values converge monotonically to 0. In the case of $f'(x) < 0$ we find

$$\frac{1}{8}f'(n) < \frac{1}{8}f'(\xi_n) < -\frac{1}{8}[f'(\eta_n) - f'(\xi_n) + f'(\eta_{n+1}) - f'(\xi_{n+1}) + \cdots] < 0.$$

This proves for $f(x) = \frac{1}{x}$ the existence of

$$\lim_{n\to\infty}\left(\frac{1}{1} + \frac{1}{2} + \frac{1}{3} + \cdots + \frac{1}{n} - \log n\right) = C$$

(Euler's constant) and furnishes the inequalities

$$\frac{1}{2n} - \frac{1}{8n^2} < \frac{1}{1} + \frac{1}{2} + \frac{1}{3} + \cdots + \frac{1}{n} - \log n - C < \frac{1}{2n}.$$

For $f(x) = -\log x$ we find

$$\log n! = (n + \tfrac{1}{2})\log n - n + 1 - s + \varepsilon_n,$$

where $s$ is a constant and $0 < \varepsilon_n < \frac{1}{8n}$. Stirling's formula [**205**] states that $1 - s = \log\sqrt{2\pi}$.

**19.** Cf. solution **18**. The sum mentioned there,

$$\tfrac{1}{8}[f'(\eta_1) - f'(\xi_1) + f'(\eta_2) - f'(\xi_2) + \cdots + f'(\eta_{n-1}) - f'(\xi_{n-1})],$$

is positive in this case. Furthermore

$$f'(\eta_1) - f'(\xi_2) < 0,\ f'(\eta_2) - f'(\xi_3) < 0, \ldots,\ f'(\eta_{n-2}) - f'(\xi_{n-1}) < 0,$$

$$f'(\xi_1) > f'(1), \quad f'(\eta_{n-1}) < f'(n).$$

**19.1.** Cf. **5**. The relation

$$\lim_{n\to\infty}(H_n - \log n) = \lim_{n\to\infty}(H_{2n} - \log 2n)$$

implies

$$\log 2 = \lim_{n\to\infty}(H_{2n} - H_n).$$

**19.2.** [G. Pólya; see Research Papers in Statistics, Festschrift for J. Neyman. New York: Wiley & Sons 1966, pp. 259—261.] Use notation **5**. From

$$\lim_{n\to\infty}(2H_n - 2\log n) = 2C,$$

$$\lim_{n\to\infty}(H_{n^2} - \log n^2) = C$$

follows that $C$ is the limit of

$$\begin{aligned} 2H_n - H_{n^2} &= \frac{2}{1} \qquad\qquad + \frac{4}{4} \qquad \cdots + \frac{2n}{n^2} \\ &\quad - \frac{1}{1} - \frac{1}{2} - \frac{1}{3} - \frac{1}{4} - \frac{1}{5} - \cdots - \frac{1}{n^2} \\ &= \frac{1}{1} - \frac{1}{2} - \frac{1}{3} + \frac{3}{4} - \frac{1}{5} - \cdots + \frac{2n-1}{n^2}, \end{aligned}$$

which coincides with the partial sum of the proposed series that has $1 + 2 + 3 + \cdots + (2n - 1)$ terms. It still must be shown that the limit of this subsequence of partial sums is the sum of the series. (Easy.)

**20.** [Cf. e.g. l.c. **9**.] Let $f(x)$ be monotone increasing. [Otherwise consider $-f(x)$.] Then

$$\int_0^{1-\frac{1}{n}} f(x)\,dx \leqq \frac{f\left(\frac{1}{n}\right) + f\left(\frac{2}{n}\right) + \cdots + f\left(\frac{n-1}{n}\right)}{n} \leqq \int_{\frac{1}{n}}^{1} f(x)\,dx.$$

The condition that the function is monotone increasing is essential only in the neighbourhood of the singular points.

**21.** We can assume that the function $f(x)$ increases [**20**] and that $f(x) \geqq 0$ [otherwise decompose $f(x) = \frac{f(x) + |f(x)|}{2} + \frac{f(x) - |f(x)|}{2}$ and examine the two terms separately]. Let $\varepsilon > 0$, $\eta$ be chosen such that $0 < \eta < 1$, $\int_{1-\eta}^{1} f(x)\,dx < \varepsilon$. Then we have

$$\lim_{n\to\infty} \frac{1}{n} \sum_{\nu=1}^{[(1-\eta)n]} \varphi\left(\frac{\nu}{n}\right) f\left(\frac{\nu}{n}\right) = \int_0^{1-\eta} \varphi(x)\, f(x)\,dx.$$

If, on the other hand, $M$ denotes the least upper bound of $|\varphi(x)|$ we can write

$$\left| \frac{1}{n} \sum_{\nu=[(1-\eta)n]+1}^{n-1} \varphi\left(\frac{\nu}{n}\right) f\left(\frac{\nu}{n}\right) \right| \leqq \frac{M}{n} \sum_{\nu=[(1-\eta)n]+1}^{n-1} f\left(\frac{\nu}{n}\right) \leqq M \int_{1-\eta}^{1} f(x)\,dx \leqq M\varepsilon.$$

**22.** In **20** set $f(x) = x^{\alpha-1}$.

**23.**

$$a_n = 1^{\alpha-1}(n-1)^{\beta-1} + 2^{\alpha-1}(n-2)^{\beta-1} + \cdots + (n-1)^{\alpha-1}\, 1^{\beta-1}$$

$$= n^{\alpha+\beta-1} \sum_{\nu=1}^{n-1} \frac{1}{n}\left(\frac{\nu}{n}\right)^{\alpha-1}\left(1 - \frac{\nu}{n}\right)^{\beta-1} \sim n^{\alpha+\beta-1} \int_0^1 x^{\alpha-1}\,(1-x)^{\beta-1}\,dx.$$

Cf. **20**.

**24.**

$$f(x) = \frac{1}{x} - \frac{1}{1-x}.$$

**25.** Let $f(x)$ be monotone decreasing and finite for $x = 1$. The inequality [**20**]

$$\int_{\frac{1}{n}}^{1} f(x)\,dx \leqq \frac{f\left(\frac{1}{n}\right) + f\left(\frac{2}{n}\right) + \cdots + f\left(\frac{n-1}{n}\right)}{n}$$

implies that the left hand side is bounded, i.e. that

$$\lim_{\varepsilon\to+0}\int_{\varepsilon}^{1} f(x)\,dx$$

is finite.

**26.** Let $f(x)$ be monotone decreasing. Then

$$\frac{1}{n}f\left(\frac{2n-1}{2n}\right)+\int_{\frac{1}{2n}}^{\frac{2n-1}{2n}} f(x)\,dx \le \frac{1}{n}\sum_{\nu=1}^{n} f\left(\frac{2\nu-1}{2n}\right) \le \frac{1}{n}f\left(\frac{1}{2n}\right)+\int_{\frac{1}{2n}}^{\frac{2n-1}{2n}} f(x)\,dx,$$

and a fortiori

$$2\int_{\frac{2n-1}{2n}}^{1} f(x)\,dx+\int_{\frac{1}{2n}}^{\frac{2n-1}{2n}} f(x)\,dx \le \frac{1}{n}\sum_{\nu=1}^{n} f\left(\frac{2\nu-1}{2n}\right) \le 2\int_{0}^{\frac{1}{2n}} f(x)\,dx+\int_{\frac{1}{2n}}^{\frac{2n-1}{2n}} f(x)\,dx.$$

Similarly

$$\lim_{n\to\infty}\frac{2}{n}\sum_{\nu=1}^{\left[\frac{n}{2}\right]} f\left(\frac{2\nu-1}{n}\right)=\int_0^1 f(x)\,dx.$$

**27.** Cf. **28** for $f(x)=x^{\alpha-1}$.

**28.**

$$\frac{1}{n}\sum_{\nu=1}^{n-1}(-1)^{\nu-1} f\left(\frac{\nu}{n}\right)=\frac{2}{n}\sum_{\nu=1}^{\left[\frac{n}{2}\right]} f\left(\frac{2\nu-1}{n}\right)-\frac{1}{n}\sum_{\nu=1}^{n-1} f\left(\frac{\nu}{n}\right).$$

[**20** and solution **26**.]

**29.** Cf. **26**.

**30.** Since $f(x)$ is monotone and $\lim\limits_{x\to\infty} f(x)=0$, $f(x)$ cannot change its sign. Assume that $f(x)$ is positive and monotone decreasing. Then we have

$$\int_h^{(m+1)h} f(x)\,dx \le h\big(f(h)+f(2h)+\cdots+f(mh)\big) \le \int_0^{mh} f(x)\,dx,$$

and so for $m\to\infty$

$$\int_h^{\infty} f(x)\,dx \le h\sum_{n=1}^{\infty} f(nh) \le \int_0^{\infty} f(x)\,dx.$$

The condition that $f(x)$ be monotone is essential only for large $x$ and in the neighborhood of $x=0$.

**31.** By setting $f(x) = e^{-x}x^{\alpha-1}$, $e^{-h} = t$ in **30** we get

$$\Gamma(\alpha) = \lim_{t\to 1-0}\left(\log\frac{1}{t}\right)^{\alpha}(1^{\alpha-1}t + 2^{\alpha-1}t^2 + 3^{\alpha-1}t^3 + \cdots)$$

$$= \lim_{t\to 1-0}(1-t)^{\alpha}\sum_{n=1}^{\infty} n^{\alpha-1}t^n = \lim_{n\to\infty}\frac{n^{\alpha-1}n!}{\alpha(\alpha+1)\cdots(\alpha+n-1)} \quad [\text{I } \mathbf{89}].$$

**32.** Introduce in **30** $f(x) = e^{-x}\left(\frac{1}{1-e^{-x}} - \frac{1}{x}\right)$, $e^{-h} = t$. We have

$$h\sum_{n=1}^{\infty}\frac{e^{-nh}}{nh} = \log\frac{1}{1-e^{-h}}.$$

**33.** In **30** put $f(x) = \frac{e^{-x}}{1+e^{-x}}$, $e^{-h} = t$ and notice that

$$\int_0^{\infty}\frac{e^{-x}}{1+e^{-x}}\,dx = \int_0^1\frac{dy}{1+y} = \log 2.$$

Or apply **32** and note

$$\sum_{n=1}^{\infty}\frac{t^n}{1+t^n} = \sum_{n=1}^{\infty}\frac{t^n}{1-t^n} - 2\sum_{n=1}^{\infty}\frac{t^{2n}}{1-t^{2n}}.$$

**34.** The statement follows from

$$\int_0^{\infty}\frac{xe^{-x}}{1-e^{-x}}\,dx = \int_0^{\infty}x\left(\sum_{n=1}^{\infty}e^{-nx}\right)dx = \sum_{n=1}^{\infty}\frac{1}{n^2}.$$

We can also argue in the following manner: we have [VIII **49**, VIII **65**]

$$\sum_{n=1}^{\infty} n^{\alpha}\frac{t^n}{1-t^n} = \sum_{n=1}^{\infty}\sigma_{\alpha}(t)\,t^n,$$

$$\frac{1}{1-t}\sum_{n=1}^{\infty}\sigma_{\alpha}(n)\,t^n = \sum_{n=1}^{\infty}\left(\sigma_{\alpha}(1) + \sigma_{\alpha}(2) + \cdots + \sigma_{\alpha}(n)\right)t^n.$$

Noticing **45** we now apply I **88**.

**35.** In **30** set $f(x) = e^{-x^2}$, $e^{-h^2} = t$, respectively $f(x) = e^{-x^{\alpha}}$, $e^{-h^{\alpha}} = t$.

**36.** The limit is $\pi$. Introduce $f(x) = \frac{2}{1+x^2}$, $h^{-1} = t$ in **30**. Observe the formula

$$\frac{1}{t} + \frac{2t}{t^2+1^2} + \frac{2t}{t^2+2^2} + \cdots + \frac{2t}{t^2+n^2} + \cdots = \pi\,\frac{e^{\pi t} + e^{-\pi t}}{e^{\pi t} - e^{-\pi t}}$$

[Hurwitz-Courant, pp. 122—123.]

**37.** The proposition follows from **30**: $f(x) = \log(1 + x^{-\alpha})$, $h = t^{-\frac{1}{\alpha}}$. Transformation of variables and partial integration lead to

$$\int_0^\infty \log(1 + x^{-\alpha})\, dx = \int_0^\infty \frac{u^{-\frac{1}{\alpha}}}{1+u}\, du = \frac{\pi}{\sin\frac{\pi}{\alpha}}.$$

**38.** Apply **30** with $f(x) = \log(1 - 2x^{-2}\cos 2\varphi + x^{-4})$. Write $t = \frac{\pi}{h} e^{i\varphi}$, use the formula for $\frac{\sin t}{t}$ and square the absolute value on both sides:

$$\prod_{n=1}^\infty \left(1 - \frac{2\cos 2\varphi}{n^2h^2} + \frac{1}{n^4h^4}\right) = \frac{h^2}{4\pi^2}\left[e^{\frac{2\pi}{h}\sin\varphi} + e^{-\frac{2\pi}{h}\sin\varphi} - 2\cos\left(\frac{2\pi}{h}\cos\varphi\right)\right].$$

**39.**

$$\sum_{n=0}^\infty (aq^n - aq^{n+1}) \log aq^{n+1} < \int_0^a \log x\, dx < \sum_{n=0}^\infty (aq^n - aq^{n+1}) \log aq^n$$

$$= a\log a + \frac{aq\log q}{1-q} \to a\log a - a$$

for $q \to 1$. A more general proposition can be deduced in analogy to **30**.

**40.** Taking **58** into account we obtain

$$\sum_{\nu=0}^n \binom{n}{\nu}^k \sim 2^{kn}\left(\frac{2}{\pi}\right)^{\frac{k}{2}} n^{-\frac{k-1}{2}} \sum_{\nu=0}^n e^{-2k\left(\frac{\nu-\frac{n}{2}}{\sqrt{n}}\right)^2} \cdot \frac{1}{\sqrt{n}} \sim 2^{kn}\left(\frac{2}{\pi}\right)^{\frac{k}{2}} n^{-\frac{k-1}{2}} \int_{-\infty}^\infty e^{-2kx^2}\, dx$$

$$= 2^{kn}\left(\frac{2}{\pi}\right)^{\frac{k}{2}} n^{-\frac{k-1}{2}} \left(\frac{\pi}{2k}\right)^{\frac{1}{2}}.$$

For more details see e.g. Jordan: Cours d'Analyse, Vol. 2, 3rd Ed. Paris: Gauthier-Villars 1913, pp. 218–221.

**41.** [For problems **41**–**47** cf. G. Pólya: Arch. Math. Phys. Ser. 3, Vol. 26, pp. 196–201 (1917).]

**42.**

$$\lim_{n\to\infty} \frac{1}{n}\sum_{\nu=1}^n \left(\frac{n}{\nu} - \left[\frac{n}{\nu}\right]\right) = \int_0^1 \left(\frac{1}{x} - \left[\frac{1}{x}\right]\right) dx = \lim_{n\to\infty} \int_{\frac{1}{n}}^1 \left(\frac{1}{x} - \left[\frac{1}{x}\right]\right) dx$$

$$= 1 - \lim_{n\to\infty}\left(1 + \frac{1}{2} + \frac{1}{3} + \cdots + \frac{1}{n} - \log n\right) = 1 - C,$$

where $C$ is Euler's constant.—Define

$$\Phi(\alpha) = \int_0^1 \frac{1 - x^\alpha}{1 - x}\, dx = \frac{\Gamma'(\alpha + 1)}{\Gamma(\alpha + 1)} + C$$

and note that the relation

$$\int_0^1 \alpha\, d\Phi(\alpha) = \Phi(1) - \int_0^1 \Phi(\alpha)\, d\alpha = 1 - C$$

holds; i.e. the operations of taking the limit and computing the mean value can be interchanged [**44**].

**43.** [Cf. Cesàro, Problem: Nouv. Annls Math. Ser. 3, Vol. 2, p. 239 (1883).]

$$\lim_{n\to\infty} \frac{1}{n} \sum_{\nu=1}^{n} \left(1 - \left[\frac{n}{\nu}\right]\frac{\nu}{n}\right) = 1 - \int_0^1 \left[\frac{1}{x}\right] x\, dx$$

$$= 1 - \sum_{n=1}^{\infty} \frac{n}{2}\left(\frac{1}{n^2} - \frac{1}{(n+1)^2}\right) = 1 - \frac{\pi^2}{12}.$$

**44.** [G. L. Dirichlet: Werke, Vol. 2. Berlin: G. Reimer 1897, p. 97—104; cf. also G. Pólya, l.c. **41**, p. 197 and Nachr. Akad. Wiss. Göttingen 1917, pp. 149—159.] We are dealing with [VIII **4**]

$$\lim_{n\to\infty} \frac{1}{n} \sum_{\nu=1}^{n} \left(\left[\frac{n}{\nu}\right] - \left[\frac{n}{\nu} - \alpha\right]\right) = \int_0^1 \left(\left[\frac{1}{x}\right] - \left[\frac{1}{x} - \alpha\right]\right) dx$$

$$= \lim_{n\to\infty} \int_{\frac{1}{n}}^{1} \left(\left[\frac{1}{x}\right] - \left[\frac{1}{x} - \alpha\right]\right) dx = \lim_{n\to\infty} \sum_{\nu=1}^{n-1} \left(\frac{1}{\nu} - \frac{1}{\nu + \alpha}\right)$$

$$= 1 - \frac{1}{1+\alpha} + \frac{1}{2} - \frac{1}{2+\alpha} + \cdots = \int_0^1 \frac{1 - x^\alpha}{1 - x}\, dx.$$

If $\alpha = \frac{1}{2}$ we obtain **41**.

**45.** [G. Pólya, l.c. **41**, pp. 199—200.] We assume at first that $\alpha > 1$. Then

$$(\alpha + 1)\int_0^1 \left[\frac{1}{x}\right] x^\alpha\, dx = \sum_{n=1}^{\infty} n \int_{\frac{1}{n+1}}^{\frac{1}{n}} (\alpha + 1)\, x^\alpha\, dx = 1 + \frac{1}{2^{\alpha+1}} + \frac{1}{3^{\alpha+1}} + \cdots$$

$$= \zeta(\alpha + 1).$$

The total variation [cf. solution **9**] of $\left[\frac{1}{x}\right] x^\alpha = f(x)$ is

$$\left(f(1) - f(\tfrac{1}{2}+0)\right) + \left(f(\tfrac{1}{2}-0) - f(\tfrac{1}{2}+0)\right) + \left(f(\tfrac{1}{2}-0) - f(\tfrac{1}{3}+0)\right) + \cdots$$
$$= 1(1^{-\alpha} - 2^{-\alpha}) + 2^{-\alpha} + 2(2^{-\alpha} - 3^{-\alpha}) + 3^{-\alpha} + \cdots = 2\zeta(\alpha) - 1$$

whence the two statements follow [**9**]. The limit relation holds also for $\alpha = 1$. If $0 < \alpha < 1$ we have to examine $\left(\frac{1}{x} - \left[\frac{1}{x}\right]\right) x^\alpha$ and use **22**.

**46.** [G. Pólya, l.c. **41**, pp. 200–201.] We write $f(x) = \frac{1}{x} - \left[\frac{1}{x}\right]$; according to **42** $\int_0^1 f(x)\,dx = 1 - C$; furthermore, the total variation of $f(x)$ in the interval $\left(\frac{1}{m}, 1\right)$ is $2(m-1)$. We have

$$\left|\frac{1}{n}\sum_{\nu=1}^{n}\left[\frac{n}{\nu}\right] - \sum_{\nu=1}^{n}\frac{1}{\nu} + 1 - C\right| = \left|\int_0^1 f(x)\,dx - \frac{1}{n}\sum_{\nu=1}^{n} f\left(\frac{\nu}{n}\right)\right|$$
$$\leqq \left|\int_{\frac{m}{n}}^{1} f(x)\,dx - \frac{1}{n}\sum_{\nu=m+1}^{n} f\left(\frac{\nu}{n}\right)\right| + \sum_{\nu=1}^{m}\int_0^{\frac{1}{n}}\left|f\left(\frac{\nu-1}{n}+x\right) - f\left(\frac{\nu}{n}\right)\right| dx.$$

Since $\frac{m}{n} > \frac{1}{m}$ the first term is not larger than $\frac{2(m-1)}{n}$ [**9**]; the second is smaller than $\frac{m}{n}$.

**47.**

$$\sum_{\nu=1}^{n}(O_\nu - E_\nu) = \sum_{\nu=1}^{n}(-1)^{\nu-1}\left[\frac{n}{\nu}\right]$$
$$= \sum_{\nu=1}^{n}(-1)^{\nu-1}\left(\left[\frac{n}{\nu}\right] - \frac{n}{\nu}\right) + n\sum_{\nu=1}^{n}\frac{(-1)^{\nu-1}}{\nu}.$$

The first sum on the left hand side divided by $n$ converges to 0 as $n \to \infty$ [**28**].

**48.** Consequence of the definition of the definite integral.

**49.** Special case of **20** for $f(x) = \log x$. With regard to $\int_0^1 \log x\,dx$ cf. **39**.

**50.** Put $c = \frac{a}{d}$. Then

$$\frac{G_n}{A_n} = \frac{\sqrt[n]{\frac{c}{n}\,\frac{c+1}{n}\,\frac{c+2}{n}\cdots\frac{c+n-1}{n}}}{\frac{c}{n} + \frac{n-1}{2n}}, \qquad [\mathbf{29}].$$

**51.** $A_n = \frac{2^n}{n+1}$. Moreover

$$\binom{n}{0}\binom{n}{1}\binom{n}{2}\cdots\binom{n}{n} = \frac{n!^{n+1}}{(1!\,2!\,3!\cdots n!)^2} = \prod_{\nu=1}^{n} (n+1-\nu)^{n+1-2\nu}$$

$$= \prod_{\nu=1}^{n} \left(\frac{n+1-\nu}{n+1}\right)^{n+1-2\nu},$$

because $\sum_{\nu=1}^{n} (n+1-2\nu) = 0$. Making appropriate use of **20** we obtain

$$\lim_{n\to\infty} \frac{1}{n} \log G_n = \lim_{n\to\infty} \frac{1}{n} \sum_{\nu=1}^{n} \left(1 - \frac{2\nu}{n+1}\right) \log\left(1 - \frac{\nu}{n+1}\right)$$

$$= \int_0^1 (1-2x) \log(1-x)\, dx = \frac{1}{2}.$$

**52.** In **48** set $f(x) = 1 - 2r\cos x + r^2 = |r - e^{ix}|^2$, $a = 0$, $b = 2\pi$. The identity

$$r^n - 1 = \prod_{\nu=1}^{n} (r - e^{2\pi i\nu/n})$$

implies

$$f_{1n} f_{2n} \cdots f_{nn} = (r^n - 1)^2.$$

If $r = 1$, $f_{nn}$ must be omitted (it vanishes); notice **20**.

**53.** [G. Szegö, Problem: Arch. Math. Phys. Ser. 3, Vol. 25, p. 196 (1917). Solved by J. Mahrenholz: Arch. Math. Phys. Ser. 3, Vol. 28, pp. 79–80 (1920).] According to the hypothesis $e^{-ix}(e^{i\xi} - r)$ is real, i.e. the arguments of $e^{ix}$ and $e^{i\xi} - r$ are equal or differ by $\pi$. Since $\xi$ is the number closest to $x$ with this property $e^{ix}$ and $e^{i\xi} - r$ have the same argument, which means that $e^{i\xi}$ is the point of intersection of the ray from $r$ parallel to the vector $e^{ix}$ with the unit circle. If, therefore, $0 \leqq x < \pi$ and $\xi'$ is the argument that belongs so to $x + \pi$ as $\xi$ to $x$, then $e^{i\xi}$, $r$, and $e^{i\xi'}$ are on the same line. Thus we obtain by elementary geometry

$$|e^{i\xi'} - r|^2 \, |e^{i\xi} - r|^2 = (1 - 2r\cos\xi' + r^2)(1 - 2r\cos\xi + r^2) = (1 - r^2)^2,$$

hence

$$\frac{1}{2\pi} \int_0^{\pi} [\log(1 - 2r\cos\xi + r^2) + \log(1 - 2r\cos\xi' + r^2)]\, dx$$

$$= \frac{1}{2\pi} \int_0^{\pi} \log(1 - r^2)^2\, dx = \log(1 - r^2).$$

**54.** We have [Maclaurin series]

$$|\log(1+x) - x| \leqq x^2 \quad \text{for} \quad |x| \leqq \tfrac{1}{2}.$$

Assume $|f(x)| < M$. Whenever $\delta_n M \leqq \frac{1}{2}$

$$\left|\sum_{\nu=1}^{n} \log(1 + f_{\nu n}\delta_n) - \sum_{\nu=1}^{n} f_{\nu n}\delta_n\right| \leqq \delta_n \sum_{\nu=1}^{n} f_{\nu n}^2 \delta_n.$$

The sum on the right hand side converges to an integral. Cf. **67**. Other subdivisions of the interval $[a, b]$ may be considered instead of the subdivision by points in arithmetic progression, and we can choose any point in a given subinterval as the point where we evaluate the function. There is an obvious analogy between the integral as a limit of sums and the limit of products considered.

**55.** According to **54**:

$$\lim_{n\to\infty} \prod_{\nu=1}^{n} \frac{1 + \frac{\nu}{n}\frac{1}{n}}{1 - \frac{\nu}{n}\frac{1}{n}} = \frac{e^{\int_0^1 x\,dx}}{e^{-\int_0^1 x\,dx}} = e.$$

**56.** The product in question is

$$= \frac{1 \cdot 3 \cdot 5 \cdots (2n-1)\,\alpha^n \cdot 2^n}{[(n+1)\alpha - 1]\,[(n+2)\alpha - 1] \cdots (2n\alpha - 1)}$$

$$= \frac{(n+1)\alpha}{(n+1)\alpha - 1} \cdot \frac{(n+2)\alpha}{(n+2)\alpha - 1} \cdots \frac{(n+n)\alpha}{(n+n)\alpha - 1}$$

$$= \frac{1}{\left(1 - \frac{1}{n}\frac{1}{\left(1+\frac{1}{n}\right)\alpha}\right)\left(1 - \frac{1}{n}\frac{1}{\left(1+\frac{2}{n}\right)\alpha}\right)\cdots\left(1 - \frac{1}{n}\frac{1}{\left(1+\frac{n}{n}\right)\alpha}\right)}$$

$$\to \frac{1}{e^{-\frac{1}{\alpha}\int_0^1 \frac{dx}{1+x}}} \quad [\mathbf{54}].$$

The particular case of $\alpha = 2$, i.e.

$$\frac{1}{2} \cdot \frac{3}{2} \cdot \frac{5}{6} \cdot \frac{7}{6} \cdot \frac{9}{10} \cdot \frac{11}{10} \cdots = \frac{1}{\sqrt{2}}$$

follows also from the product representation of $\cos x$ at the point $x = \frac{\pi}{4}$. [Euler: Opera Omnia, Ser. 1, Vol. 17. Leipzig and Berlin: B. G. Teubner 1915, p. 419 (distorted by a misprint).]

**57.**

$$\frac{a + \nu d}{b + \nu d} = 1 + \frac{\alpha - \beta}{\delta} \frac{1}{1 + \frac{\beta + \nu\delta}{n}} \frac{\delta}{n}.$$

Make use of the remark to solution **54** for $f(x) = \frac{\alpha-\beta}{\delta}\frac{1}{1+x}$ on the interval $[0, \delta]$.

**58.** Let $n$ be even, $n = 2m$, $\frac{\nu - m}{\sqrt{m}} \to \lambda\sqrt{2}$, moreover $\lambda \geqq 0$, $\nu > m$. Then we get

$$\frac{\binom{2m}{\nu}}{\binom{2m}{m}} = \frac{m}{m+1}\frac{m-1}{m+2}\cdots\frac{m-(\nu-m-1)}{m+(\nu-m)}$$

$$= \frac{1}{1+\frac{1}{\sqrt{m}}\frac{1}{\sqrt{m}}}\,\frac{1-\frac{1}{\sqrt{m}}\frac{1}{\sqrt{m}}}{1+\frac{2}{\sqrt{m}}\frac{1}{\sqrt{m}}}\cdots\frac{1-\frac{\nu-m-1}{\sqrt{m}}\frac{1}{\sqrt{m}}}{1+\frac{\nu-m}{\sqrt{m}}\frac{1}{\sqrt{m}}}$$

$$\to \frac{e^{-\int_0^{\lambda\sqrt{2}} x\,dx}}{e^{\int_0^{\lambda\sqrt{2}} x\,dx}} = e^{-2\lambda^2} \quad [\mathbf{54}].$$

Notice that $\binom{2m}{m} \sim \frac{2^{2m}}{\sqrt{m\pi}}$ [**202**], and that, furthermore,

$$\frac{\binom{2m+1}{\nu}}{\binom{2m+1}{m+1}} = \frac{m+1}{\nu}\,\frac{\binom{2m}{\nu-1}}{\binom{2m}{m}}.$$

**59.**

$$\prod_{\nu=1}^{n}\left|\frac{2n-\nu}{z-\nu}\right|^2 = \prod_{\nu=1}^{n}\frac{(2n-\nu)^2}{4n^2-4n\nu\cos\frac{t}{\sqrt{n}}+\nu^2} = \prod_{\nu=1}^{n}\frac{1}{1+\frac{8n\nu}{(2n-\nu)^2}\sin^2\frac{t}{2\sqrt{n}}}$$

$$\sim \prod_{\nu=1}^{n}\frac{1}{1+\frac{8n\nu}{(2n-\nu)^2}\frac{t^2}{4n}} \sim e^{-\int_0^1 \frac{2xt^2}{(2-x)^2}dx}.$$

We can justify the substitution of $\frac{t}{2\sqrt{n}}$ for $\sin\frac{t}{2\sqrt{n}}$ by expanding $\sin\frac{t}{2\sqrt{n}}$ in powers of $\frac{t}{\sqrt{n}}$ and taking the logarithm of the product in the same manner as in **54**.

**60.** We write for the second difference in question

$$F(b,d) - F(b,c) - F(a,d) + F(a,c) = \mathop{\Delta^2}_{R} F(x,y),$$

thus

$$\underset{R}{\Delta^2} F(x, y) = \sum_{\mu=1}^{m} \sum_{\nu=1}^{n} \underset{R_{\mu\nu}}{\Delta^2} F(x, y).$$

The mean value theorem of the differential calculus applied to

$$G_\nu(x) = F(x, y_\nu) - F(x, y_{\nu-1})$$

leads to

$$\underset{R_{\mu\nu}}{\Delta^2} F(x, y) = G_\nu(x_\mu) - G_\nu(x_{\mu-1}) = (x_\mu - x_{\mu-1}) \left(F'_x(\xi_\mu, y_\nu) - F'_x(\xi_\mu, y_{\nu-1})\right)$$

$$= (x_\mu - x_{\mu-1}) (y_\nu - y_{\nu-1}) f(\xi_\mu, \eta_\nu), \quad x_{\mu-1} < \xi_\mu < x_\mu, \quad y_{\nu-1} < \eta_\nu < y_\nu.$$

Cf. **7**.

**61.** Multiplying the determinant in question by its complex conjugate row by row we find

$$\left|1 + \varepsilon^{\lambda-\mu} + \varepsilon^{2(\lambda-\mu)} + \cdots + \varepsilon^{(n-1)(\lambda-\mu)}\right|_{\lambda,\mu=0,1,\ldots,n-1} = n^n.$$

On the other hand we have (Vandermonde's determinant)

$$\prod_{j<k}^{0,1,\ldots,n-1} |\varepsilon^j - \varepsilon^k|^2 = \prod_{j<k}^{0,1,\ldots,n-1} \left(2 \sin \frac{j-k}{n} \pi\right)^2.$$

Hence we conclude by comparison

$$\sum_{j<k}^{0,1,\ldots,n-1} \frac{\pi^2}{n^2} \log \left|\sin\left(\frac{j\pi}{n} - \frac{k\pi}{n}\right)\right| = \frac{\pi^2}{2n} \log n - \left(1 - \frac{1}{n}\right) \frac{\pi^2}{2} \log 2.$$

Argument **20**!

**62.** Cf. **54**.

**63.** We denote the expression in question by $\Pi_n$. Then the inequality used in solution **54** yields for sufficiently large $n$

$$\left|\log \Pi_n - \frac{1}{n^2} \sum_{\mu=1}^{n} \sum_{\nu=1}^{n} f\left(\frac{\mu}{n}, \frac{\nu}{n}\right)\right| \le \frac{1}{n^4} \sum_{\nu=1}^{n} \left[\sum_{\mu=1}^{n} f\left(\frac{\mu}{n}, \frac{\nu}{n}\right)\right]^2.$$

Cauchy's inequality [**80**] provides an upper bound for the right hand side:

$$n^{-4} \sum_{\nu=1}^{n} n \sum_{\mu=1}^{n} \left[f\left(\frac{\mu}{n}, \frac{\nu}{n}\right)\right]^2.$$

This expression converges to 0 as $n \to \infty$. Therefore

$$\lim_{n \to \infty} \Pi_n = e^{\int_0^1 \int_0^1 f(x,y)\,dx\,dy}.$$

**64.** [Cf. G. Pólya: Math. Ann. Vol. 74, pp. 204—208 (1913).] Subdivide the space by the three sequences of planes

$$x, y, z = \cdots, \quad -\frac{5}{2n}, \quad -\frac{3}{2n}, \quad -\frac{1}{2n}, \quad \frac{1}{2n}, \quad \frac{3}{2n}, \quad \frac{5}{2n}, \ldots$$

into cubes of volume $\frac{1}{n^3}$. The expression in I **30**, with $s = [n\sigma]$, gives the number of cubes whose center is in $\mathfrak{D}$. This number multiplied by $n^{-3}$ converges to the given integral.

**65.** We have

$$a_n = \sum_{\nu_1+\nu_2+\cdots+\nu_p=n}\sum\cdots\sum \nu_1^{\alpha_1-1}\nu_2^{\alpha_2-1}\cdots\nu_p^{\alpha_p-1}$$

$$= \sum_{\nu_1+\nu_2+\cdots+\nu_{p-1}\leq n}\sum\cdots\sum \nu_1^{\alpha_1-1}\nu_2^{\alpha_2-1}\cdots\nu_{p-1}^{\alpha_{p-1}-1}(n-\nu_1-\nu_2-\cdots-\nu_{p-1})^{\alpha_p-1},$$

thus

$$\frac{a_n}{n^{\alpha_1+\alpha_2+\cdots+\alpha_p-1}} = \frac{1}{n^{p-1}}\sum_{\nu_1+\nu_2+\cdots\nu_{p-1}\leq n}\sum\cdots\sum\left(\frac{\nu_1}{n}\right)^{\alpha_1-1}\left(\frac{\nu_2}{n}\right)^{\alpha_2-1}\cdots\left(\frac{\nu_{p-1}}{n}\right)^{\alpha_{p-1}-1}$$

$$\cdot\left(1-\frac{\nu_1}{n}-\frac{\nu_2}{n}-\cdots-\frac{\nu_{p-1}}{n}\right)^{\alpha_p-1}.$$

Cf. **23**.

**66.** We use the same notation as in solution **65**. According to solution **31** we have for $t \to 1-0$

$$f_k(t) \sim \Gamma(\alpha_k)\,(1-t)^{-\alpha_k}, \qquad k = 1, 2, \ldots, p.$$

Introducing

$$F(z) = \sum_{n=1}^{\infty} n^{\alpha_1+\alpha_2+\cdots+\alpha_p-1}z^n$$

we obtain

$$F(t) \sim \Gamma(\alpha_1+\alpha_2+\cdots+\alpha_p)\,(1-t)^{-(\alpha_1+\alpha_2+\cdots+\alpha_p)};$$

$$\lim_{t\to 1-0}\frac{f_1(t)\,f_2(t)\cdots f_p(t)}{F(t)} = \frac{\Gamma(\alpha_1)\,\Gamma(\alpha_2)\cdots\Gamma(\alpha_p)}{\Gamma(\alpha_1+\alpha_2+\cdots+\alpha_p)}.$$

On the other hand this limit is, according to I **85** and **65**, equal to the integral in question. The $p$-fold Dirichlet-Jordan integral can be easily computed with the help of this relation, cf. E. T. Whittaker and G. N. Watson, p. 258.

**67.** The term in question is

$$\delta_n^p \sum_{1 \le \nu_1 < \nu_2 < \cdots < \nu_p \le n} \sum \cdots \sum f_{\nu_1 n} f_{\nu_2 n} \cdots f_{\nu_p n}$$

$$\to \int\int_{a \le x_1 \le x_2 \le \cdots \le x_p \le b} \cdots \int f(x_1) f(x_2) \cdots f(x_p)\, dx_1\, dx_2 \cdots dx_p$$

$$= \frac{1}{p!} \int_a^b \int_a^b \cdots \int_a^b f(x_1) f(x_2) \cdots f(x_p)\, dx_1\, dx_2 \cdots dx_p.$$

Moreover [I **62**]

$$\prod_{\nu=1}^{n} (1 + z f_{\nu n} \delta_n) \ll (1 + zM\delta_n)^n \ll e^{zM\delta_n n} = e^{zM(b-a)}.$$

This inequality combined with the above proved proposition implies [I **180**]

$$\lim_{n\to\infty} \prod_{\nu=1}^{n} (1 + z f_{\nu n} \delta_n) = 1 + \frac{z}{1!} \int_a^b f(x)\, dx + \frac{z^2}{2!} \left( \int_a^b f(x)\, dx \right)^2 + \cdots$$

$$+ \frac{z^p}{p!} \left( \int_a^b f(x)\, dx \right)^p + \cdots = e^{z \int_a^b f(x) dx}.$$

This new solution of **54** illustrates well the limit operations which lead to Fredholm's solution of integral equations. Conversely **67** can be deduced from **54** with the help of I **179**.

**68.** Multiplication by rows leads to a determinant $P$ of order $m$ with the general term

$$\sum_{\nu=1}^{n} f_{\nu n}^{(\lambda)} \varphi_{\nu n}^{(\mu)} \sim \frac{n}{b-a} \int_a^b f_\lambda(x)\, \varphi_\mu(x)\, dx; \qquad \lambda, \mu = 1, 2, \ldots, m.$$

On the other hand we have ($n \ge m$)

$$P = \sum_{1 \le \nu_1 < \nu_2 < \cdots < \nu_m \le n} \begin{vmatrix} f_{\nu_1 n}^{(1)} & f_{\nu_2 n}^{(1)} & \cdots & f_{\nu_m n}^{(1)} \\ f_{\nu_1 n}^{(2)} & f_{\nu_2 n}^{(2)} & \cdots & f_{\nu_m n}^{(2)} \\ \cdots & \cdots & \cdots & \cdots \\ f_{\nu_1 n}^{(m)} & f_{\nu_2 n}^{(m)} & \cdots & f_{\nu_m n}^{(m)} \end{vmatrix} \cdot \begin{vmatrix} \varphi_{\nu_1 n}^{(1)} & \varphi_{\nu_2 n}^{(1)} & \cdots & \varphi_{\nu_m n}^{(1)} \\ \varphi_{\nu_1 n}^{(2)} & \varphi_{\nu_2 n}^{(2)} & \cdots & \varphi_{\nu_m n}^{(2)} \\ \cdots & \cdots & \cdots & \cdots \\ \varphi_{\nu_1 n}^{(m)} & \varphi_{\nu_2 n}^{(m)} & \cdots & \varphi_{\nu_m n}^{(m)} \end{vmatrix}.$$

If $\nu_1, \nu_2, \ldots, \nu_m$ assume independently all the values $1, 2, \ldots, n$ we obtain $m!\,P$. The sum established in this way is $\sim \left(\frac{n}{b-a}\right)^m$-times the $m$-fold integral exhibited in the problem.

**69.** [For a fuller account of the subject studied in the present Chap. 2 see G. H. Hardy, J. E. Littlewood and G. Pólya: Inequalities.

Cambridge: Cambridge University Press 1952.] The statement follows as a limit relation from the proposition on the arithmetic, geometric and harmonic mean [**48**].

When we pass to the limit ">" becomes "≧", and so our proof is not suitable for specifying the cases in which equality is attained. For an essential remark on such cases see **109**.

**70.** [J. L. W. V. Jensen: Acta Math. Vol. 30, p. 175 (1906).] The proof is analogous to Cauchy's proof for the inequality between the arithmetic, geometric and harmonic means given in the footnote on p. 64 (which deals with the case $\varphi(t) = \log t$). First the statement is proved for $n = 2^k$ ($k$ integer), then it is extended to arbitrary $n$.

**71.** [J. L. W. V. Jensen, l.c. **70**.] Using a similar notation as in **48** we get for each $n$

$$\varphi\left(\frac{f_{1n} + f_{2n} + \cdots + f_{nn}}{n}\right) \leqq \frac{\varphi(f_{1n}) + \varphi(f_{2n}) + \cdots + \varphi(f_{nn})}{n}.$$

Let $n$ increase to infinity and notice **124, 110**.

**72.** Let $t_1, t_2$ be two arbitrary points on $[m, M]$, $t_1 < t_2$. Then

$$\varphi(t_1) = \varphi\left(\frac{t_1 + t_2}{2}\right) + \frac{t_1 - t_2}{2}\varphi'\left(\frac{t_1 + t_2}{2}\right) + \frac{(t_1 - t_2)^2}{8}\varphi''(\tau_1),$$

$$\varphi(t_2) = \varphi\left(\frac{t_1 + t_2}{2}\right) + \frac{t_2 - t_1}{2}\varphi'\left(\frac{t_1 + t_2}{2}\right) + \frac{(t_1 - t_2)^2}{8}\varphi''(\tau_2);$$

where $t_1 < \tau_1 < \frac{t_1 + t_2}{2}$ and $\frac{t_1 + t_2}{2} < \tau_2 < t_2$. Hence

$$\varphi(t_1) + \varphi(t_2) - 2\varphi\left(\frac{t_1 + t_2}{2}\right) > 0,$$

provided $\varphi''(t) > 0$ on $[m, M]$.

**73.** [**72**.]

**74.** [J. L. W. V. Jensen, l.c. **70**.] In the case where the $p_\nu$ are integers the proposition is a consequence of **70** where $p_1$ points coincide with $t_1$, $p_2$ points with $t_2, \ldots, p_n$ points with $t_n$. Then we extend the proposition to rational $p_\nu$; for arbitrary $p_\nu$ we need the continuity of $\varphi(t)$ [**124**].

**75.** [J. L. W. V. Jensen, l.c. **70**.] Introducing

$$f_{\nu n} = f\left(x_1 + \nu\frac{x_2 - x_1}{n}\right), \quad p_{\nu n} = p\left(x_1 + \nu\frac{x_2 - x_1}{n}\right), \quad \nu = 1, 2, \ldots, n,$$

we obtain according to **73**

$$\varphi\left(\frac{p_{1n}f_{1n} + p_{2n}f_{2n} + \cdots + p_{nn}f_{nn}}{p_{1n} + p_{2n} + \cdots + p_{nn}}\right)$$
$$\leqq \frac{p_{1n}\varphi(f_{1n}) + p_{2n}\varphi(f_{2n}) + \cdots + p_{nn}\varphi(f_{nn})}{p_{1n} + p_{2n} + \cdots + p_{nn}}.$$

Let $n$ increase to infinity.

**76.** [O. Hölder: Nachr. Akad. Wiss. Göttingen 1889, p. 38.] We put

$$\frac{p_1t_1 + p_2t_2 + \cdots + p_nt_n}{p_1 + p_2 + \cdots + p_n} = M,$$

then

$$\varphi(t_\nu) = \varphi(M) + (t_\nu - M)\,\varphi'(M) + \frac{(t_\nu - M)^2}{2}\varphi''(\tau_\nu),$$

thus

$$\frac{p_1\varphi(t_1) + p_2\varphi(t_2) + \cdots + p_n\varphi(t_n)}{p_1 + p_2 + \cdots + p_n} = \varphi(M) + \frac{\sum_{\nu=1}^{n} p_\nu \frac{(t_\nu - M)^2}{2}\varphi''(\tau_\nu)}{\sum_{\nu=1}^{n} p_\nu} > \varphi(M),$$

provided that at least one of the $t_k$'s is different from $M$.

**77.** [Cf. G. Pólya, Problem: Arch. Math. Phys. Ser. 3, Vol. 21, pp. 370–371 (1913).] Analogous to **76**.

**78.** In **76** set $\varphi(t) = -\log t$, and $t \log t$ resp., $M > m > 0$; furthermore replace $a_\nu$ by $\frac{1}{a_\nu}$, $\nu = 1, 2, \ldots, n$.

**79.** In **77** put: $\varphi(t) = -\log t$, and $t\log t$ resp., $M > m > 0$; then replace $f(x)$ by $\frac{1}{f(x)}$.

**80.** First proof:

$$\sum_{\nu=1}^{n} a_\nu^2 \sum_{\nu=1}^{n} b_\nu^2 - \left(\sum_{\nu=1}^{n} a_\nu b_\nu\right)^2 = \sum_{i<k}^{1,2,\ldots,n} (a_ib_k - a_kb_i)^2 \geqq 0.$$

Second proof: If $\lambda$ and $\mu$ denote real variables the quadratic form

$$(\lambda a_1 + \mu b_1)^2 + (\lambda a_2 + \mu b_2)^2 + \cdots + (\lambda a_n + \mu b_n)^2$$
$$= A\lambda^2 + 2B\lambda\mu + C\mu^2 \geqq 0.$$

Provided that $\lambda^2 + \mu^2 > 0$ the case of equality presents itself only then when there is a particular set of $\lambda$, $\mu$ for which $\lambda a_\nu + \mu b_\nu = 0$, $\nu = 1, 2, \ldots, n$. Therefore $AC - B^2$ is positive or 0 as asserted.

**81.** By taking the limit in **80**: Writing $f_{\nu n} = f\left(x_1 + \nu\frac{x_2 - x_1}{n}\right)$, $g_{\nu n} = g\left(x_1 + \nu\frac{x_2 - x_1}{n}\right)$, we obtain [**80**]

$$\left(\frac{f_{1n}g_{1n} + f_{2n}g_{2n} + \cdots + f_{nn}g_{nn}}{n}\right)^2 \leqq \frac{f_{1n}^2 + f_{2n}^2 + \cdots + f_{nn}^2}{n} \cdot \frac{g_{1n}^2 + g_{2n}^2 + \cdots + g_{nn}^2}{n}.$$

Let $n$ increase to infinity. It is also possible to adapt both methods used in **80** to the present problem; as to the first method cf. **68**.

**81.1.** See **81.3**.

**81.2.** See **81.4**.

**81.3.** Let $\Sigma$ stand for $\sum_{\nu=1}^{n}$, set

$$\Sigma a_\nu = A, \quad \Sigma b_\nu = B, \ldots, \quad \Sigma l_\nu = L,$$

and use **78**:

$$\begin{aligned}(A^\alpha B^\beta \cdots L^\lambda)^{-1} \sum a_\nu^\alpha b_\nu^\beta \cdots l_\nu^\lambda &= \sum \left(\frac{a_\nu}{A}\right)^\alpha \left(\frac{b_\nu}{B}\right)^\beta \cdots \left(\frac{l_\nu}{L}\right)^\lambda \\ &\leqq \sum \left(\alpha \frac{a_\nu}{A} + \beta \frac{b_\nu}{B} + \cdots + \lambda \frac{l_\nu}{L}\right) \\ &= \alpha + \beta + \cdots + \lambda = 1.\end{aligned}$$

**81.4.** From **81.3** by a passage to the limit or by analogy. Analogy may be better: It may allow us to discuss the case of equality.

**82.** Let $t \neq 0$ and introduce $a_\nu^t = A_\nu$, $\nu = 1, 2, \ldots, n$. Because of **78** we have

$$\begin{aligned}t^2 \frac{\varphi'(t)}{\varphi(t)} = &\frac{A_1 \log A_1 + A_2 \log A_2 + \cdots + A_n \log A_n}{A_1 + A_2 + \cdots + A_n} \\ &- \log \frac{A_1 + A_2 + \cdots + A_n}{n} > 0.\end{aligned}$$

We find

$$\varphi(-\infty) = \min(a), \quad \varphi(-1) = \mathfrak{H}(a), \quad \varphi(0) = \mathfrak{G}(a), \quad \varphi(1) = \mathfrak{A}(a),$$
$$\varphi(+\infty) = \max(a).$$

Thus we have a new proof of the proposition on the relation between the arithmetic, geometric and harmonic means.

**83.** Assume $t \neq 0$ and set $[f(x)]^t = F(x)$. Proposition **79** implies

$$t^2 \frac{\Psi'(t)}{\Psi(t)} = \frac{\int_{x_1}^{x_2} F(x) \log F(x)\, dx}{\int_{x_1}^{x_2} F(x)\, dx} - \log \left(\frac{1}{x_2 - x_1} \int_{x_1}^{x_2} F(x)\, dx\right) \geqq 0.$$

(Or taking the limit in **82**.)

$$\Psi(-1) = \mathfrak{H}(f), \quad \Psi(0) = \mathfrak{G}(f), \quad \Psi(1) = \mathfrak{A}(f).$$

Let $M$ denote the maximum of $f(x)$ on $[x_1, x_2]$ and $\delta$ the length of a subinterval of $[x_1, x_2]$ in which $f(x) > M - \varepsilon$. Then we have for $t > 0$

$$(M - \varepsilon) \left(\frac{\delta}{x_2 - x_1}\right)^{\frac{1}{t}} \leqq \Psi(t) \leqq M,$$

i.e. $\Psi(\infty) = \lim_{t\to\infty} \Psi(t) = M$; $\Psi(-\infty)$ is found in a similar way. This proposition contains therefore a new proof for **69**.

**84.** [H. Minkowski; cf. e.g. Hardy, Littlewood and Pólya, l.c. **69**, p. 21.] First proof: We assume $0 \leqq t \leqq 1$ and define

$\varphi(t) = \prod_{\nu=1}^{n} [ta_\nu + (1-t)\, b_\nu]^{\frac{1}{n}}$. Then [**80**]

$$\frac{\varphi''(t)}{\varphi(t)} = \frac{1}{n^2}\left(\sum_{\nu=1}^{n} \frac{a_\nu - b_\nu}{ta_\nu + (1-t)\, b_\nu}\right)^2 - \frac{1}{n}\sum_{\nu=1}^{n}\left(\frac{a_\nu - b_\nu}{ta_\nu + (1-t)\, b_\nu}\right)^2 < 0,$$

unless $a_\nu = \lambda b_\nu$, $\nu = 1, 2, \ldots, n$.

Second proof: With $\log b_\nu - \log a_\nu = t_\nu$ the inequality becomes

$$\frac{\log(1+e^{t_1}) + \log(1+e^{t_2}) + \cdots + \log(1+e^{t_n})}{n} \geqq \log\left(1 + e^{\frac{t_1+t_2+\cdots+t_n}{n}}\right).$$

$\log(1+e^t)$, however, is convex [**73**].

Third proof: Particular case of **81.3**:

$$n = 2, \quad \alpha = \beta = \cdots = \lambda.$$

Also particular case of **90**, $k = 0$.

**85.** [Cf. W. Blaschke: Arch. Math. Phys. Ser. 3, Vol. 24, p. 281 (1916).] We define

$$\varphi(t) = e^{\frac{1}{x_2-x_1}\int_{x_1}^{x_2} \log[tf(x)+(1-t)g(x)]dx}, \qquad 0 \leqq t \leqq 1;$$

Schwarz's inequality implies [**81**]

$$\frac{\varphi''(t)}{\varphi(t)} = \left(\frac{1}{x_2-x_1}\int_{x_1}^{x_2} \frac{f(x)-g(x)}{tf(x)+(1-t)\, g(x)}\, dx\right)^2$$
$$- \frac{1}{x_2-x_1}\int_{x_1}^{x_2}\left(\frac{f(x)-g(x)}{tf(x)+(1-t)\, g(x)}\right)^2 dx \leqq 0.$$

(Or take the limit in **84**.) Particular case of **91**.

**86.** By repeated application of **85** to the functions

$$p_1 f_1(x),\ p_2 f_2(x),\ \ldots,\ p_m f_m(x).$$

**87.** By definition

$$l_k = \text{least upper bound of } \sum_{\nu=1}^{n} \sqrt{(x^{(\nu)} - x^{(\nu-1)})^2 + [f_k(x^{(\nu)}) - f_k(x^{(\nu-1)})]^2}$$

for all possible subdivisions of the interval $[x_1, x_2]$

$$(x_1 = x^{(0)} < x^{(1)} < x^{(2)} < \cdots < x^{(n)} = x_2).$$

Since $\sqrt{c^2+t^2}$ is convex [**73**] we get for an arbitrary subdivision

$$\frac{p_1l_1+p_2l_2+\cdots+p_ml_m}{p_1+p_2+\cdots+p_m} \geqq \sum_{\nu=1}^{n} \frac{\sum_{k=1}^{m} p_k \sqrt{(x^{(\nu)}-x^{(\nu-1)})^2+[f_k(x^{(\nu)})-f_k(x^{(\nu-1)})]^2}}{\sum_{k=1}^{m} p_k}$$

$$\geqq \sum_{\nu=1}^{n} \sqrt{(x^{(\nu)}-x^{(\nu-1)})^2+[F(x^{(\nu)})-F(x^{(\nu-1)})]^2}\,.$$

**88.** Put

$$p_\nu^{(n)} = \int_{(\nu-1)\frac{2\pi}{n}}^{\nu\frac{2\pi}{n}} p(\xi)\,d\xi,\ \nu=1,2,\ldots,n;\quad F_n(x) = \frac{\sum_{\nu=1}^{n} p_\nu^{(n)} f\left((\nu-1)\frac{2\pi}{n}+x\right)}{\sum_{\nu=1}^{n} p_\nu^{(n)}}\,.$$

Then the sequence $F_n(x)$ converges to $F(x)$,

$$\lim_{n\to\infty} F_n(x) = F(x)\,,$$

uniformly in $x$, $0 \leqq x \leqq 2\pi$, because

$$\left|\sum_{\nu=1}^{n} \int_{(\nu-1)\frac{2\pi}{n}}^{\nu\frac{2\pi}{n}} p(\xi)\left[f(\xi+x)-f\left((\nu-1)\frac{2\pi}{n}+x\right)\right]d\xi\right| < \omega_n \int_0^{2\pi} p(\xi)\,d\xi\,,$$

where $\omega_n$ denotes the maximum of the oscillation of $f(x)$ in an interval of length $\frac{2\pi}{n}$ [p.77]. Hence $\mathfrak{G}(F) = \lim_{n\to\infty} \mathfrak{G}(F_n)$; because of **86** $\mathfrak{G}(F_n) \geqq \mathfrak{G}(f)$.

**89.** Using the same notation as in **88** we establish that $\lim_{n\to\infty} F_n(x) = F(x)$ uniformly in $x$, $0 \leqq x \leqq 2\pi$:

$$\left|\sum_{\nu=1}^{n} \int_{(\nu-1)\frac{2\pi}{n}}^{\nu\frac{2\pi}{n}} p(\xi)\left[f(\xi+x)-f\left((\nu-1)\frac{2\pi}{n}+x\right)\right]d\xi\right|$$

$$\leqq \max(p) \int_0^{\frac{2\pi}{n}} \left[\sum_{\nu=1}^{n} \left|f\left((\nu-1)\frac{2\pi}{n}+\xi+x\right)-f\left((\nu-1)\frac{2\pi}{n}+x\right)\right|\right] d\xi.$$

Thus

$$|F_n(x)-F(x)| \leqq \frac{2\pi}{n} \frac{\max(p)}{\int_0^{2\pi} p(\xi)\,d\xi} V\,,$$

where $V$ denotes the total variation of $f(x)$ on $[0, 2\pi]$. Since the length $l$ of the arc of $f\left((\nu-1)\frac{2\pi}{n}+x\right)$, $0 \leqq x \leqq 2\pi$, is the same for each $\nu$, the length $L_n$ of the arc of $F_n(x)$ can be estimated by $L_n \leqq l$. Besides, the limit relation

$$\sum_{\alpha=1}^{s} \sqrt{(x^{(\alpha)} - x^{(\alpha-1)})^2 + [F(x^{(\alpha)}) - F(x^{(\alpha-1)})]^2}$$

$$= \lim_{n\to\infty} \sum_{\alpha=1}^{s} \sqrt{(x^{(\alpha)} - x^{(\alpha-1)})^2 + [F_n(x^{(\alpha)}) - F_n(x^{(\alpha-1)})]^2} \leqq l$$

holds for any arbitrary subdivision of the interval $[0, 2\pi]$,

$$0 = x^{(0)} < x^{(1)} < \cdots < x^{(s-1)} < x^{(s)} = 2\pi.$$

A particularly interesting special case of this problem is due to F. Lukács: The arc lengths of Fejér's means of the Fourier series of $f(x)$ [**134**] cannot be larger than the length of the arc of $y = f(x)$ over the interval $[0, 2\pi]$. (The "jumps" have to be included in the length of the arc, also $|f(+0) - f(2\pi - 0)|$.)

**90.** Assume $\varkappa \neq 0$ [**84**]. Put $A_\nu = ta_\nu + (1-t)\,b_\nu$, $0 \leqq t \leqq 1$, $\nu = 1, 2, \ldots, n$ and $\varphi(t) = \mathfrak{M}_\varkappa(A)$. The second derivative is

$$\varphi''(t) = (\varkappa - 1)\,(A_1^\varkappa + A_2^\varkappa + \cdots + A_n^\varkappa)^{\frac{1}{\varkappa}-2} \cdot$$
$$\{(A_1^\varkappa + A_2^\varkappa + \cdots + A_n^\varkappa)\,[(a_1 - b_1)^2 A_1^{\varkappa-2} + (a_2 - b_2)^2 A_2^{\varkappa-2}$$
$$+ \cdots + (a_n - b_n)^2 A_n^{\varkappa-2}]$$
$$- ((a_1 - b_1)\,A_1^{\varkappa-1} + (a_2 - b_2)\,A_2^{\varkappa-1} + \cdots + (a_n - b_n)\,A_n^{\varkappa-1})^2\}.$$

The quantity in curly brackets is always positive unless $a_\nu = \lambda b_\nu$ [**80**]. Thus $\operatorname{sgn} \varphi''(t) = \operatorname{sgn}(\varkappa - 1)$. Therefore we have

$$2\varphi\left(\frac{1}{2}\right) \leqq \text{ or } \geqq \varphi(0) + \varphi(1)$$

according as $\varkappa \geqq 1$ or $\varkappa \leqq 1$. If $\varkappa = 2$, $\mathfrak{M}_\varkappa$ represents the distance of the point $a_1, a_2, \ldots, a_n$ in the $n$-dimensional space $R^n$ from the origin. The proposition states in this case that one side of a triangle is shorter than the sum of the other two sides.

**91.** In **90** put $a_\nu = f\left(x_1 + \nu \frac{x_2 - x_1}{n}\right)$, $\quad b_\nu = g\left(x_1 + \nu \frac{x_2 - x_1}{n}\right)$, $\nu = 1, 2, \ldots, n$ and let $n$ become infinite.

**92.** [For the special case $a_\nu b_\nu = 1$, $\nu = 1, 2, \ldots, n$, cf. P. Schweitzer: Mat. phys. lap. Vol. 23, pp. 257–261 (1914).] We rearrange the numbers $a_\nu$ so that $a_1 \leqq a_2 \leqq \cdots \leqq a_n$. To determine the maximum it is then sufficient to consider values $b_1 \geqq b_2 \geqq \cdots \geqq b_n$. (If $b_\nu < b_\mu$, $\nu < \mu$, we interchange $b_\nu$ and $b_\mu$: $b_\nu^2 + b_\mu^2 = b_\mu^2 + b_\nu^2$ and $a_\mu b_\mu + a_\nu b_\nu \geqq a_\mu b_\nu + a_\nu b_\mu$.)

We may also assume that not all the $a_\nu$'s are equal, nor all the $b_\nu$'s, i.e.

$$a_n b_1 - a_1 b_n = (a_n - a_1)b_1 + a_1(b_1 - b_n) > 0.$$

If $n > 2$ the numbers $u_1, u_2, \ldots, u_{n-1}, v_1, v_2, \ldots, v_{n-1}$ are defined by the equations

$$a_\nu^2 = u_\nu a_1^2 + v_\nu a_n^2, \quad b_\nu^2 = u_\nu b_1^2 + v_\nu b_n^2, \quad \nu = 2, 3, \ldots, n-1.$$

We find $u_\nu \geqq 0$, $v_\nu \geqq 0$ and $a_\nu b_\nu \geqq u_\nu a_1 b_1 + v_\nu a_n b_n$ [**80**]; $u_\nu = 0$ if and only if $a_\nu = a_{\nu+1} = \cdots = a_n$, $b_\nu = b_{\nu+1} = \cdots = b_n$ and $v_\nu = 1$. In a similar way $v_\nu = 0$ implies $u_\nu = 1$, etc. If $u_\nu > 0$, $v_\nu > 0$ then $a_\nu b_\nu > u_\nu a_1 b_1 + v_\nu a_n b_n$. Thus the expression in question is

$$\leqq \frac{(pa_1^2 + qa_n^2)(pb_1^2 + qb_n^2)}{(pa_1b_1 + qa_nb_n)^2},$$

where $1 + u_2 + u_3 + \cdots + u_{n-1} = p$, $v_2 + v_3 + \cdots + v_{n-1} + 1 = q$. The inequality becomes an equality if and only if the $u_\nu$'s and $v_\nu$'s are 0 or 1, $p$, $q$, are integers, $a_1 = a_2 = \cdots = a_p$, $a_{p+1} = a_{p+2} = \cdots = a_n$, $b_1 = b_2 = \cdots = b_p$, $b_{p+1} = b_{p+2} = \cdots = b_n$. The last expression is

$$= 1 + pq\left(\frac{a_nb_1 - a_1b_n}{pa_1b_1 + qa_nb_n}\right)^2 \leqq 1 + pq\left(\frac{a_nb_1 - a_1b_n}{2\sqrt{pa_1b_1qa_nb_n}}\right)^2$$

$$= \left(\frac{\sqrt{\dfrac{a_nb_1}{a_1b_n}} + \sqrt{\dfrac{a_1b_n}{a_nb_1}}}{2}\right)^2;$$

it is an equality if and only if $pa_1b_1 = qa_nb_n$. If we replace $a_1, a_n, b_1, b_n$ by $a, A, B, b$, in the term on the right hand side it does not decrease.

**93.** [For the special case $a = b$, $A = B$ see J. Kürschák: Mat. phys. lap. Vol. 23, p. 378 (1914).] In **92** define

$$a_\nu = f\left(x_1 + \nu\frac{x_2 - x_1}{n}\right), \quad b_\nu = g\left(x_1 + \nu\frac{x_2 - x_1}{n}\right), \quad \nu = 1, 2, \ldots, n.$$

Then let $n$ increase to infinity.

**93.1.** Define

$$\Sigma a_\nu^r = R.$$

Then

$$\frac{(\Sigma a_\nu^s)^{\frac{1}{s}}}{(\Sigma a_\nu^r)^{\frac{1}{r}}} = \frac{(\Sigma a_\nu^s)^{\frac{1}{s}}}{R^{\frac{1}{r}}} = \left[\Sigma\left(\frac{a_\nu^r}{R}\right)^{\frac{s}{r}}\right]^{\frac{1}{s}}$$

$$< \left(\Sigma\frac{a_\nu^r}{R}\right)^{\frac{1}{s}} = 1.$$

Whereas several foregoing problems of this chapter were arranged in pairs, each problem about sums followed by a companion problem about integrals, the present problem has no such companion. (Why is that so? Whereas in former cases, e.g. **81.3**, multiplying each $\Sigma$ by $\frac{1}{n}$ leaves the relation in question unchanged, this is not so in the present case.)

**94.** [G. Pólya, Problem: Arch. Math. Phys. Ser. 3, Vol. 28, p. 174 (1920).] Assume that $f(x)$ is not constant. We show that the quadratic form

$$Q(x, y) = \int_0^1 f(t)\,[(2a + 1)\,t^{2a}x^2 + 2(a + b + 1)\,t^{a+b}xy + (2b + 1)\,t^{2b}y^2]\,dt$$
$$= Ax^2 + 2Bxy + Cy^2$$

is indefinite, i.e. $AC - B^2 < 0$. Integration by parts leads to

$$\int_0^1 t^k f(t)\,dt = \left(\frac{t^{k+1}}{k+1} f(t)\right)_0^1 - \int_0^1 \frac{t^{k+1}}{k+1}\,df(t) = \frac{f(1)}{k+1} - \int_0^1 \frac{t^{k+1}}{k+1}\,df(t), \quad k > 0,$$

provided that $f(1) = \lim_{t \to 1-0} f(t)$ is finite. Thus

$$Q(x, y) = f(1)\,(x + y)^2 - \int_0^1 (t^a x + t^b y)^2\,t\,df(t).$$

We have $Q(1, 1) > 0$, $Q(1, -1) < 0$. If $f(1) = \infty$ we can also establish by careful manipulation [**112**] that $Q(1, -1) < 0$.

**94.1.** Assume that the solid is a polyhedron and let $p_\nu$, $q_\nu$ and $r_\nu$ stand for the areas of the orthogonal projections of one of its faces onto the three coordinate planes, respectively. Extending the sums over all faces we have

$$S = \Sigma(p_\nu^2 + q_\nu^2 + r_\nu^2)^{\frac{1}{2}},$$
$$S \leqq \Sigma(p_\nu + q_\nu + r_\nu) = 2(P + Q + R),$$
$$S \geqq [(\Sigma p_\nu)^2 + (\Sigma q_\nu)^2 + (\Sigma r_\nu)^2]^{\frac{1}{2}} = 2[P^2 + Q^2 + R^2]^{\frac{1}{2}},$$

see **93.1** and **90**. A cube attains the upper bound for $S$ and a regular octahedron the lower bound provided that the edges of the former and the diagonals of the latter are parallel to the coordinate axes; and these bounds remain attained even if we apply to these solids arbitrary dilatations parallel to the axes.

If the solid has a smooth (differentiable) surface, integrals should be used instead of sums.

**94.2.** By symmetry

$$\iint \xi^2\,d\omega = \iint \eta^2\,d\omega = \iint \zeta^2\,d\omega = \frac{1}{3}\iint (\xi^2+\eta^2+\zeta^2)\,d\omega = \frac{4\pi}{3}.$$

(1) By **80**

$$E = \iint (b^2c^2\xi^2 + c^2a^2\eta^2 + a^2b^2\zeta^2)^{\frac{1}{2}}(\xi^2+\eta^2+\zeta^2)^{\frac{1}{2}}\,d\omega$$
$$\geqq \iint (bc\xi^2 + ca\eta^2 + ab\zeta^2)\,d\omega = \frac{4\pi}{3}(bc+ca+ab).$$

(2) We may, and shall, assume that $c$ is the shortest semiaxis.

$$E = ab\iint\left[1 - \frac{a^2-c^2}{a^2}\xi^2 - \frac{b^2-c^2}{b^2}\eta^2\right]^{\frac{1}{2}} d\omega$$
$$\leqq ab\iint\left[1 - \tfrac{1}{2}\left(\frac{a^2-c^2}{a^2}\xi^2 + \frac{b^2-c^2}{b^2}\eta^2\right)\right] d\omega$$
$$\leqq \frac{4\pi}{3}(a^2+b^2+c^2).$$

To obtain the last line work backwards and use that $c \leqq a$, $c \leqq b$.

(3) The more elementary **94.1** yields better estimates when $c$ is near 0. For other approximations to the area of the ellipsoid see G. Pólya: Publicaciones del Instituto de Mat. Rosario Vol. 5, pp. 51—61 (1943).

**94.3.** Lower bound in **94.2** and the theorem of the means.

**95.** [G. Pólya, Problem: Arch. Math. Phys. Ser. 3, Vol. 26, p. 65 (1917). Solved by G. Szegö: Arch. Math. Phys. Ser. 3, Vol. 28, pp. 81—82 (1920).] [For a generalization see G. Pólya and M. Schiffer: Journal d'Analyse math. Vol. 3, p. 323 (1953/54).]

**95.1.** Consider the polynomial in $u$

$$w = (a^2+u)(b^2+u)(c^2+u).$$

By the theorem of the means

$$w^{-\frac{1}{3}} \leqq \frac{1}{3}\left(\frac{1}{a^2+u} + \frac{1}{b^2+u} + \frac{1}{c^2+u}\right) = \frac{1}{3w}\frac{dw}{du}$$

and so

$$\frac{1}{2}\int_0^\infty w^{-\frac{1}{2}}\,du \leqq \frac{1}{2}\int_0^\infty w^{-\frac{1}{6}}\frac{w'}{3w}\,du = \left[-w^{-\frac{1}{6}}\right]_{u=0}^{\infty} = (abc)^{-\frac{1}{3}}.$$

A comparison with **94.3** may suggest generalizations. Cf. G. Pólya and G. Szegö: Isoperimetric Inequalities in Mathematical Physics. Princeton: Princeton University Press 1951.

**95.2.** [E. Laguerre: Oeuvres, Vol. 1. Paris: Gauthier-Villars 1898, p. 93.] Let the roots be $x, x_1, x_2, \ldots, x_{n-1}$. Then, by **80**,

$$(a_1 - x)^2 = (x_1 + x_2 + \cdots + x_{n-1})^2 \leqq (n-1)(x_1^2 + x_2^2 + \cdots + x_{n-1}^2)$$
$$= (n-1)(a_1^2 - 2a_2 - x^2).$$

The bounds proposed are the two roots of the quadratic equation that we obtain hence by considering the case of equality.

**95.3.** [See G. Pólya: Numerische Math. Vol. 11, pp. 315—319 (1968).] By **93.1**

$$s_n^{\frac{1}{n}} = (\Sigma \gamma_\nu^{-n})^{\frac{1}{n}} > (\Sigma \gamma_\nu^{-n-1})^{\frac{1}{n+1}} = s_{n+1}^{\frac{1}{n+1}}.$$

By **80**

$$s_n^2 = (\Sigma \gamma_\nu^{-(n-1)/2} \gamma_\nu^{-(n+1)/2})^2$$
$$\leqq \Sigma \gamma_\nu^{-n+1} \, \Sigma \gamma_\nu^{-n-1} = s_{n-1} s_{n+1}.$$

Obviously

$$s_{n+1} = \Sigma \gamma_\nu^{-n} \gamma_\nu^{-1} < \gamma^{-1} \Sigma \gamma_\nu^{-n} = \gamma^{-1} s_n.$$

The rest is more obvious.

**95.4.** From **95.3**

$$\frac{s_3}{s_4} \frac{s_2}{s_3} < \left(\frac{s_1}{s_2}\right)^2, \; \left(\frac{s_2}{s_4}\right)^{1/2} < \frac{s_1}{s_2}; \; \frac{s_7}{s_8} \frac{s_6}{s_7} \frac{s_5}{s_6} \frac{s_4}{s_5} < \left(\frac{s_3}{s_4} \frac{s_2}{s_3}\right)^2, \; \left(\frac{s_4}{s_8}\right)^{1/4} < \left(\frac{s_2}{s_4}\right)^{1/2};$$

and so on. Also directly from **81.1**

$$s_{2n} = \sum (\gamma_\nu^{-n})^{2/3} (\gamma_\nu^{-4n})^{1/3} < s_n^{2/3} \cdot s_{4n}^{1/3}, \qquad \left(\frac{s_{2n}}{s_{4n}}\right)^{1/2n} < \left(\frac{s_n}{s_{2n}}\right)^{1/n}.$$

Obviously

$$s_{2n} = \Sigma \gamma_\nu^{-n} \gamma_\nu^{-n} < \gamma^{-n} s_n.$$

The result is useful in certain applications of Graeffe's method, see l.c. **95.3**.

**95.5.** [See Hardy, Littlewood and Pólya, l.c. **69**, p. 163, theor. 218.]

$$2A = \int_0^{2\pi} r^2 \, d\varphi.$$

By **81.2**

$$2\pi = \int_0^{2\pi} \left(\frac{1}{r}\right)^{2/3} (r^2)^{1/3} \, d\varphi \leqq \left(\int_0^{2\pi} \frac{d\varphi}{r}\right)^{2/3} \left(\int_0^{2\pi} r^2 d\varphi\right)^{1/3} = F^{2/3} (2A)^{1/3}.$$

The case of equality, and so the minimum of $F$, is attained when $r$ is constant.

**96.** [Cf. I. Schur: Sber. Berlin. Math. Ges. Vol. 22, pp. 16—17 (1923).] Suppose that $x_1, x_2, \ldots, x_n$ are positive. Since $\log x$ is concave on any positive interval

$$\log y_\mu \geqq a_{\mu 1} \log x_1 + a_{\mu 2} \log x_2 + \cdots + a_{\mu n} \log x_n, \quad \mu = 1, 2, \ldots, n.$$

Add these inequalities.

**97.** [G. Pólya, Problem: Arch. Math. Phys. Ser. 3, Vol. 20, p. 272 (1913). Solved by G. Szegö: Arch. Math. Phys. Ser. 3, Vol. 22, pp. 361—362 (1914).]

**98.** [Cf. E. Steinitz, Problem: Arch. Math. Phys. Ser. 3, Vol. 19, p. 361 (1912). Solved by G. Pólya: Arch. Math. Phys. Ser. 3, Vol. 21, p. 290 (1913).] If $x$ is an integer $g(x) = 0$; if $x$ is not an integer $g(x) = 1$. If $x$ is rational $G(x) = 0$; if $x$ is irrational $G(x) = 1$. Any lower sum of $G(x)$ is 0, any upper sum over the interval $[a, b]$ is $b - a$. The function $G(x)$ is integrable over no interval.

**99.** If $x$ is irrational $f(x) = 0$ and if $h$ converges to 0, $x + h$ is either irrational and so $f(x + h) = 0$, or rational, $x + h = \frac{p}{q}$, and so $f(x + h) = \frac{1}{q}$; $\frac{1}{q}$ converges to 0 as $h \to 0$. If $x = \frac{p}{q}$, rational, $f(x) = \frac{1}{q}$, and $x + h$ irrational, $f(x + h) - f(x) = -\frac{1}{q}$. Any lower sum is 0. We now divide the interval $[0, 1]$ into $k^3$ equal parts. Since there are at most $1 + 2 + \cdots + (k - 1) = \frac{k(k-1)}{2}$ positive proper fractions with denominator $\leqq k$, the upper sum is $< \frac{k(k-1)}{2} \frac{1}{k^3} + \frac{2}{k^3} + \frac{1}{k} \cdot 1$.

**100.** If we call $\Omega_\nu$ the oscillation of $f(x)$ in the interval $[x_{\nu-1}, x_\nu]$ and if $|\varphi(x)| < M$ we get

$$\left| \sum_{\nu=1}^{n} f(y_\nu) \varphi(\eta_\nu) (x_\nu - x_{\nu-1}) - \sum_{\nu=1}^{n} f(\eta_\nu) \varphi(\eta_\nu) (x_\nu - x_{\nu-1}) \right| < M \sum_{\nu=1}^{n} \Omega_\nu (x_\nu - x_{\nu-1}).$$

The last sum converges to 0.

**101.** Let $n$ denote a positive integer, $\delta < \frac{b-a}{n}$ and $\Omega_\nu$ the oscillation of $\varphi(x)$ on the interval $\left[a + (\nu - 1)\frac{b-a}{n}, a + \nu \frac{b-a}{n}\right]$, $\nu = 1, 2, \ldots, n$. Suppose that $\varphi(x) < M$. Then

$$\int_a^b |\varphi(x + \delta) - \varphi(x)| \, dx < \frac{b-a}{n} \sum_{\nu=1}^{n-1} (\Omega_\nu + \Omega_{\nu+1}) + 2M \frac{b-a}{n}.$$

**102.** Construct the lower sum $L$ and the upper sum $U$ that belong to the subdivision $a = x_0 < x_1 < \cdots < x_n = b$ (as described on p. 46):

$$U = \sum_{\nu=1}^{n} M_\nu(x_\nu - x_{\nu-1}), \qquad L = \sum_{\nu=1}^{n} m_\nu(x_\nu - x_{\nu-1}).$$

By a proper choice of the subdivision we can attain that $U - L < \varepsilon$. Now we define $\Psi(x)$ as follows: $\Psi(x) = M_\nu$ on $[x_{\nu-1}, x_\nu)$, $\nu = 1, 2, \ldots, n-1$, $\Psi(x) = M_n$ on $[x_{n-1}, x_n]$. We define $\psi(x)$ similarly using $m_\nu$. Then

$$\int_a^b \Psi(x)\,dx = U, \qquad \int_a^b \psi(x)\,dx = L.$$

The only condition imposed on the subdividing points is that the maximal length of the subintervals $[x_{\nu-1}, x_\nu]$, $\nu = 1, 2, \ldots, n$, converges to 0 as $n$ increases. Therefore these points can be chosen equidistant, forming an arithmetic progression. The functions $\Psi(x)$ and $\psi(x)$ constructed in the described way are continuous on the right; they could be defined continuous on the left instead.

**103.** Define $\Psi(x)$ and $\psi(x)$ as in solution **102**. Then the total variation of $\Psi(x)$ is

$$|M_2 - M_1| + |M_3 - M_2| + \cdots + |M_n - M_{n-1}|$$

and that of $\psi(x)$ is

$$|m_2 - m_1| + |m_3 - m_2| + \cdots + |m_n - m_{n-1}|.$$

Both are not larger than the total variation of $f(x)$ because $f(x)$ assumes on $[x_{\nu-1}, x_\nu]$ values which are arbitrarily close to $M_\nu$ and $m_\nu$.

**104.** Let $\nu$ be an integer, $\nu = 1, 2, \ldots, n$; in the first half of the interval $\left[\frac{\nu-1}{n}, \frac{\nu}{n}\right]$ we have $s(nx) = +1$, in the second half $s(nx) = -1$. Thus

$$\int_0^1 f(x)\,s(nx)\,dx = \int_0^{\frac{1}{2n}} \sum_{\nu=1}^{n} \left\{f\left(\frac{\nu-1}{n} + y\right) - f\left(\frac{\nu-1}{n} + y + \frac{1}{2n}\right)\right\} dy.$$

The absolute value of the expression between the curly brackets is smaller than the oscillation of $f(x)$ on $\left[\frac{\nu-1}{n}, \frac{\nu}{n}\right]$.

**105.** [Riemann: Werke. Leipzig: B. G. Teubner 1876, p. 240; E. W. Hobson: The Theory of Functions of a Real Variable & The Theory of Fourier's Series, 2nd Ed., Vol. II. New York: Dover Publication 1957, pp. 514—515.] We may choose $a = 0$, $b = 2\pi$.

$$\int_0^{2\pi} f(x) \sin nx\, dx$$

$$= \int_0^{\frac{\pi}{n}} \sin ny \sum_{\nu=1}^{n} \left\{ f\left((\nu-1)\frac{2\pi}{n} + y\right) - f\left((\nu-1)\frac{2\pi}{n} + y + \frac{\pi}{n}\right)\right\} dy.$$

The absolute value of the expression in the curly brackets is smaller than the oscillation of $f(x)$ on $\left[(\nu-1)\frac{2\pi}{n}, \nu\frac{2\pi}{n}\right]$.

**106.** [L. Fejér: J. reine angew. Math. Vol. 138, p. 27 (1910).] We may choose $a = 0$, $b = 2\pi$.

$$\int_0^{2\pi} f(x)\,|\sin nx|\,dx = \sum_{\nu=1}^{n} \int_{(\nu-1)\frac{2\pi}{n}}^{\nu\frac{2\pi}{n}} f(x)\,|\sin nx|\,dx = \sum_{\nu=1}^{n} f_{\nu n} \int_{(\nu-1)\frac{2\pi}{n}}^{\nu\frac{2\pi}{n}} |\sin nx|\,dx,$$

where $f_{\nu n}$ denotes a value between the least upper and the greatest lower bound of $f(x)$ on $\left[(\nu-1)\frac{2\pi}{n}, \nu\frac{2\pi}{n}\right]$.

**107.** If the points of discontinuity of the bounded function $f(x)$ have only finitely many accumulation points on $[a, b]$ it is possible to find finitely many intervals of arbitrary small length outside of which $f(x)$ is continuous. Such a function is therefore integrable. The function in question has the points of discontinuity $x = \frac{1}{2}, \frac{1}{3}, \ldots, \frac{1}{n}, \ldots$; and also $x = 0$ if $\alpha = 0$.

**108.** According to Riemann's criterion it is possible to find in an arbitrary subinterval $[a_0, b_0]$ of $[a, b]$ a smaller subinterval $[a_1, b_1]$, $b_1 - a_1 < \frac{b_0 - a_0}{2}$, in which the oscillation of $f(x)$ is smaller than $\frac{1}{2}$. Iterating this procedure we obtain a sequence of intervals $[a_n, b_n]$, $n = 1, 2, \ldots$, $b_n - a_n < \frac{b_0 - a_0}{2^n}$, on which the oscillation of $f(x)$ is smaller than $\frac{1}{2^n}$. The point $\alpha = \lim_{n\to\infty} a_n = \lim_{n\to\infty} b_n$ lies in $[a_0, b_0]$ and $f(x)$ is continuous at $\alpha$.

**109.** Assume that $f(\xi) = 0$ at each point $\xi$ of continuity of $f(x)$, $a < \xi < b$.

$$\int_a^b f(x)^2\,dx = \lim_{n\to\infty} \sum_{\nu=1}^{n} f(\xi_\nu)^2 (x_\nu - x_{\nu-1}),$$

where $x_{\nu-1} < \xi_\nu < x_\nu$ and the maximal length of the subintervals $[x_{\nu-1}, x_\nu]$ converges to 0 as $n \to \infty$. According to **108** $\xi_\nu$ can be chosen such that $f(x)$ is continuous at $x = \xi_\nu$, and so $f(\xi_\nu) = 0$. Let, on the other hand, be $f(\xi) \neq 0$ at the point of continuity $\xi$, $a < \xi < b$. Then we have for $\delta$, $\delta > 0$ and sufficiently small, $f(x)^2 > \frac{f(\xi)^2}{2}$ whenever $|x - \xi| < \delta$, and therefore

$$\int_a^b f(x)^2\,dx \geqq \int_{\xi-\delta}^{\xi+\delta} f(x)^2\,dx \geqq \delta f(\xi)^2 > 0.$$

**110.** Assume that $\varepsilon$, $\eta$ are given, $\varepsilon, \eta > 0$ and that $\delta$ is such that $|\varphi(y_1) - \varphi(y_2)| < \varepsilon$ whenever $|y_1 - y_2| < \delta$. Since $f(x)$ is integrable, a subdivision of $[a, b]$ can be found for which the total length of the subintervals where the oscillation of $f(x)$ is $\geqq \delta$ is $< \eta$. On the other subintervals the oscillation of $\varphi[f(x)]$ is at most $\varepsilon$.

**111.** [Cf. C. Carathéodory: Vorlesungen über reelle Funktionen. Leipzig and Berlin: B. G. Teubner 1918, pp. 379–380.] Let $f(x)$ be defined as in **99** and $G(x)$ as in **98** and

$$\varphi(y) = \begin{cases} 1 & \text{for } y = 0 \\ 0 & \text{for } y \gtrless 0. \end{cases}$$

Then $\varphi[f(x)] = G(x)$.

**112.** Assume that $f(x)$ is non-increasing. We obtain for $0 < x < \frac{1}{2}$

$$\int_{\frac{x}{2}}^{x} \zeta^a f(\zeta)\,d\zeta \geqq f(x) \int_{\frac{x}{2}}^{x} \zeta^a d\zeta = x^{a+1} f(x)\, \frac{1 - \left(\frac{1}{2}\right)^{a+1}}{a+1}$$

$$\int_{x}^{2x} \zeta^a f(\zeta)\,d\zeta \leqq f(x) \int_{x}^{2x} \zeta^a\,d\zeta = x^{a+1} f(x)\, \frac{2^{a+1} - 1}{a+1}.$$

In the case of $a = -1$ the last factors, both positive, must be replaced by log 2.

**113.** Change the variable of integration in **112**: substitute $\frac{1}{x}$ for $x$. Or prove the statement directly in a similar manner as **112**.

**114.** The integral over $[0, \varepsilon]$ exists if and only if the integral of $x^\alpha \left(1 - \frac{x^2}{2}\right)^{x^\beta}$ or of $x^\alpha e^{-\frac{1}{2}x^{\beta+2}}$ converges, i.e. certainly for $\beta < -2$; for $\beta \geqq -2$ if and only if $\alpha > -1$. The integral over $[\omega, \infty)$, $\omega > 0$, is convergent for $\beta \leqq 0$ if and only if $\alpha < -1$. If $\beta > 0$ and $n$ an integer the two following integrals can be compared as $n \to \infty$:

$$\int_{n\pi}^{(n+1)\pi} x^\alpha |\cos x|^{x^\beta} dx \sim (n\pi)^\alpha \int_0^\pi |\cos x|^{(x+n\pi)^\beta} dx.$$

The second integral lies between

$$\int_0^\pi |\cos x|^{(n\pi)^\beta} dx \quad \text{and} \quad \int_0^\pi |\cos x|^{[(n+1)\pi]^\beta} dx,$$

which increase like $n^{-\frac{\beta}{2}}$ [**202**]. For the integral to converge we must have $\alpha - \frac{\beta}{2} < -1$. Combining all these results we find that the integral in question is convergent if and only if either $\alpha < -1$, $\beta < -2$ or if $-1 < \alpha < \frac{\beta}{2} - 1$.

**114.1.** Let $a_1, a_2, a_3, \ldots$ be a sequence of positive numbers such that $\sum_{n=1}^\infty a_n$ converges and $\sum_{n=1}^\infty a_n^\beta$ diverges for $\beta < 1$, moreover $a_n < \frac{1}{2}$. We may choose e.g.

$$a_n = \frac{1}{n (\log n)^2}, \qquad n = 3, 4, \ldots$$

Set for $n = 1, 2, 3, \ldots$

$$f(x) = \begin{cases} a_n & \text{for } n - 1 \leqq x < n - a_n^2, \\ \frac{1}{a_n} & \text{for } n - a_n^2 \leqq x < n. \end{cases}$$

Observe that

$$\int_{n-1}^n f(x)\, dx < 2a_n, \qquad \int_{n-1}^n [f(x)]^\alpha dx > \tfrac{1}{2} a_n^\alpha + a_n^{2-\alpha}.$$

**115.** The limit function is properly integrable over every finite interval $[-\omega, \omega]$ and $\lim_{n\to\infty} \int_{-\omega}^{\omega} f_n(x)\, dx = \int_{-\omega}^{\omega} f(x)\, dx$. Since $|f(x)| \leqq F(x)$ the integral $\int_{-\infty}^{\infty} f(x)\, dx$ exists.

We have

$$\left| \int_{-\infty}^{\infty} f(x)\, dx - \int_{-\infty}^{\infty} f_n(x)\, dx \right| \leqq \left| \int_{-\infty}^{-\omega} f(x)\, dx \right| + \left| \int_{\omega}^{\infty} f(x)\, dx \right|$$
$$+ \int_{-\infty}^{-\omega} F(x)\, dx + \int_{\omega}^{\infty} F(x)\, dx + \left| \int_{-\omega}^{\omega} f(x)\, dx - \int_{-\omega}^{\omega} f_n(x)\, dx \right|.$$

The first four terms on the right hand side are arbitrarily small for sufficiently large $\omega$ and the same is true for the last term if $n$ is sufficiently large while $\omega$ is fixed.

**116.** Set $\nu = \frac{n}{2} + \lambda_n \sqrt{n}$, $\lambda_n \to \lambda$. VI **31** implies

$$\frac{\sqrt{n}}{2^n}\binom{n}{\nu} = \frac{1}{2\pi}\int_{-\pi\sqrt{n}}^{\pi\sqrt{n}} \left(\cos\frac{x}{2\sqrt{n}}\right)^n \cos\lambda_n x\, dx.$$

In **115** we put:

$$f_n(x) = \begin{cases} \frac{1}{2\pi}\left(\cos\frac{x}{2\sqrt{n}}\right)^n \cos\lambda_n x & \text{for } |x| \leqq \pi\sqrt{n} \\ 0 & \text{for } |x| > \pi\sqrt{n}. \end{cases}$$

We have $\lim\limits_{n\to\infty} f_n(x) = f(x) = \frac{1}{2\pi} e^{-\frac{x^2}{8}} \cos\lambda x$. To find a function $F(x)$ as described in **115** we proceed as follows: Since

$$\frac{\log\cos x}{x^2}$$

is continuous and negative on $\left(0, \frac{\pi}{2}\right)$,

$$\lim_{x\to+0} \frac{\log\cos x}{x^2} = -\frac{1}{2}, \quad \lim_{x\to\frac{\pi}{2}-0} \frac{\log\cos x}{x^2} = -\infty,$$

there exists a (absolute) constant $K$, $K > 0$, such that

$$\frac{\log\cos x}{x^2} < -K, \quad \cos x \leqq e^{-Kx^2}, \quad 0 \leqq x \leqq \frac{\pi}{2}.$$

Thus we can pick $F(x) = \frac{1}{2\pi} e^{\frac{-Kx^2}{4}}$.

**117.** Put $\sum\limits_{\nu=1}^{n} a_\nu e^{-\nu y} = P_n(y)$. Then

$$\int_0^\infty P_n(y)\, y^{s-1}\, dy = (a_1 1^{-s} + a_2 2^{-s} + \cdots + a_n n^{-s})\,\Gamma(s) = D_n(s)\,\Gamma(s).$$

According to **115** it is sufficient to find $F(x)$ such that $|P_n(y)\, y^{s-1}| < F(y)$, whereby $\int_0^\infty F(y)\, dy$ exists. Partial summation yields

$$P_n(y) = \sum_{\nu=1}^{n-1} D_\nu(\sigma)\,[\nu^\sigma e^{-\nu y} - (\nu+1)^\sigma e^{-(\nu+1)y}] + n^\sigma D_n(\sigma)\, e^{-ny}.$$

The function $x^\sigma e^{-xy}$ has its maximum, $(\sigma e^{-1})^\sigma y^{-\sigma}$, at $x = \frac{\sigma}{y}$; thus $|P_n(y)| < Ay^{-\sigma}$, $|P_n(y)\, y^{s-1}| < Ay^{s-\sigma-1}$, $A$ independent of $n$ and $y$. Set $F(y) = Ay^{s-\sigma-1}$ for $0 < y < 1$; $F(y) = Be^{-y}$ for $y \geqq 1$, $B > 0$, $B$ independent of $y$.

**118.** We have with $\omega > 0$

$$\left|\int_{-\infty}^{\infty} f(x) \sin nx\, dx\right| \leqq \int_{-\infty}^{-\omega} |f(x)|\, dx + \int_{\omega}^{\infty} |f(x)|\, dx + \left|\int_{-\omega}^{\omega} f(x) \sin nx\, dx\right|.$$

The first two integrals on the right converge to 0 as $\omega \to \infty$, the third integral converges to 0 as $n \to \infty$, $\omega$ fixed [**105**]. The proof of the second part of the problem runs similarly:

$$\left|\int_{-\infty}^{\infty} f(x)\, |\sin nx|\, dx - \frac{2}{\pi}\int_{-\infty}^{\infty} f(x)\, dx\right|$$

$$\leqq \left(1 + \frac{2}{\pi}\right)\int_{-\infty}^{-\omega} |f(x)|\, dx + \left(1 + \frac{2}{\pi}\right)\int_{\omega}^{\infty} |f(x)|\, dx$$

$$+ \left|\int_{-\omega}^{\omega} f(x)\, |\sin nx|\, dx - \frac{2}{\pi}\int_{-\omega}^{\omega} f(x)\, dx\right| \qquad [\mathbf{106}].$$

**118.1.** [G. Pólya, Problem: Jber. deutsch. Math. Verein. Vol. 40, 2. Abt., p. 81 (1931). Solved by G. Szegö: Jber. deutsch. Math. Verein. Vol. 43, 2. Abt., pp. 17—20 (1934).]

**119.** (1) Reduction to two functions of two variables possible: Let $\varphi(x, y)$ make different values $u = \varphi(x, y)$ correspond to different pairs of numbers $x, y$ (e.g. a one to one correspondence between the $xy$-plane and the number line $u$). For a given function $f(x, y, z)$, $\psi(u, z)$ is then found in the following way:

If $u^*$ does not belong to the range of $\varphi(x, y)$, $\psi(u^*, z)$ is chosen arbitrarily, e.g. $\psi(u^*, z) = 1$.

If $u^*$ does belong to the range of $\varphi(x, y)$, it corresponds to a *unique* pair $x^*, y^*$, thus $u^* = \varphi(x^*, y^*)$; in this case we choose $\psi(u^*, z) = f(x^*, y^*, z)$.

Then it is true for all $x, y, z$: $\psi(\varphi(x, y), z) = f(x, y, z)$.

(2) Representation impossible, e.g. for the function
$f(x, y, z) = yz + zx + xy$.

Arbitrarily many pairs $x_1, y_1$; $x_2, y_2$; ...; $x_n, y_n$ exist so that a given continuous function $\varphi(x, y)$ assumes the same value for all of them. (They are points of the same "level line".) If $\varphi(x, y) =$ const. the statement is evident. If $\varphi(x, y)$ is not a constant and if e.g. $\varphi(x', y') = 1$, $\varphi(x''', y''') = 3$ there exists on each circular arc connecting $x', y'$ with $x''', y'''$ an intermediate point $x'', y''$ for which $\varphi(x'', y'') = 2$.

If $f(x, y, z) = \psi(\varphi(x, y), z)$ as proposed, $\varphi(x, y)$ continuous, there are arbitrarily many pairs $x_1, y_1; x_2, y_2; \ldots; x_n, y_n$ for which

$$f(x_1, y_1, z) = f(x_2, y_2, z) = \cdots = f(x_n, y_n, z)$$

identically in $z$. If

$$(x_1 + y_1) z + x_1 y_1 = (x_2 + y_2) z + x_2 y_2$$

is to hold identically in $z$ the two equations

$$x_1 + y_1 = x_2 + y_2, \quad x_1 y_1 = x_2 y_2$$

must be satisfied. Eliminating $y_1$ we find

$$x_1^2 - (x_2 + y_2) x_1 + x_2 y_2 = 0.$$

If $x_1 \neq x_2$, then $x_1 = y_2$, $y_1 = x_2$. This means that to a given pair $x_1, y_1$ there exists exactly one different pair on the "level line" and not arbitrarily many as the proposed form asks for.

$yz + zx + xy$, being a symmetric function, cannot be represented in the form $\psi(\varphi(y, z), x)$ or $\psi(\varphi(z, x), y)$ either.

In the same manner one can show that the continuous function $xy + yz + z$ cannot be written with the help of two continuous functions of two variables boxed in each other, however it can be done with three such functions:

$$xy + yz + z = (x + z) y + z = S\{P[S(x, z), y], z\}.$$

(Notation **119**a.)

It is easier to discuss similar questions for narrower classes of functions [**119**a]. We cannot represent an analytic function of three variables that does not satisfy some algebraic partial differential equation by boxing into each other a finite number of analytic functions of two variables. Cf. the 13th of the "Mathematical Problems" by D. Hilbert: Nachr. Akad. Wiss. Göttingen 1900, p. 280.

**119a.** We write $f(x, y, z) = yz + zx + xy$ and denote the partial derivatives by subscripts.

(1) The relation $f(x, y, z) = \varphi\{\psi[\chi(x, y), z], z\}$ for all values $x, y, z$ for which $f_y = z + x \neq 0$ implies

$$\frac{\partial}{\partial z}\left(\frac{f_x}{f_y}\right) = \frac{\partial}{\partial z}\left(\frac{\varphi_\psi \psi_\chi \chi_x}{\varphi_\psi \psi_\chi \chi_y}\right) = 0,$$

thus $f_{xz} f_y - f_x f_{yz} = 0$; $yz + zx + xy$ does not satisfy this equation.

(2) **First proof:** Taking the derivative with respect to $x, y, z$ and eliminating $\varphi_\psi$ and $\varphi_\chi$ we deduce from the relation

$$f(x, y, z) = \varphi[\psi(x, z), \chi(y, z)]$$

that

$$\begin{vmatrix} f_x & \psi_x & 0 \\ f_y & 0 & \chi_y \\ f_z & \psi_z & \chi_z \end{vmatrix} = 0.$$

We can assume that $\psi_x \not\equiv 0$, $\chi_y \not\equiv 0$. If $\psi_x \neq 0$ and $\chi_y \neq 0$ we may write

$$\frac{\psi_z}{\psi_x} = -v, \qquad \frac{\chi_z}{\chi_y} = -u,$$

$v = v(x, z)$, $u = u(y, z)$, thus $v_y = u_x = 0$. Moreover

$$f_y u + f_x v + f_z = 0.$$

This leads to a contradiction: Put $F = f_y u + f_x v + f_z$, $f = yz + zx + xy$. Then

$$F - \frac{\partial F}{\partial x} f_y - \frac{\partial F}{\partial y} f_x + \frac{\partial^2 F}{\partial x\, \partial y} f_x f_y = -2z.$$

**Second proof:** Take the derivatives with respect to $x, y, z$ and set $z = -y$ in the three corresponding equations. The first equation implies then $0 = \varphi_\psi \psi_x$. We can assume $\psi_x \not\equiv 0$ so that for the special values mentioned we have $\varphi_\psi = 0$, as long as $\psi_x \neq 0$. I.e.

$$f_y = x - y = \varphi_\chi \chi_y, \qquad f_z = x + y = \varphi_\chi \chi_z.$$

If $x \neq y$, then $\varphi_\chi \neq 0$, $\chi_y \neq 0$ and

$$\frac{x + y}{x - y} = \frac{\chi_z}{\chi_y}.$$

The right hand side depends on $y$ and $z = -y$ only: contradiction.

(3) Differentiation of $f(x, y, z) = \varphi\{\psi[\chi(x, y), z], x\}$ implies

$$\begin{aligned} f_x &= \varphi_\psi \psi_\chi \chi_x + \varphi_x, \\ f_y &= \varphi_\psi \psi_\chi \chi_y, \\ f_z &= \varphi_\psi \psi_z, \\ f_{xy} &= \cdots + \varphi_\psi \chi_x \chi_y \psi_{\chi\chi}, \\ f_{yy} &= \cdots + \varphi_\psi \chi_y^2 \psi_{\chi\chi}. \end{aligned}$$

In the last two equations the terms containing $\psi_\chi$ are not written down.

We set $z = -x$ and obtain $\varphi_\psi \psi_\chi \chi_y = 0$. Since we may assume $\chi_y \neq 0$ and because of the third equation we conclude for $x + y \neq 0$ and $\chi_y \neq 0$ that $\psi_\chi = 0$. Thus

$$1 = \varphi_\psi \chi_x \chi_y \psi_{\chi\chi}, \qquad 0 = \varphi_\psi \chi_y^2 \psi_{\chi\chi},$$

for the special values mentioned above. These two equations contain a contradiction because, according to the second equation, at least one of the three functions involved has to vanish.

**120.** No. Example: $f(x) = x^3$; $\xi = 0$ is a point of inflection.

**121.** [Cf. G. Pólya: Tôhoku Mat. J. Vol. 19, p. 3 (1921).] Denote the least upper bound of $|f'(x)|$ on $[a, b]$ by $M$. Then $M > 0$ and

$$f(x) = f'(\xi)(x - a) \leqq M(x - a) \quad \text{for} \quad a \leqq x \leqq \frac{a+b}{2},$$

$$f(x) = f'(\eta)(x - b) \leqq M(b - x) \quad \text{for} \quad \frac{a+b}{2} \leqq x \leqq b,$$

$a < \xi < x$, $x < \eta < b$. It is not possible that both inequalities hold identically because such a function would cease to be differentiable at $x = \frac{a+b}{2}$. Thus

$$\int_a^b f(x)\,dx < M \int_a^{\frac{a+b}{2}} (x - a)\,dx + M \int_{\frac{a+b}{2}}^b (b - x)\,dx = M\frac{(b-a)^2}{4}.$$

**122.** [W. Blaschke; Problem: Arch. Math. Phys. Ser. 3, Vol. 25, p. 273 (1917).] According to Taylor's theorem we can write for $x \gtrless x_0$

$$f(x) - f(x_0) = (x - x_0) f'(x_0) + \frac{(x - x_0)^2 f''(\bar{x})}{2}, \qquad x_0 - r < \bar{x} < x_0 + r.$$

Integration and the first law of the mean for integrals yield

$$\int_{x_0-r}^{x_0+r} [f(x) - f(x_0)]\,dx = \int_{x_0-r}^{x_0+r} \frac{(x - x_0)^2}{2} f''(\bar{x})\,dx = f''(\xi) \int_{x_0-r}^{x_0+r} \frac{(x - x_0)^2}{2}\,dx,$$

because the function $f''(x)$, being a derivative, assumes all the values between its greatest lower and its least upper bound.

**122.1.** The function $f(x)$ is linear; method of **122.2.**

**122.2.** Introduce the abbreviations

$$f(a) = A, \; f(x) = X, \; f(b) = B,$$

and take for $(u, v)$ first $(a, x)$, then $(x, b)$, then $(a, b)$:

$$(x - a)\sqrt{AX} + (b - x)\sqrt{XB} = (b - a)\sqrt{AB},$$

and hence

$$f(x) = X = \frac{1}{(cx + d)^2},$$

where $c$ and $d$ are appropriate constants. Convince yourself by performing an elementary integration that a function of this form satisfies the imposed condition.

**122.3.** By the method of **122.2**

$$f(x) = \frac{1}{\sqrt{|cx + d|}}.$$

**123.** First proof: Suppose that the series $\sum_{n=0}^{\infty} p_n e^{nx} = f(x)$ is convergent on the interval $(a, b)$; then it can be differentiated arbitrarily often in that interval:

$$f'(x) = \sum_{n=0}^{\infty} n p_n e^{nx}, \quad f''(x) = \sum_{n=0}^{\infty} n^2 p_n e^{nx},$$

$$f(x) f''(x) - [f'(x)]^2 = \sum_{m=0}^{\infty} \sum_{n=0}^{\infty} \tfrac{1}{2} (m - n)^2 p_m p_n e^{(m+n)x} > 0 \quad \textbf{[72]}.$$

Second proof. Proposition **80** implies for $x_1 < x_2$, $a_\nu = \sqrt{p_\nu e^{\frac{\nu x_1}{2}}}$, $b_\nu = \sqrt{p_\nu e^{\frac{\nu x_2}{2}}}$

$$\left[f\left(\frac{x_1 + x_2}{2}\right)\right]^2 < f(x_1) f(x_2).$$

**124.** [Cf. J. L. W. V. Jensen, l.c. **70**, pp. 187—190.] Suppose that the function $\varphi(x)$ is convex on the interval $[a, b]$ and that $\varphi(x) < G$.

$$\varphi\left(\frac{x_1 + x_2 + \cdots + x_n}{n}\right) \leqq \frac{\varphi(x_1) + \varphi(x_2) + \cdots + \varphi(x_n)}{n}$$

implies that for $x_1 = x_2 = \cdots = x_m = x + n\,\delta$, $x_{m+1} = \cdots = x_n = x$, $x$ an arbitrary point of $(a, b)$, $|\delta|$ sufficiently small, $m < n$,

$$\frac{\varphi(x + n\delta) - \varphi(x)}{n} \geqq \frac{\varphi(x + m\delta) - \varphi(x)}{m}.$$

Substituting $-\delta$ for $\delta$ and remembering that

$\varphi(x + m\,\delta) - \varphi(x) \geqq \varphi(x) - \varphi(x - m\,\delta)$ we obtain

$$(*) \quad \begin{cases} \dfrac{\varphi(x + n\delta) - \varphi(x)}{n} \geqq \dfrac{\varphi(x + m\delta) - \varphi(x)}{m} \\ \qquad \geqq \dfrac{\varphi(x) - \varphi(x - m\delta)}{m} \geqq \dfrac{\varphi(x) - \varphi(x - n\delta)}{n}. \end{cases}$$

Hence for $m = 1$

$$\frac{G-\varphi(x)}{n} \geqq \varphi(x+\delta)-\varphi(x) \geqq \varphi(x)-\varphi(x-\delta) \geqq \frac{\varphi(x)-G}{n}.$$

Let $\delta$ converge to 0 and $n$ increase to infinity in such a way that $x \pm n\,\delta$ remains in $(a, b)$. Then the continuity of $\varphi(x)$ is established. Assume $\delta > 0$ and replace $\delta$ by $\frac{\delta}{n}$ in (*):

$$\frac{\varphi(x+\delta)-\varphi(x)}{\delta} \geqq \frac{\varphi\left(x+\frac{m}{n}\delta\right)-\varphi(x)}{\frac{m}{n}\delta}$$

$$\geqq \frac{\varphi(x)-\varphi\left(x-\frac{m}{n}\delta\right)}{\frac{m}{n}\delta} \geqq \frac{\varphi(x)-\varphi(x-\delta)}{\delta}.$$

Since $\varphi(x)$ is continuous:

$$\left.\begin{aligned} &\frac{\varphi(x+\delta)-\varphi(x)}{\delta} \geqq \frac{\varphi(x+\delta')-\varphi(x)}{\delta'} \\ &\qquad \geqq \frac{\varphi(x)-\varphi(x-\delta')}{\delta'} \geqq \frac{\varphi(x)-\varphi(x-\delta)}{\delta} \end{aligned}\right\} 0 < \delta' < \delta.$$

As $\delta \to 0$ the first term converges to a limit $l_+$ and the last to a limit $l_-$, $l_+ \geqq l_-$.

**125.** [G. Pólya, Problem: Arch. Math. Phys. Ser. 3, Vol. 24, p. 283 (1916).] The values which $y = f(x)$ assumes on an interval of length $l$ fill a closed interval of length $L$:

$$L = \max|f(x_1)-f(x_2)| = \max\left|\int_{x_1}^{x_2} f'(x)\,dx\right| \leqq l \max|f'(x)|.$$

Let $\varepsilon > 0$. About each point $x$ where $f'(x) = 0$ construct the largest interval for which $|f'(x)| \leqq \varepsilon$. We denote the length of the interval by $l_x$. The values $f(x)$ is assuming on such an interval cover at most an interval of length $l_x\varepsilon$. The points of $M$ ($y = f(x)$ with $f'(x) = 0$, $a \leqq x \leqq b$) are enclosed in countably many intervals of total length $\leqq \varepsilon(b-a)$. This proof shows in addition that the set $M$ has measure 0.

**126.** [U. Dini; cf. C. Carathéodory, l.c. **111**, pp. 176–177; Hille, Vol. II, p. 78.] It is sufficient to consider the following case:
$f_1(x) \geqq f_2(x) \geqq \cdots \geqq f_n(x) \geqq \cdots$, $f_n(x)$ continuous, $\lim\limits_{n\to\infty} f_n(x) = 0$, $0 \leqq x \leqq 1$. If the convergence were not uniform, infinitely many points $x_n$ would exist for which $f_n(x_n) > a > 0$, $0 \leqq x_n \leqq 1$, $a$ independent of $n$. Let $\xi$ denote an accumulation point of the $x_n$'s and $m$ be such that

$f_m(\xi) < a$. Determine a neighbourhood of $\xi$ in which $f_m(x) < a$, thus $f_n(x) < a$ for $n \geqq m$. There are infinitely many $x_n$'s in this neighbourhood: contradiction.

**127.** [Cf. G. Pólya, Problem: Arch. Math. Phys. Ser. 3, Vol. 28, p. 174 (1920).] The limit function is monotone too, say monotone increasing. Subdivide the interval of convergence of the sequence $f_n(x)$, $n = 1, 2, 3, \ldots$, into subintervals $[x_{\nu-1}, x_\nu]$, $\nu = 1, 2, \ldots, N$, so small that $f(x_\nu) - f(x_{\nu-1}) < \varepsilon$, moreover choose $n$ so large that $|f_n(x_\nu) - f(x_\nu)| < \varepsilon$ for every $\nu$. Then we have for $x_{\nu-1} \leqq x \leqq x_\nu$

$$f(x_{\nu-1}) - \varepsilon < f_n(x_{\nu-1}) \leqq f_n(x) \leqq f_n(x_\nu) < f(x_\nu) + \varepsilon$$

thus $|f_n(x) - f(x)| < 2\varepsilon$; we have used the hypothesis that $f_n(x)$ is increasing.

**128.** Obvious.

**129.** Assume $a < x < b$. (It is obvious what has to be changed in the proof below to accommodate the cases $x = a$ and $x = b$.)

$$\int_a^b p_n(t)\, f(t)\, dt - f(x) = \int_{x-\varepsilon}^{x+\varepsilon} p_n(t)\, [f(t) - f(x)]\, dt + \int_a^{x-\varepsilon} + \int_{x+\varepsilon}^{b}.$$

The absolute value of the first term on the right is $< \delta \int_{x-\varepsilon}^{x+\varepsilon} p_n(t)\, dt \leqq \delta$ whenever $|f(t) - f(x)| < \delta$ for $x - \varepsilon \leqq t \leqq x + \varepsilon$. The absolute values of the other two terms are smaller than

$$2 \max_{a \leqq t \leqq b} |f(t)| \left( \int_a^{x-\varepsilon} p_n(t)\, dt + \int_{x+\varepsilon}^{b} p_n(t)\, dt \right).$$

Thus the condition is sufficient.—We define

$$f(t) = \begin{cases} 0 & \text{for } x - \varepsilon + \eta \leqq t \leqq x + \varepsilon - \eta, \\ 1 & \text{for } a \leqq t \leqq x - \varepsilon \text{ and } x + \varepsilon \leqq t \leqq b, \\ \text{linear for } x - \varepsilon \leqq t \leqq x - \varepsilon + \eta \text{ and } x + \varepsilon - \eta \leqq t \leqq x + \varepsilon, \end{cases}$$

$0 < \eta < \varepsilon$; $f(t)$ is continuous. From

$$\int_a^b p_n(t)\, f(t)\, dt - f(x) = \int_{x-\varepsilon}^{x-\varepsilon+\eta} p_n(t)\, f(t)\, dt + \int_{x+\varepsilon-\eta}^{x+\varepsilon} p_n(t)\, f(t)\, dt + \int_a^{x-\varepsilon} p_n(t)\, dt$$
$$+ \int_{x+\varepsilon}^{b} p_n(t)\, dt \to 0$$

follows that the condition is necessary because all the terms are positive.

**130.** The statement of **129** can be extended to the case $b = \infty$, $x = \infty$. In **129** put $n = \frac{1}{\varepsilon}$, $p_n(t) = \varepsilon e^{-\varepsilon t}$; then

$$\lim_{\varepsilon \to 0} \varepsilon \int_0^\omega e^{-\varepsilon t}\, dt = 0, \qquad \varepsilon \int_0^\infty e^{-\varepsilon t}\, dt = 1,$$

where $\omega$ is positive and independent of $\varepsilon$.

**131.** The substitution of $e^t$ for $t$ changes the interval of integration $[0, \infty)$ into $(-\infty, \infty)$. The new integral can be split into two parts, over $(-\infty, 0]$ and $[0, \infty)$ resp. We examine the two parts separately. If the integral

$$\int_0^\infty e^{\lambda t} f(e^t)\, e^t dt = \int_0^\infty e^{\lambda t} \varphi(t)\, dt$$

converges for $\lambda = \beta$ it converges for any smaller $\lambda$, $\lambda = \beta - \varepsilon$, $\varepsilon > 0$: writing $e^{\beta t}\varphi(t) = \psi(t)$ we obtain

$$\int_0^\omega e^{-\varepsilon t} \psi(t)\, dt = e^{-\varepsilon \omega} \int_0^\omega \psi(t)\, dt + \varepsilon \int_0^\omega e^{-\varepsilon t} \left( \int_0^t \psi(\tau)\, d\tau \right) dt,$$

hence for $\omega \to \infty$

$$\int_0^\infty e^{-\varepsilon t} \psi(t)\, dt = \varepsilon \int_0^\infty e^{-\varepsilon t} \left( \int_0^t \psi(\tau)\, d\tau \right) dt.$$

This integral is, as a function of $\varepsilon$, continuous on the right [**130**].

**132.** Cf. **128** and the first part of the proof of **129**. The number $\delta$ defined there can be made arbitrarily small with $\varepsilon$, independently of $x$ (uniform continuity).

**133.** [E. Landau: Rend. Circ. Mat. Palermo Vol. 25, pp. 337–345 (1908).] Apply **132**: $a = 0$, $b = 1$,

$$p_n(x, t) = \frac{[1 - (x - t)^2]^n}{\int_0^1 [1 - (x - t)^2]^n\, dt}.$$

For $0 < \varepsilon \leqq x \leqq 1 - \varepsilon$

$$\int_0^{x-\varepsilon} p_n(x, t)\, dt + \int_{x+\varepsilon}^1 p_n(x, t)\, dt < \frac{(1 - \varepsilon^2)^n}{\int_{x-\varepsilon}^{x+\varepsilon} [1 - (x - t)^2]^n\, dt} = \frac{(1 - \varepsilon^2)^n}{\int_{-\varepsilon}^{\varepsilon} (1 - t^2)^n dt}.$$

According to **201**, **202** this expression converges to 0 because

$$\int_{-\varepsilon}^{\varepsilon} (1 - t^2)^n\, dt \sim \sqrt{\frac{\pi}{n}} \sim \frac{2}{1} \cdot \frac{2}{3} \cdot \frac{4}{5} \cdots \frac{2n}{2n + 1}.$$

**134.** [L. Fejér: Math. Ann. Vol. 58, pp. 51–69 (1904).] Apply **132**: $a = 0$, $b = 2\pi$, $p_n(x, t) = \frac{1}{2n\pi}\left(\frac{\sin n\frac{x-t}{2}}{\sin\frac{x-t}{2}}\right)^2$. According to VI **18** the $p_n$'s are "normed", i.e.

$$\int_0^{2\pi} p_n(x, t)\, dt = 1;$$

also for $0 < \varepsilon \leqq x \leqq 2\pi - \varepsilon$

$$\int_0^{x-\varepsilon} p_n(x, t)\, dt + \int_{x+\varepsilon}^{2\pi} p_n(x, t)\, dt < \frac{1}{n \sin^2 \frac{\varepsilon}{2}}.$$

Since the integrand is periodic, any interval of length $< 2\pi$ can be considered as an interior subinterval.

**135.** Cf. **133**.

**136.** Cf. **134**.

**137.** We may assume that $f(x)$ is non-negative and [**102**] that it is piecewise constant. Such a function can be obtained by addition and multiplication by positive constants of functions with the following properties

$$f(x) = \begin{cases} 1 \text{ on a subinterval } [\alpha, \beta] \text{ of } [a, b],\ a \leqq \alpha < \beta \leqq b \\ 0 \text{ outside of } [\alpha, \beta]. \end{cases}$$

For simplicity's sake let $a < \alpha$, $\beta < b$.

We define for sufficiently small $\eta$, $\eta > 0$,

$$f_\eta(x) = \begin{cases} 1 + \eta, \text{ if } \alpha \leqq x \leqq \beta; \\ \eta, \text{ if } a \leqq x \leqq \alpha - \eta \text{ or } \beta + \eta \leqq x \leqq b; \\ \text{linear, if } \alpha - \eta \leqq x \leqq \alpha \text{ or } \beta \leqq x \leqq \beta + \eta. \end{cases}$$

Then $f_\eta(x) - f(x) \geqq \eta$, furthermore

$$0 < \int_a^b f_\eta(x)\, dx - \int_a^b f(x)\, dx = \eta(\alpha - \eta - a) + \eta(b - \beta - \eta) + \eta(1 + 2\eta) + \eta(\beta - \alpha),$$

i.e. arbitrarily small with $\eta$. The function $f_\eta(x)$ is everywhere continuous. If the polynomial $P(x)$ is such that

$$|f_\eta(x) - P(x)| < \eta, \text{ then } f(x) \leqq f_\eta(x) - \eta < P(x)$$

and

$$\int_a^b P(x)\,dx - \int_a^b f(x)\,dx < \eta(b-a) + \int_a^b f_\eta(x)\,dx - \int_a^b f(x)\,dx.$$

For $\alpha = a$ or $\beta = b$ some slight changes have to be made. A similar proposition holds for trigonometric polynomials if $a = 0$, $b = 2\pi$.

**138.** [M. Lerch: Acta Math. Vol. 27, pp. 345–347 (1903); E. Phragmén: Acta Math. Vol. 28, pp. 360–364 (1904).] Let $P(x)$ denote an arbitrary polynomial:

$$\int_a^b [f(x)]^2\,dx = \int_a^b f(x)\,[f(x) - P(x)]\,dx + \int_a^b f(x)\,P(x)\,dx$$

$$= \int_a^b f(x)[f(x) - P(x)]\,dx,$$

thus

$$\int_a^b [f(x)]^2\,dx \leqq \max_{a \leqq x \leqq b} |f(x) - P(x)| \cdot \int_a^b |f(x)|\,dx. \quad [\mathbf{135, 109}.]$$

A similar statement holds for trigonometric moments (Fourier constants) $\int_0^{2\pi} f(x) \begin{matrix}\cos\\ \sin\end{matrix}\, nx\,dx$, $n = 0, 1, 2, \ldots$

**139.** Determine $p(x)$ as in **137** and assume $|f(x)| \leqq M$. Then [cf. solution **138**]

$$\int_a^b [f(x)]^2\,dx \leqq M \int_a^b [f(x) - p(x)]\,dx < M\varepsilon.$$

**140.** Let $f(x)$ not vanish identically and not change sign more than $n-1$ times. According to the first condition, $\int_a^b f(x)\,dx = 0$, there exist numbers $x_1, x_2, \ldots, x_k$, $a < x_1 < x_2 < \cdots < x_k < b$, with the following property: $f(x)$ does not vanish identically in any of the subintervals $(a, x_1), (x_1, x_2), \ldots, (x_{k-1}, x_k), (x_k, b)$; $f(x)$ does not change sign in any of these subintervals (V, Chap. 1, § 2) but the signs alternate for consecutive subintervals. Therefore $f(x)\,(x - x_1)\,(x - x_2) \cdots (x - x_k)$ has the same sign for the entire interval $(a, b)$ and does not vanish identically. According to the hypothesis we would have in the case $k \leqq n - 1$

$$\int_a^b f(x)\,(x - x_1)\,(x - x_2) \cdots (x - x_k)\,dx = 0$$

i.e. [**109**] $f(x)\,(x - x_1)\,(x - x_2) \cdots (x - x_k) \equiv 0$, $f(x) \equiv 0$: contradiction.

**141.** [Cf. A. Hurwitz: Math. Ann. Vol. 57, pp. 425—446 (1903).] Suppose that $f(0) > 0$ and that $f(x)$ changes sign $2k < 2n + 2$ times in the interval $(0, 2\pi)$. Let $x_1, x_1', x_2, x_2', \ldots, x_k, x_k'$,

$$0 < x_1 < x_1' < x_2 < x_2' < \cdots < x_k < x_k' < 2\pi,$$

be the points where the changes of sign occur. In analogy to solution **140** we form

$$f(x) \sin\frac{x - x_1}{2} \sin\frac{x - x_1'}{2} \sin\frac{x - x_2}{2} \sin\frac{x - x_2'}{2} \cdots \sin\frac{x - x_k}{2} \sin\frac{x - x_k'}{2}.$$

Note that

$$\sin\frac{x-\alpha}{2}\sin\frac{x-\beta}{2} = \frac{1}{2}\cos\frac{\alpha-\beta}{2} - \frac{1}{2}\cos\left(x - \frac{\alpha+\beta}{2}\right)$$

($\alpha, \beta$ constant, $0 < \alpha < 2\pi$, $0 < \beta < 2\pi$) changes sign in the interval $(0, 2\pi)$ only at $x = \alpha$ and $x = \beta$. Use VI **10**. If $f(0) = 0$ consider $f(x + a)$ with $f(a) \neq 0$.

**142.** [L.c. **138.**] Writing $\int_0^x e^{-k_0 t} \varphi(t)\, dt = \Phi(x)$ we obtain for $k > k_0$

$$J(k) = [\Phi(x) e^{-(k-k_0)x}]_0^\infty + (k - k_0) \int_0^\infty \Phi(x) e^{-(k-k_0)x} dx.$$

The relations

$$e^{-\alpha x} = y, \quad \Phi\left(\frac{1}{\alpha} \log\frac{1}{y}\right) = \psi(y), \quad \psi(0) = J(k_0) = 0,$$

define $\psi(y)$ as a continuous function on the interval $[0, 1]$, furthermore

$$\int_0^1 \psi(y) y^{n-1} dy = 0, \qquad n = 1, 2, \ldots$$

Hence [**138**] $\psi(y) \equiv 0$, $\Phi(x) \equiv 0$, $\varphi(x) \equiv 0$.

**143.** [M. Lerch, communicated by M. Plancherel.] If $s_0$ were a zero of the $\Gamma$-function $s_0 + m$, $m = 1, 2, \ldots$, would be zeros too [functional equation]. Let $m$ be so large that $\Re(s_0 + m)$ is positive, put $s = s_0 + m + 1 = \sigma + it$, $\sigma > 1$. The equation

$$\int_0^\infty e^{-nx} x^{\sigma-1} \cos(t \log x)\, dx = \Re \int_0^\infty e^{-nx} x^{s-1} dx = \Re \frac{\Gamma(s)}{n^s} = 0, n = 1, 2, 3, \ldots$$

would imply [**142**] that $x^{\sigma-1} \cos(t \log x) \equiv 0$: contradiction.

**144.** For $f(x) = 1, x, x^2$ cf. I **40**. For $f(x) = e^x$:

$$K_n(x) = \sum_{\nu=0}^{n} e^{\frac{\nu}{n}} \binom{n}{\nu} x^\nu (1-x)^{n-\nu} = \left(e^{\frac{1}{n}} x + 1 - x\right)^n = \left[1 + \left(e^{\frac{1}{n}} - 1\right)x\right]^n.$$

**145.** We have [I **40**]

$$\sum_{\nu=0}^{n} (\nu - nx)^2 \binom{n}{\nu} x^\nu (1-x)^{n-\nu} = nx(1-x),$$

thus

$$n^{\frac{3}{2}} \sum{}^{II} \leqq nx(1-x) \leqq \frac{n}{4}.$$

**146.** [Cf. S. Bernstein: Communic. Soc. Math. Charkow Ser. 2, Vol. 13, pp. 1–2 (1912).] We define $\varepsilon_n(x) = \max \left| f(x) - f\left(\frac{\nu}{n}\right) \right|$ for all $\nu$ for which $\left| \frac{\nu}{n} - x \right| \leqq n^{-\frac{1}{4}}$. $\lim\limits_{n\to\infty} \varepsilon_n(x) = 0$ uniformly in $x$, i.e. $\varepsilon_n(x) < \varepsilon_n$, $\lim\limits_{n\to\infty} \varepsilon_n = 0$. Moreover

$$f(x) - K_n(x) = \sum_{\nu=0}^{n} \left[ f(x) - f\left(\frac{\nu}{n}\right) \right] \binom{n}{\nu} x^\nu (1-x)^{n-\nu}.$$

According to **145** we have the inequality

$$|f(x) - K_n(x)| < \varepsilon_n \sum{}^{I} + 2M \sum{}^{II} < \varepsilon_n + \frac{M}{2} n^{-\frac{1}{2}},$$

where $|f(x)| < M$ on $[0, 1]$.

**147.** [Cf. J. Franel: Math. Ann. Vol. 52, pp. 529–531 (1899).] Obvious for $0 \leqq r < r_1$. For $r_m \leqq r < r_{m+1}$ the right hand side is equal to

$$mf(r) - 1[f(r_2) - f(r_1)] - 2[f(r_3) - f(r_2)] - \cdots - (m-1)[f(r_m) - f(r_{m-1})] - m[f(r) - f(r_m)].$$

But this exactly equals the expression on the left hand side. Indeed, the formula we have proved is the formula for "partial integration":

$$\int_0^r f(t)\, dN(t) = N(r) f(r) - \int_0^r N(t) f'(t)\, dt.$$

**148.** If $r_{n-k-1} < r_{n-k} = \cdots = r_n = r_{n+1} = \cdots = r_{n+l} < r_{n+l+1}$ (possibly $k = 0$ or $l = 0$; $r_0 = 0$) then

$$\frac{N(r_n - 0) + 1}{r_n} = \frac{n-k}{r_n} \leqq \frac{n}{r_n} \leqq \frac{n+l}{r_n} = \frac{N(r_n)}{r_n}, \quad \lim_{n\to\infty} \frac{1}{r_n} = 0.$$

If $r_m \leqq r < r_{m+1}$ we find $N(r) = m$ and

$$\frac{m+1}{r_{m+1}} - \frac{1}{r_{m+1}} = \frac{m}{r_{m+1}} < \frac{N(r)}{r} \leqq \frac{m}{r_m}.$$

Analogously in the second case.

**149.** Cf. **148** and I **113**.

**150.** [E. Landau: Bull. Acad. Belgique 1911, pp. 443—472. Cf. G. Pólya: Nachr. Akad. Wiss. Göttingen 1917, pp. 149—159.] We assume $c > 1$ and choose $m$ so large that $1 < c < 2^m$. Then

$$1 < \frac{L(cr)}{L(r)} < \frac{L(2^m r)}{L(r)} = \frac{L(2r)}{L(r)} \frac{L(2^2 r)}{L(2r)} \cdots \frac{L(2^m r)}{L(2^{m-1} r)} \to 1.$$

If $c < 1$ we replace $c$ by $\frac{1}{c}$ and use the slowly increasing function $L(cr)$.

**151.** Mathematical induction shows that for positive integral $k$

$$\lim_{r\to\infty} \frac{\log_k 2r}{\log_k r} = 1.$$

For $k = 1$ the value of the limit is obvious; if $k > 1$ we have for sufficiently large $r$

$$1 < \frac{\log_k (2r)}{\log_k r} < \frac{\log_k r^2}{\log_k r} = \frac{\log_{k-1} (2 \log r)}{\log_{k-1} (\log r)} .$$

**152.** It is sufficient to prove

$$\lim_{m\to\infty} \frac{\log L(2^m)}{m} = 0.$$

This relation is a consequence of the inequality $\frac{L(2^m)}{L(2^{m-1})} < 1 + \delta$, where $\delta > 0$ is arbitrary and $m$ sufficiently large, $m > M(\delta)$.

**153.** [**149, 152.**]

**154.** $\frac{N(cr)}{N(r)} \sim c^\lambda \frac{L(cr)}{L(r)}$.

**155.** Any function that is continuous on the left and piecewise constant can be obtained as linear combination of the functions

$$f(x) = \begin{cases} 1 \text{ for } 0 < x \leqq \gamma, \gamma > 0 \\ 0 \text{ for all the other values of } x. \end{cases}$$

For such functions the proposition reads

$$\lim_{r\to\infty} \frac{N(\gamma r)}{N(r)} = \gamma^\lambda \qquad [\mathbf{154}].$$

Cf. **156** for other functions.

**156.** We bound $f(x)$ by two functions that are piecewise constant and continuous on the left

$$\psi(x) \leqq f(x) \leqq \Psi(x)$$

so that

$$\int_0^{c^\lambda} \Psi\left(x^{\frac{1}{\lambda}}\right) dx - \int_0^{c^\lambda} \psi\left(x^{\frac{1}{\lambda}}\right) dx < \varepsilon$$

$\varepsilon > 0$, $\varepsilon$ arbitrary [**102**]. Then

$$\frac{1}{N(r)} \sum_{r_n \le cr} \psi\left(\frac{r_n}{r}\right) \le \frac{1}{N(r)} \sum_{r_n \le cr} f\left(\frac{r_n}{r}\right) \le \frac{1}{N(r)} \sum_{r_n \le cr} \Psi\left(\frac{r_n}{r}\right).$$

lim inf and lim sup of the middle term lie therefore [**155**] between $\int_0^{c^\lambda} \psi\left(x^{\frac{1}{\lambda}}\right) dx$ and $\int_0^{c^\lambda} \Psi\left(x^{\frac{1}{\lambda}}\right) dx$, which differ from $\int_0^{c^\lambda} f\left(x^{\frac{1}{\lambda}}\right) dx$ by less than $\varepsilon$.

**157.** According to **147** we have

$$\frac{1}{N(r)} \sum_{r_n \le r} \left(\frac{r_n}{r}\right)^{\alpha-\lambda} = \frac{r^{\lambda-\alpha}}{N(r)} \left(\frac{N(r)}{r^{\lambda-\alpha}} + (\lambda-\alpha) \int_0^r N(t)\, t^{-\lambda+\alpha-1} dt\right)$$

$$\sim 1 + \frac{(\lambda-\alpha) r^{-\alpha}}{L(r)} \int_0^r L(t)\, t^{\alpha-1}\, dt.$$

For $0 < c < 1$

$$\frac{r^{-\alpha} L(cr)}{L(r)} \int_{cr}^r t^{\alpha-1}\, dt < \frac{r^{-\alpha}}{L(r)} \int_0^r L(t)\, t^{\alpha-1}\, dt < r^{-\alpha} \int_0^r t^{\alpha-1}\, dt = \frac{1}{\alpha};$$

lim inf and lim sup of the middle term lie therefore between $\frac{1-c^\alpha}{\alpha}$ and $\frac{1}{\alpha}$ where $c$ is arbitrarily small.

**158.** Proposition **147** implies that

$$\sum_{r < r_n \le R} r_n^{-\alpha-\lambda} = N(R)\, R^{-\alpha-\lambda} - N(r)\, r^{-\alpha-\lambda} + (\alpha+\lambda) \int_r^R N(t)\, t^{-\alpha-\lambda-1}\, dt,$$

hence [**152, 153**]

$$\frac{1}{N(r)} \sum_{r_n > r} \left(\frac{r_n}{r}\right)^{-\alpha-\lambda} = -1 + \frac{(\alpha+\lambda) r^{\alpha+\lambda}}{N(r)} \int_r^\infty N(t)\, t^{-\alpha-\lambda-1}\, dt$$

$$\sim -1 + \frac{(\alpha+\lambda) r^\alpha}{L(r)} \int_r^\infty L(t)\, t^{-\alpha-1}\, dt.$$

Let $0 < \varepsilon < 2^\alpha - 1$; for sufficiently large $r$ the inequality $L(2r) < L(r)\,(1+\varepsilon)$ holds, hence $L(2^\nu r) < L(r)\,(1+\varepsilon)^\nu$ for $\nu = 1, 2, 3, \ldots$; consequently

$$\frac{1}{\alpha} < \frac{r^\alpha}{L(r)} \int_r^\infty L(t)\, t^{-\alpha-1}\, dt < \frac{r^\alpha}{L(r)} \sum_{\nu=1}^\infty L(2^\nu r) \int_{2^{\nu-1} r}^{2^\nu r} t^{-\alpha-1}\, dt <$$

$$\frac{1}{\alpha} \sum_{\nu=1}^\infty (1+\varepsilon)^\nu \left(2^{-\alpha(\nu-1)} - 2^{-\alpha\nu}\right) = \frac{1+\varepsilon}{\alpha} \frac{2^\alpha - 1}{2^\alpha - 1 - \varepsilon}.$$

**159.** [As to **159**—**161** cf. G. Pólya: Math. Ann. Vol. 88, pp. 173—177 (1923).] Generalization of **155**—**158**. As in **156** bound $f(x)$ by two functions $\psi(x)$ and $\Psi(x)$ that coincide with $-x^{\alpha-\lambda}$ and $x^{\alpha-\lambda}$ respectively on $[0, \delta)$, with $f(x)$ on $[\delta, \omega)$ and with $-x^{-\alpha-\lambda}$ and $x^{-\alpha-\lambda}$ respectively on $[\omega, \infty)$. We have [**157**]

$$\lim_{r\to\infty} \frac{1}{N(r)} \sum_{r_n \leq \delta r} \left(\frac{r_n}{r}\right)^{\alpha-\lambda} = \lim_{r\to\infty} \frac{\delta^{\alpha-\lambda} N(\delta r)}{N(r)} \frac{1}{N(\delta r)} \sum_{r_n \leq \delta r} \left(\frac{r_n}{\delta r}\right)^{\alpha-\lambda} = \delta^\alpha \frac{\lambda}{\alpha}$$

and [**158**]

$$\lim_{r\to\infty} \frac{1}{N(r)} \sum_{r_n > \omega r} \left(\frac{r_n}{r}\right)^{-\alpha-\lambda} = \lim_{r\to\infty} \frac{\omega^{-\alpha-\lambda} N(\omega r)}{N(r)} \frac{1}{N(\omega r)} \sum_{r_n > \omega r} \left(\frac{r_n}{\omega r}\right)^{-\alpha-\lambda} = \omega^{-\alpha} \frac{\lambda}{\alpha}.$$

lim sup and lim inf of $\frac{1}{N(r)} \sum_{n=1}^{\infty} f\left(\frac{r_n}{r}\right)$ are therefore bounded by

$$-\delta^\alpha \frac{\lambda}{\alpha} + \int_{\delta^\lambda}^{\omega^\lambda} f\left(x^{\frac{1}{\lambda}}\right) dx - \omega^{-\alpha} \frac{\lambda}{\alpha} \quad \text{and} \quad \delta^\alpha \frac{\lambda}{\alpha} + \int_{\delta^\lambda}^{\omega^\lambda} f\left(x^{\frac{1}{\lambda}}\right) dx + \omega^{-\alpha} \frac{\lambda}{\alpha}.$$

Let $\delta$ converge to 0 and $\omega$ increase to infinity.

**160.** Let $f(x)$ be decreasing, $\beta > \lambda$. There exist [I **115**] infinitely many values of $n$ such that

$$\frac{r_\mu}{r_n} < \left(\frac{\mu}{n}\right)^{\frac{1}{\beta}}, \qquad \mu = 1, 2, \ldots, n-1.$$

Choose a number $r$, $r_{n-1} < r < r_n$ such that the inequalities

$$\frac{r_\mu}{r} < \left(\frac{\mu}{n}\right)^{\frac{1}{\beta}}, \qquad \mu = 1, 2, \ldots, n-1,$$

are satisfied too. Thus, since $f(x)$ is decreasing,

$$f\left(\frac{r_\mu}{r}\right) \geq f\left[\left(\frac{\mu}{n}\right)^{\frac{1}{\beta}}\right], \qquad \mu = 1, 2, \ldots, n-1.$$

We have $N(r) = n - 1$, consequently

$$\frac{1}{N(r)} f\left(\frac{r_\mu}{r}\right) \geq \frac{1}{n-1} f\left[\left(\frac{\mu}{n}\right)^{\frac{1}{\beta}}\right] \geq \frac{n}{n-1} \int_{\frac{\mu}{n}}^{\frac{\mu+1}{n}} f\left(x^{\frac{1}{\beta}}\right) dx,$$

$$\frac{1}{N(r)} \sum_{\mu=1}^{n-1} f\left(\frac{r_\mu}{r}\right) = \frac{1}{N(r)} \sum_{r_\mu \leq r} f\left(\frac{r_\mu}{r}\right) \geq \frac{n}{n-1} \int_{\frac{1}{n}}^{1} f\left(x^{\frac{1}{\beta}}\right) dx.$$

The integral $\int_0^1 f\left(x^{\frac{1}{\beta}}\right) dx$ is a continuous function of $\beta$ [**131**]. Similar arguments apply if $f(x)$ is increasing. Replace $f(x)$ by $-f(x)$.

**161.** Let $0 < \alpha < \lambda < \beta$. By making use of I **116** we establish similarly as in **160** the existence of arbitrarily large $n$ for which

$$\frac{1}{N(r_n)} \sum_{k=1}^{\infty} f\left(\frac{r_k}{r_n}\right) \leqq \int_0^1 f\left(x^{\frac{1}{\alpha}}\right) dx + \int_1^{\infty} f\left(x^{\frac{1}{\beta}}\right) dx.$$

Choose $\alpha$ and $\beta$ sufficiently close to $\lambda$ [**131**].

**162.** [For **162**–**166** cf. H. Weyl: Nachr. Akad. Wiss. Göttingen 1914, pp. 235–236; Math. Ann. Vol. 77, pp. 313–315 (1916).] If $f(x) = 1$ on the subinterval $[\alpha, \beta]$ of $[0, 1]$ and $f(x) = 0$ otherwise the equation (*) on p. 88 leads to the condition $\lim_{n\to\infty} \frac{\nu_n(\alpha, \beta)}{n} = \beta - \alpha$. Now suppose that the condition is satisfied. To begin we note that it does not matter whether the subinterval $\alpha, \beta$ is open, halfopen or closed. The relation (*) holds for any function that is constant ($\neq 0$) on a subinterval and vanishes outside this subinterval, therefore (*) holds also for any linear combination $c_1 f_1(x) + c_2 f_2(x) + \cdots + c_l f_l(x)$ of such functions $f_\nu(x)$, $c_\nu$ constant, i.e. for any piecewise constant function. If $f(x)$ is properly integrable there exist piecewise constant functions [**102**, with $a = 0$, $b = 1$], $\psi(x)$ and $\Psi(x)$ such that

$$\frac{\psi(x_1) + \psi(x_2) + \cdots + \psi(x_n)}{n} \leqq \frac{f(x_1) + f(x_2) + \cdots + f(x_n)}{n}$$
$$\leqq \frac{\Psi(x_1) + \Psi(x_2) + \cdots + \Psi(x_n)}{n}.$$

The first and the last expression converge to $\int_0^1 \psi(x)\,dx$ and $\int_0^1 \Psi(x)\,dx$ resp. and both are arbitrarily close to $\int_0^1 f(x)\,dx$. The weaker conditions

$$\lim_{n\to\infty} \frac{\nu_n(0, \beta)}{n} = \beta, \qquad 0 < \beta < 1$$

or

$$\lim_{n\to\infty} \frac{\nu_n(\beta, 1)}{n} = 1 - \beta, \qquad 0 < \beta < 1$$

are also sufficient. It will even do if one of these conditions is satisfied for a set of $\beta$-values everywhere dense on $(0, 1)$.

**163.** The condition is obviously necessary. As in **162** the interval may be open, half-open or closed. We replace the functions $\psi(x)$ and

$\Psi(x)$ in **102**, if $a > 0$ by *piecewise linear functions* for which the extensions of the single line segments pass through the origin ($y = kx$). As in **162** we now can establish (*) for a properly integrable function on [0, 1] that vanishes in an interval containing the origin, e.g.

$$\lim_{n\to\infty} \frac{\nu_n(\beta,1)}{n} = 1 - \beta, \quad 0 < \beta < 1 \qquad \text{[solution 162]}.$$

**164.** Cf. **162**. Instead of **102** use **137**.

**165.** Cf. **162**. Instead of **102** use **137**.

**166.** The condition of **165** is satisfied because

$$e^{2\pi i k x_1} + e^{2\pi i k x_2} + \cdots + e^{2\pi i k x_n} = \sum_{\nu=1}^{n} e^{2\pi i k \nu\theta} = e^{2\pi i k\theta} \frac{e^{2\pi i k n\theta} - 1}{e^{2\pi i k\theta} - 1}.$$

**167.** In the notation of **166** we have to examine

$$\lim_{n\to\infty} \frac{f(x_1) + f(x_2) + \cdots + f(x_n)}{n}, \quad f(x) = \varphi(a\theta - \theta d + xd),$$

where $\varphi(y) = 1$ or 0 depending on whether the integer next to $y$ is on the right or on the left of $y$. (If $y = n$, $n + \frac{1}{2}$, $n$ integer, we choose, e.g. $\varphi(y) = \frac{1}{2}$.) **166** implies that the limit is

$$= \int_0^1 f(x)\,dx = \int_0^1 \varphi(y)\,dy = \frac{1}{2}.$$

The proposition is also valid for arithmetic progressions of higher order [I **128**] as can be proved by more elaborate methods [H. Weyl, l.c. **162**, p. 326]. The result might be interpreted as an expression of a certain degree of "irregularity" of the sequence $\varepsilon_1, \varepsilon_2, \varepsilon_3, \ldots, \varepsilon_n, \ldots$ [cf. R. v. Mises: Math. Z. Vol. 5, p. 57 (1919).]

**168.** [E. Hecke: Abh. math. Sem. Hamburg Vol. 1, pp. 57–58 (1922).] According to I **88** we have

$$\lim_{r=1-0} (1 - r) \sum_{n=1}^{\infty} a_n r^n = \lim_{n\to\infty} \frac{a_1 + a_2 + \cdots + a_n}{n},$$

provided that the limit on the right hand side exists. For

$$a_n = (n\theta - [n\theta])\, e^{2\pi i n\alpha}$$

this limit becomes, according to **166**,

$$= \int_0^1 x e^{2\pi i q x}\,dx = \frac{1}{2\pi i q}.$$

**169.** [E. Steinitz, Problem: Arch. Math. Phys. Ser. 3, Vol. 19, p. 361 (1912). Solved by G. Pólya: Arch. Math. Phys. Ser. 3, Vol. 21, p. 290

(1913).] The limit is the function $f(x)$ defined in **99**. For irrational $x$ cf. **166**; easier in the case of rational $x$.

**170.** We obtain $10^n\theta - [10^n\theta]$ by multiplying the decimal fraction by $10^n$ and omitting all the digits to the left of the decimal point. Let $\alpha = \alpha_1\alpha_2 \cdots \alpha_k$ represent a finite decimal fraction. Choose $n$ so that $10^n\theta - [10^n\theta]$ starts with the digits $\alpha_1, \alpha_2, \ldots, \alpha_k$ and that $r$ zeros follow. Then

$$|10^n\theta - [10^n\theta] - \alpha| < \frac{1}{10^{k+r}}.$$

**171.** The Taylor series of $e$ is

$$e = 1 + \frac{1}{1!} + \frac{1}{2!} + \cdots + \frac{1}{n!} + \frac{e^{\theta_n}}{(n+1)!}, \quad 0 < \theta_n < 1,$$

hence $n!e = n! + \frac{n!}{1!} + \frac{n!}{2!} + \cdots + 1 + \frac{e^{\theta_n}}{n+1}$. For $n \geqq 2$ we have $\frac{e^{\theta_n}}{n+1} < \frac{e}{n+1} < 1$, thus $n!e - [n!e] = \frac{e^{\theta_n}}{n+1} < \frac{3}{n+1}$.

**172.** [Communicated by H. Prüfer; H. Weyl proved, l.c. **162**, that the set in question is everywhere dense, even equidistributed on the interval [0, 1].] For $r = 1$ the statement follows from **166**. Assume $r > 1$ and that $a_r$ is irrational [otherwise we omit the highest rational terms which are periodic mod 1]. If the set had only a finite number of limit points this would be true also of the remainder mod 1 of the numbers

$$P(n+r-1) - \binom{r-1}{1} P(n+r-2) + \binom{r-1}{2} P(n+r-3) - \cdots$$
$$+ (-1)^{r-1} P(n) = a_r r! \left(n + \frac{r-1}{2}\right) + a_{r-1}(r-1)!.$$

This, however, contradicts **166**.

**173.** Let $k$ be a positive integer. Then

$$e^{2\pi i k x_1} + e^{2\pi i k x_2} + \cdots + e^{2\pi i k x_n}$$

is bounded [**166**]. Partial summation shows that

$$\alpha_1 e^{2\pi i k x_1} + \alpha_2 e^{2\pi i k x_2} + \cdots + \alpha_n e^{2\pi i k x_n}$$

is bounded too [**165**].

**174.** [L. Fejér.] Let $N(x)$ denote the counting function of the sequence $g(1), g(2), \ldots, g(n), \ldots$ and let $t = \gamma(x)$ be the inverse function of $x = g(t)$. Then $N(x) = [\gamma(x)]$. The conditions (1)–(4) imposed on $g(t)$ imply the following properties of $\gamma(x)$ for $x \geqq g(1)$: $\gamma(x)$ is continuously differentiable, $\gamma(x)$ is monotone increasing to infinity as $x \to \infty$, $\gamma'(x)$ is monotone increasing to infinity as $x \to \infty$, while $\frac{\gamma'(x)}{\gamma(x)} \to 0$.

Since $\frac{x}{\gamma(x)} < 2\frac{x-\frac{x}{2}}{\gamma(x)-\gamma\left(\frac{x}{2}\right)} = \frac{2}{\gamma'(x_1)}, \frac{x}{2} < x_1 < x$, we conclude from the above that $\frac{x}{\gamma(x)} \to 0$ as $x \to \infty$; furthermore $\frac{\gamma(x+\varepsilon)}{\gamma(x)} \to 1$, $\varepsilon$ fixed or bounded, as $x$ increases.

Let $0 < \alpha < 1$. It is sufficient to prove [**162**] that

$$\frac{\sum_{k=1}^{m-1}(N(k+\alpha)-N(k))+N(m+\lambda_n)-N(m)}{N(m+x_n)}, \quad m=[g(n)], \lambda_n = \min(x_n, \alpha)$$

converges to $\alpha$ as $n \to \infty$. We replace $N(x)$ by $\gamma(x)$. In view of the properties of $\gamma(x)$ we find that this proposition is equivalent to

$$\lim_{m\to\infty}\frac{1}{\gamma(m)}\sum_{k=1}^{m}\left(\gamma(k+\alpha)-\gamma(k)\right)=\alpha.$$

According to **19** this quotient is, with $g(1) = x_0$, equal to

$$\frac{\gamma(m+\alpha)-\gamma(m)}{2\gamma(m)}+\frac{1}{\gamma(m)}\int_{x_0}^{m}\left(\gamma(x+\alpha)-\gamma(x)\right)dx+O\left(\frac{\gamma'(m)}{\gamma(m)}\right)$$

$$=\frac{\alpha}{\gamma(m)}\int_{x_0}^{m}\gamma'(\xi)\,dx+o(1),$$

$x < \xi = \xi(x) < x + \alpha$. Since $\gamma'(x)$ is montone we have

$$\gamma(m)-\gamma(x_0)=\int_{x_0}^{m}\gamma'(x)\,dx<\int_{x_0}^{m}\gamma'(\xi)\,dx<\int_{x_0}^{m}\gamma'(x+\alpha)\,dx$$

$$=\gamma(m+\alpha)-\gamma(x_0+\alpha).$$

**175.** Special case of **174**: $g(t) = at^\sigma$.

**176.** Special case of **174**: $g(t) = a(\log t)^\sigma$.

**177.** Assume $0 < \varrho \leqq 1$. We write

$$s_n = |\sin 1^\sigma\xi| + |\sin 2^\sigma\xi| + \cdots + |\sin n^\sigma\xi|.$$

**174** implies for

$$g(t)=\frac{\xi}{2\pi}t^\sigma, \quad f(x)=|\sin 2\pi x|,$$

$$\lim_{n\to\infty}\frac{s_n}{n}=\int_0^1|\sin 2\pi x|\,dx=\frac{2}{\pi}.$$

Thus ($s_0 = 0$)

$$\sum_{\nu=1}^{n}\frac{s_\nu - s_{\nu-1}}{\nu^\varrho}=\sum_{\nu=1}^{n-1}s_\nu\left(\frac{1}{\nu^\varrho}-\frac{1}{(\nu+1)^\varrho}\right)+\frac{s_n}{n^\varrho}\to+\infty.$$

**178.** [J. Franel: Vjschr. Naturf. Ges. Zürich Vol. 62, p. 295 (1917).] Write in decimal notation

$$\sqrt{n} = c^{(n)}.\ c_1^{(n)} c_2^{(n)} c_3^{(n)} \cdots.$$

(We exlude the case where all the $c_j^{(n)}$'s are equal to 9 for $j$ larger than a certain given index.) In **175** we put $\sigma = \frac{1}{2}$, $a = 10^{j-1}$; then

$$x_n = 10^{j-1}\sqrt{n} - [10^{j-1}\sqrt{n}] = 0.\ c_j^{(n)} c_{j+1}^{(n)} c_{j+2}^{(n)} \cdots,$$

i.e. $c_j^{(n)} = [10x_n]$. We find $c_j^{(n)} = g$ if and only if $\frac{g}{10} \leqq x_n < \frac{g+1}{10}$. This means

$$\lim_{n\to\infty} \frac{\nu_g(n)}{n} = \int\limits_{\frac{g}{10}}^{\frac{g+1}{10}} dx = \frac{1}{10}.$$

**179.** Using the notation of solution **174** we have to prove

$$\lim \frac{\sum\limits_{k=1}^{m-1} (N(k+\alpha) - N(k)) + N(m+\lambda_n) - N(m)}{N(m+x_n)} = \int\limits_0^\alpha K(x,\xi)\,dx,$$

as $n \to \infty$, $m \to \infty$, $x_n \to \xi$, with $N(x) = [q^x]$. We replace $N(x)$ by $q^x$ in the expression in question, which is justified because $mq^{-m} \to 0$, and obtain

$$\frac{\sum\limits_{k=1}^{m-1} q^k(q^\alpha - 1) + q^m(q^{\lambda_n} - 1)}{q^{m+x_n}} \to \frac{q^\alpha - 1}{q-1} q^{-\xi} + (q^\lambda - 1)\, q^{-\xi},\ \lambda = \min(\xi, \alpha).$$

The last expression is $= \int\limits_0^\alpha K(x,\xi)\,dx$ as can be easily checked by integration, or better by differentiation with respect to $\alpha$.

**180.** The function

$$\varphi(\xi) = \int\limits_0^1 f(x)\,K(x,\xi)\,dx = \frac{\log q}{q-1} q^{-\xi}\left(\int\limits_0^1 f(x)\,q^x\,dx + (q-1)\int\limits_0^\xi f(x)\,q^x\,dx\right)$$

is continuous in the interval $[0, 1]$, $\varphi(0) = \varphi(1)$; $\varphi(\xi)$ is constant if and only if $f(x)$ assumes the same value at all its points of continuity (differentiate). As $a \to \infty$, $q$ converges to 1, as $a \to 0$, $q$ tends to infinity. We find accordingly

$$\lim_{q\to\infty} \int\limits_0^1 f(x)\,K(x,\xi)\,dx = \int\limits_0^1 f(x)\,dx.$$

As $a$ increases to infinity $J(a; f)$ is reduced to a point, $\int_0^1 f(x)\,dx$; the distribution is almost uniform for large $a$'s [**176**]. If $0 < \xi < 1$ we find in addition

$$\lim_{q\to\infty} \int_0^1 f(x)\,K(x,\xi)\,dx = f(\xi)$$

for any point of continuity $\xi$ [**132**]. If $\xi$ is a point of ordinary discontinuity (jump) of $f(x)$ the limit is $f(\xi - 0)$. The limit is $f(1-0)$ for $\xi = 0$ and $\xi = 1$. If $f(x)$ is e.g. of bounded variation $J(a; f)$ approaches the entire range of the function $f(x)$ on $(0, 1)$ as $a \to 0$. The jumps of $f(x)$ at points of ordinary discontinuity are included. For small values of $a$ the distribution is almost like the distribution in **182**.

**181.** [J. Franel, l.c. **178**, pp. 285–295.] We denote the common logarithm of $n$ by Log $n$ and write it as a decimal fraction

$$\text{Log}\, n = \frac{\log n}{\log 10} = c^{(n)}.\ c_1^{(n)}\, c_2^{(n)}\, c_3^{(n)} \cdots.$$

In **179** and **180** set $a = \frac{10^{j-1}}{\log 10}$. Then

$$x_n = 10^{j-1}\,\text{Log}\, n - [10^{j-1}\text{Log}\, n] = 0.\ c_j^{(n)} c_{j+1}^{(n)} c_{j+2}^{(n)} \cdots,$$

i.e. $c_j^{(n)} = [10 x_n]$. Thus we are concerned with the range of the continuous function

$$\varphi(\xi) = \int_{\frac{g}{10}}^{\frac{g+1}{10}} K(x,\xi)\,dx$$

on the interval $[0, 1]$; $K(x, \xi)$ is defined as in **179** and $q = e^{\frac{\log 10}{10^{j-1}}} = 10^{10^{1-j}}$.

**182.** Let [solution **174**]

$$m = [g(n)], \quad x = g(t), \quad t = \gamma(x), \quad N(x) = [\gamma(x)].$$

Because of the conditions (1)–(4) the function $\gamma(x)$ is continuously differentiable for $x \geqq g(1)$, monotone increasing to infinity as $x \to \infty$, moreover $\gamma'(x) \to \infty$, $\frac{\gamma'(x)}{\gamma(x)} \to \infty$, whence $\frac{\gamma(x-\varepsilon)}{\gamma(x)} \to 0$ as $x \to \infty$ and also $\frac{N(x-\varepsilon)}{N(x)} \to 0$ if $\varepsilon$ fixed, $\varepsilon > 0$ or if $\varepsilon$ has a lower positive bound as $x$ increases. Assume that $0 < \xi < 1$, $f(x)$ continuous at $x = \xi$, $\varepsilon > 0$, $\varepsilon$ arbitrary, $\delta > 0$ such that $|f(x) - f(\xi)| < \varepsilon$ whenever $|x - \xi| < \delta$. We choose $n$ so that $|x_n - \xi| < \frac{\delta}{2}$ [I **101**] and find

$$\left|\frac{f(x_1) + f(x_2) + \cdots + f(x_n)}{n} - f(\xi)\right|$$

$$\leqq \frac{|f(x_1) - f(\xi)| + |f(x_2) - f(\xi)| + \cdots + |f(x_n) - f(\xi)|}{n} < 2M\frac{N(m+\xi-\delta)}{n} + \varepsilon,$$

provided $|f(x)| < M$. According to the hypothesis we have, however,

$$\frac{N(m+\xi-\delta)}{n} < \frac{N(m+\xi-\delta)}{N\left(m+\xi-\frac{\delta}{2}\right)} \to 0 \quad \text{for} \quad m \to \infty.$$

Now suppose that $\xi$ is a jump point of $f(x)$ and that again $|x_n - \xi| < \frac{\delta}{2}$. Then

$$-2M\frac{N(m+\xi-\delta)}{n} + \mu(\delta) < \frac{f(x_1)+f(x_2)+\cdots+f(x_n)}{n}$$
$$< 2M\frac{N(m+\xi-\delta)}{n} + M(\delta);$$

where $\mu(\delta)$ and $M(\delta)$ denote the greatest lower and the least upper bound of $f(x)$ in the interval $|x-\xi| < \delta$. These inequalities imply that the limit points lie between $f(\xi - 0)$ and $f(\xi + 0)$. Now we prove that the limit points cover the entire interval between $f(\xi - 0)$ and $f(\xi + 0)$: Since $f(x)$ is integrable there exist points of continuity of $f(x)$ arbitrarily close to $\xi$ [**108**]. Let $\xi'$ and $\xi''$ be two such points, $0 < \xi' < \xi < \xi'' < 1$ and let $x_{n'}$ and $x_{n''}$ ($n'$, $n''$ integers) be two sequences such that $x_{n'} \to \xi'$ and $x_{n''} \to \xi''$, $[g(n')] = [g(n'')]$. We set $f(x_1) + f(x_2) + \cdots + f(x_n) = nF_n$, thus $F_{n'} \to f(\xi')$, $F_{n''} \to f(\xi'')$. Moreover we have for each $n$

$$|F_n - F_{n+1}| = \left|\frac{F_n}{n+1} - \frac{f(x_{n+1})}{n+1}\right| < \frac{2M}{n+1}.$$

The sequence $F_{n'}, F_{n'+1}, F_{n'+2}, \ldots, F_{n''}$ is "slowly increasing" or "slowly decreasing" in the sense of solution I **100**. It is easy to adapt the proof to the case $\xi = 0$ or $\xi = 1$.

**183.** Special case of **182**: $g(t) = a (\log t)^\sigma$.

**184.** The proposition follows from **182**, **183** for

$$g(t) = 10^{j-1}\sqrt{\operatorname{Log} t} = \frac{10^{j-1}}{\sqrt{\log 10}}\sqrt{\log t},\ f(x) = 1 \text{ for } [10x] = g, \text{ otherwise}$$

$f(x) = 0$. Cf. **178**, **181**.

**185.** [H. Weyl, l.c. **162**, pp. 319–320.] It is sufficient to prove the proposition for $f(x_1, x_2, \ldots, x_p) = e^{2\pi i(k_1x_1+k_2x_2+\cdots+k_px_p)}$, where $k_1, k_2, \ldots, k_p$ are integers at least one of which does not vanish [**165**]. Introduce

$$k_1a_1 + k_2a_2 + \cdots + k_pa_p = a,\quad k_1\theta_1 + k_2\theta_2 + \cdots + k_p\theta_p = \theta;$$

then

$$\frac{1}{t}\int_0^t e^{2\pi i(a+\theta t)}\,dt = e^{2\pi i a}\frac{e^{2\pi i\theta t} - 1}{2\pi i\theta t} \to 0.$$

**186.** Special case of **185**: $p = 2$, $f(x_1, x_2) = 1$, if $\alpha_1 \leqq x_1 \leqq \alpha_2$, $\beta_1 \leqq x_2 \leqq \beta_2$, otherwise $f(x_1, x_2) = 0$ in the unit square.

**187.** [Cf. D. König and A. Szücs: Rend. Circ. Mat. Palermo Vol. 36, pp. 79—83 (1913).] We may assume that the motion in question takes place in the square $0 \leqq x \leqq \frac{1}{2}$, $0 \leqq y \leqq \frac{1}{2}$. By reflection in the lines $x = \frac{1}{2}$, $y = \frac{1}{2}$ three domains besides $\mathfrak{f}$ are obtained. The four domains form a subdomain $\mathfrak{f}^*$ of the unit square $0 \leqq x \leqq 1$, $0 \leqq y \leqq 1$. Translating $\mathfrak{f}^*$ by an integer number of units parallel to the axes we construct an infinite number of domains as in **186**. The original zigzag motion in the square $0 \leqq x \leqq \frac{1}{2}$, $0 \leqq y \leqq \frac{1}{2}$ can be replaced by a rectilinear motion. Special case of **185**: $p = 2$, $f(x_1, x_2) = 1$ if $x_1, x_2$ is in $\mathfrak{f}^*$, otherwise $f(x_1, x_2) = 0$ in the unit square.

**188.** [G. Pólya: Nachr. Akad. Wiss. Göttingen 1918, pp. 28—29.] We put $\psi(y) = e^{2\pi iky}$, $k$ positive integer, in VIII **35**. Then we obtain, using the same notation, $g(n) = 0$ for $n > k$. I.e.

$$\left|\sum_{(r,n)=1} \psi\left(\frac{r}{n}\right)\right| = \left|\sum_{t|n;\, t\leqq k} \mu\left(\frac{n}{t}\right) g(t)\right| \leqq |g(1)| + |g(2)| + \cdots + |g(k)|$$

for any value of $n$ [**165**]; and $\varphi(n) \to \infty$ cf. VIII **264**.

**189.** Using the notation of **188** we set in I **70**

$$a_n = f\left(\frac{r_{1n}}{n}\right) + f\left(\frac{r_{2n}}{n}\right) + \cdots + f\left(\frac{r_{\varphi n}}{n}\right),$$

$$b_n = \varphi(n) \qquad n = 1, 2, 3, \ldots$$

**190.** Apply **137** to $x^{-p} f(x)$. Cf. **40** for the special functions $a_0 x^p + a_1 x^{p+1} + \cdots + a_l x^{p+l}$, where $a_0, a_1, \ldots, a_l$ are constants. The result generalizes a well known theorem in probability theory. Cf. e.g. A. A. Markoff: Wahrscheinlichkeitsrechnung. Leipzig: G. B. Teubner 1912, pp. 33—34. H. Cramér: Mathematical Methods of Statistics. Princeton: Princeton University Press 1966, p. 214.

**191.** We denote the highest coefficient of $P_n(x)$ by $k_n = \frac{(2n)!}{2^n (n!)^2}$ [solution VI **84**]. Then

$$\left(1 + \frac{x_{1n}}{\lambda}\right)\left(1 + \frac{x_{2n}}{\lambda}\right) \cdots \left(1 + \frac{x_{nn}}{\lambda}\right) = k_n^{-1}(-\lambda)^{-n} P_n(-\lambda) = k_n^{-1} \lambda^{-n} P_n(\lambda)$$

[**49, 203**].

**192.** We find [**52**]

$$\log \frac{\lambda + \sqrt{\lambda^2 - 1}}{2\lambda} = \frac{1}{\pi} \int_0^\pi \log\left(1 + \frac{\cos\vartheta}{\lambda}\right) d\vartheta\,.$$

In I **179** put

$$x = \frac{1}{\lambda},\ a_{nk} = \frac{(-1)^{k-1}}{k} \left(\frac{x_{1n}^k + x_{2n}^k + \cdots + x_{nn}^k}{n} - \frac{1}{\pi} \int_0^\pi \cos^k \vartheta\, d\vartheta\right),\ A = 2.$$

**193.** [Cf. G. Szegö: Math. természettud. Ért. Vol. 36, p. 531 (1918).] Follows from **192** in analogy to **164**.

**194.** Special case of **193**: $f(x) = 1$ if $\alpha \leqq x \leqq \beta$, otherwise $f(x) = 0$ in the interval $[-1, 1]$.

**195.** We can assume $a_1 > a_2 > \cdots > a_l$. Then

$$p_1 a_1^n + p_2 a_2^n + p_3 a_3^n + \cdots + p_l a_l^n$$

$$= p_1 a_1^n \left[1 + \frac{p_2}{p_1}\left(\frac{a_2}{a_1}\right)^n + \frac{p_3}{p_1}\left(\frac{a_3}{a_1}\right)^n + \cdots + \frac{p_l}{p_1}\left(\frac{a_l}{a_1}\right)^n\right].$$

The expression in square brackets converges to 1. A different proof follows from **196** and I **68**. Cf. **82**.

**196.** [Solution **195**.]

**197.** Let $f(x) = c(x - a_1)(x - a_2) \cdots (x - a_l)$, $c \neq 0$. Because of

$$\frac{f'(x)}{f(x)} = \frac{1}{x - a_1} + \frac{1}{x - a_2} + \cdots + \frac{1}{x - a_l}$$

we find

$$c_n = a_1^{-n-1} + a_2^{-n-1} + \cdots + a_l^{-n-1}, \qquad n = 0, 1, 2, \ldots$$

Use **195**, **196**. (**95.3**, somewhat amplified, so that it applies also to finite sums, is much more informative. See also III **242**.)

**198.** Let $f(x)$ attain its maximum at $x = \xi$, $a \leqq \xi \leqq b$, $\varepsilon > 0$, $\delta$ be positive and so small that

$$f(\xi) - \varepsilon < f(x) \leqq f(\xi)$$

whenever $|x - \xi| < \delta$, $a \leqq x \leqq b$. Then

$$[f(\xi) - \varepsilon]^n \int_{\xi-\delta}^{\xi+\delta} \varphi(x)\, dx \leqq \int_a^b \varphi(x)\, [f(x)]^n\, dx \leqq [f(\xi)]^n \int_a^b \varphi(x)\, dx.$$

If $\xi - \delta < a$ the lower limit $\xi - \delta$ of the integration is replaced by $a$, if $\xi + \delta > b$ the upper limit $\xi + \delta$ is replaced by $b$. Take the $n$-th root, let $n$ increase to infinity and then let $\varepsilon$ converge to 0. Cf. **83**.

**199.** First proof [P. Csillag]: Let $0 < \varepsilon < M = \max f(x)$. Then

$$\int_a^b \varphi(x)\, [f(x)]^{n+1}\, dx \geqq \int_{f(x) \geqq M-\varepsilon} \varphi(x)\, [f(x)]^{n+1}\, dx \geqq (M - \varepsilon) \int_{f(x) \geqq M-\varepsilon} \varphi(x)\, [f(x)]^n\, dx,$$

$$\int_{f(x) \geqq M-\varepsilon} \varphi(x)\, [f(x)]^n\, dx \geqq \int_{f(x) \geqq M-\frac{\varepsilon}{2}} \varphi(x)\, [f(x)]^n\, dx \geqq C\left(M - \frac{\varepsilon}{2}\right)^n,$$

where the positive constant $C$ is independent of $n$. Consequently

$$M \geqq \frac{\int_a^b \varphi(x)\,[f(x)]^{n+1}dx}{\int_a^b \varphi(x)\,[f(x)]^n dx} \geqq (M-\varepsilon)\,\frac{\int\limits_{f(x)\geqq M-\varepsilon} \varphi(x)\,[f(x)]^n dx}{\int\limits_{f(x)\geqq M-\varepsilon} \varphi(x)\,[f(x)]^n dx + \int\limits_{f(x)<M-\varepsilon} \varphi(x)\,[f(x)]^n dx}$$

As $n \to \infty$ the last quotient converges to 1 because

$$\frac{\int\limits_{f(x)<M-\varepsilon} \varphi(x)\,[f(x)]^n dx}{\int\limits_{f(x)\geqq M-\varepsilon} \varphi(x)\,[f(x)]^n dx} \leqq \frac{(M-\varepsilon)^n \int_a^b \varphi(x)\,dx}{C\left(M-\frac{\varepsilon}{2}\right)^n}.$$

The proof almost involves the concept of the Lebesgue integral.

Second proof: We write

$$I_n = \int_a^b \varphi(x)\,[f(x)]^n\,dx = \int_a^b \sqrt{\varphi(x)}[f(x)]^{\frac{n-1}{2}} \cdot \sqrt{\varphi(x)}\,[f(x)]^{\frac{n+1}{2}}\,dx,$$

then [**81**]

$$I_n^2 \leqq I_{n-1} I_{n+1}, \qquad n = 1, 2, 3, \ldots$$

The sequence $\frac{I_{n+1}}{I_n}$ is therefore monotone increasing. The value of the limit follows from **198** and I **68.1**.

**200.** By introducing the new variable $\sqrt{kn}(x-\xi) = t$ we transform the integral into

$$\frac{1}{\sqrt{kn}} \int\limits_{-\sqrt{kn}(\xi-a)}^{\sqrt{kn}(b-\xi)} e^{-t^2}\,dt.$$

As $n \to \infty$ this integral converges to $\int_{-\infty}^{\infty} e^{-t^2}\,dt = \sqrt{\pi}$.

**201.** [Laplace: Théorie analytique des probabilités, Vol. 1, Part 2, Chap. 1; Oeuvres, Vol. 7, p. 89. Paris: Gauthier-Villars 1886. G. Darboux: J. Math. Pures Appl. Ser. 3, Vol. 4, pp. 5–56, 377–416 (1878). T. J. Stieltjes, Ch. Hermite: Correspondence d'Hermite et de Stieltjes, Vol. 2. Paris: Gauthier-Villars 1905, p. 185, 315–317, 333. H. Lebesgue: Annls Fac. Sci. Univ. Toulouse Ser. 3, Vol. 1, pp. 119–128 (1909). H. Burkhardt: Sber. bayer. Akad. Wiss. 1914, pp. 1–11. O. Perron: Sber. bayer. Akad. Wiss. 1917, pp. 191–219.] Let $\varepsilon > 0$, $\delta$ be positive and so small that $a < \xi - \delta < \xi + \delta < b$ and

$$\varphi(\xi) - \varepsilon < \varphi(x) < \varphi(\xi) + \varepsilon, \quad h''(\xi) - \varepsilon < h''(x) < h''(\xi) + \varepsilon < 0,$$

whenever $\xi - \delta < x < \xi + \delta$. Then

$$\int_a^b \varphi(x)\, e^{n[h(x)-h(\xi)]}\, dx = \int_{\xi-\delta}^{\xi+\delta} \varphi(x)\, e^{n[h(x)-h(\xi)]}\, dx + O(\alpha^n)$$

$$= \varphi(\xi') \int_{\xi-\delta}^{\xi+\delta} e^{\frac{n}{2}(x-\xi)^2 h''(\xi'')}\, dx + O(\alpha^n);$$

where $0 < \alpha < 1$, $\alpha$ depends on $\varepsilon$ but not on $n$, $\xi - \delta < \xi' < \xi + \delta$, $\xi - \delta < \xi'' < \xi + \delta$. The first term on the right hand side lies between

$$[\varphi(\xi) - \varepsilon] \int_{\xi-\delta}^{\xi+\delta} e^{\frac{n}{2}(x-\xi)^2[h''(\xi)-\varepsilon]}\, dx \quad \text{and} \quad [\varphi(\xi) + \varepsilon] \int_{\xi-\delta}^{\xi+\delta} e^{\frac{n}{2}(x-\xi)^2[h''(\xi)+\varepsilon]}\, dx.$$

According to **200** these bounds are asymptotically equal to

$$[\varphi(\xi) - \varepsilon]\sqrt{-\frac{2\pi}{[h''(\xi)-\varepsilon]\, n}} \quad \text{and} \quad [\varphi(\xi) + \varepsilon]\sqrt{-\frac{2\pi}{[h''(\xi)+\varepsilon]\, n}} \qquad \text{resp.}$$

The theorem is also true if $n$ increases continuously to infinity.

**202.** [Wallis' formula.]

$$\frac{1 \cdot 3 \cdots (2n-1)}{2 \cdot 4 \cdots 2n} = \frac{1}{\pi} \int_0^\pi \sin^{2n} x\, dx.$$

Special case of **201**:

$a = 0$, $b = \pi$, $\varphi(x) = \frac{1}{\pi}$, $f(x) = \sin^2 x$, $\xi = \frac{\pi}{2}$. (Follows also from **205.**)

**203.**

$$P_n(\lambda) = \frac{1}{2\pi} \int_{-\pi}^{\pi} (\lambda + \sqrt{\lambda^2 - 1} \cos x)^n\, dx.$$

Special case of **201**:

$$a = -\pi, \quad b = \pi, \quad \varphi(x) = \frac{1}{2\pi}, \quad f(x) = \lambda + \sqrt{\lambda^2 - 1} \cos x, \qquad \xi = 0.$$

**204.**

$$i^{-\nu} J_\nu(it) = \frac{1}{2\pi} \int_0^{2\pi} e^{-t\cos x} \cos \nu x\, dx.$$

Special case of **201**:

$$a = 0, \quad b = 2\pi, \quad \varphi(x) = \frac{1}{2\pi} \cos \nu x, \quad f(x) = e^{-\cos x}, \quad \xi = \pi, \quad n = t.$$

**205.** [Stirling's formula.]

$$\Gamma(n+1) = n^{n+1} \int_0^\infty (e^{-x} x)^n\, dx.$$

Special case of **201**:

$$a = 0,\ b = \infty,\ \varphi(x) = 1,\ f(x) = e^{-x}x,\ \xi = 1.$$

For more accurate approximates see **18, I 167, I 155.**

**206.** According to **205** we have

$$\binom{nk+l}{n} = \frac{\Gamma(nk+l+1)}{\Gamma(n+1)\,\Gamma(nk-n+l+1)} \sim \left(\frac{nk}{nk-n}\right)^l \frac{\Gamma(nk+1)}{\Gamma(n+1)\,\Gamma(nk-n+1)}$$

$$\sim \left(\frac{k}{k-1}\right)^l \left(\frac{nk}{e}\right)^{nk} \left(\frac{e}{n}\right)^n \left(\frac{e}{nk-n}\right)^{nk-n} \sqrt{\frac{2\pi nk}{2\pi n \cdot 2\pi(nk-n)}}.$$

**207.** The substitution of $tx$ for $x$ transforms the integral into

$$t^\alpha \int_{t^{-1}}^{\infty} x^{\alpha-1} \left(\frac{e}{x}\right)^{tx} dx.$$

Let $t$ be so large that $t^{-1} < \frac{1}{2}$. The integral

$$t^\alpha \int_{t^{-1}}^{\frac{1}{2}} x^{\alpha-1} \left(\frac{e}{x}\right)^{tx} dx$$

can be disregarded because the function $f(x) = \left(\frac{e}{x}\right)^x$ increases on the interval $\left[0, \frac{1}{2}\right]$, i.e. $f(x) \leqq \sqrt{2e} < e$ on $\left[0, \frac{1}{2}\right]$. Apply **201** to the integral

$$\int_{\frac{1}{2}}^{\infty} x^{\alpha-1}\left(\frac{e}{x}\right)^{tx} dx:$$

$$a = \frac{1}{2},\ b = \infty,\ \varphi(x) = x^{\alpha-1},\ f(x) = \left(\frac{e}{x}\right)^x,\ \xi = 1,\ n = t.$$

**208.** Substitute $\tau^{-\frac{1}{1-\alpha}}(1+x)$ for $x$. There results

$$\tau^{-\frac{1}{1-\alpha}} \exp\left(\frac{1-\alpha}{\alpha}\tau^{-\frac{\alpha}{1-\alpha}}\right) \int_{-1}^{\infty} \exp\left(\tau^{-\frac{\alpha}{1-\alpha}} \frac{(1+x)^\alpha - 1 - \alpha x}{\alpha}\right) dx.$$

Special case of **201**:

$$a = -1,\ b = \infty,\ \varphi(x) = 1,\ h(x) = \frac{(1+x)^\alpha - 1 - \alpha x}{\alpha},$$

$$\xi = 0, \qquad n = \tau^{-\frac{\alpha}{1-\alpha}}.$$

**209.** Substitute $e^{-1}t^{\frac{1}{\alpha}}(1+x)$ for $x$. There results

$$e^{-1}t^{\frac{1}{\alpha}} \exp\left(e^{-1}\alpha t^{\frac{1}{\alpha}}\right) \int_{-1}^{\infty} \exp\left\{e^{-1}\alpha t^{\frac{1}{\alpha}}\,[x - (1+x)\log(1+x)]\right\} dx.$$

Special case of **201**:

$$a = -1,\ b = \infty,\ \varphi(x) = 1,\ h(x) = x - (1 + x) \log (1 + x),$$

$$\xi = 0,\ n = e^{-1} \alpha t^{\frac{1}{\alpha}}.$$

**210.** We put $\eta = n^{-\frac{1}{2}+\varepsilon}$, $0 < \varepsilon < \frac{1}{6}$. The integral in question becomes [**205**]

$$= \sqrt{\frac{n}{2\pi}}\left[1 + O\left(\frac{1}{n}\right)\right] \int\limits_{-1}^{\alpha n^{-\frac{1}{2}} + \beta n^{-1}} [e^{-x}(1 + x)]^n \, dx$$

$$= \sqrt{\frac{n}{2\pi}}\left[1 + O\left(\frac{1}{n}\right)\right] \int\limits_{-\eta}^{\alpha n^{-\frac{1}{2}} + \beta n^{-1}} [e^{-x}(1 + x)]^n \, dx + O\left(\sqrt{n}\, e^{-\frac{1}{2} n^{2\varepsilon}}\right).$$

Since the function $e^{-x}(1 + x)$ increases for $x < 0$ the integral over $[-1, -\eta]$ is smaller than $[e^{\eta}(1 - \eta)]^n$. We now expand the integrand on the remaining interval

$$e^{-x}(1 + x) = e^{-\frac{x^2}{2} + \frac{x^3}{3} - \frac{x^4}{4}(1+\theta x)^{-4}},\quad 0 < \theta = \theta(x) < 1.$$

The factor $e^{-n\frac{x^4}{4}(1+\theta x)^{-4}}$ is of the form $1 + O(n^{-1+4\varepsilon})$. The integral becomes therefore

$$\sqrt{\frac{n}{2\pi}}\,[1 + O(n^{-1+4\varepsilon})] \int\limits_{-\eta}^{\alpha n^{-\frac{1}{2}} + \beta n^{-1}} e^{-n\left(\frac{x^2}{2} - \frac{x^3}{3}\right)} dx + O\left(\sqrt{n}\, e^{-\frac{1}{2} n^{2\varepsilon}}\right).$$

We have

$$e^{n\frac{x^3}{3}} = 1 + n\frac{x^3}{3} + O(n^{-1+6\varepsilon}).$$

Since $\int\limits_{-\eta}^{\alpha n^{-\frac{1}{2}} + \beta n^{-1}} e^{-n\frac{x^2}{2}} dx$ is of order $n^{-\frac{1}{2}}$ the $O$-term of $e^{n\frac{x^3}{3}}$ yields a contribution of $O(n^{-1+6\varepsilon})$. Hence

$$\sqrt{\frac{n}{2\pi}}\,[1 + O(n^{-1+4\varepsilon})] \int\limits_{-\eta}^{\alpha n^{-\frac{1}{2}} + \beta n^{-1}} e^{-n\frac{x^2}{2}}\left(1 + n\frac{x^3}{3}\right) dx + O(n^{-1+6\varepsilon})$$

$$= \frac{1}{\sqrt{2\pi}} \int\limits_{-n^{\varepsilon}}^{\alpha + \beta n^{-\frac{1}{2}}} e^{-\frac{x^2}{2}}\left(1 + \frac{x^3}{3\sqrt{n}}\right) dx + O(n^{-1+6\varepsilon}) = \frac{1}{\sqrt{2\pi}} \int\limits_{-\infty}^{\alpha + \beta n^{-\frac{1}{2}}} e^{-\frac{x^2}{2}}\left(1 + \frac{x^3}{3\sqrt{n}}\right) \times dx + O(n^{-1+6\varepsilon})$$

$$= \frac{1}{\sqrt{2\pi}} \int\limits_{-\infty}^{\alpha} e^{-\frac{x^2}{2}} dx + \frac{1}{\sqrt{2\pi}} \int\limits_{\alpha}^{\alpha + \beta n^{-\frac{1}{2}}} e^{-\frac{x^2}{2}} dx + \frac{1}{\sqrt{2\pi n}} \int\limits_{-\infty}^{\alpha + \beta n^{-\frac{1}{2}}} e^{-\frac{x^2}{2}} \frac{x^3}{3}\, dx + O(n^{-1+6\varepsilon}).$$

**211.** [Cf. A. de Moivre: The Doctrine of Chances, 2nd Ed. London 1738, pp. 41–42.] We have

$$K_n(x) = \frac{1}{n!}\int_0^x e^{-x}x^n\,dx = 1 - e^{-x}\left(1 + \frac{x}{1!} + \frac{x^2}{2!} + \cdots + \frac{x^n}{n!}\right).$$

Therefore $x_n$ is the only positive root of the transcendental equation $K_n(x) = 1 - \lambda$. According to **210** we find for arbitrary $n$-free $\alpha$ and $\beta$

$$K_n(n + \alpha\sqrt{n} + \beta) = A + \frac{B}{\sqrt{n}} + o\left(\frac{1}{\sqrt{n}}\right),$$

where $A$ and $B$ have the same meaning as in **210**. Determine $\alpha$ and $\beta$ so that $A = 1 - \lambda$ and $B = 0$. Then $x_n - (n + \alpha\sqrt{n} + \beta)$ must converge to zero. If, on the contrary, we had $x_n - (n + \alpha\sqrt{n} + \beta) > c > 0$ for infinitely many $n$ we could conclude that

$$1 - \lambda = K_n(x_n) > K_n(n + \alpha\sqrt{n} + \beta + c) = A + \frac{B'}{\sqrt{n}} + o\left(\frac{1}{\sqrt{n}}\right),$$

where $B'$ depends on $\alpha$ and $\beta + c$ as $B$ depends on $\alpha$ and $\beta$. In particular $B' = B + \frac{1}{\sqrt{2\pi}}ce^{-\frac{\alpha^2}{2}} = \frac{1}{\sqrt{2\pi}}ce^{-\frac{\alpha^2}{2}} > 0$. Since $A = 1 - \lambda$ the last inequality is impossible. In a similar way we show that

$$x_n - (n + \alpha\sqrt{n} + \beta) < -c < 0$$

cannot hold for infinitely many $n$.

**212.** By a similar computation as in **201**: Instead of **200** consider the formula

$$\int_a^{\xi + \frac{\alpha}{\sqrt{n}}} e^{-kn(x-\xi)^2}\,dx \sim \frac{1}{\sqrt{2kn}}\int_{-\infty}^{\alpha\sqrt{2k}} e^{-\frac{t^2}{2}}\,dt,$$

where $a$ is real, $k > 0$, $a$, $k$ fixed. **212** does not completely imply **210**.

**213.** By a similar argument as in **201** we find that the integral in question is

$$\sim \int_{\xi-\delta}^{\xi+\varepsilon_n} \varphi(x)\,e^{nh(x)}\,dx = \varphi(\xi')\,e^{nh(\xi)}\int_{\xi-\delta}^{\xi+\varepsilon_n} e^{n[h(x)-h(\xi)]}\,dx,$$

where $\varepsilon_n = \alpha n^{-1}\log n + \beta n^{-1}$, $\delta$ a positive constant, $\delta$ so small and $n$ so large that $\varphi(x)$ is continuous, $h(x)$ twice continuously differentiable and $h'(x) > 0$ in the interval of integration; we have $\xi - \delta < \xi' < \xi + \varepsilon_n$. Put $\eta_n = n^{-3/4}$ and choose $n$ so large that $\varepsilon_n < \eta_n < \delta$. On $[\xi - \delta, \xi - \eta_n]$ the integrand is of the order of $e^{-n\eta_n h'} = e^{-n^{1/4}h'}$ where $h'$ denotes a positive lower bound of $h'(x)$. On the remaining interval expand up to terms of second order,

$$\int_{\xi-\eta_n}^{\xi+\varepsilon_n} e^{nh'(\xi)(x-\xi)+\frac{n}{2}(x-\xi)^2h''(\xi'')}\,dx, \qquad \xi-\eta_n<\xi''<\xi+\varepsilon_n.$$

Herein $h''(\xi'')$ is bounded and $n(x-\xi)^2 \leqq n\eta_n^2 = n^{-\frac{1}{3}}$, furthermore

$$\int_{\xi-\eta_n}^{\xi+\varepsilon_n} e^{nh'(\xi)(x-\xi)}\,dx = \frac{e^{n\varepsilon_n h'(\xi)} - e^{-n\eta_n h'(\xi)}}{nh'(\xi)} \sim \frac{e^{(\alpha\log n+\beta)h'(\xi)}}{nh'(\xi)}.$$

**214.** A change of variables leads to

$$\frac{e^{-n}n^{n+1}}{n!}\int_0^{\xi+\frac{\alpha\log n}{n}+\frac{\beta}{n}} (e^{1+x}x)^n\,dx \qquad \textbf{[205, 213]}.$$

**215.** According to solution **211** we have

$$\frac{1}{n!}\int_0^{x_n} e^{-x}x^n\,dx = \frac{1}{n!}\int_0^{x_n} e^{x}x^n\,dx = 1.$$

Determine the constants $\alpha$ and $\beta$ in **214** so that $A=0$, $B=1$. Then $x_n-(\xi n+\alpha\log n+\beta)$ must converge to 0 [solution **211**].

**216.** Assume $\frac{g(x)}{x}<G=\text{const.}$ for $x>1$. We split the integral

$$a_n = \frac{e^{-n}n^{n+1}}{n!}\int_0^\infty \left(e^{1-x+x\frac{g(nx)}{nx}}x\right)^n dx$$

into four parts corresponding to the intervals $(0,\varepsilon)$, $(\varepsilon, 1-\varepsilon)$, $(1-\varepsilon, 1+\varepsilon)$, $(1+\varepsilon,\infty)$ where $\varepsilon$ does not depend on $n$, $0<\varepsilon<\frac{1}{2}$, $\varepsilon<\gamma$ and so small that in the first interval $xe^{1-x+G\varepsilon}<1$. In the second and fourth interval $xe^{1-x}<1$; choose $\delta=\delta(\varepsilon)$ so small that we have even $xe^{1-x+\delta x}<1$ and $n$ so large, $n>N=N(\varepsilon)$, that $\frac{g(nx)}{nx}<\delta$ and $n\varepsilon>1$. Then, except on the third interval, the integrand is $O(\theta^n)$, $0<\theta<1$, $\theta$ independent of $x$ and $n$, $\theta=\theta(\varepsilon)$. The mean value theorem of integrals [in addition note **205**] implies that

$$a_n = e^{g(n\xi)}\frac{e^{-n}n^{n+1}}{n!}\int_{1-\varepsilon}^{1+\varepsilon}(e^{1-x}x)^n\,dx + O(\sqrt{n}\theta^n),\ 1-\varepsilon<\xi<1+\varepsilon.$$

Hence

$$\frac{\log a_n}{g(n)} = \frac{g(n\xi)}{g(n)} + o(1).$$

Now $\lim_{n\to\infty} \frac{g(\alpha n)}{g(n)}$ exists and the convergence is uniform on $1-\varepsilon \leqq \alpha \leqq 1+\varepsilon$. The limit is arbitrarily close to 1 if $\varepsilon$ is sufficiently small.

**217.** The integral in question can be written in the following way:

$$\frac{n!\,2^{2n}}{(2n-1)(2n-2)(2n-3)\cdots n}\frac{1}{\sqrt{n}}\int_{-\pi\sqrt{n}}^{\pi\sqrt{n}} 2^{2n\left(\cos\frac{x}{\sqrt{n}}-1\right)}\prod_{\nu=1}^{n}\left|\frac{2n-\nu}{2n\,e^{\frac{ix}{\sqrt{n}}}-\nu}\right| dx.$$

Apply **115**. The limit of the integrand is $e^{-x^2}$ [**59**], the convergence is uniform on any finite interval as a supplement to the proof of **59** shows. For an appropriate $F(x)$ in the sense of **115** cf. G. Pólya: Nachr. Akad. Wiss. Göttingen 1920, pp. 6–7.—For the generalized Laplace formula cf. also R. v. Mises: Math. Z. Vol. 4, p. 9 (1919).

**217.1.** The real-valued functions $\varphi(x, y)$ and $h(x, y)$ are defined in the (bounded or unbounded) region $\mathfrak{R}$ and satisfy there the following conditions:

(1) $\varphi(x, y)\, e^{nh(x,y)}$ is absolutely integrable in $\mathfrak{R}$ for $n = 0, 1, 2, \ldots$

(2) The function $h(x, y)$ attains its maximum in $\mathfrak{R}$ at a single point $\xi, \eta$, and in the region $\mathfrak{R}'$ that remains when we exclude from $\mathfrak{R}$ a closed neighbourhood of $\xi, \eta$ the upper bound of $h(x, y)$ is less than $h(\xi, \eta)$. Moreover, the second partial derivatives $h_{xx}$, $h_{xy}$ and $h_{yy}$ exist and are continuous in a certain neighbourhood of $\xi, \eta$ and at the point $\xi, \eta$

$$h_{xx} < 0,\quad h_{xx}h_{yy} - h_{xy}^2 > 0.$$

(3) $\varphi(x, y)$ is continuous at the point $\xi, \eta$ and $\varphi(\xi, \eta) \neq 0$.

The proof is closely analogous to the proof of **201**.

**218.** [Cf. G. Pólya, Problem: Arch. Math. Phys. Ser. 3, Vol. 24, p. 282 (1916).] Let $x_n$, $x_n > n$, be the point at which the function attains its maximum $M_n$:

$$\frac{M_n}{n!} = \frac{\sqrt{x_n}(x_n-1)(x_n-2)\cdots(x_n-n)}{n!}a^{-x_n} = \frac{x_n-n}{\sqrt{x_n}}\binom{x_n}{n}a^{-x_n},$$

$$\frac{1}{2x_n}+\frac{1}{x_n-1}+\frac{1}{x_n-2}+\cdots+\frac{1}{x_n-n} = \log a.$$

In view of **16** we derive from this that $x_n = \left(n+\frac{1}{2}\right)b + \varepsilon_n$ where $b = (1-a^{-1})^{-1}$, $\lim_{n\to\infty}\varepsilon_n = 0$. We have $b > 1$. Thus we obtain for $\varepsilon$ positive and $n$ sufficiently large

$$\binom{(n+\frac{1}{2})b-\varepsilon}{n} < \binom{x_n}{n} < \binom{(n+\frac{1}{2})b+\varepsilon}{n}.$$

According to **206** these two bounds are asymptotically equal to

$$\frac{(b-1)^n}{\sqrt{2\pi n}}\left(\frac{b}{b-1}\right)^{(n+\frac{1}{2})b+\frac{1}{2}-\varepsilon} \quad \text{and} \quad \frac{(b-1)^n}{\sqrt{2\pi n}}\left(\frac{b}{b-1}\right)^{(n+\frac{1}{2})b+\frac{1}{2}+\varepsilon} \quad \text{resp.}$$

**219.** Analogous to **218**, use **17**.

**220.** We put $f(n, x) = |Q_n(x)|\, a^{-x}$, then we have for $m - 1 \leqq x < m$, $m$ positive integer, $a \geqq 1$,

$$f(m-1, x) \geqq f(m, x) \geqq f(m+1, x) \geqq \cdots$$

[III **12**] so that the least upper bound of $f(n, x)$ for fixed positive $x$ and variable $n$ is attained when $n < x$. Cf. **218**.

**221.** [**219**, **220**.]

**222.** The point $x_n$ where $M_n$ is attained is determined by $n = x_n + a\mu x_n^{\mu}$. We have $n > x_n$, $\lim\limits_{n\to\infty} \frac{x_n}{n} = 1$, $\lim\limits_{n\to\infty} \frac{n - x_n}{n^{\mu}} = a\mu$. Consequently $x_n = n - a\mu n^{\mu} + o(n^{\mu})$, $\log x_n = \log n - a\mu n^{\mu-1} + o(n^{\mu-1})$, thus

$$\log \frac{M_n}{n!} = n \log x_n - x_n - a x_n^{\mu} - n \log \frac{n}{e} + o(n^{\mu}) = -a n^{\mu} + o(n^{\mu}).$$

***223.** Let

$$\varphi(x_1) = f(x_1, y_1), \quad \varphi(x_2) = f(x_2, y_2).$$

By the definition of $\varphi(x)$

$$\varphi(x_1) \geqq f(x_1, y_2),$$

and by the continuity of $f(x, y)$

$$f(x_1, y_2) \geqq f(x_2, y_2) - \varepsilon$$

when $|x_1 - x_2|$ is sufficiently small. Hence

$$\varphi(x_1) \geqq \varphi(x_2) - \varepsilon$$

and we can interchange $x_1$ and $x_2$ in the foregoing argument.

***224.** Let

$$\max_y \left(\min_x f(x, y)\right) = f(x_1, y_1), \quad \min_x \left(\max_y f(x, y)\right) = f(x_2, y_2).$$

In view of the first operation (the inner one, written on the right)

$$f(x_1, y_1) \leqq f(x_2, y_1) \leqq f(x_2, y_2).$$

***225.**

$$\max_y f(x, y) = 1, \quad \min_x f(x, y) = \begin{cases} 1 - (y - 1)^2 & \text{when } y \geqq 2, \\ 1 - (3 - y)^2 & \text{when } y \leqq 2. \end{cases}$$

$$\min_x \max_y f(x, y) = 1, \quad \max_y \min_x f(x, y) = 0$$

***226.** By using **198** complete the following outline: The quantity whose limit is desired is

$$\sim \left[ \int_a^{a'} \varphi(x)^{-n} \, dx \right]^{-1/n} \sim \min \varphi(x).$$

## Part Three

# Functions of One Complex Variable

**1.** $z + \bar{z} = 2x, \quad z - \bar{z} = 2iy, \quad z\bar{z} = r^2.$

**2.** The open right half-plane; the closed right half-plane; the open horizontal strip bounded by the lines $y = a$ and $y = b$ parallel to the $x$-axis; the closed sector between the two rays which form the angles $\alpha$ and $\beta$ resp. with the positive $x$-axis; the imaginary axis; the circle with center $z_0$ and radius $R$; the open disk and the closed disk resp. with center $z_0$ and radius $R$; the closed annulus between the two circles with radii $R$ and $R'$ and centred at the origin; the circle with center at $z = \frac{R}{2}$ and radius $\frac{R}{2}$.

**3.** The ellipse, and the domain bounded by the ellipse, with foci $z = a$ and $z = b$ and the semimajor axis $k$ if $|a - b| \leq k$. (If $|a - b| = k$ the ellipse degenerates into a segment.) If $k < |a - b|$ no point $z$ satisfies the condition.

**4.** Let $z_1$ and $z_2$ denote the two roots of the equation $z^2 + az + b = 0$. The region in question is the interior of the curve $|z - z_1|\,|z - z_2| = R^2$ with "foci" $z_1$ and $z_2$. The curve is the locus of all points for which the product of the distances to $z_1$ and $z_2$ is constant, equal to $R^2$. It consists of two pieces for $R \leq \frac{|z_1 - z_2|}{2}$ and of one piece for $R > \frac{|z_1 - z_2|}{2}$. If $R = \frac{|z_1 - z_2|}{2}$ the curve is called lemniscate.

**5.** The condition in question is equivalent to

$$|z - a|^2 \lesseqqgtr |1 - \bar{a}z|^2 \quad \text{or to} \quad (1 - |a|^2)(|z|^2 - 1) \lesseqqgtr 0.$$

The first set is the open disk $|z| < 1$; the second set is the unit circle $|z| = 1$; the third set is the exterior $|z| > 1$ of the closed unit disk. The value of the expression in question is $|a|^{-1}$ for $z = \infty$, thus $z = \infty$ belongs to the third set.

**6.** The condition in question is equivalent to

$$|a - z|^2 \lesseqgtr |\bar{a} + z|^2 \qquad \text{or to} \qquad -\Re(a + \bar{a})\, z \lesseqgtr 0.$$

Since $a + \bar{a}$ is real and positive the condition means $\Re z \gtreqless 0$. The first set is the open right half-plane; the third set is the open left half-plane; the second set is the imaginary axis. (The value of the expression in question is for $z = \infty$ equal to $-1$; $z = \infty$ belongs to the second set.)

**7.** With the notation $a = \gamma + i\,\delta$, $\frac{z_1}{z_2} = x + iy$, the equation reads

$$\alpha(x^2 + y^2) + 2(\gamma x + \delta y) + \beta = 0.$$

**8.** Suppose that a wheel of radius $a$ is rolling on the real axis. Let $P$ be a point on the wheel at a distance $b$ from its center. The point $z_1$ moves on a straight line, the path of the center of the wheel; the point $z_2$ describes the path of the point $P$ if the wheel would rotate around the origin but not slide. The point $z = z_1 + z_2$ describes a prolate, a regular, or a curtate cycloid according as $a >, =,$ or $< b$.

**9.** The point describes an epicycloid.

**10.**

$$\frac{dz}{dt} = \frac{dr}{dt} e^{i\vartheta} + ire^{i\vartheta} \frac{d\vartheta}{dt},$$

$$\frac{d^2z}{dt^2} = \frac{d^2r}{dt^2} e^{i\vartheta} + 2i \frac{dr}{dt} \frac{d\vartheta}{dt} e^{i\vartheta} - re^{i\vartheta} \left(\frac{d\vartheta}{dt}\right)^2 + ire^{i\vartheta} \frac{d^2\vartheta}{dt^2}$$

$$= \left[\frac{d^2r}{dt^2} - r\left(\frac{d\vartheta}{dt}\right)^2\right] e^{i\vartheta} + \frac{ie^{i\vartheta}}{r} \frac{d}{dt}\left(r^2 \frac{d\vartheta}{dt}\right).$$

The coefficient of $e^{i\vartheta}$ is the radial component, the coefficient of $ie^{i\vartheta}$ is the component perpendicular to the radius.

**11.** In the annulus

$$\Re_n: \qquad n < |z| < n + 1, \qquad n = 0, 1, 2, \ldots$$

the inequalities

$$1 < \left|\frac{z}{1!}\right| < \left|\frac{z^2}{2!}\right| < \cdots < \left|\frac{z^{n-1}}{(n-1)!}\right| < \left|\frac{z^n}{n!}\right| > \left|\frac{z^{n+1}}{(n+1)!}\right| > \cdots,$$

are satisfied; i.e. in $\Re_n$ the absolute value of the $n$-th term is larger than the absolute value of any other term. On the common border of $\Re_n$ and $\Re_{n+1}$ the absolute values of the $n$-th and the $n + 1$-st term are equal, all the others are smaller. In general: let

$$a_0 + a_1 z + a_2 z^2 + \cdots + a_n z^n + \cdots, \qquad a_0 \neq 0,$$

be an everywhere convergent infinite power series. Then the $z$-plane can be divided by concentric circles in such a way that in each annulus the absolute value of a certain term is largest (maximal term). The subscripts of these terms increase as $z$ proceeds from one annulus to the next larger one. [I **119**, I **120**.]

**12.** The circles

$$\mathfrak{C}_n: \qquad |z-n| = n+1, \qquad n = 0, 1, 2, \ldots,$$

are tangent to one another at $z = -1$ where they are perpendicular to the $x$-axis. They intersect with the $x$-axis also at the points $z = 2n+1$. $\mathfrak{C}_{n+1}$ contains $\mathfrak{C}_n$. In the crescent shaped region $\mathfrak{R}_n$ inside $\mathfrak{C}_n$ and outside $\mathfrak{C}_{n-1}$ the inequalities

$$\left|\frac{z}{1}\right| > 1, \left|\frac{z-1}{2}\right| > 1, \ldots, \left|\frac{z-n+1}{n}\right| > 1, \left|\frac{z-n}{n+1}\right| < 1, \left|\frac{z-n-1}{n+2}\right| < 1, \ldots$$

are satisfied. Hence

$$1 < \left|\binom{z}{1}\right| < \left|\binom{z}{2}\right| < \cdots < \left|\binom{z}{n-1}\right| < \left|\binom{z}{n}\right| > \left|\binom{z}{n+1}\right| > \cdots,$$

i.e. the absolute value of the $n$-th term is larger than any other in $\mathfrak{R}_n$. On the common boundary of $\mathfrak{R}_n$ and $\mathfrak{R}_{n+1}$ the $n$-th and the $n+1$-st terms have the same absolute value, all the other terms have smaller absolute values. (At $z = -1$ all the terms have the absolute value 1.)

The regions $\mathfrak{R}_n$ together with their boundaries cover the entire half-plane $\mathfrak{R}z > -1$, including $z = -1$. If $\mathfrak{R}z \leqq -1$, $z \neq -1$, the absolute values increase monotonically, there does not exist a largest term.

**13.** The lemniscates

$$\mathfrak{L}_n: \qquad |z^2 - n^2| = n^2, \qquad n = 1, 2, 3, \ldots$$

are tangent to each other and to the straight lines $\mathfrak{R}z = \pm\mathfrak{I}z$ at the point $z = 0$. They intersect the real axis at the points $\pm n\sqrt{2}$. $\mathfrak{L}_{n+1}$ contains $\mathfrak{L}_n$. The inequalities

$$\left|1 - \frac{z^2}{1^2}\right| > 1, \; \left|1 - \frac{z^2}{2^2}\right| > 1, \; \ldots, \; \left|1 - \frac{z^2}{n^2}\right| > 1,$$

$$\left|1 - \frac{z^2}{(n+1)^2}\right| < 1, \; \left|1 - \frac{z^2}{(n+2)^2}\right| < 1, \ldots$$

hold in the "double crescent" shaped region $\mathfrak{R}_n$ between $\mathfrak{L}_{n+1}$ and $\mathfrak{L}_n$ ($\mathfrak{R}_0$ is the region bounded by $\mathfrak{L}_1$). Therefore the inequalities

$$|P_0(z)| < |P_1(z)| < |P_2(z)| < \cdots < |P_{n-1}(z)| < |P_n(z)| > |P_{n+1}(z)| > \cdots,$$

are satisfied in $\mathfrak{R}_n$, i.e. $|P_n(z)|$ is in $\mathfrak{R}_n$ larger than the absolute value of any other partial product. On the common boundary of $\mathfrak{R}_n$ and $\mathfrak{R}_{n+1}$ the absolute values of the $n$-th and $n+1$-st partial products are equal and larger than any other. (At $z = 0$ all vanish.) The $\mathfrak{R}_n$'s together with their boundaries fill out the region consisting of the two angles

$$-\frac{\pi}{4} < \arg z < \frac{\pi}{4}, \frac{3\pi}{4} < \arg z < \frac{5\pi}{4}$$

and the point $z = 0$. Outside this region and this point $P_n(z)$ increases with $n$ monotonically.

**14.** We put $\vartheta = \arg \int\limits_a^b f(t)\, e^{i\varphi(t)}\, dt$. Then

$$\left|\int\limits_a^b f(t)\, e^{i\varphi(t)}\, dt\right| = e^{-i\vartheta} \int\limits_a^b f(t) e^{i\varphi(t)}\, dt = \int\limits_a^b f(t) \cos\ [\varphi(t) - \vartheta]\, dt < \int\limits_a^b f(t)\, dt,$$

except if $\varphi(t) \equiv \vartheta$ (mod $2\pi$) at all points of continuity of $\varphi(t)$.

**15.** [K. Löwner: Math. Ann. Vol. 89, p. 120 (1923).] It is sufficient to prove $\mathfrak{R}(4P^2 - 2Q) \leqq 3$. [Replace $\varphi(t)$ by $\varphi(t) + \frac{\vartheta}{2}$ where $\vartheta = \arg(4P^2 - 2Q)$, cf. solution **14**.] Now we have [II **81**]

$$\mathfrak{R}(4P^2 - 2Q) = 4\left(\int\limits_0^\infty e^{-t} \cos \varphi(t)\, dt\right)^2 - 4\left(\int\limits_0^\infty e^{-t} \sin \varphi(t)\, dt\right)^2 -$$

$$- 2\int\limits_0^\infty e^{-2t} \cos 2\varphi(t)\, dt \leqq$$

$$\leqq 4\left(\int\limits_0^\infty e^{-t} |\cos \varphi(t)|\, dt\right)^2 - 2\int\limits_0^\infty e^{-2t} \cos 2\varphi(t)\, dt \leqq$$

$$\leqq 4\int\limits_0^\infty e^{-t} \cos^2 \varphi(t)\, dt - 2\int\limits_0^\infty e^{-2t} \cos 2\varphi(t)\, dt =$$

$$= 4\int\limits_0^\infty (e^{-t} - e^{-2t}) \cos^2 \varphi(t)\, dt + 1 \leqq 4\int\limits_0^\infty (e^{-t} - e^{-2t})\, dt + 1 = 3.$$

If $\mathfrak{R}(4P^2 - 2Q) = 3$ then $\cos^2 \varphi(t) = 1$ and $\cos \varphi(t)$ has the same sign at all the points of continuity of $\varphi(t)$, i.e. $\varphi(t) \equiv 0$ or $\varphi(t) \equiv \pi$ (mod $2\pi$).

**16.** The function $p_1 z^{-1} + p_2 z^{-2} + \cdots + p_n z^{-n}$ is monotone decreasing from $\infty$ to 0 as $z$ is positive and increasing; therefore it assumes the value 1 at exactly one positive point $\zeta$. Consequently

$$z^n - p_1 z^{n-1} - p_2 z^{n-2} - \cdots - p_n > 0 \text{ or } \leqq 0$$

according as $z > \zeta$ or $z \leqq \zeta$.

**17.** We have

$$|z_0|^n = |a_1 z_0^{n-1} + a_2 z_0^{n-2} + \cdots + a_n|$$
$$\leqq |a_1|\,|z_0|^{n-1} + |a_2|\,|z_0|^{n-2} + \cdots + |a_n|,$$

hence, according to **16**, $|z_0| \leqq \zeta$.

**18.** Apply **17** to $a_n^{-1} z^n P(z^{-1})$.

**19.** The two test polynomials considered in **17** and **18** are, in this case, identical, namely $z^n - |c|$.

***20.** By **16** and **17** it is enough to ascertain the sign of

$$M^n - |a_1| M^{n-1} - |a_2| M^{n-2} - \cdots - |a_n|$$
$$\geqq M^n - c_1 M M^{n-1} - c_2 M^2 M^{n-2} - \cdots - c_n M^n \geqq 0.$$

***21.** From **20**. Substitute for $c_k$

$$\frac{1}{n}, \frac{\binom{n}{k}}{2^n - 1}, 2^{-k}$$

respectively and add a little remark to **20** in the last case.

**22.** [G. Eneström: Öfvers. K. Vetensk. Akad. Förh. 1893, pp. 405–415; Tôhoku Math. J. Vol. 18, pp. 34–36 (1920); S. Kakeya: Tôhoku Math. J. Vol. 2, pp. 140–142 (1912); A. Hurwitz: Tôhoku Math. J. Vol. 4, p. 89 (1913).] We find for $|z| \leqq 1$, $z \neq 1$,

$$|(1 - z)(p_0 + p_1 z + p_2 z^2 + \cdots + p_n z^n)|$$
$$= |p_0 - (p_0 - p_1) z - (p_1 - p_2) z^2 - \cdots - (p_{n-1} - p_n) z^n - p_n z^{n+1}|$$
$$\geqq p_0 - |(p_0 - p_1) z + (p_1 - p_2) z^2 + \cdots + p_n z^{n+1}|$$
$$> p_0 - (p_0 - p_1 + p_1 - p_2 + \cdots + p_n) = 0,$$

because $(p_0 - p_1) z, (p_1 - p_2) z^2, \ldots, p_n z^{n+1}$ cannot have all the same argument. (Unless $z \geqq 0$, in which case the proposition is trivial.) A weaker statement, $<$ instead of $\leqq$, follows immediately from **17**.

**23.** We replace $z$ by $\frac{z}{\varrho}$ and $\frac{\varrho}{z}$ respectively and choose $\varrho$, $\varrho > 0$ (in the second case we first multiply by $z^n$), so that **22** can be applied.

**24.** We call the polynomial in question $f(z)$. For $\Re z \geqq 0$, $|z| > 1$, we have $\Re \frac{1}{z} \geqq 0$, therefore

$$\left|\frac{f(z)}{z^n}\right| \geqq \left|a_n + \frac{a_{n-1}}{z}\right| - \frac{a_{n-2}}{|z|^2} - \frac{a_{n-3}}{|z|^3} - \cdots - \frac{a_0}{|z|^n}$$
$$> \Re\left(a_n + \frac{a_{n-1}}{z}\right) - \frac{9}{|z|^2} - \frac{9}{|z|^3} - \cdots$$
$$\geqq 1 - \frac{9}{|z|^2 - |z|}.$$

The last expression is $\geqq 0$ if $|z| \geqq r$, where $r$ is the positive root of the equation $r^2 - r = 9$, $r = \frac{1}{2}\left((1 + \sqrt{37}\right)$, $3 < r < 4$. The polynomial corresponding to the number 109, $9 + z^2$, has the roots $\pm 3i$.

**25.** [Ch. Hermite and Ch. Biehler; cf. Laguerre: Oeuvres, Vol. 1. Paris: Gauthier-Villars 1898, p. 109.] We write

$$P(z) = U(z) + iV(z) = a_0(z - z_1)(z - z_2) \cdots (z - z_n), \; a_0 \neq 0;$$

let $x$ be a root of $V(x) = 0$ or $U(x) = 0$. Hence

$$U(x) + iV(x) = U(x) - iV(x) \text{ or } U(x) + iV(x) = -[U(x) - iV(x)],$$

that is

$$a_0(x - z_1)(x - z_2) \cdots (x - z_n) = \pm \bar{a}_0(x - \bar{z}_1)(x - \bar{z}_2) \cdots (x - \bar{z}_n).$$

Such an equation, however, is possible only for real $x$. Assume $\Im x > 0$; then $|x - z_\nu| < |x - \bar{z}_\nu|$ for $\nu = 1, 2, \ldots, n$. By the same token $x$ cannot lie in the open lower halfplane.

**26.** [Cf. I. Schur: J. reine angew. Math. Vol. 147, p. 230 (1917).] Put $P(z) = a_0(z - z_1)(z - z_2) \cdots (z - z_n)$, $a_0 \neq 0$. Let $x$ be a root of $P(z) + P^*(z) = 0$. Then $|P(x)| = |P^*(x)|$; since $P^*(z) = \bar{a}_0(1 - \bar{z}_1 z)(1 - \bar{z}_2 z) \cdots (1 - \bar{z}_n z)$ we have

$$\prod_{\nu=1}^{n} |x - z_\nu| = \prod_{\nu=1}^{n} |1 - \bar{z}_\nu x|.$$

Such an equation can hold only for $|x| = 1$. Assume $|x| < 1$: then [**5**] $|x - z_\nu| < |1 - \bar{z}_\nu x|$ for all $\nu$, thus the first product is smaller than the second. By the same token it is impossible that $|x| > 1$. We argue analogously in the case of $P(z) + \gamma P^*(z)$, $|\gamma| = 1$.

**27.** [M. Fekete.] Define $\gamma = \lambda P(a) + \mu P(b)$, $0 < \lambda < 1$, $\lambda + \mu = 1$. If all the zeros of $P(z) - \gamma = a_0(z - z_1)(z - z_2) \cdots (z - z_n)$ were outside the domain in question, the inequality

$$-\frac{\pi}{n} < \arg \frac{a - z_\nu}{b - z_\nu} < \frac{\pi}{n}, \qquad \nu = 1, 2, \ldots, n$$

would hold, thus

$$-\pi < \arg \frac{P(a) - \gamma}{P(b) - \gamma} < \pi, \quad \text{in contradiction to} \quad \frac{P(a) - \gamma}{P(b) - \gamma} = -\frac{\mu}{\lambda}.$$

**28.** We assume that the straight line mentioned is the imaginary axis and $\Re z_\nu > 0$ for all $\nu$ (this can always be achieved by multiplication with a suitable $e^{i\alpha}$); then we have also $\Re \frac{1}{z_\nu} > 0$ and

$$\Re(z_1 + z_2 + \cdots + z_n) > 0, \qquad z_1 + z_2 + \cdots + z_n \neq 0,$$

$$\Re\left(\frac{1}{z_1} + \frac{1}{z_2} + \cdots + \frac{1}{z_n}\right) > 0, \qquad \frac{1}{z_1} + \frac{1}{z_2} + \cdots + \frac{1}{z_n} \neq 0.$$

The conclusion holds also in the case where all the points are in a closed half-plane determined by the straight line unless all the points lie on the line itself.

**29.** Cf. **28**.

**30.** We have $m_1(z_1 - z) + m_2(z_2 - z) + \cdots + m_n(z_n - z) = 0$. Apply **29** to the points $m_\nu(z_\nu - z)$. If $m_\nu(z_\nu - z)$ lies on one side of a straight line $l'$ through the origin, $z_\nu$ lies on the corresponding side of the straight line $l$ through $z$ and parallel to $l'$.

**31.** [Gauss: Werke, Vol. 3, p. 112. Göttingen: Ges. d. Wiss. 1886; Vol. 8, p. 32, 1900; Ch. F. Lucas: C. R. Acad. Sci. (Paris) Sér. A–B, Vol. 67, pp. 163–164 (1868); Vol. 106, pp. 121–122 (1888). Cf. also L. Fejér: C. R. Acad. Sci. (Paris) Sér. A–B, Vol. 145, p. 460 (1907), and Math. Ann. Vol. 65, p. 417 (1907).] First solution: The vector determined by the complex number $\frac{1}{\overline{z} - \overline{a}}$ represents a force directed from $a$ to $z$ the magnitude of which is inversely proportional to the distance. Let $z_1, z_2, \ldots, z_n$ denote the zeros of $P(z)$ and let $z$ be a zero of $P'(z)$ different from $z_1, z_2, \ldots, z_n$. Then

$$\frac{P'(z)}{P(z)} = \sum_{\nu=1}^{n} \frac{1}{z - z_\nu} = 0, \quad \text{i.e.} \quad \frac{1}{\overline{z} - \overline{z}_1} + \frac{1}{\overline{z} - \overline{z}_2} + \cdots + \frac{1}{\overline{z} - \overline{z}_n} = 0,$$

this means that $z$ represents an *equilibrium position* of a material point subjected to repellent forces exerted by the points $z_1, z_2, \ldots, z_n$ and inversely proportional to the distance. If $z$ were outside the smallest convex polygon that contains the $z_\nu$'s the resultant of the several forces acting on $z$ could not vanish: there could be no equilibrium. (**315**.)

Second solution: Using the same notation as before we have

$$\frac{z - z_1}{|z - z_1|^2} + \frac{z - z_2}{|z - z_2|^2} + \cdots + \frac{z - z_n}{|z - z_n|^2} = 0,$$

thus

$$z = m_1 z_1 + m_2 z_2 + \cdots + m_n z_n, \quad m_1 + m_2 + \cdots + m_n = 1,$$

where the $\nu$-th "mass" $m_\nu$ is proportional to $\frac{1}{|z - z_\nu|^2}$, $\nu = 1, 2, \ldots, n$.

**32.** [L. Fejér, O. Toeplitz.] Let $\zeta$ denote an arbitrary interior point of the convex polygon in question. Then

$$\zeta = \lambda_1 z_1 + \lambda_2 z_2 + \cdots + \lambda_n z_n, \quad \lambda_1 > 0, \quad \lambda_2 > 0, \quad \ldots, \quad \lambda_n > 0,$$
$$\lambda_1 + \lambda_2 + \cdots + \lambda_n = 1,$$

thus for $\zeta \neq z_\nu$

$$\frac{m_1}{\zeta - z_1} + \frac{m_2}{\zeta - z_2} + \cdots + \frac{m_n}{\zeta - z_n} = 0,$$

$$m_\nu = \frac{\lambda_\nu |\zeta - z_\nu|^2}{\lambda_1 |\zeta - z_1|^2 + \lambda_2 |\zeta - z_2|^2 + \cdots + \lambda_n |\zeta - z_n|^2}, \qquad \nu = 1, 2, \ldots, n.$$

Approximate $m_\nu$ by the rational numbers $\frac{p_\nu}{P}$, $\nu = 1, 2, \ldots, n$, $p_1 + p_2 + \cdots + p_n = P$, and remember that the roots of algebraic equations are continuous functions of the coefficients. The derivative of the polynomial $\prod_{\nu=1}^{n} (z - z_\nu)^{p_\nu}$ has a zero arbitrarily close to $\zeta$. Since $\zeta$ lies inside or on the border of at least one of the triangles determined by three of the points $z_\nu$ it is sufficient for this problem to know that the zeros of a polynomial of degree 2 are continuous functions of the coefficients, which is obvious. [Remark due to A. and R. Brauer.]

**33.** [M. Fujiwara: Tôhoku Math. J. Vol. 9, pp. 102—108 (1916); T. Takagi: Proc. Phys. Math. Soc. Japan Ser. 3, Vol. 3, pp. 175—179 (1921).] Set $P(z) = a_0(z - z_1)(z - z_2) \cdots (z - z_n)$ and let $z$ denote a point at which

$$P(z) - cP'(z) = 0, \qquad P(z) \neq 0.$$

Hence

$$\text{(1)} \qquad \frac{P'(z)}{P(z)} - \frac{1}{c} = \frac{1}{z - z_1} + \frac{1}{z - z_2} + \cdots + \frac{1}{z - z_n} - \frac{1}{c} = 0.$$

Introducing

$$m_1 = \frac{1}{|z - z_1|^2}, \quad m_2 = \frac{1}{|z - z_2|^2}, \quad \ldots, \quad m_n = \frac{1}{|z - z_n|^2},$$

$$M = \frac{1}{|c|^2(m_1 + m_2 + \cdots + m_n)}$$

we can write (1) as

$$\text{(2)} \qquad z = \frac{m_1 z_1 + m_2 z_2 + \cdots + m_n z_n}{m_1 + m_2 + \cdots + m_n} + Mc.$$

The first term on the right hand side of the equation represents the center of gravity of a certain mass distribution at the points $z_1, z_2, \ldots, z_n$, that means a point inside the smallest convex polygon containing all the points $z_\nu$. The second term represents a vector parallel to the vector $c$. Hence the statement follows.—Cf. V **114**.

**34.** [T. J. Stieltjes: Acta Math. Vol. 6, pp. 321–326 (1885); G. Pólya: C. R. Acad. Sci. (Paris), Vol. 155, p. 767 1–769 (1912).] Let $z_\nu$, $\nu = 1, 2, \ldots, n$ denote the zeros of $P(z)$ and assume $A(z_\nu) \neq 0$. Then $P'(z_\nu) \neq 0$ because otherwise the differential equation for $P(z)$ would imply that $P''(z_\nu) = 0$ and repeated differentiation would show that $P(z)$ is identically zero. The equation

$$\frac{P''(z_\nu)}{2P'(z_\nu)} + \frac{B(z_\nu)}{A(z_\nu)} = 0,$$

$$\frac{1}{z_\nu - z_1} + \frac{1}{z_\nu - z_2} + \cdots + \frac{1}{z_\nu - z_{\nu-1}} + \frac{1}{z_\nu - z_{\nu+1}} + \cdots + \frac{1}{z_\nu - z_n}$$

$$+ \frac{\varrho_1}{z_\nu - a_1} + \frac{\varrho_2}{z_\nu - a_2} + \cdots + \frac{\varrho_p}{z_\nu - a_p} = 0$$

implies [**31**] that $z_\nu$ lies in the interior of the smallest convex polygon that contains the points $z_1, z_2, \ldots, z_{\nu-1}, z_{\nu+1}, \ldots, z_n, a_1, a_2, \ldots, a_p$ (on the line segment that contains all these points). Consider now the smallest convex polygon that encloses $z_1, z_2, \ldots, z_n, a_1, a_2, \ldots, a_p$. Only the $a_\nu$'s and no $z_\nu$ different from the zeros of $A(z)$ can lie on the polygon.

**35.** [L. J. W. V. Jensen: Acta Math. Vol. 36, p. 190 (1913); J. v. Sz. Nagy: Jber. deutsch. Math. Verein. Vol. 31, p. 239–240 (1922).] We denote the zeros of $f(z)$ by $z_1, z_2, \ldots, z_n$ and assume

$$\frac{1}{z - z_1} + \frac{1}{z - z_2} + \cdots + \frac{1}{z - z_n} = 0, \quad z \neq \bar{z}, \quad z \neq z_\nu, \quad \nu = 1, 2, \ldots, n.$$

Because of the pairwise symmetry of the zeros we have

$$\sum_{\nu=1}^{n} \Im\left(\frac{1}{z - z_\nu} + \frac{1}{z - \bar{z}_\nu}\right) = 0.$$

The formula

$$z = x + iy, \quad z_0 = x_0 + iy_0, \quad \Im\left(\frac{1}{z - z_0} + \frac{1}{z - \bar{z}_0}\right) = 2y \frac{y_0^2 - (x - x_0)^2 - y^2}{|(z - z_0)(z - \bar{z}_0)|^2},$$

shows that the above equation can not hold when $z$ is outside all the circles described.

**36.** Writing $z_n = x_n + iy_n$ we find $|z_n| \leq \frac{x_n}{\cos\alpha}$. Therefore the convergence of $\sum_{n=1}^{\infty} x_n$ implies the convergence of $\sum_{n=1}^{\infty} |z_n|$. The converse is obvious.

**37.** Set $z_n = x_n + iy_n$. The hypothesis implies in turn the convergence of

$$\sum_{n=1}^{\infty} x_n, \quad \sum_{n=1}^{\infty} \Re z_n^2 = \sum_{n=1}^{\infty} (x_n^2 - y_n^2), \quad 2\sum_{n=1}^{\infty} x_n^2 - \sum_{n=1}^{\infty} (x_n^2 - y_n^2) = \sum_{n=1}^{\infty} |z_n|^2.$$

**38.** Example: $z_n = e^{2\pi i n\theta} (\log (n+1))^{-1}$, $\theta$ irrational, $\sum_{\nu=1}^{n} e^{2\pi i \nu k\theta}$ is bounded as $n \to \infty$ [solution II **166**, Knopp, p. 315].

**39.** We assume that all the numbers $z_n$ are different from zero and that they are arranged according to increasing magnitude, $0 < |z_1| \leqq |z_2| \leqq |z_3| \leqq \cdots$. We enclose each $z_\nu$, $\nu = 1, 2, \ldots, m$, in a circle with center $z_\nu$ and radius $\frac{\delta}{2}$. These circles have no common inner points and are completely contained in $|z| \leqq |z_m| + \frac{\delta}{2}$. Therefore

$$m\pi \frac{\delta^2}{4} < \pi \left( |z_m| + \frac{\delta}{2} \right)^2, \quad \text{i.e.} \quad \limsup_{m\to\infty} \frac{\log m}{\log |z_m|} \leqq 2 \qquad [\text{I } \mathbf{113}].$$

**40.** The expression in question is

$$= n^{i\alpha} \sum_{\nu=1}^{n} \left(\frac{\nu}{n}\right)^{i\alpha} \frac{1}{n}.$$

The first factor is everywhere dense on the unit circle [I **101**]; the second is a sum of rectangles and converges to

$$\int_0^1 x^{i\alpha}\, dx = \frac{1}{1+i\alpha}.$$

**41.** $$\lim_{n\to\infty} |z_n|^2 = \left(1 + \frac{1}{1^2}\right)\left(1 + \frac{1}{2^2}\right)\left(1 + \frac{1}{3^2}\right) \cdots \left(1 + \frac{1}{n^2}\right) \cdots$$

$$= \left(\frac{\sin \pi x}{\pi x}\right)_{x=i} = \frac{e^\pi - e^{-\pi}}{2\pi}.$$

Writing $i + n = \sqrt{1+n^2}\, e^{2\pi i \vartheta_n}$, $0 < \vartheta_n < \frac{1}{4}$, we have $\tan 2\pi\vartheta_n = \frac{1}{n}$; thus the series $\vartheta_1 + \vartheta_2 + \vartheta_3 + \cdots + \vartheta_n + \cdots$ diverges and $\lim_{n\to\infty} \vartheta_n = 0$. We find

$$\arg z_n = 2\pi(\vartheta_1 + \vartheta_2 + \cdots + \vartheta_n - [\vartheta_1 + \vartheta_2 + \cdots + \vartheta_n]).$$

According to I **101** the limit points of $z_n$ cover the entire circle

$$|z| = \left(\frac{e^\pi - e^{-\pi}}{2\pi}\right)^{\frac{1}{2}}.$$

**42.**
$$r_n^2 = |z_n|^2 = \left(1+\frac{1}{1}\right)\left(1+\frac{1}{2}\right)\cdots\left(1+\frac{1}{n}\right) = \frac{2}{1}\cdot\frac{3}{2}\cdots\frac{n+1}{n} = n+1,$$
$$|z_{n+1}-z_n| = \left|\frac{iz_n}{\sqrt{n+1}}\right| = 1, \frac{r_n - r_{n-1}}{\varphi_n - \varphi_{n-1}} = \frac{\sqrt{n+1}-\sqrt{n}}{\arctan\frac{1}{\sqrt{n}}} \sim n\left(\sqrt{1+\frac{1}{n}}-1\right)$$
$$= \frac{1}{2} - \frac{1}{8n} + \cdots.$$
Hence $r_n \sim \frac{1}{2}\varphi_n$ [I **70**].

**43.** According to II **59** and II **202** the absolute value of the expression in question converges to $e^{-t^2}$. Therefore it is sufficient to prove
$$2n \sin\frac{t}{\sqrt{n}} \log 2 - \sum_{\nu=1}^{n} \arctan \frac{2n \sin\frac{t}{\sqrt{n}}}{2n\cos\frac{t}{\sqrt{n}} - \nu} = o(1),$$
where $-\frac{\pi}{2} < \arctan x < \frac{\pi}{2}$. Let $n > t^2$, then [I **142**]
$$\left|\frac{2n\sin\frac{t}{\sqrt{n}}}{2n\cos\frac{t}{\sqrt{n}} - \nu}\right| \leqq \frac{2\sqrt{n}\,|t|}{2n\left(1-\frac{t^2}{2n}\right)-\nu} \leqq \frac{2\sqrt{n}\,|t|}{n-t^2} = O(n^{-\frac{1}{2}}).$$
Thus $\arctan x = x - \frac{x^3}{3} + \cdots$ can be replaced by $x$. The statement now follows from
$$\log 2 - \frac{1}{n}\sum_{\nu=1}^{n} \frac{1}{2-\left[\frac{\nu}{n} + 2\left(1-\cos\frac{t}{\sqrt{n}}\right)\right]} = O(n^{-1}).$$
This estimate can be justified by a slight extension of II **10**: if we replace in the sum $\Delta_n$ the term $f\left(a+\nu\frac{b-a}{n}\right)$ by $f\left(a+\nu\frac{b-a}{n}+\varepsilon_n\right)$ where $\varepsilon_n = O(n^{-1})$ we still obtain $\Delta_n = O(n^{-1})$. (We assume that the interval of definition of $f(x)$ contains the interval $a-\varepsilon_N \leqq x \leqq b+\varepsilon_N$, $N < n$, and that $f(x)$ is bounded on this interval.)

**44.** Assume that $\lim_{n\to\infty} \sigma_n = \sigma$, $\zeta_n < K$ and $\lim_{n\to\infty} z_n = z$, $|z_n| < M$, $|z| \leqq M$, $K$ and $M$ independent of $n$. The series
$$\alpha = a_0 + a_1 + a_2 + \cdots + a_n + \cdots$$
and
$$\beta = a_0 z_0 + a_1 z_1 + a_2 z_2 + \cdots + a_n z_n + \cdots$$

are absolutely convergent; let $\varepsilon$ be an arbitrary positive number and $N = N(\varepsilon)$ be such that $|z_n - z| < \varepsilon$ whenever $n > N$. We define $w = (\sigma - \alpha) z + \beta$ and deduce from

$$w_n = \left(\sigma_n - \sum_{\nu=0}^{n} a_\nu\right) z + \sum_{\nu=0}^{n} a_\nu z_\nu + \sum_{\nu=0}^{n} (a_{n\nu} - a_\nu)(z_\nu - z)$$

that

$$|w - w_n| < M|\sigma - \sigma_n| + M\left|\sum_{\nu=n+1}^{\infty} a_\nu\right| + \left|\sum_{\nu=n+1}^{\infty} a_\nu z_\nu\right| + 2M \sum_{\nu=0}^{N} |a_{n\nu} - a_\nu| + 2K\varepsilon.$$

**45.** [Cf. I. Schur: J. reine angew. Math. Vol. 151, pp. 100–101 (1921); F. Mertens: J. reine angew. Math. Vol. 79, pp. 182–184 (1875).] We denote by $V_n$ and $W_n$ the partial sums of the series $\sum_{n=0}^{\infty} v_n$ and $\sum_{n=0}^{\infty} (u_0 v_n + u_1 v_{n-1} + \cdots + u_n v_0)$. The two sums are related by the equation

$$W_n = u_n V_0 + u_{n-1} V_1 + \cdots + u_0 V_n, \qquad n = 0, 1, 2, \ldots$$

In order that a convergent sequence $V_n$ generates a convergent sequence $W_n$ it is necessary that the sums $|u_n| + |u_{n-1}| + \cdots + |u_0|$ are bounded, i.e. the series $u_0 + u_1 + u_2 + \cdots$ must be absolutely convergent. In this case the conditions (1) and (2) (cf. p. 111) are automatically satisfied. Therefore the absolute convergence of the series

$$u_0 + u_1 + u_2 + \cdots + u_n + \cdots$$

is the desired necessary and sufficient condition.

**46.** [Cf. I. Schur, l.c. **45**, pp. 103–104; T. J. Stieltjes: Nouv. Annls Math. Ser. 3, Vol. 6, pp. 210–213 (1887).] Let $V_n$ and $W_n$ be the partial sums of the series $\sum_{n=1}^{\infty} v_n$ and $\sum_{n=1}^{\infty} \left(\sum_{t/n} u_t v_{\frac{n}{t}}\right)$. Now [VIII **81**]

$$W_n = u_1 V_n + u_2 V_{\left[\frac{n}{2}\right]} + u_3 V_{\left[\frac{n}{3}\right]} + \cdots + u_n V_{\left[\frac{n}{n}\right]}, \quad n = 1, 2, 3, \ldots$$

The coefficient of $V_k$ is equal to the sum of those $u_l$'s for which $\left[\frac{n}{l}\right] = k$.

Set $\nu = [\sqrt{n}]$. For those values of $l$ which are less or equal to $\nu$ the coefficient of $V_{\left[\frac{n}{l}\right]}$ is precisely $u_l$. This follows from the fact that for $2 \leqq l \leqq \nu$

$$\frac{n}{l-1} - \frac{n}{l} = \frac{n}{l(l-1)} > \frac{n}{l^2} \geqq 1.$$

In order that the sum of the absolute values of the coefficients in the $n$-th row be bounded the same must hold for $|u_1| + |u_2| + \cdots + |u_\nu|$, that is, $u_1 + u_2 + \cdots + u_n + \cdots$ is absolutely convergent. Hence the validity of the other conditions follows, and so the absolute convergence of the series $u_1 + u_2 + \cdots + u_n + \cdots$ is the desired necessary and sufficient condition.

**47.** [R. Dedekind, cf. Knopp, p. 315; Hadamard: Acta Math. Vol. 27, pp. 177–183 (1903).—Cf. I. Schur, l.c. **45**, pp. 104–105.] Put

$$A_n = a_0 + a_1 + a_2 + \cdots + a_n, \quad B_n = \gamma_0 a_0 + \gamma_1 a_1 + \gamma_2 a_2 + \cdots + \gamma_n a_n,$$

then

$$B_n = \sum_{\nu=0}^{n-1} (\gamma_\nu - \gamma_{\nu+1}) A_\nu + \gamma_n A_n.$$

**48.** [Cf. G. Pólya, Problem: Arch. Math. Phys. Ser. 3, Vol. 24, p. 282 (1916). Solved by S. Sidon: Arch. Math. Phys. Ser. 3, Vol. 26, p. 68 (1917).] Set

$$cu_0 = z_0, \quad u_0 + u_1 + \cdots + u_{n-1} + cu_n = z_n, \quad n = 1, 2, 3, \ldots,$$
$$u_0 + u_1 + \cdots + u_{n-1} + u_n = w_n, \quad n = 0, 1, 2, \ldots,$$
$$\sum_{n=0}^{\infty} u_n \zeta^n = U(\zeta), \quad \sum_{n=0}^{\infty} z_n \zeta^n = Z(\zeta), \quad \sum_{n=0}^{\infty} w_n \zeta^n = W(\zeta).$$

By comparing the coefficients we find that

$$W(\zeta) = \frac{U(\zeta)}{1-\zeta}, \quad Z(\zeta) = \frac{U(\zeta)}{1-\zeta} + (c-1)\, U(\zeta),$$

and hence

$$W(\zeta) = \frac{Z(\zeta)}{c + (1-c)\,\zeta}.$$

Assume $c \neq 0$. Compare the coefficients in the last equation:

$$w_n = \frac{(c-1)^n}{c^{n+1}} z_0 + \frac{(c-1)^{n-1}}{c^n} z_1 + \cdots + \frac{c-1}{c^2} z_{n-1} + \frac{1}{c} z_n, \quad n = 0, 1, 2, \ldots$$

According to the criterion on p. 111 $\sum_{n=0}^{\infty} \frac{(c-1)^n}{c^{n+1}}$ must be absolutely convergent. This is the case if and only if $\left|\frac{c-1}{c}\right| < 1$, i.e. if and only if $\Re c > \frac{1}{2}$.

**49.** [I. Schur: Math. Ann. Vol. 74, pp. 453–456 (1913.)] The case $c = -k$, $k$ positive integer, can be excluded from the start; example: $u_n = \binom{n}{k-1}$, $u_n + c\,\frac{u_0 + u_1 + \cdots + u_n}{n+1} = 0$ for $k \geqq 2$, $u_n = \log(n+1)$

for $k=1$ [I **69**]. The case $c=0$ is obvious. We put $u_n + c\dfrac{u_0+u_1+\cdots+u_n}{n+1} = z_n$, $u_n = w_n$ and, as on p. 111,

$$w_n = a_{n0}z_0 + a_{n1}z_1 + \cdots + a_{nn}z_n.$$

Multiplying the relation

$$(n+1)\,z_n - nz_{n-1} = (n+1)\,w_n - nw_{n-1} + cw_n, \quad n = 1, 2, 3, \ldots$$

by $\dfrac{\Gamma(n+c+1)}{\Gamma(n+1)}$ and adding the first $n$ equations we obtain

$$\frac{\Gamma(n+c+2)}{\Gamma(n+1)} w_n - \Gamma(c+2)\,w_0 = \sum_{\nu=1}^{n} \frac{\Gamma(\nu+c+1)}{\Gamma(\nu+1)} [(\nu+1)\,z_\nu - \nu z_{\nu-1}]$$

$$= \frac{\Gamma(n+c+1)}{\Gamma(n+1)} (n+1)\,z_n - c\sum_{\nu=1}^{n-1} \frac{\Gamma(\nu+c+1)}{\Gamma(\nu+1)} z_\nu - \Gamma(c+2)\,z_0,$$

i.e.

$$w_n = \frac{n+1}{n+c+1} z_n - c\frac{\Gamma(n+1)}{\Gamma(n+c+2)} \sum_{\nu=0}^{n-1} \frac{\Gamma(\nu+c+1)}{\Gamma(\nu+1)} z_\nu.$$

For fixed $\nu$ one finds $a_{n\nu} \sim -c\dfrac{\Gamma(\nu+c+1)}{\Gamma(\nu+1)} n^{-c-1}$ [I **155**]. Thus $\lim\limits_{n\to\infty} a_{n\nu}$ exists if and only if $\Re c > -1$. Assume $\Re c > -1$ and put $u_0 = u_1 = u_2 = \cdots = \dfrac{1}{1+c}$, then $z_n = 1$, $w_n = \dfrac{1}{1+c}$, thus

$$a_{n0} + a_{n1} + a_{n2} + \cdots + a_{nn} = \frac{1}{1+c}.$$

With $\Re c = \gamma$ we have

$$\left|\frac{\Gamma(n+c+1)}{\Gamma(n+1)}\right| < An^\gamma, \qquad \left|\frac{\Gamma(n+1)}{\Gamma(n+c+2)}\right| < Bn^{-\gamma-1},$$

where $A$ and $B$ are constants, independent of $n$. Hence

$$|a_{n0}| + |a_{n1}| + \cdots + |a_{n,n-1}| < |c|\,ABn^{-\gamma-1} \sum_{\nu=0}^{n-1} \nu^\gamma$$

$$\to |c|\,AB\int_0^1 x^\gamma dx = \frac{|c|\,AB}{1+\gamma} \qquad \text{[II 22]}.$$

The desired necessary and sufficient condition is therefore $\Re c > -1$.

**50.** Set $a_n = \alpha_n + i\beta_n$, $\alpha_n$, $\beta_n$ real. The relation

$$\sum_{\nu=1}^{n} |\nu^s - (\nu+1)^s| = \sum_{\nu=1}^{n} \nu^\sigma \left|1 - \left(1+\frac{1}{\nu}\right)^s\right| = O(n^\sigma)$$

(binomial series) and a generalisation of the proof of I **75** imply $\lim_{n\to\infty} (a_1 + a_2 + \cdots + a_n)\, n^{-\sigma} = 0$, i.e.

$$\lim_{n\to\infty} (\alpha_1 + \alpha_2 + \cdots + \alpha_n)\, n^{-\sigma} = \lim_{n\to\infty} (\beta_1 + \beta_2 + \cdots + \beta_n)\, n^{-\sigma} = 0.$$

Now I **92** can be applied to both power series

$$\alpha_1 t + \alpha_2 t^2 + \cdots + \alpha_n t^n + \cdots, \quad \beta_1 t + \beta_2 t^2 + \cdots + \beta_n t^n + \cdots,$$

hence

$$\lim_{t\to 1-0} (1-t)^\sigma (\alpha_1 t + \alpha_2 t^2 + \cdots + \alpha_n t^n + \cdots)$$
$$= \lim_{t\to 1-0} (1-t)^\sigma (\beta_1 t + \beta_2 t^2 + \cdots + \beta_n t^n + \cdots) = 0.$$

**51.** Whenever the four subseries consisting of the terms in the four quadrants ($\Re z \geqq 0$, $\Im z \geqq 0$, etc.) converge, the series converges absolutely.

**52.** By successive bisection: Assume that all the terms $z_{r_1}, z_{r_2}, \ldots$ lie in the sector $\vartheta_1 \leqq \arg z \leqq \vartheta_2$ and that $|z_{r_1}| + |z_{r_2}| + \cdots$ diverges. Construct the two subseries with terms in $\vartheta_1 \leqq \arg z \leqq \frac{\vartheta_1 + \vartheta_2}{2}$ and in $\frac{\vartheta_1 + \vartheta_2}{2} \leqq \arg z \leqq \vartheta_2$ resp. At least one of the two is divergent.

**53.** Choose a finite number of terms $z_m = x_m + iy_m$ from each of the successive sectors

$$\left(-\frac{\pi}{2}, \frac{\pi}{2}\right), \left(-\frac{\pi}{4}, \frac{\pi}{4}\right), \ldots, \left(-\frac{\pi}{2^n}, \frac{\pi}{2^n}\right), \ldots$$

The different sectors should contribute different terms so that the points $z_{r_h}, z_{r_{h+1}}, \ldots, z_{r_{h+k}}$ that correspond to $\left(-\frac{\pi}{2^n}, \frac{\pi}{2^n}\right)$ are in this sector and that

$$1 < x_{r_1} + x_{r_2} + \cdots + x_{r_{h+k}} < 2.$$

**54.** [More on this topic: P. Lévy: Nouv. Annls Math. Sér. 4, Vol. 5, pp. 506–511 (1905); E. Steinitz: J. reine angew. Math. Vol. 143, pp. 128–175 (1913).] Let the direction of the positive real axis be the direction of accumulation (which can always be arranged by multiplication with a suitable $e^{i\alpha}$) and $z_{r_1} + z_{r_2} + \cdots$ be the subseries chosen in **53**; apply I **134** to the real part, I **133** to the imaginary part.

**55.** $z = x + iy$, $z^2 = x^2 - y^2 + 2ixy$ are analytic, $|z| = \sqrt{x^2 + y^2}$ and $\bar{z} = x - iy$ however are not analytic.

**55.1.** Cauchy-Riemann equations.

**55.2.** Obvious from the Cauchy-Riemann equations once the existence of the second derivatives is known.

**55.3.** As **55.1**. The result is useful in some problems of mathematical physics, see e.g. G. Pólya and G. Szegö: Isoperimetric Inequalities in Mathematical Physics. Princeton: Princeton University Press 1951, p. 95.

**55.4.** [**55.2**]. Useful in the same manner as **55.3**.

**55.5.** [**55.3**, **55.4**.]

**56.** (1) Put $f(x+iy) = u + iv$, so $u$ is known and

$$f(x+iy) = u(x,y) + i\left(\int_0^x v_x'(x,0)\,dx + \int_0^y v_y'(x,y)\,dy\right)$$

$$= u(x,y) + i\left(-\int_0^x u_y'(x,0)\,dx + \int_0^y u_x'(x,y)\,dy\right)$$

is uniquely determined.

(2) We seek the function $f(z)$ among those functions that are conjugate complex for conjugate complex values of $z$:

$$f(z) + f(\bar{z}) = 2u(x,y) = \frac{(z+\bar{z})(1+z\bar{z})}{1+z^2+\bar{z}^2+z^2\bar{z}^2} = \frac{z}{1+z^2} + \frac{\bar{z}}{1+\bar{z}^2},$$

hence $f(z) = \dfrac{z}{1+z^2}$, because the function is uniquely determined according to (1) and because rational functions of $z$ are analytic. [Hurwitz-Courant, p. 47, p. 282.]

**57.** First solution: Express $\omega$ in terms of $x$ and $y$, verify $\dfrac{\partial^2\omega}{\partial x^2} + \dfrac{\partial^2\omega}{\partial y^2} = 0$ and determine $\mathfrak{I} f(z)$ by integration.

Second solution: $\pi - \mathfrak{I}\log(z-a) = \pi + \mathfrak{R} i \log(z-a)$ is the angle under which the real axis from $a$ to $\infty$ is seen from $z$. Thus

$$f(z) = \pi + i\log(z-a) - [\pi + i\log(z-b)] + ic = ic + i\log\frac{z-a}{z-b},$$

where $c$ stands for a real constant.

**58.** Put $f(x+iy) = u(x,y) + iv(x,y)$. We denote the partial derivatives of $u$ and $v$ by $u_x, u_y, u_{xx}, \ldots$ Differentiation yields

$$\frac{\partial^2}{\partial x^2}(u^2+v^2) = 2(u_x^2 + v_x^2 + uu_{xx} + vv_{xx}),$$

$$\frac{\partial^2}{\partial y^2}(u^2+v^2) = 2(u_y^2 + v_y^2 + uu_{yy} + vv_{yy}),$$

[**55.1**, **55.2**].

**59.** Cf. solution **58**; in addition note that

$$(uu_x + vv_x)^2 + (uu_y + vv_y)^2 = (u^2+v^2)(u_x^2+v_x^2)$$
$$= |f(x+iy)|^2\,|f'(x+iy)|^2.$$

**60.** By comparison of similar triangles:

$$x:\xi = y:\eta = 1:1-\zeta, \quad \text{thus} \quad x+iy = \frac{\xi+i\eta}{1-\zeta},$$

$$x^2+y^2 = \frac{\xi^2+\eta^2}{(1-\zeta)^2} = \frac{2}{1-\zeta} - 1,$$

$$\xi = \frac{2x}{x^2+y^2+1}, \quad \eta = \frac{2y}{x^2+y^2+1}, \quad \zeta = \frac{x^2+y^2-1}{x^2+y^2+1}.$$

**61.** We have in turn [**60**]

$$P: x+iy = \frac{\xi+i\eta}{1-\zeta}, \quad P': \xi,\eta,\zeta, \quad P'': \xi, -\eta, -\zeta;$$

therefore

$$u+iv = \frac{\xi-i\eta}{1+\zeta} = \frac{(x-iy)(1-\zeta)}{1+\zeta} = \frac{x-iy}{x^2+y^2} = \frac{1}{x+iy}.$$

**62.** In the case of Mercator's projection: straight lines parallel to the axes on the unrolled cylinder, generatrices and directrices on the cylinder itself. This property together with the condition of conformity determines Mercator's projection completely (cf. e.g. E. Goursat: Cours d'analyse mathématique, Vol. 2, 3rd Ed. Paris: Gauthier-Villars 1918, p. 58). In the case of stereographic projection: rays from the origin and concentric circles with center at the origin.

**63.** $x+iy = \dfrac{\xi+i\eta}{1-\zeta} = \dfrac{\cos\varphi e^{i\theta}}{1-\sin\varphi} = \tan\left(\dfrac{\varphi}{2}+\dfrac{\pi}{4}\right)e^{i\theta}$,

hence

$$x+iy = e^{u+iv}.$$

**64.** With $w = u+iv$ we have

$$|z| = e^u, \quad \arg z = v.$$

The curves in question are concentric circles, centred at the origin, and rays perpendicular to the circles. [**62**, **63**.]

**65.** Comparison of the real and imaginary parts of $w = u+iv = z^2 = (x+iy)^2$ yields

$$u = x^2-y^2, \qquad v = 2xy.$$

The curves $u =$ const. and $v =$ const. form two families of hyperbolas. Since the mapping is conformal the images in the $z$-plane of the straight lines $u =$ const. and $v =$ const. in the $w$-plane are orthogonal.

**66.** Put $z = x+iy$, $w = u+iv$; then $x = u^2-v^2$, $y = 2uv$; elimination of $v$ and $u$ resp. yields the parabolas $y^2 = 4u^2(u^2-x)$ as images of the lines $u =$ const. and the parabolas $y^2 = 4v^2(v^2+x)$ as images of the lines $v = $ const. All these parabolas have the common axis $y = 0$ and the focus $x = y = 0$. Two orthogonal parabolas pass through every point $z$, $z \neq 0$, of the $z$-plane.

**67.** Let $z = x + iy$, $w = u + iv$. Then

$$u + iv = \frac{e^{i(x+iy)} + e^{-i(x+iy)}}{2}, \quad \text{thus } u = \frac{e^y + e^{-y}}{2} \cos x,\ v = \frac{e^{-y} - e^y}{2} \sin x;$$

The lines $x = $ const. are mapped onto the hyperbolas $\frac{u^2}{\cos^2 x} - \frac{v^2}{\sin^2 x} = 1$,

and the lines $y = $ const. onto the ellipses $\frac{u^2}{\left(\frac{e^y + e^{-y}}{2}\right)^2} + \frac{v^2}{\left(\frac{e^{-y} - e^y}{2}\right)^2} = 1.$

They have the common foci $w = -1$, $w = 1$. The two families of curves are perpendicular to each other (confocal conics).

**68.** Elimination of $v$ and $u$ respectively from $x = u + e^u \cos v$ and $y = v + e^u \sin v$ leads to

$$x - u = e^u \cos\left(y - \sqrt{e^{2u} - (x-u)^2}\right), \quad y - v = e^{x-(y-v)\cot v} \sin v.$$

The line $v = 0$ is transformed into $y = 0$ and the line $v = \pi$ into the twice covered line segment $y = \pi$, $-\infty < x \leqq -1$.

**69.** The image of the square is bounded by the two rays $\arg w = \varepsilon$ and $\arg w = -\varepsilon$ and the two circles $|w| = e^{a+\varepsilon}$ and $|w| = e^{a-\varepsilon}$. The area is therefore $\varepsilon(e^{2a+2\varepsilon} - e^{2a-2\varepsilon})$. The ratio in question is

$$\lim_{\varepsilon\to+0} \frac{e^{2a+2\varepsilon} - e^{2a-2\varepsilon}}{4\varepsilon} = e^{2a}.$$

**70.**

$$\int_{x_1}^{x_2}\int_{y_1}^{y_2} |f'(z)|^2 dx\, dy = \int_{x_1}^{x_2}\int_{y_1}^{y_2} |\sin(x+iy)|^2\, dx\, dy.$$

Because of

$$|\sin(x+iy)|^2 = \sin(x+iy)\sin(x-iy) = -\tfrac{1}{2}\cos 2x + \tfrac{1}{4}(e^{2y} + e^{-2y})$$

the integral is

$$\frac{x_2 - x_1}{8}(e^{2y_2} - e^{2y_1} - e^{-2y_2} + e^{-2y_1}) - \frac{y_2 - y_1}{4}(\sin 2x_2 - \sin 2x_1).$$

For $x_1 = 0$, $x_2 = \frac{\pi}{2}$, $y_1 = 0$, $y_2 = y$ we obtain one quarter of the area of an ellipse with semi-axes

$$a = \frac{e^y + e^{-y}}{2}, \quad b = \frac{e^y - e^{-y}}{2}.$$

**71.** $f'(z) = 2z$. On circles with center at the origin, i.e. $|z| = $ const. and on rays from the origin, i.e. $\arg z = $ const.

**72.** When $0 < a \leqq \pi$ the image of the square is the simply covered region bounded by the two circles with radii $e^a$ and $e^{-a}$ and the two rays $\arg w = -a$ and $\arg w = a$. When $a > \pi$ either part or all of the region is covered several times. If $a = n\pi$ the image is covered exactly $n$ times, except certain points of the real axis that are covered only $n - 1$ times.

**73.** The intersection of the ray $\arg w = \alpha$ with the circle $|w| = 1$; observe the vertical line segment within the disk $|z| \leqq r$ that intersects as many of the parallels $\Im z = \alpha + 2k\pi$, $k = 0, \pm 1, \pm 2, \ldots$ as possible: it lies on $\Re z = 0$, i.e. on the image of $|w| = 1$ (VIII **16**; $N(r, a, \alpha)$ assumes its largest possible value for $r$, $\alpha$ fixed if $\log a = 0$.)

**74.** Assume $z_1 \neq z_2$, $|z_1| < 1$, $|z_2| < 1$. Then

$$z_2^2 + 2z_2 + 3 - (z_1^2 + 2z_1 + 3) = (z_2 - z_1)(z_2 + z_1 + 2) \neq 0.$$

**75.** If $z = re^{i\vartheta}$, $r > 0$, $0 < \vartheta < \pi$, then $w = Re^{i\theta} = r^2 e^{2i\vartheta}$. $R = r^2$, $\theta = 2\vartheta$, and so $R > 0$, $0 < \theta < 2\pi$. If, on the other hand, $R$ and $\theta$ are given $r$, and $\vartheta$ are completely determined.

**76.** The function in question is schlicht on the closed unit disk $|z| \leqq 1$, where furthermore $|w| \leqq 1$ [**5**]. The inverse function is

$$z = \frac{a + e^{-i\alpha} w}{1 + \bar{a} e^{-i\alpha} w},$$

and so the dependence of $z$ on $e^{-i\alpha} w$ is of the same nature as the dependence of $w$ on $z$. Hence each value $w$, $|w| \leqq 1$, is assumed. The locus of the points with constant linear enlargement is given by the relation

$$\frac{1 - |a|^2}{|1 - \bar{a}z|^2} = \text{const}.$$

If $a \neq 0$ these are certain arcs of circles centred at $\frac{1}{\bar{a}}$ (reflection of $a$ with respect to the unit circle).

**77.** [Cf. A. Winternitz: Monatsh. Math. Vol. 30, p. 123 (1920).] According to the hypothesis we have $\left|\frac{z - a}{1 - \bar{a}z}\right| = \text{const}$. along the circle $C$, i.e. $a$ and $\frac{1}{\bar{a}}$ (0 and $\infty$ if $a = 0$) are the pair of harmonic points common to $C$ and the unit circle. Let $z_0$ denote the center of $C$ and $r$ be its radius, $z_0 \neq 0$, $r < 1 - |z_0|$. Then $a$, $|a| < 1$, satisfies the quadratic equation

$$(a - z_0)\left(\frac{1}{a} - \bar{z}_0\right) = r^2 \quad \text{or} \quad (|a| - |z_0|)\left(\frac{1}{|a|} - |\bar{z}_0|\right) = r^2,$$

$\arg a = \arg z_0$; $\alpha$ arbitrary.

**78.** $w = a\dfrac{z - i}{z + i}$ where $a$ is constant, $|a| = 1$.

**79.** Write $z = re^{i\vartheta}$. Then $w$ becomes

$$w = \frac{\frac{1}{r} + r}{2} \cos\vartheta - i \frac{\frac{1}{r} - r}{2} \sin\vartheta.$$

The circles $|z| = r$, $0 < r < 1$, are mapped onto confocal ellipses with semiaxes $\frac{\frac{1}{r} + r}{2}$ and $\frac{\frac{1}{r} - r}{2}$, the common foci are $w = +1$, $w = -1$. The rays $\vartheta = \text{const.}$ are transformed into confocal hyperbolas with the same foci, $w = +1$, $w = -1$. The two families of conics are orthogonal to each other. If $|z| = 1$, $z = e^{i\vartheta}$, then $w = \cos\vartheta$. Consequently, if $z$ describes the unit circle then $w$ describes the segment $-1 \leqq w \leqq 1$ twice.

**80.** The function $w = kz + \frac{1}{kz}$, $0 < kr_1 < kr_2 < 1$ maps the annulus in question onto the region bounded by the two ellipses with foci $w = -2$ and $w = 2$ and major semiaxes $kr_1 + \frac{1}{kr_1}$ and $kr_2 + \frac{1}{kr_2}$ respectively. Put

$$k = \frac{a_1 - \sqrt{a_1^2 - 1}}{r_1} = \frac{a_2 - \sqrt{a_2^2 - 1}}{r_2}.$$

**81.** The function $w = -\frac{z + \frac{1}{z}}{2}$, $z = re^{i\vartheta}$ maps the upper half of the unit disk into the upper half-plane [**79**]:

$$w = -\frac{\frac{1}{r} + r}{2} \cos\vartheta + i \frac{\frac{1}{r} - r}{2} \sin\vartheta; \quad \mathfrak{J}w > 0 \quad \text{for} \quad 0 < \vartheta < \pi.$$

The linear enlargement

$$\left| \frac{\frac{1}{z^2} - 1}{2} \right| = \tfrac{1}{2}$$

for the points $z = x + iy$ for which

$$|x^2 - y^2 - 1 + 2ixy|^2 = |x^2 - y^2 + 2ixy|^2, \quad \text{i.e.} \quad x^2 - y^2 = \tfrac{1}{2}.$$

They define an equilateral hyperbola which intersects the real axis at the points $z = \pm \frac{1}{\sqrt{2}}$. The rotation becomes

$$\arg \frac{\frac{1}{z^2} - 1}{2} = \pm \frac{\pi}{2}$$

when $\mathfrak{R} \frac{1}{z^2} = 1$, i.e. $r^2 = \cos 2\vartheta$. These points lie on the lemniscate $\left| z - \frac{1}{\sqrt{2}} \right| \left| z + \frac{1}{\sqrt{2}} \right| = \frac{1}{2}$.

**82.** The auxiliary function $\zeta = -\frac{z + \frac{1}{z}}{2}$ maps the region in question onto the upper half-plane $\Im\zeta > 0$ [**81**]. The origin $z = 0$ corresponds to $\zeta = \infty$, $z = i$ to $\zeta = 0$ and $z = \pm 1$ to $\zeta = \mp 1$. The function $w = \frac{1}{\zeta^2}$ transforms the upper half-plane $\Im\zeta > 0$, note **75**, into the $w$-plane cut open along the non-negative real axis. The point $\zeta = \infty$ corresponds to $w = 0$, $\zeta = 0$ to $w = \infty$, $\zeta = \pm 1$ to $w = 1$. The mapping function having the required properties is therefore given by

$$w = \left(\frac{2}{z + \frac{1}{z}}\right)^2 = \frac{4z^2}{(1 + z^2)^2}.$$

The images of $z = \pm 1$ are both at $w = 1$ but on different sides of the cut.

**83.** $\arg w = \frac{2\pi}{\beta - \alpha}(\arg z - \alpha)$, i.e. $0 < \arg w < 2\pi$.

**84.** The first auxiliary function $\zeta = (e^{-i\alpha} z)^{\frac{\pi}{\beta - \alpha}}$ maps the circular sector onto the upper half of the disk $|\zeta| < 1$. The second auxiliary function $s = -\frac{\zeta + \frac{1}{\zeta}}{2}$ [**81**] maps the disk onto the upper half-plane $\Im s > 0$. Apply **78**.

**85.** Follows from

$$u - iv = \frac{\partial}{\partial x}[\varphi(x, y) + i\psi(x, y)] = \frac{1}{i}\frac{\partial}{\partial y}[\varphi(x, y) + i\psi(x, y)]$$

by separation of real and imaginary parts.

**86.** They are the images in the $z$-plane of the lines $\Re f = \text{const.}$ and $\Im f = \text{const.}$ parallel to the axes in the $f$-plane under the conformal mapping $f = f(z)$.

**87.** Follows from **85** by virtue of the Cauchy-Riemann differential equation $\frac{\partial u}{\partial x} + \frac{\partial v}{\partial y} = 0$. Also the function $\psi(x, y)$ satisfies Laplace's differential equation.

**88.** With $u \cos\tau + v \sin\tau = \Re(u - iv)\, e^{i\tau}$ the integral in question becomes

$$= \Re \int_L \frac{df}{dz} e^{i\tau} ds = \Re \int_L \frac{df}{dz} dz = \Re[f(z_2) - f(z_1)]$$

where $dz$ denotes the directed line element with modulus $ds$ and argument $\tau$.

**89.** $u \sin\tau - v \cos\tau = \Im(u - iv)\, e^{i\tau}$; cf. **88**.

**90.** The third equation is identical with the second Cauchy-Riemann differential equation. We get the first two by differentiation, keeping the first Cauchy-Riemann differential equation in mind:

$$\frac{1}{\varrho}\frac{\partial p}{\partial x} = -u\frac{\partial u}{\partial x} - v\frac{\partial v}{\partial x} = -u\frac{\partial u}{\partial x} - v\frac{\partial u}{\partial y},$$

$$\frac{1}{\varrho}\frac{\partial p}{\partial y} = -u\frac{\partial u}{\partial y} - v\frac{\partial v}{\partial y} = -u\frac{\partial v}{\partial x} - v\frac{\partial v}{\partial y}.$$

**91.** The vector $\overline{w} = \frac{1}{r}e^{i\vartheta}$ forms the angle $\vartheta$ with the positive real axis, its modulus is $\frac{1}{r}$. The functions in question are up to a constant

$$f(z) = \log z, \quad \varphi(x, y) = \log r = \log\sqrt{x^2 + y^2}, \quad \psi(x, y) = \vartheta = \arctan\frac{y}{x}.$$

The level lines are concentric circles around the origin, the stream lines are rays perpendicular to these circles.

**92.**

$$\varphi_2 - \varphi_1 = \log r_2 - \log r_1 = \log\frac{r_2}{r_1}, \quad \psi' - \psi = 2\pi,$$

$$\frac{\frac{1}{4\pi}(\psi' - \psi)}{\varphi_2 - \varphi_1} = \frac{1}{2\log\frac{r_2}{r_1}}.$$

**93.** The amplitude of $\overline{w}$ is equal to $\vartheta + \frac{\pi}{2}$, the modulus is $\frac{1}{r}$. Furthermore we have (up to an additive constant)

$$f(z) = -i\log z, \quad \varphi(x, y) = \vartheta = \arctan\frac{y}{x},$$

$$\psi(x, y) = -\log r = -\log\sqrt{x^2 + y^2}.$$

The level and stream lines are the stream and level lines resp. of **91**. The potential $\varphi$ is infinitely multivalued.

**94.** According to **93** the field of force is described (up to a real constant factor) by

$$w = \frac{2i}{z^2 - 1}, \text{ thus } f(z) = i\log\frac{z-1}{z+1}, \quad \psi = \log\left|\frac{z-1}{z+1}\right|, \quad -\varphi = \arg\frac{z-1}{z+1}.$$

The level lines are circles through the points $z = -1$ and $z = +1$, the stream lines are circles too, namely the ones with respect to which the points $z = -1$ and $z = +1$ are mirror images of each other (circles of Apollonius).

**95.** We are looking for the points $z$ for which [**93**]

$$-\frac{i\lambda_1}{z-z_1}-\frac{i\lambda_2}{z-z_2}-\cdots-\frac{i\lambda_n}{z-z_n}=0,$$

i.e.

$$\frac{\lambda_1}{z-z_1}+\frac{\lambda_2}{z-z_2}+\cdots+\frac{\lambda_n}{z-z_n}=0;$$

the positive numbers $\lambda_1, \lambda_2, \ldots, \lambda_n$ are proportional to the intensities of the currents. Cf. **31**, in particular the first solution given.

**96.** The vector field is generated by an analytic function $f(z)$ for which $\Re f=$ const. on the given ellipses. The function [**80**]

$$z=kZ+\frac{1}{kZ}, \quad 2kZ=z-\sqrt{z^2-4}$$

maps the region bounded by the two ellipses onto the annulus $r_1<|Z|<r_2$. The semi-axes and the radii are related by

$$\frac{r_1}{r_2}=\frac{a_1-\sqrt{a_1^2-1}}{a_2-\sqrt{a_2^2-1}}, \quad k=\frac{a_1-\sqrt{a_1^2-1}}{r_1}=\frac{a_2-\sqrt{a_2^2-1}}{r_2},$$

provided that $a_1>a_2$ and the positive roots are chosen. The problem is now reduced to **91**: $w=\frac{1}{Z}$ defines a vector field in the $Z$-plane for which the concentric circles around the origin $Z=0$ are level lines, i.e. $\Re\int\frac{dZ}{Z}=$ const. Consequently the same is true for $|Z|=$ const.

$$\int\frac{dZ}{dz}\frac{1}{Z}dz=-\int\frac{dz}{\sqrt{z^2-4}}$$

along the given ellipses in the $z$-plane, i.e.

$$w=-\frac{1}{\sqrt{z^2-4}}, \quad f(z)=\log\left(z-\sqrt{z^2-4}\right).$$

The stream lines are confocal hyperbolas, the level lines confocal ellipses with foci $-2$ and 2. The relations between the potentials are

$$\psi'-\psi=2\pi, \quad \varphi_2-\varphi_1=\log\frac{r_2}{r_1}=\log\frac{a_2-\sqrt{a_2^2-1}}{a_1-\sqrt{a_1^2-1}}.$$

The capacity is

$$\frac{1}{2\log\dfrac{a_2-\sqrt{a_2^2-1}}{a_1-\sqrt{a_1^2-1}}}.$$

**97.** Put [**93**]

$$w=-\frac{i}{z}; \quad \psi_1=-\log a, \quad \psi_2=-\log b, \quad \varphi_1=\alpha, \quad \varphi_2=\beta.$$

The resistance is (the sign is not important) equal to

$$-\frac{\beta-\alpha}{\log b-\log a}.$$

**98.** According to **85** the unit circle $|z|=1$ is a stream line. If the constant value of the conjugate potential along the unit circle and on the real axis inside the vector field is assumed to be zero the function $f=f(z)$ transforms the unit circle $|z|=1$ into a segment of the real axis. In view of **79** set

$$f(z)=k\left(z+\frac{1}{z}\right)+k_0, \qquad k, k_0 \text{ real}.$$

Since $\bar{w}=1$ for $z=\infty$ we have $k=1$, i.e.

$$w=1-\frac{1}{z^2}.$$

**99.** The stagnation points are $z=\pm 1$; $\bar{w}$ assumes the same values at each pair of points that are symmetric with respect to the origin. Therefore the resultant total pressure on the pillar vanishes [cf. **90**]. The pressure is minimal or maximal when $\left|1-\frac{1}{z^2}\right|$ is maximal or minimal resp., i.e. for $z=\pm i$ and $z=\pm 1$ resp. Rotation of all the vectors through 90° generates a field of force which admits the following interpretation: A homogeneous electrostatic field is disturbed by a circular, insulated wire perpendicular to the direction of the field. (The most simple example of electrostatic influence.)

**100.** [G. Kirchhoff: Vorlesungen über Mechanik, 4th Ed. 1897, pp. 303—307; A. Sommerfeld: Mechanics of Deformable Bodies. New York: Academic Press 1950, pp. 215—217.] Supplementary continuity condition: the boundary of the wake (the stagnant water) stretches to infinity where $|w|=1$; since $|w|$ is constant on the entire boundary it has to be equal to 1. The *direction* of $\bar{w}$ is known along the boundary segments $AB$, $AD$ (barrier) and the *magnitude* of $\bar{w}$ is known along the boundary lines $BC$, $DC$ (along the wake), $\bar{w}$ is completely known at the four points $A$, $B$, $C$, $D$. Noticing that either the direction or the magnitude of $\bar{w}$ is *constant* on the respective parts of the boundary we find a half-circle in the $\bar{w}$-plane as image of the boundary of the field of flow. We fix the constant contained in $f$ [p. 123] so that $f=0$ corresponds to the stagnation point $z=0$. Then the left and right "banks" of the positive real axis of the $f$-plane correspond to the streamlines $ABC$ and $ADC$, respectively. The $f$-plane cut along the positive real axis corresponds to the whole field of flow; it is not possible that only a subregion of the $f$-plane so cut should correspond to the field of flow because $w = \frac{df}{dz} \sim i$ as $z \to \infty$, thus $f \sim iz$. The same point, but associated with the left or right bank of the cut in the $f$-plane, corresponds to the two points $B$ and $D$, respectively, because of symmetry. Note that a dilatation of the $f$-plane with $\bar{w}$ retained, causes the same dilatation of the $z$-plane because $df = w\,dz$. Now we dilate the $f$-plane so that $f = 1$ becomes the image of $z = \pm l$. In this way we attribute a numerical value to $l$.—If a one to one relationship can be established between the corresponding parts of the $z$-, $\bar{w}$- and $f$-planes we can find out [cf. **188**] whether the interior is to the left or to the right as one moves in the direction given by $ABCDA$ (see the last line given in the table). In the following diagrams the points in the different planes corresponding to $A$ are also called $A$; $B$, $C$, $D$ are used similarly.

| | $z$ | $\bar{w}$ | $w$ | $f$ |
|---|---|---|---|---|
| $A$ | $0$ | $0$ | $0$ | $0$ |
| $B$ | $l$ | $1$ | $1$ | $1$ |
| $C$ | $\infty$ | $-i$ | $i$ | $\infty$ |
| $D$ | $-l$ | $-1$ | $-1$ | $1$ |
| region lies to the | left | right | left | left |

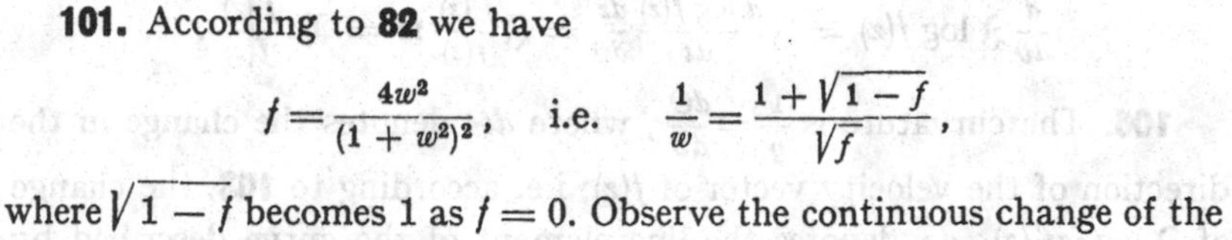

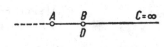

Velocity plane ($\bar{w}$-plane) (w-plane) Potential plane ($f$-plane)

**101.** According to **82** we have

$$f = \frac{4w^2}{(1+w^2)^2}, \quad \text{i.e.} \quad \frac{1}{w} = \frac{1+\sqrt{1-f}}{\sqrt{f}},$$

where $\sqrt{1-f}$ becomes 1 as $f = 0$. Observe the continuous change of the value of $w$ on the two banks of the cut in the $f$-plane. Hence

$$z = \int_0^f \frac{df}{w} = 2\sqrt{f} + \sqrt{f}\sqrt{1-f} + \arcsin\sqrt{f},$$

$$z = x + iy = 2\sqrt{f} + \frac{\pi}{2} - i\left[f\sqrt{1-\frac{1}{f}} - \log\left(\sqrt{f} + \sqrt{f-1}\right)\right].$$

The first formula for $z$ is to be used when $0 < f < 1$, the second when $f > 1$ (positive roots); it furnishes the boundary of the wake:

$z = l = 2 + \frac{\pi}{2}$ for $f = 1$; $x \sim 2\sqrt{f}$, $y \sim -f$, thus the width of the wake is $2x \sim 4\sqrt{|y|}$ at a great distance from the barrier.

**102.** If the pressure is $p = c - \frac{1}{2}|w|^2$ at the point $z$ [**90**] it is in particular equal to $p_1 = c - \frac{1}{2}$ on the boundary and therefore everywhere in the wake. The total pressure desired is

$$\int_{-l}^{+l} (p - p_1)\, dz = \int_{-l}^{+l} \tfrac{1}{2}(1 - |w|^2)\, dz = \int_0^l (1 - w^2)\, dz$$

$$= \int_0^1 (1 - w^2) \frac{df}{w} = \int_0^1 4\sqrt{1 - f}\, d\sqrt{f} = \pi.$$

**103.** Put $z = re^{i\vartheta}$. Since the sign of the angular velocity is positive, $\vartheta$ is increasing, i.e. $z$ describes the circle in the positive sense. We have $\frac{dz}{d\vartheta} = iz$ and the velocity vector in question is

$$\frac{df(z)}{d\vartheta} = \frac{df(z)}{dz} \frac{dz}{d\vartheta} = izf'(z).$$

**104.** Let $\omega$ denote the angle through which the vector $w$ (radius vector) has to be rotated in the positive sense to fall into the direction of the vector $izf'(z)$ (tangential vector, **103**). Then the distance in question is given by

$$|f(z)| \sin \omega = |f(z)| \frac{\Im \frac{izf'(z)}{f(z)}}{\left|\frac{izf'(z)}{f(z)}\right|} = \frac{\Re zf'(z)\overline{f(z)}}{|zf'(z)|}.$$

**105.** The amplitude of the vector in question is $\Im \log f(z)$. The angular velocity is therefore, with $z = re^{i\vartheta}$,

$$\frac{d}{d\vartheta} \Im \log f(z) = \Im \frac{d \log f(z)}{dz} \frac{dz}{d\vartheta} = \Im \frac{f'(z)}{f(z)} iz = \Re z \frac{f'(z)}{f(z)}.$$

**106.** The curvature is $\frac{1}{\varrho} = \frac{d\Theta}{dS}$, where $d\Theta$ denotes the change in the direction of the velocity vector of $f(z)$, i.e. according to **103**, the change of $\Im \log izf'(z)$; $dS$ denotes the line element of the curve described by $f(z)$. According to **103** we have $\frac{dS}{d\vartheta} = |izf'(z)|$; consequently, with $z = re^{i\vartheta}$,

$$\frac{1}{\varrho} = \frac{\frac{d\Theta}{d\vartheta}}{\frac{dS}{d\vartheta}} = \frac{\frac{d}{d\vartheta} \Im \log izf'(z)}{|izf'(z)|} = \frac{\Im \frac{d \log zf'(z)}{dz} \frac{dz}{d\vartheta}}{|zf'(z)|}.$$

The curvature is positive or negative according as the velocity vector turns in the positive or the negative sense.

**107.**

| | concave | convex |
|---|---|---|
| to the left | + | − |
| to the right | − | + |

If $w = z^n + a$, $n \gtrless 0$, then $\frac{1}{\varrho} = \frac{\operatorname{sgn} n}{r^n}$. The four possibilities are already shown by the special cases $r = 1$, $|a| > 1$, $n = 1$, or $n = -1$, where $w = 0$ is chosen as reference point.

**108.** [**106**, **107**.]

**109.** The angular velocity of the vector $w = f(z)$ is always positive. [**105**.]

**110.** The proposition follows from **108** and **109** or from the following consideration: the angle between $dw$, $w = f(z)$, and the positive real axis is given by the argument of $izf'(z)$ [**103**]. Convexity means that this argument is always changing in the same sense, which property coincides with the definition of the image of the circle under $w = zf'(z)$ as being star-shaped.

**111.** [Thekla Lukács.] Let $a$ and $b$ be two points with respect to which the image of the circle $|z| = r$ is star-shaped. Then

$$\Re z \frac{f'(z)}{f(z) - a} > 0, \qquad \Re z \frac{f'(z)}{f(z) - b} > 0,$$

i.e.

$$\Re z f'(z)\, \overline{f(z) - a} > 0, \qquad \Re z f'(z)\, \overline{f(z) - b} > 0.$$

Hence for $\lambda > 0, \mu > 0, \lambda + \mu = 1$

$$\Re z f'(z)\, \overline{f(z) - (\lambda a + \mu b)} > 0, \qquad \text{thus} \quad \Re z \frac{f'(z)}{f(z) - (\lambda a + \mu b)} > 0.$$

The proposition can also be easily proved by elementary geometric reasoning.

**112.** $h(\varphi) = \Re \bar{a} e^{i\varphi} = |a| \cos(\varphi - \alpha)$.

**113.** (1) The velocity vector [**103**] $izf'(z)$ forms the angle $\varphi + \frac{\pi}{2}$ with the positive real axis: $\varphi = \arg zf'(z) = \Im \log zf'(z)$.

(2)

$$h(\varphi) = \frac{\Re z f'(z)\, \overline{f(z)}}{|zf'(z)|}$$

[**104**.]; note the sign.

**114.**

$$\varphi = \Im \log \frac{z}{1+z} = \Im \log \frac{e^{\frac{i\vartheta}{2}}}{2\cos\frac{\vartheta}{2}} = \frac{\vartheta}{2}, \qquad z = e^{i\vartheta} \qquad [\mathbf{113}].$$

**115.** For $\frac{\pi}{4} \leqq \varphi \leqq \frac{3\pi}{4}$ the support function is identical with the one of the point $\pi i$ [**112**]. Outside this sector we find on the basis of **113**

$$\varphi = \Im \log \frac{2z}{\sqrt{1+z^2}} = \Im \log \frac{2e^{\frac{i\vartheta}{2}}}{\sqrt{2\cos\vartheta}} = \frac{\vartheta}{2}.$$

**116.** $\varphi = \arg z f'(z) = \arg \frac{w}{1-w}$, $h(\varphi) = \left|\frac{w}{1-w}\right| \Re(1-w)$ if the line of support is a tangent to the boundary; $h(\varphi) = \cos\varphi$ if the line of support passes through the corner $w = 1$.

**117.** $\frac{1}{2\pi}\int\limits_0^{2\pi} e^{i(k-l)\vartheta}\, d\vartheta = \frac{1}{2\pi}\int\limits_0^{2\pi} e^{in\vartheta} d\vartheta = 0$ or $1$ depending on whether the integer $n = k - l$ is different from zero or not.

**118.** Follows from the expansion

$$f(re^{i\vartheta}) = a_0 + a_1 r e^{i\vartheta} + a_2 r^2 e^{2i\vartheta} + \cdots + a_n r^n e^{in\vartheta} + \cdots$$

by integration and application of **117**.

**119.** Apply **118** to $\log f(z)$ and consider the real part.

**120.** [Jensen's Formula, cf. **175**.] The geometric mean of the single factor $|z - z_\nu|$ can be computed with the help of II **52**, the geometric mean of $|f^*(z)|$ with the help of **119**.

**121.** For $f(z) \neq 0$ in $|z| \leqq r$ the means are equal: $\mathfrak{g}(r) = \mathfrak{G}(r) = |f(0)|$ [**119**]. If $f(z)$ is the product $f(z) = f_1(z)\, f_2(z)$ of two functions $f_1(z)$ and $f_2(z)$ that are regular for $|z| \leqq r$, its mean is equal to the product of the corresponding means. It is therefore sufficient to examine the particular case $f(z) = (z - z_0)$, $|z_0| \leqq r$. According to II **52**

$$\mathfrak{G}(r) = e^{\frac{1}{2\pi}\int\limits_0^{2\pi} \log|re^{i\vartheta} - z_0| d\vartheta} = \max(r, |z_0|) = r,$$

$$\mathfrak{g}(r) = e^{\frac{1}{\pi r^2}\int\limits_0^r\int\limits_0^{2\pi} \log|\varrho e^{i\vartheta} - z_0|\varrho\, d\varrho\, d\vartheta} = e^{\frac{2}{r^2}\int\limits_0^r \max(\log\varrho, \log|z_0|)\varrho\, d\varrho} = e^{\log r - \frac{1}{2} + \frac{|z_0|^2}{2r^2}}.$$

**122.** [Cf. M. A. Parseval: Mém. par. divers savants Vol. 1, pp. 639—648 (1805); A. Gutzmer: Math. Ann. Vol. 32, pp. 596—600 (1888).]

$$\frac{1}{2\pi}\int\limits_0^{2\pi} f(re^{i\vartheta})\, \overline{f(re^{i\vartheta})}\, d\vartheta = \sum_{k=0}^{\infty}\sum_{l=0}^{\infty} r^{k+l} \frac{1}{2\pi}\int\limits_0^{2\pi} e^{i(k-l)\vartheta}\, d\vartheta \cdot a_k \bar{a}_l \qquad [\mathbf{117}].$$

**123.** We set $P(z) = x_0 + x_1 z + x_2 z^2 + \cdots + x_n z^n$ with arbitrary complex coefficients $x_0, x_1, \ldots, x_n$. Then

$$\frac{1}{2\pi}\int_0^{2\pi} |f(e^{i\vartheta}) - P(e^{i\vartheta})|^2 \, d\vartheta$$

$$= |a_0 - x_0|^2 + |a_1 - x_1|^2 + \cdots + |a_n - x_n|^2 + |a_{n+1}|^2 + |a_{n+2}|^2 + \cdots.$$

This expression becomes a minimum if and only if the first $n + 1$ terms vanish.

**124.** [As to **124**—**127** cf. L. Bieberbach: Rend. Circ. Mat. Palermo Vol. 38, pp. 98—112 (1914); Sber. Berlin Math. Ges. 1916, pp. 940—955 and T. Carleman: Math. Z. Vol. 1, pp. 208—212 (1918).] Particular case of **125**; replace $r$ and $R$ by 0 and $r$ respectively.

**125.** With $w = f(x + iy) = u + iv$ the area in question becomes

$$F = \iint\limits_{r^2 \leq x^2 + y^2 \leq R^2} du\, dv = \iint\limits_{r^2 \leq x^2 + y^2 \leq R^2} \left|\frac{\partial(u, v)}{\partial(x, y)}\right| dx\, dy = \int_r^R \int_0^{2\pi} \left|\frac{\partial(u, v)}{\partial(x, y)}\right| \varrho\, d\varrho\, d\vartheta.$$

We have (cf. p. 117)

$$\frac{\partial(u, v)}{\partial(x, y)} = \frac{\partial u}{\partial x}\frac{\partial v}{\partial y} - \frac{\partial u}{\partial y}\frac{\partial v}{\partial x} = \left(\frac{\partial u}{\partial x}\right)^2 + \left(\frac{\partial v}{\partial x}\right)^2 = \left|\frac{\partial(u + iv)}{\partial x}\right|^2 = |f'(z)|^2,$$

hence [cf. **122**]

$$F = \int_r^R \int_0^{2\pi} |f'(\varrho e^{i\vartheta})|^2 \varrho\, d\varrho\, d\vartheta = 2\pi \int_r^R \left(\sum_{n=-\infty}^{\infty} n^2 |a_n|^2 \varrho^{2n-1}\right) d\varrho$$

$$= \pi \sum_{n=-\infty}^{\infty} n |a_n|^2 (R^{2n} - r^{2n}).$$

**126.** Particular case of **127**. As to the orientation cf. **188** or **190**.— What does the result suggest when $c = 0$? [**124**.]

**127.** The area is given as the sum of elementary triangles provided with a sign, bounded by the arcs of the curve $L$ and radii. The sign is positive or negative depending on whether the point $w = 0$ is to the left or to the right of the bounding oriented arc. It coincides with the sign of $\sin \omega$ introduced in solution **104**. Thus the area is

$$\left.\begin{aligned} &= \tfrac{1}{2}\int_0^{2\pi} |izf'(z)|\,|f(z)| \sin\omega\, d\vartheta = \tfrac{1}{2}\int_0^{2\pi} \Re zf'(z)\, \overline{f(z)}\, d\vartheta \\ &= \tfrac{1}{2}\Re \int_0^{2\pi} \sum_{k=-\infty}^{\infty} k a_k r^k e^{ik\vartheta} \sum_{l=-\infty}^{\infty} \bar{a}_l r^l e^{-il\vartheta}\, d\vartheta, \end{aligned}\right\} z = re^{i\vartheta}.$$

**128.** According to **124** we have

$$4\int_0^r \frac{J(\varrho)}{\varrho}\,d\varrho = 2\pi \sum_{n=1}^{\infty} |a_n|^2 r^{2n} \qquad [\mathbf{122}].$$

**129.** [Cf. K. Löwner and Ph. Frank: Math. Z. Vol. 3, p. 84 (1919).] The density function of the mass distribution in question is proportional to $|\varphi'(z)|^{-1}$, because $\int |\varphi'(z)|^{-1}\,|dw|$ over an arbitrary arc of $L$ gives the length of the corresponding arc of the circle $|z| = r$. Therefore

$$\xi \int_L |\varphi'(z)|^{-1}\,|dw| = \int_L \varphi(z)\,|\varphi'(z)|^{-1}|dw|, \text{ i.e. } \xi \int_0^{2\pi} r\,d\vartheta = \int_0^{2\pi} \varphi(re^{i\vartheta})\,r\,d\vartheta.$$

**130.** The volume is given by

$$\int_0^r \int_0^{2\pi} |f(\varrho e^{i\vartheta})|^2\,\varrho\,d\varrho\,d\vartheta = 2\pi \int_0^r \left(\sum_{n=0}^{\infty} |a_n|^2 \varrho^{2n+1}\right) d\varrho,$$

where $z = \varrho e^{i\vartheta}$.

**131.** [J. L. W. V. Jensen: Acta Math. Vol. 36, p. 195 (1912).]

$$\cos\gamma = \left[1 + \left(\frac{\partial\zeta}{\partial x}\right)^2 + \left(\frac{\partial\zeta}{\partial y}\right)^2\right]^{-\frac{1}{2}}, \quad \tan^2\gamma = \left(\frac{\partial\zeta}{\partial x}\right)^2 + \left(\frac{\partial\zeta}{\partial y}\right)^2.$$

The Cauchy-Riemann differential equations together with the relation $\zeta = u^2 + v^2$ imply

$$\frac{1}{4}\tan^2\gamma = \left(u\frac{\partial u}{\partial x} + v\frac{\partial v}{\partial x}\right)^2 + \left(u\frac{\partial u}{\partial y} + v\frac{\partial v}{\partial y}\right)^2 = (u^2 + v^2)\left[\left(\frac{\partial u}{\partial x}\right)^2 + \left(\frac{\partial v}{\partial x}\right)^2\right].$$

**132.** Let $z_0$ be a point with horizontal tangent plane. According to **131** we have either $f(z_0) = 0$ or $f(z_0) \neq 0$, $f'(z_0) = 0$. As to the first case one has to note that an analytic function has only isolated zeros. In the second case we consider $f'(z_0) = f''(z_0) = \cdots = f^{(l-1)}(z_0) = 0$, $f^{(l)}(z_0) \neq 0$ [$l \geqq 2$, saddle point of order $l - 1$], thus

$$f(z_0 + h) = f(z_0) + \frac{f^{(l)}(z_0)}{l!} h^l + \cdots,$$

$$\zeta = |f(z_0 + h)|^2 = |f(z_0)|^2 + f(z_0)\overline{\frac{f^{(l)}(z_0)}{l!}}\,\overline{h}^l + \overline{f(z_0)}\frac{f^{(l)}(z_0)}{l!} h^l + \cdots$$
$$= |f(z_0)|^2 + A\,|h|^l \cos(l\varphi - \alpha) + \cdots,$$

where $h = |h|\,e^{i\varphi}$; $A$, $\alpha$ are real constants, $A \gtrless 0$, determined by $f(z_0)$ and $f^{(l)}(z_0)$. The $2l$ values of $\varphi$ that correspond to the $2l$ directions of the branches joining at $z_0$ are

$$\varphi = \frac{\alpha}{l} + \frac{2k-1}{2l}\pi, \qquad k = 1, 2, \ldots, 2l.$$

The sign of $A \cos (l\varphi - \alpha)$ is alternately positive and negative between these directions.

**133.** Suppose that the polynomial in question is given by $a_0(z-\alpha_1)(z-\alpha_2)\cdots(z-\alpha_n)$, $\alpha_1, \alpha_2, \ldots, \alpha_n$ real. Then

$$\zeta = |a_0|^2 \prod_{\nu=1}^{n} [(x-\alpha_\nu)^2 + y^2] \geqq |a_0|^2 \prod_{\nu=1}^{n} (x-\alpha_\nu)^2, \qquad \frac{\partial^2 \zeta}{\partial y^2} > 0.$$

**134.** **122** implies

$$|f(z_0)|^2 + \left|\frac{f'(z_0)}{1!}\right|^2 r^2 + \left|\frac{f''(z_0)}{2!}\right|^2 r^4 + \cdots + \left|\frac{f^{(n)}(z_0)}{n!}\right|^2 r^{2n} + \cdots$$
$$= \frac{1}{2\pi}\int_0^{2\pi} |f(z_0 + re^{i\vartheta})|^2 \, d\vartheta \leqq M^2,$$

i.e. $|f(z_0)| \leqq M$. In the case of equality we must have

$$f'(z_0) = f''(z_0) = \cdots = f^{(n)}(z_0) = \cdots = 0$$

which means $f(z) \equiv f(z_0)$, constant.

**135.** The continuous function $|f(z)|$ has to attain its maximum in the closed domain $\mathfrak{D}$. According to **134** $|f(z)|$ cannot assume its largest value at an interior point $z_0$ of $\mathfrak{D}$.

**136.** The piece cut off from the modular graph by an arbitrary cylinder perpendicular to the $x, y$-plane has its highest point on the boundary unless the modular graph is parallel to the $x, y$-plane. There are no "peaks" in an "analytic landscape".

**137.** Geometric interpretation of the proposition that a polynomial $(z-z_1)(z-z_2)\cdots(z-z_n)$ always assumes its maximum on the boundary of any domain [**135**].

**138.** The function $\frac{1}{f(z)}$ is regular in $\mathfrak{D}$ [**135**].

**139.** Assume $|z_\nu| < R$, $\nu = 1, 2, \ldots, n$. The function

$$f(z) = \frac{1}{R}\sqrt[n]{(R^2 - \bar{z}_1 z)(R^2 - \bar{z}_2 z)\cdots(R^2 - \bar{z}_n z)}$$

has a regular branch for $|z| \leqq R$, furthermore $f(z) \neq 0$ for $|z| \leqq R$. According to **135** and **138** $|f(0)| = R$ must lie between the maximum and the minimum of $|f(z)|$ on $|z| = R$. We have for $|z| = R$ [**5**]

$$|f(z)| = \sqrt[n]{|z-z_1|\,|z-z_2|\cdots|z-z_n|}.$$

The only exception occurs in the case $f(z) \equiv$ const., i.e. if $z_1 = z_2 = \cdots = z_n = 0$.

**140.** We find for $|z| = R$

$$R \leq \max \left| \frac{1}{R} \frac{(R^2 - \bar{z}_1 z) + (R^2 - \bar{z}_2 z) + \cdots + (R^2 - \bar{z}_n z)}{n} \right|$$

$$\leq \max \left( \frac{1}{R} \frac{|R^2 - \bar{z}_1 z| + |R^2 - \bar{z}_2 z| + \cdots + |R^2 - \bar{z}_n z|}{n} \right)$$

$$= \max \frac{|z - z_1| + |z - z_2| + \cdots + |z - z_n|}{n} \qquad [\mathbf{5}].$$

The inequality becomes an equality only if $z_1 + z_2 + \cdots + z_n = 0$ and all the $z_\nu$'s have the same argument, i.e. $z_\nu = 0$, $\nu = 1, 2, \ldots, n$. Notice the particular case $n = 4$, $z_1 = 1$, $z_2 = i$, $z_3 = -1$, $z_4 = -i$, $R > 1$. The arithmetic mean of the *projections* of the distances in question into the diameter through $P$ is already equal to $R$.

**141.** The function

$$\frac{1}{R^2 - \bar{z}_1 z} + \frac{1}{R^2 - \bar{z}_2 z} + \cdots + \frac{1}{R^2 - \bar{z}_n z}$$

is regular for $|z| \leq R$. Hence [**135**] for $|z| = R$

$$\min \frac{n}{\sum_{\nu=1}^{n} \frac{1}{|z - z_\nu|}} = \min \frac{n}{\sum_{\nu=1}^{n} \frac{R}{|R^2 - \bar{z}_\nu z|}} \leq \frac{n}{\max \left| \sum_{\nu=1}^{n} \frac{R}{R^2 - \bar{z}_\nu z} \right|} \leq R.$$

Note the special case $n = 3$, $z_1 = e^{i\vartheta_1} = 1$, $z_2 = e^{i\vartheta_2} = e^{i\frac{2\pi}{3}}$, $z_3 = e^{i\vartheta_3} = e^{i\frac{4\pi}{3}}$, $R \geq 5$. Then we have for $z = Re^{i\vartheta}$, $\nu = 1, 2, 3$,

$$\frac{1}{|z - z_\nu|} = \frac{1}{\sqrt{R^2 + 1 - 2R\cos(\vartheta - \vartheta_\nu)}} = \sum_{k=0}^{\infty} \frac{P_k[\cos(\vartheta - \vartheta_\nu)]}{R^{k+1}},$$

where $P_k(x)$ denotes the $k$-th Legendre polynomial [VI, § 11]; $P_0(\cos\vartheta) = 1$, $P_1(\cos\vartheta) = \cos\vartheta$, $P_2(\cos\vartheta) = \frac{1}{4} + \frac{3}{4}\cos 2\vartheta$. Hence [VI **91**]

$$\frac{1}{3} \sum_{\nu=1}^{3} \frac{1}{|z - z_\nu|} = \frac{1}{R} + \frac{1}{4R^3} +$$

$$+ \sum_{k=3}^{\infty} \frac{P_k[\cos(\vartheta - \vartheta_1)] + P_k[\cos(\vartheta - \vartheta_2)] + P_k[\cos(\vartheta - \vartheta_3)]}{3R^{k+1}}$$

$$> \frac{1}{R} + \frac{1}{4R^3} - \sum_{k=3}^{\infty} \frac{1}{R^{k+1}} = \frac{1}{R} + \frac{R - 5}{4R^3(R - 1)} \geq \frac{1}{R}.$$

**142.** If $f(z) \neq 0$ everywhere inside the level line the absolute value $|f(z)|$ attains its maximum and its minimum on the boundary according to **135** and **138**. This implies that $|f(z)|$ must be constant in the interior, and so $f(z) \equiv$ const. Geometric interpretation: Since there is no peak inside a closed level line on the modular graph there must be at least one pit unless the modular surface is a horizontal plane.

**143.** At least one zero must lie inside any closed line [**142**] along which the absolute value of the polynomial $(z - z_1)(z - z_2)\cdots(z - z_n)$ is constant. There are only $n$ zeros.

**144.** The theorem is not valid for $f(z) =$ const. Thus we may assume $f(z_0) \neq 0$. If there is a saddle point on the circle $|z| = r$ the projection of at least one of the sectors with points above the saddle point mentioned in **132** protrudes into the interior of $|z| \leqq r$. Therefore $z_0$ cannot be a saddle point, i.e. $f'(z_0) \neq 0$.

We put $f(z) = w$ and consider the image of the circle $|z| = r$ in the $w$-plane. The point farthest from $w = 0$ is $w_0 = f(z_0)$, the curve has a definite tangent at $w_0$ because $f'(z_0) \neq 0$. The tangent, that is the vector $iz_0f'(z_0)$ [**103**], is perpendicular to the vector $w_0 = f(z_0)$ (obvious for geometric reasons). Hence $\frac{iz_0f'(z_0)}{f(z_0)}$ is purely imaginary. In the neighbourhood of $w_0$ the side of the image curve turned towards the origin corresponds, according to the hypothesis, to the side of the circle turned towards the origin. Therefore $\frac{iz_0f'(z_0)}{f(z_0)}$ must be positive imaginary.

**145.** [Cf. A. Pringsheim: Sber. bayer. Akad. Wiss. 1920, p. 145; 1921, p. 255.]

$$2\sum_{\nu=1}^{n}\frac{a(\omega^\nu - \omega^{\nu-1})}{a(\omega^\nu + \omega^{\nu-1})} = 2\sum_{\nu=1}^{n}\frac{\omega^{\frac12} - \omega^{-\frac12}}{\omega^{\frac12} + \omega^{-\frac12}} = 2ni\tan\frac{\pi}{n} \to 2\pi i.$$

**146.** [A. Pringsheim, l.c. **145**.] We define

$$\zeta_\nu^{(k)} = \begin{cases} \dfrac{1}{k+1}\left(z_{\nu-1}^k + z_{\nu-1}^{k-1}z_\nu + \cdots + z_\nu^k\right) & \text{for } k = 0, 1, 2, \ldots, \\ -\dfrac{1}{k+1}\left(z_{\nu-1}^{k+1}z_\nu^{-1} + z_{\nu-1}^{k+2}z_\nu^{-2} + \cdots + z_{\nu-1}^{-1}z_\nu^{k+1}\right) & \text{for } k = -2, -3, \ldots; \end{cases}$$

then

$$\zeta_1^{(k)}(z_1 - z_0) + \zeta_2^{(k)}(z_2 - z_1) + \cdots + \zeta_n^{(k)}(z_n - z_{n-1}) = \frac{z_n^{k+1} - z_0^{k+1}}{k+1} = 0.$$

The total length of $L$ is called $l$, $R$ is the largest, $r$ the smallest distance of $L$ from the origin $z = 0$; assume $R > 1$. The quantity

$$|z_1 - z_0| + |z_2 - z_1| + |z_3 - z_2| + \cdots + |z_n - z_{n-1}|$$

is the length of an inscribed polygon, thus $\leqq l$. The points $z_1, z_2, \ldots, z_n$ can be chosen so that for a given $\delta$, $\delta > 0$, and sufficiently large $n$,

$|z_\nu - z_{\nu-1}| \leqq \delta$, $\nu = 1, 2, \ldots, n$. Assume $k \geqq 0$, then

$$|(\zeta_\nu^{(k)} - z_\nu^k)(z_\nu - z_{\nu-1})| = \frac{1}{k+1}|(z_{\nu-1}^k - z_\nu^k) + z_\nu(z_{\nu-1}^{k-1} - z_\nu^{k-1}) + \cdots||z_\nu - z_{\nu-1}|$$

$$\leqq \frac{1}{k+1}[k + (k-1) + \cdots + 1 + 0] R^{k-1} |z_\nu - z_{\nu-1}|^2,$$

consequently

$$\left|\sum_{\nu=1}^{n} z_\nu^k (z_\nu - z_{\nu-1})\right| \leqq \frac{k}{2} R^{k-1} \sum_{\nu=1}^{n} |z_\nu - z_{\nu-1}|^2 \leqq \frac{k}{2} R^{k-1} l \delta.$$

The case $R \leqq 1$ is similar. If $k \leqq -2$ we have to use $r$ instead of $R$ for the estimate.

**147.** [Cf. G. N. Watson: Complex Integration and Cauchy's Theorem. Cambr. Math. Tracts No. 15, p. 66 (1914).] The interior of the given ellipse is described by the inequality

$$x^2 - xy + y^2 + x + y < 0.$$

Only one, $z_0 = -\frac{1}{\sqrt{2}} - \frac{i}{\sqrt{2}}$, of the four poles, $\pm\frac{1}{\sqrt{2}} \pm \frac{i}{\sqrt{2}}$, of the integrand lies inside the ellipse. Therefore

$$\oint \frac{dz}{1+z^4} = \frac{2\pi i}{4z_0^3} = \frac{\pi}{2\sqrt{2}}(-1+i).$$

**148.**

$$4i \int_0^{\frac{\pi}{2}} \frac{x\, d\vartheta}{x^2 + \sin^2\vartheta} = \int_{-\pi}^{\pi} \frac{ix\, d\vartheta}{\sin^2\vartheta + x^2} = \int_{-\pi}^{\pi} \frac{d\vartheta}{\sin\vartheta - ix} = \oint \frac{2dz}{z^2 + 2zx - 1},$$

with $z = e^{i\vartheta}$. The integral is taken along the circle $|z| = 1$ which includes only the pole at $z_0 = -x + \sqrt{1+x^2}$. Hence

$$\frac{4\pi i}{z_0 + z_0^{-1}} = \frac{2\pi i}{\sqrt{1+x^2}}.$$

**149.** [G. Pólya, Problem: Arch. Math. Phys. Ser. 3, Vol. 24, p. 84 (1916). Solved by J. Mahrenholz: Arch. Math. Phys. Ser. 3, Vol. 26, p. 66 (1917).]

$$\int_0^{2\pi} \frac{(1 + 2\cos\vartheta)^n e^{in\vartheta}}{1 - r - 2r\cos\vartheta} d\vartheta = \frac{1}{i} \oint \frac{(1 + z + z^2)^n}{(1-r)z - r(1+z^2)} dz,$$

the integral is taken along the circle $|z| = 1$. If $r \gtrless 0$, $-1 < r < \frac{1}{3}$, we have $\left|\frac{1}{r} - 1\right| > 2$, therefore the quadratic equation

$(1 - r)\,z - r(1 + z^2) = 0$ has its two roots separated by the unit circle. Let $\varrho$ be the root inside the unit circle, i.e. $|\varrho| < 1$, $(1 - r)\,z - r(1 + z^2) = -r(z - \varrho)\left(z - \frac{1}{\varrho}\right)$. We obtain

$$2\pi\left[\frac{(1 + z + z^2)^n}{-r\left(z - \frac{1}{\varrho}\right)}\right]_{z=\varrho} = \frac{2\pi}{r\left(\frac{1}{\varrho} - \varrho\right)}\left(\frac{\varrho}{r}\right)^n;$$

$\varrho$ is real, $\varrho = \dfrac{1 - r - \sqrt{1 - 2r - 3r^2}}{2r}$.

**150.** With the notation $z = x + iy$ the integral can be written

$$\frac{1}{2i}\oint\frac{z\,dz - \bar z\,d\bar z + z\bar z(\bar z\,dz - z\,d\bar z)}{1 + z^2 + \bar z^2 + z^2\bar z^2} = \frac{1}{2i}\oint\frac{(1 + \bar z^2)\,z\,dz - (1 + z^2)\,\bar z\,d\bar z}{(1 + z^2)(1 + \bar z^2)}$$

$$= \Im\oint\frac{z\,dz}{1 + z^2} = \Im\,2\pi i = 2\pi.$$

The foci of the ellipse are the poles of the integrand.

**151.** We have for $\omega > 0$

$$\int_0^\omega x^{s-1}e^{-x}\,dx + \omega^s i\int_0^{\frac{\pi}{2}} e^{is\vartheta - \omega e^{i\vartheta}}\,d\vartheta - e^{\frac{i\pi s}{2}}\int_0^\omega x^{s-1}e^{-ix}\,dx = 0.$$

Let $\omega$ converge to $+\infty$. Then the first integral converges to $\Gamma(s)$ and the third to the integral in question. The modulus of the second integral is

$$< A\int_0^{\frac{\pi}{2}} e^{-\omega\cos\vartheta}\,d\vartheta = A\int_0^{\frac{\pi}{2}-\varepsilon} e^{-\omega\cos\vartheta}\,d\vartheta + A\int_{\frac{\pi}{2}-\varepsilon}^{\frac{\pi}{2}} e^{-\omega\cos\vartheta}\,d\vartheta < \frac{A\pi}{2}e^{-\omega\sin\varepsilon} + A\varepsilon,$$

where $A = e^{\frac{\pi}{2}|\Im s|}$, $0 < \varepsilon < \frac{\pi}{2}$. By setting $\varepsilon = \frac{1}{\omega^k}$, $\Re s < k < 1$, we see that the second term converges to 0.

**152.** The integral is

$$= \frac{1}{n}\int_0^\infty x^{\frac{1}{n}-2}\sin x\,dx.$$

We conclude from **151** that for real $s$, $0 < s < 1$,

$$\int_0^\infty x^{s-1}\sin x\,dx = \Gamma(s)\sin\frac{\pi s}{2}.$$

This integral converges for $-1 < \Re s < 1$ and the right hand side is regular for the same $s$ values. Hence the formula holds for $-1 < \Re s < 1$.

**153.** [Cf. Correspondance d'Hermite et de Stieltjes, Vol. 2. Paris: Gauthier-Villars 1905, p. 337; cf. G. H. Hardy: Mess. Math. Vol. 46, pp. 175—182 (1917).]

$$\int_0^\infty e^{-x^\mu e^{i\alpha}} x^n\,dx = \frac{1}{\mu} e^{-i\frac{(n+1)\alpha}{\mu}} \int e^{-z} z^{\frac{n+1}{\mu}-1}\,dz.$$

The last integral is taken along the ray $\arg z = \alpha$ and it is equal to

$$\int_0^{+\infty} e^{-x} x^{\frac{n+1}{\mu}-1}\,dx = \Gamma\left(\frac{n+1}{\mu}\right),$$

because the integrand, with $z = \varrho e^{i\vartheta}$, $0 \leqq \vartheta \leqq \alpha$, converges to 0 as $\frac{1}{r}$ tends to 0; the convergence is uniform in $\vartheta$. For $\alpha = \mu\pi$, $0 < \mu < \frac{1}{2}$ we get the function

$$e^{-x^\mu \cos\mu\pi} \sin(x^\mu \sin\mu\pi),$$

all the Stieltjes moments of which vanish without the function vanishing itself (no contradiction to II **138**, II **139**). According to E. Borel [Leçons sur les séries divergentes. Paris: Gauthier-Villars 1901, pp. 73—75; cf. also G. Pólya: Astronom. Nachr. Vol. 208, p. 185 (1919)] a similar statement can not be true for a function $f(x)$, $|f(x)| < e^{-k\sqrt{x}}$, $k>0$, $k=$ const. Our formula shows that $\sqrt{x}$ cannot be replaced in this theorem of Borel by a lower power, $x^\mu$, $\mu < \frac{1}{2}$, of $x$. H. Hamburger has proved [Math. Z. Vol. 4, pp. 209—211 (1919)] that $\sqrt{x}$ cannot even be replaced by $\frac{\sqrt{x}}{(\log x)^2}$ by showing that

$$\int_0^\infty \exp\left(-\frac{\pi\sqrt{x}-\log x}{(\log x)^2+\pi^2}\right) \sin\left(\frac{\sqrt{x}\log x+\pi}{(\log x)^2+\pi^2}\right) x^n dx = 0, \quad n = 0, 1, 2, \ldots$$

**154.**

$$x^{\mu+1}\int_0^{+\infty} e^{-t^\mu}\cos xt\,dt = x^\mu \int_0^{+\infty} \sin xt \cdot \mu t^{\mu-1} e^{-t^\mu}\,dt = \Im \int_0^{+\infty} e^{iz^\nu - \delta z}\,dz,$$

where the variable of integration is $z = x^\mu t^\mu$, $\mu^{-1} = \nu$ and $x^{-\mu} = \delta$. Rotate the line of integration through a small positive angle, put $\delta = 0$; then rotate the line of integration in the positive sense until $\arg z = \frac{\mu\pi}{2}$.

**155.** We replace in the integral

$$\frac{1}{2\pi i}\int_{a-iT}^{a+iT} \frac{e^{\alpha s}}{s^2}\,ds, \qquad T > a,$$

the rectilinear path of integration by the semicircle over the segment $(a - iT, a + iT)$ to the right or to the left according as $\alpha \leqq 0$ or $\alpha > 0$.

In the first case the integral does not change and is absolutely smaller than $\frac{1}{2\pi}\frac{e^{\alpha a}}{T^2}\pi T = \frac{e^{\alpha a}}{2T}$; in the second case it decreases by $\alpha$ (residue at the pole $s = 0$) and the new integral is absolutely $< \frac{1}{2\pi}\frac{e^{\alpha a}}{(T-a)^2}\pi T$. Now let $T$ increase to $+\infty$.

**156.** [The case $\lambda = 1 + e^{-1}$ is due to H. Weyl.] We have

$$\mu(t) = \frac{t^n}{n!} \quad \text{for } n \leqq t \leqq n+1,$$

thus (the interchange of summation and integration can be justified by various arguments)

$$\begin{aligned}\int_0^{+\infty} \mu(t)\, e^{-\lambda t}\, dt &= \sum_{n=0}^{\infty} \int_n^{n+1} \frac{t^n}{n!} e^{-\lambda t}\, dt \\ &= \sum_{n=0}^{\infty} \int_0^{+\infty} \frac{t^n}{n!} e^{-\lambda t} \left( \frac{1}{\pi} \int_{-\infty}^{+\infty} \frac{\sin \frac{u}{2}\, e^{i(n+\frac{1}{2}-t)u}\, du}{u} \right) dt \\ &= \frac{1}{\pi} \int_{u=-\infty}^{+\infty} \int_{t=0}^{+\infty} \frac{\sin \frac{u}{2}\, e^{\frac{iu}{2}}}{u} e^{-t\lambda} e^{-itu} \sum_{n=0}^{\infty} \frac{t^n e^{inu}}{n!}\, dt\, du \\ &= \frac{1}{2\pi i} \int_{-\infty}^{+\infty} \int_0^{+\infty} \frac{e^{iu}-1}{u} e^{t(-\lambda - iu + e^{iu})}\, dt\, du \\ &= \frac{1}{2\pi i} \int_{-\infty}^{+\infty} \frac{e^{iu}-1}{u(\lambda + iu - e^{iu})}\, du.\end{aligned}$$

This integral is equal to the sum of the residues in the upper half-plane because the integral of $\frac{e^{iu}-1}{u(\lambda + iu - e^{iu})}$ along the half-circle $u = re^{i\vartheta}$, $0 \leqq \vartheta \leqq \pi$, converges to 0 as $r$ tends to infinity. The only pole in the upper half-plane is $u = iz$ [**196**], the corresponding residue is $\frac{1}{z}$. (Cf. **215**, IV **55**.)

**157.** [Dirichlet: J. reine angew. Math. Vol. 17, p. 35 (1837); Mehler: Math. Ann. Vol. 5, p. 141 (1872).] Assume $-1 < x < 1$, $x = \cos\vartheta$, $0 < \vartheta < \pi$. Then [cf. G. Szegö: Orthogonal Pelynomials, 3rd Ed. 1967, pp. 86–90]

$$P_n(\cos\vartheta) = \frac{1}{2\pi i} \oint \frac{z^n}{\sqrt{1 - 2z\cos\vartheta + z^2}}\, dz,$$

where the (positively oriented) path of integration encloses the two singular points $e^{i\vartheta}$ and $e^{-i\vartheta}$. We may contract the path to a straight line segment from $e^{i\vartheta}$ to $e^{-i\vartheta}$ (Laplace's formula) or to the arc $-\vartheta \leqq \arg z \leqq \vartheta$ of the unit circle (Dirichlet-Mehler formula). (In both cases we should proceed carefully because the integrand becomes infinite at the endpoints, although only of the order $\frac{1}{2}$.) When $z$ goes around the singular point the integrand changes only its sign. Thus

$$P_n(\cos\vartheta) = \frac{2}{2\pi i}\int_{-1}^{1} \frac{(\cos\vartheta + i\alpha\sin\vartheta)^n\, i\sin\vartheta\, d\alpha}{\sqrt{1 - 2(\cos\vartheta + i\alpha\sin\vartheta)\cos\vartheta + (\cos\vartheta + i\alpha\sin\vartheta)^2}}$$

$$= \frac{2}{2\pi i}\int_{-\vartheta}^{\vartheta} \frac{e^{int}\cdot ie^{it}\, dt}{\sqrt{1 - 2e^{it}\cos\vartheta + e^{2it}}}.$$

We obtain the third expression either by contracting the path of integration to the arc $\vartheta \leqq \arg z \leqq 2\pi - \vartheta$ of the unit circle or by changing the variable and replacing $\vartheta$ by $\pi - \vartheta$ in the second expression. [$P_n(-\cos\vartheta) = (-1)^n P_n(\cos\vartheta)$.]

**158.** If $z$ is real and negative, any two pieces of $L$ that are symmetric to each other with respect to the real axis contribute conjugate complex values. We denote by $L_\alpha$ the negatively oriented boundary of the half-strip $\Re z > \alpha$, $-\pi < \Im z < \pi$ (in particular $L_0 = L$). If $z$ is not in $\mathfrak{H}$ and if $\alpha > 0$ the integral along each of the $L_\alpha$'s has the same value. By letting $\alpha$ increase to $+\infty$ we successively extend $E(z)$ analytically over the entire plane.

**159.** The integral along $L = L_0$ has the same value as the integral along $L_\alpha$, $\alpha \gtreqless 0$. Since $e^{e^{x+i\pi}} = e^{e^{x-i\pi}} = e^{-e^x}$ the contributions of the horizontal parts of $L_\alpha$ cancel each other. The vertical segment of $L_\alpha$ supplies

$$\frac{1}{2\pi}\int_{-\pi}^{\pi} e^{e^{\alpha+iy}}\, dy.$$

This value is independent of $\alpha$, hence $=1$, as can be seen for $\alpha \to -\infty$.

**160.** (1) Let $z$ be outside $\mathfrak{H}$. We have [**159**]

$$E(z) = -\frac{1}{z} + \frac{1}{z^2}\frac{1}{2\pi i}\int_L \left(-\zeta + \frac{\zeta^2}{\zeta - z}\right) e^{e^\zeta}\, d\zeta.$$

Evaluate the integral not along $L$ but along $L'$, an inner parallel curve to $L$ at the distance $\delta$ (boundary of the region $\Re z > \delta$, $-\pi + \delta < \Im z < \pi - \delta$) whereby $0 < \delta < \frac{\pi}{2}$. Note that the real integral

$$\int_0^\infty \xi^2 e^{-e^\xi \cos\delta}\, d\xi$$

converges.

(2) Let $z$ be in the rectangle $-1 < \Re z < 0, -\pi < \Im z < \pi$. According to the residue theorem we have

$$\int_L \frac{e^{e^\zeta}}{\zeta - z}\, d\zeta - \int_{L_{-1}} \frac{e^{e^\zeta}}{\zeta - z}\, d\zeta = 2\pi i e^{e^z}.$$

Consequently

$$E(z) = e^{e^z} - \frac{1}{z} + \frac{1}{z^2}\frac{1}{2\pi i}\int_{L_{-1}}\left(-\zeta + \frac{\zeta^2}{\zeta - z}\right) e^{e^\zeta}\, d\zeta$$

for all $z$ in the rectangle and then, by virtue of analytic continuation, in the half-strip $\mathfrak{H}$. Instead of $L_{-1}$ choose the outer parallel curve at the distance $\delta$, $0 < \delta < \frac{\pi}{2}$, as path of integration.

**161.** The numerator is

$$= 2\pi i \sum_{\nu=0}^{n} (-1)^{n-\nu} \frac{2^\nu}{\nu!\,(n-\nu)!} = \frac{2\pi i}{n!}.$$

The denominator is

$$= \int_{-\pi}^{\pi} \frac{2^{2n\cos\vartheta}}{\prod_{\nu=1}^{n} |2ne^{i\vartheta} - \nu|}\, d\vartheta \sim \frac{2\pi}{n!} \qquad \text{[II 217]}.$$

Explanation: The principal parts of both integrals stem from an arc centred at $z = 2n$ whose length is of order $\sqrt{n}$. Along it the argument of $dz$ is close to $\frac{\pi}{2}$ and the argument of $f_n(z)$ is nearly 0. [**43**.]

**162.** The number of zeros in the disk $|z| < r$ is:

$$= \frac{1}{2\pi i}\oint_{|z|=r} \frac{f'(z)}{f(z)}\, dz = \frac{1}{2\pi}\int_0^{2\pi} z\frac{f'(z)}{f(z)}\, d\vartheta = \frac{1}{2\pi}\int_0^{2\pi} \Re z\frac{f'(z)}{f(z)}\, d\vartheta, \quad z = re^{i\vartheta}.$$

**163.** The function $\frac{\omega(\zeta) - \omega(z)}{\zeta - z}$ is a polynomial of degree $n - 1$ in $z$. Furthermore

$$P(z_\nu) = \frac{1}{2\pi i}\oint_L \frac{f(\zeta)}{\zeta - z_\nu}\, d\zeta = f(z_\nu), \qquad \nu = 1, 2, \ldots, n.$$

**164.** Apply Cauchy's integral formula to both sides. The proposition is now identical with V **97**.

**165.** Let $\varepsilon > 0$. Consider that region of the $z$-plane where all the inequalities $\left|z - \frac{n\pi}{\varrho}\right| > \varepsilon$, $n = 0, \pm 1, \pm 2, \ldots$ are satisfied (the riddled plane). There exists a constant $K$ depending on $\varepsilon$ such that in the entire riddled plane

$$|\sin \varrho(x + iy)| > K^{-1} e^{\varrho|y|}.$$

It is sufficient to check this in the region $-\frac{\pi}{2\varrho} \leqq x \leqq \frac{\pi}{2\varrho}$, $|z| > \varepsilon$, $0 < \varepsilon < \frac{\pi}{2\varrho}$. The integral

$$\frac{1}{2\pi i}\oint \frac{F(\zeta)}{\sin \varrho\zeta}\frac{d\zeta}{(\zeta - z)^2}$$

along the circle $|\zeta| = \left(n + \frac{1}{2}\right)\frac{\pi}{\varrho}$ converges to 0 as $n \to \infty$ because $|F(\zeta)(\sin \varrho\zeta)^{-1}| < CK$ along the path of integration. Compute the sum of the residues. [Hurwitz-Courant, pp. 118–123.]

**166.** We substitute in **165** $G\left(z + \frac{\pi}{2\varrho}\right)$ for $F(z)$ and then $z - \frac{\pi}{2\varrho}$ for $z$:

$$-\frac{d}{dz}\left(\frac{G(z)}{\cos \varrho z}\right) = -\sum_{n=-\infty}^{\infty} \frac{\varrho(-1)^n G\left(\frac{(n+\frac{1}{2})\pi}{\varrho}\right)}{(\varrho z - (n+\frac{1}{2})\pi)^2}.$$

We now combine the terms with the subscripts $n$ and $-n-1$:

$$\frac{d}{dz}\left(\frac{G(z)}{\cos \varrho z}\right) = \sum_{n=0}^{\infty}(-1)^n G\left(\frac{(n+\frac{1}{2})\pi}{\varrho}\right)\left(\frac{\varrho}{[\varrho z - (n+\frac{1}{2})\pi]^2} + \frac{\varrho}{[\varrho z + (n+\frac{1}{2})\pi]^2}\right)$$

and integrate

$$\frac{G(z)}{\cos \varrho z} = -\sum_{n=0}^{\infty}(-1)^n G\left(\frac{(n+\frac{1}{2})\pi}{\varrho}\right)\left(\frac{1}{\varrho z - (n+\frac{1}{2})\pi} + \frac{1}{\varrho z + (n+\frac{1}{2})\pi}\right).$$

The constant of integration has to be 0 because there must be an odd function on both sides.

**167.** The functions $f(z)\log z$ and $z^k f(z)$ are regular in the domain described in the diagram.

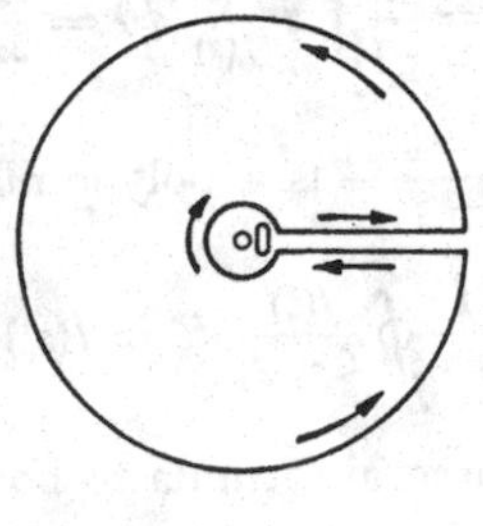


**168.** Since $|\log z - i\pi| \leqq \pi$ for $|z| = 1$ **167** implies

$$\left|\int_0^1 f(x)\,dx\right| \leqq \frac{1}{2}.$$

If $k$ is an integer we replace $f(z)$ by $z^k f(z)$; if $k$ is not an integer the second formula in **167** is used.

**169.** [D. Hilbert, cf. H. Weyl: Diss. Göttingen 1908, p. 83; F. Wiener: Math. Ann. Vol. 68, p. 361 (1910); I. Schur: J. reine angew. Math. Vol. 140, p. 16 (1911); L. Fejér and F. Riesz: Math. Z. Vol. 11, pp. 305–314 (1921).] Introduce in **168**

$$f(z) = \frac{1}{2\pi}(x_1 + x_2 z + x_3 z^2 + \cdots + x_n z^{n-1})^2, \quad k = \alpha + 1.$$

Then [**122**]

$$\int_0^{2\pi} |f(e^{i\vartheta})|\, d\vartheta = x_1^2 + x_2^2 + x_3^2 + \cdots + x_n^2 = 1$$

and

$$\int_0^1 x^k f(x)\, dx = \frac{1}{2\pi}\sum_{\lambda=1}^{n}\sum_{\mu=1}^{n} x_\lambda x_\mu \int_0^1 x^{\lambda+\mu+\alpha-1}\, dx = \frac{1}{2\pi}\sum_{\lambda=1}^{n}\sum_{\mu=1}^{n}\frac{x_\lambda x_\mu}{\lambda+\mu+\alpha}.$$

**170.** Let $L$ denote a closed continuous curve without double points, let $L$ lie completely inside $\mathfrak{R}$, and let $z$ be a point in the interior of $L$. According to Cauchy's integral theorem we have

$$f_n(z) = \frac{1}{2\pi i}\oint_L \frac{f_n(\zeta)}{\zeta - z}\, d\zeta.$$

From this and from the fact that

$$\lim_{n\to\infty} \frac{f_n(\zeta)}{\zeta - z} = \frac{f(\zeta)}{\zeta - z}$$

is uniformly valid on $L$ it follows that $f(\zeta)$ is continuous on $L$. Furthermore

$$\lim_{n\to\infty} f_n(z) = \frac{1}{2\pi i}\oint_L \frac{f(\zeta)}{\zeta - z}\, d\zeta.$$

The last function is regular inside $L$.

**171.** [T. Carleman.] We call $F_r(z)$ the area integral of $f$ over the disk with center $z$, radius $r$ and boundary circle $C_r$. The element of arc of $C_r$ is $dz = e^{i\tau}|dz|$. If $x$ varies and its increment is $\Delta x$, i.e. if the area of integration is displaced through $\Delta x$ in the direction of the $x$-axis, the change of the area per element of arc is $|dz|\,\Delta x \sin\tau$ as is obvious by geometric considerations; i.e.

$$\frac{\partial F_r(z)}{\partial x} = \oint_{C_r} f \sin\tau\, |dz|.$$

By a similar argument we obtain

$$\frac{\partial F_r(z)}{\partial y} = -\oint_{C_r} f \cos\tau\, |dz|,$$

hence by virtue of the hypothesis

$$\frac{\partial F_r(z)}{\partial x} - \frac{1}{i}\frac{\partial F_r(z)}{\partial y} = \frac{1}{i}\oint_{C_r} f\,dz = 0.$$

Therefore [p. 113] $F_r(z)$ is analytic as well as $r^{-2}F_r(z)$ and finally [**170**]

$$\lim_{r\to 0}\frac{F_r(z)}{r^2\pi} = f(z).$$

Instead of all the circles we could consider all the curves that are similar to a given closed curve without double points and similarly situated.

**172.** To show that the difference

$$\int_0^{2\pi} f(e^{i\vartheta})\,d\vartheta - 2\pi f(0) = \int_0^{2\pi} [f(e^{i\vartheta}) - f(re^{i\vartheta})]\,d\vartheta, \quad 0 \leqq r < 1 \quad [\mathbf{118}]$$

vanishes we prove that this expression can be made arbitrarily small with $1 - r$. We take each point of discontinuity as the center of an open disk of radius $\varepsilon$ so that different disks have no common point. Removing these "small" disks from the unit disk we obtain a domain in which $f(z)$ is uniformly continuous. We now split the integral into two parts: The first part consists of the integral along the arcs inside the above mentioned disks, the second part consists of the integral along the remaining piece of the circle. The first part can be made arbitrarily small with $\varepsilon$ [$f(z)$ is bounded] and then, once $\varepsilon$ is fixed, the second part can be made arbitrarily small when $1 - r$ is sufficiently small.

**173.** Cf. **174**. Cf. also **231**, Hurwitz-Courant, p. 327, and E. Hille, Vol. II, p. 361.

**174.** It is convenient to consider the following general situation: Let $\mathfrak{D}$ be a simply connected domain in the $\zeta$-plane and assume that the mapping $\psi(\zeta) = Z$ is conformal in the interior of $\mathfrak{D}$, sufficiently continuous on the boundary and that it establishes a one to one relationship between $\mathfrak{D}$ and the disk $|Z| \leqq 1$; let the point $\zeta = z$ correspond to $Z = 0$. We have $\zeta = \psi^{-1}(Z)$ where $\psi^{-1}$ denotes the inverse function of $\psi$. We "transplant" the function $f(\zeta)$, which is regular in $\mathfrak{D}$, from the $\zeta$-plane to the $Z$-plane by defining

$$f(\zeta) = f[\psi^{-1}(Z)] = F(Z).$$

We have

$$F(0) = \oint F(Z)\frac{dZ}{2\pi i Z},$$

the integral computed along the positively oriented circle $|Z| = 1$ [**172**]. The change of variable $Z = \psi(\zeta)$, for which

$$F[\psi(\zeta)] = f(\zeta), \qquad \psi(z) = 0,$$

leads to

$$f(z) = \oint \frac{f(\zeta)\, d\psi(\zeta)}{2\pi i \psi(\zeta)};$$

the integral is taken along the positively oriented boundary of $\mathfrak{D}$. **173** and **174** are particular cases:

$$\mathfrak{D}\colon |\zeta| \leqq R, \quad \psi(\zeta) = \frac{(\zeta - z) R}{R^2 - \zeta \bar{z}}, \quad \frac{\psi'(\zeta)\, d\zeta}{i\psi(\zeta)} = \frac{(R^2 - r^2)\, d\Theta}{R^2 - 2Rr \cos(\Theta - \vartheta) + r^2},$$

$$\zeta = Re^{i\Theta}, \quad z = re^{i\vartheta};$$

$$\mathfrak{D}\colon \Re\zeta \geqq 0, \quad \psi(\zeta) = \frac{z - \zeta}{\bar{z} + \zeta}, \quad \frac{\psi'(\zeta)\, d\zeta}{i\psi(\zeta)} = -\frac{2x\, d\eta}{x^2 + (\eta - y)^2},$$

$$\zeta = i\eta, \quad z = x + iy.$$

**175.** [J. L. W. V. Jensen: Acta Math. Vol. 22, pp. 359—364 (1899). Cf. E. Hille, Vol. II, pp. 189—190.] The integrand is a single valued function on $\mathfrak{R}_\varepsilon$; $\mathfrak{R}_\varepsilon$ contains the point $r = 0$ if $\varepsilon$ is sufficiently small (hypothesis) and then the value of the integral is $2\pi i \log f(0)$. We denote by $\alpha_1, \alpha_2, \ldots, \alpha_m, \beta_1, \ldots, \beta_n$ the endpoints of the paths that connect the $\varepsilon$-circles around $a_1, a_2, \ldots, a_m, b_1, \ldots, b_n$ with the unit circle $|z| = 1$ ($|\alpha_1| = \cdots = |\alpha_m| = |\beta_1| = \cdots = |\beta_n| = 1$). The loop starting at $\alpha_\mu$, turning around $a_\mu$ along the $\varepsilon$-circle and ending at $\alpha_\mu$ contributes, as $\varepsilon \to 0$,

$$-2\pi i \int_{a_\mu}^{\alpha_\mu} \frac{dz}{z} = -2\pi i \log \frac{\alpha_\mu}{a_\mu}$$

($a_\mu$ is assumed to be a simple zero). The circle $|z| = 1$ contributes $\int_0^{2\pi} \log f(e^{i\vartheta})\, i\, d\vartheta$. Take the imaginary part on both sides of the equation

$$\int_0^{2\pi} \log f(re^{i\vartheta})\, i\, d\vartheta - 2\pi i \sum_{\mu=1}^{m} \log \frac{\alpha_\mu}{a_\mu} + 2\pi i \sum_{\nu=1}^{n} \log \frac{\beta_\nu}{b_\nu} = 2\pi i \log f(0).$$

The method used in **120** leads to another proof.

**176.** Cf. **177**. Cf. also **232**.

**177.** [F. and R. Nevanlinna: Acta Soc. Sci. Fennicae Vol. 50, No. 5 (1922).] Same notation as in **174**. Consider a function $f(\zeta)$ that is meromorphic in $\mathfrak{D}$, and different from 0 and $\infty$ on the boundary of $\mathfrak{D}$ and at the inner point $z$; inside $\mathfrak{D}$ the zeros are $a_1, a_2, \ldots, a_m$ and the poles

$b_1, b_2, \ldots, b_n$. The "transplanted" function $F(Z) = f[\psi^{-1}(Z)]$ has the zeros $A_\mu = \psi(a_\mu)$ and the poles $B_\nu = \psi(b_\nu)$. Then [**175**]

$$\log |F(0)| + \sum_{\mu=1}^{m} \log \frac{1}{|A_\mu|} - \sum_{\nu=1}^{n} \log \frac{1}{|B_\nu|} = \oint \log |F(Z)| \frac{dZ}{2\pi i Z},$$

the integral taken along $|Z| = 1$. Hence by a change of variables [**174**]

$$\sum_{\mu=1}^{m} \log \frac{1}{|\psi(a_\mu)|} - \sum_{\nu=1}^{n} \log \frac{1}{|\psi(b_\nu)|} = \oint \log |f(\zeta)| \frac{d\psi(\zeta)}{2\pi i \psi(\zeta)} - \log |f(z)|,$$

the integral taken along the boundary of $\mathfrak{D}$. The signs on the left hand side are in evidence: $|\psi(a_\mu)| < 1$, $|\psi(b_\nu)| < 1$; $\frac{d\psi(\zeta)}{2\pi i \psi(\zeta)}$ is positive. Suppose that $f(\zeta)$ is variable and $|f(z)|$ is fixed; then the essence of the formula can be roughly expressed as follows: zeros in the interior increase, poles in the interior decrease the moduli of the boundary values.

**176**, **177** are a consequence of Jensen's general formula as **173** and **174** follow from the generalized formula given in **174**.—Also the proof of **120** can be suitably generalized [**174**]. —The condition that $f(z)$ be different from 0 on the boundary can always be weakened [**120**], that $f(z)$ be regular on the boundary can be weakened in many cases.

**178.** [Cf. F. Nevanlinna: C. R. Acad. Sci. (Paris) Sér. A—B, Vol. 175, p. 676 (1922); T. Carleman: Ark. Mat., Astron. Fys. Vol. 17, No. 9, p. 5 (1923).] Put a circle of radius $\varepsilon$, $\varepsilon$ sufficiently small, around $a_\mu$ and connect the $\varepsilon$-circle with the half-circle $|z| = R$, $-\frac{\pi}{2} \leq \arg z \leq \frac{\pi}{2}$ so that the connecting paths do not intersect. Removal of the $\varepsilon$-circles and the connecting paths reduces $\mathfrak{D}$ to a simply connected region $\mathfrak{R}_\varepsilon$; its boundary is the appropriate path of integration. The loop around $a_\mu$ starting and ending at $z_\mu$, $|z_\mu| = R$, contributes

$$-2\pi \left( \frac{1}{a_\mu} - \frac{a_\mu}{R^2} - \frac{1}{z_\mu} + \frac{z_\mu}{R^2} \right).$$

Take the real part of the expression

$$\int_{-\frac{\pi}{2}}^{+\frac{\pi}{2}} \log f(Re^{i\vartheta}) \frac{2 \cos \vartheta}{R} \, d\vartheta + \left( \int_R^r + \int_{-r}^{-R} \right) \left( -\frac{1}{y^2} + \frac{1}{R^2} \right) \log f(iy) \, dy$$

$$+ \int_{+\frac{\pi}{2}}^{-\frac{\pi}{2}} \log f(re^{i\vartheta}) \left( \frac{e^{-i\vartheta}}{r} + \frac{re^{i\vartheta}}{R^2} \right) d\vartheta - 2\pi \sum_{\mu=1}^{m} \left( \frac{1}{a_\mu} - \frac{a_\mu}{R^2} - \frac{1}{z_\mu} + \frac{z_\mu}{R^2} \right) = 0.$$

The formula is so simple because the differential $\left(\frac{1}{z^2}+\frac{1}{R^2}\right)\frac{dz}{i}$ is always real on the half-circle $|z| = R$, $\Re z > 0$ and the function $\frac{1}{z} - \frac{z}{R^2}$ is purely imaginary.

**179.** We introduce $a_0 = |a_0| e^{i\gamma} \neq 0$, $z - z_\nu = r_\nu e^{i\varphi_\nu}$, thus

$$U(z) + iV(z) = a_0(z - z_1)(z - z_2) \cdots (z - z_n)$$
$$= |a_0| e^{i\gamma} r_1 e^{i\varphi_1} \cdot r_2 e^{i\varphi_2} \cdots r_n e^{i\varphi_n};$$

when $z = x$ is real and increasing from $-\infty$ to $+\infty$ all the arguments increase from $-\pi$ to 0, hence $\arctan \frac{V(x)}{U(x)} = \gamma + \varphi_1 + \varphi_2 + \cdots + \varphi_n$ increases by $n\pi$. The quotient $\frac{V(x)}{U(x)} = \tan(\gamma + \varphi_1 + \varphi_2 + \cdots + \varphi_n)$ therefore assumes the value 0 $n$ times and becomes infinite $n$ times. This method shows in addition that the zeros of $U(x)$ and $V(x)$ alternate.

**180.** Obvious, because the argument changes continuously.

**181.** Consequence of the argument principle: Compare the zeros and poles of the factors with the zeros and poles of the product and take into account the points that are special (pole or zero) for both $\varphi(z)$ and $\psi(z)$.—Directly: $\arg f(z) = \arg \varphi(z) + \arg \psi(z)$.

**182.** A polynomial is the product of linear factors $z - z_0$, where $z_0$ denotes a zero. According to **181** it is sufficient to prove the proposition for a linear function $z - z_0$. The transformation $w = z - z_0$ translates the curve $L$ through the vector $-z_0$. If $z_0$ is inside $L$ the point $w = 0$ is inside the image of $L$, the winding number is equal to 1. In the other case the winding number is equal to 0.

**183.** If $f(z)$ is regular in $\mathfrak{D}$ with the possible exception of finitely many poles and if it is different from 0 on the boundary $L$ of $\mathfrak{D}$ then it can be written as a product, $f(z) = R(z)\,\varphi(z)$, of a rational function $R(z)$ and a regular function $\varphi(z)$ without zeros in $\mathfrak{D}$, that is a function $\varphi(z)$ to which the restricted principle of the argument as stated in the problem applies. [**181**, **182**.]

**184.** [A. Hurwitz: Math. Ann. Vol. 57, p. 444 (1903). Cf. also Ch. Sturm: J. Math. Pures Appl. Vol. 1, p. 431 (1836).] The number of zeros has to be $\leq 2n$ by virtue of VI **14**. The winding number of the curve described by $P(z)$, where $z = re^{i\vartheta}$, $r > 0$ and $\vartheta$ varies from 0 to $2\pi$, is at least $m$ because $P(z)$ has a zero of order $m$ at the origin. Hence it intersects the imaginary axis at least $2m$ times. Consider $r = 1 - \varepsilon$, $\varepsilon > 0$ and suitably chosen, or $r = 1$ depending on whether $P(z)$ has zeros on the circle $|z| = 1$ or not. An essentially different proof follows from II **141**.

**185.** The polynomial $P(z) = a_0 + a_1 z + a_2 z^2 + \cdots + a_n z^n$ has $n$ zeros inside the unit circle $|z| = 1$ [**22**]. Hence the winding number of the image of the unit circle under $w = P(z)$ is equal to $n$. Apply **180** to the positive and to the negative part of the imaginary axis.

**186.** [A. Ostrowski.] Imagine in the $w$-plane a circle of radius $M = \max\limits_{z \text{ on } L} |f(z)|$ and center $w = a$. This circle does not contain the point $w = 0$. The image of $L$ under the mapping $w = a - f(z)$ lies inside the above mentioned circle, therefore its winding number is 0.

**187.** For $z = iy$, $-\frac{1}{2} \leqq y \leqq \frac{1}{2}$, $w = e^{\pi iy} - e^{-\pi iy} = 2i \sin \pi y$. For $z = x \pm \frac{1}{2}i$, $x \geqq 0$, $w = \pm i(e^{\pi x} + e^{-\pi x})$. Thus, when $z$ moves along the boundary in the positive direction $w$ describes the imaginary axis from $+i\infty$ to $-i\infty$. Let $z = x + iy$, $x$ fixed, and $-\frac{1}{2} \leqq y \leqq \frac{1}{2}$. The decomposition

$$w = e^{\pi x} \cdot e^{\pi iy} - e^{-\pi x} \cdot e^{-\pi iy} = (e^{\pi x} - e^{-\pi x}) \cos \pi y + i(e^{\pi x} + e^{-\pi x}) \sin \pi y$$

shows that $w$ describes the right half of the ellipse centred at the origin and with semi-axes $e^{\pi x} - e^{-\pi x}$ and $e^{\pi x} + e^{-\pi x}$ as $z$ moves on the line $\Re z = x$, $-\frac{1}{2} \leqq \Im z \leqq \frac{1}{2}$.

Therefore, if $w_0$ is an arbitrary point in the right half-plane $\Re w > 0$, $x$ can be chosen so large that the following situation is met: As $z$ moves on the boundary of the rectangular region

$$0 < \Re z < x, \quad -\tfrac{1}{2} < \Im z < \tfrac{1}{2}$$

the winding number of the curve described by $w - w_0$ in the $w$-plane is equal to 1. Cf. also **188**.

**188.** Let $w_0$ denote an arbitrary point in the interior of the image $C$ of the circle $|z| = r$. The translation given by the vector $-w_0$ moves $C$ to $C'$; i.e. $C'$ is the image of the circle $|z| = r$ under the mapping $w = f(z) - w_0$. According to the hypothesis its winding number must be $+1$ or $-1$. Since $f(z) - w_0$ is regular the winding number has to be non-negative [argument principle], thus it is $+1$. The function $f(z) - w_0$ has, therefore, exactly one zero in the disk $|z| < r$. In an analogous way we show that a point $w_0$ outside $C$ cannot belong to the range of $f(z)$ in $|z| \leqq r$ (the corresponding winding number is 0).

**189.** We map the circular sector, $-\frac{\pi}{4} \leqq \arg z \leqq \frac{\pi}{4}$, $|z| \leqq r$, by means of $w = \int\limits_0^z e^{-\frac{1}{2}x^2}\, dx$ into the $w$-plane. The image of the two radii can be obtained by rotating through 45°, and reflecting and rotating, respectively, the right half of the Cornu spiral, described by Sommerfeld, l.c. The image of the bounding circular arc is near the point $w=\sqrt{\frac{\pi}{2}}$ [IV **189**]. We choose $r$ so small that the image of the arc $z=re^{i\vartheta}$, $-\frac{\pi}{4}\leqq\vartheta\leqq\frac{\pi}{4}$, intersects the images of the radii only at the points that correspond to $z=re^{i\frac{\pi}{4}}$ and $z=re^{-i\frac{\pi}{4}}$. We now consider the image of the boundary of the above mentioned sector outside the circle $|z|=r$. It has the winding number 0. Hence $w\neq0$ when $-\frac{\pi}{4}<\arg z<\frac{\pi}{2}$. Different proof in V **178**.

**190.** The integral

$$\frac{1}{2\pi i}\oint\frac{f'(z)\,dz}{f(z)-a}$$

is an integer. Therefore it cannot change if the path of integration is deformed continuously. The curves in question can be transformed into each other by a continuous deformation.

**191.** Let $f(z)\not\equiv0$, then $f(z)\neq0$ on $L$. The function $\log f(z)=\log R+i\theta$ is regula at every point of $L$. Now [clear when $\partial v$ and $\partial s$ coincide with $\partial x$ and $\partial y$, respectively, cf. p. 113]

$$\frac{\partial\log R}{\partial v}=\frac{\partial\Theta}{\partial s},$$

where $\frac{\partial}{\partial v}$ denotes the differentiation in the direction of the outer normal and $\frac{\partial}{\partial s}$ in the direction of the positive tangent at the point $z$ of $L$. The derivative on the left hand side is positive [maximum principle] hence the derivative on the right hand side has to be positive too. The image of $L$ is a circle (possibly described several times in the same sense).

**192.** [B. Riemann: Werke. Leipzig: B. G. Teubner 1876, pp. 106—107; H. M. Macdonald: Proc. Lond. Math. Soc. Vol. 29, pp. 576—577 (1898); cf. also G. N. Watson: Proc. Lond. Math. Soc. Ser. 2, Vol. 15, pp. 227—242 (1916); Whittaker and Watson, p. 121.] We assume that $f'(z)=Rie^{i\Theta}\frac{d\Theta}{dz}\neq0$ on $L$ (same notation as in **191**). Let the winding numbers of the curves described by $f(z)$ and $f'(z)$ be denoted by $W$ and $W'$ resp. ($W$ is the winding number of a circle which may be described several times in the same sense). Thus $2\pi(W'-W)$ is equal to the change of the argument of $\frac{d\Theta}{dz}$ as $z$ describes the curve $L$. Let $ds=|dz|$ be the line element of $L$, $\frac{d\Theta}{dz}=\frac{d\Theta}{ds}\frac{ds}{dz}$. The first factor is always real, therefore only the change of the argument of the second factor is important. The argument of $\frac{ds}{dz}=\frac{|dz|}{dz}$ is equal to the negative value of the argument of

$dz$. The change in direction of $dz$, i.e. the change in direction of the tangent vector along a closed curve without double points, is $2\pi$; this can be easily seen in the case of a polygon. Hence $W' - W = -1$. If $f'(z)$ vanishes on $L$ we consider a level line inside but sufficiently close to $L$. The geometric formulation of the proposition can also be verified directly; cf. H. M. Macdonald, l.c., or MPR, Vol. 1, pp. 163–164.

**193.** The function $f(z)$ is not a constant because $f'(z) \neq 0$; $f(z)$ has exactly one zero in the interior of $\mathfrak{D}$ [**192**]. Therefore $w = f(z)$ describes a (circular) curve with the winding number 1 as $z$ describes the boundary of $\mathfrak{D}$. The winding number of the path described by $f(z) - w_0$ is 1 or 0 depending on whether $|w_0| <$ or $|w_0| >$ the constant modulus of $f(z)$ on the boundary of $\mathfrak{D}$.

**194.** [E. Rouché: J. de l'Ec. Pol. Vol. 39, p. 217 (1862).] We may assume that $f(z)$ and $\varphi(z)$ are regular on $L$ because $f(z)$ and $f(z) + \varphi(z)$ are different from 0 and $|f(z)| > |\varphi(z)|$ sufficiently close to $L$. The function $1 + \frac{\varphi(z)}{f(z)}$ has on $L$ a positive real part and so, as $z$ describes $L$, the change of its argument is equal to 0. Furthermore $f(z) + \varphi(z) = f(z)\left(1 + \frac{\varphi(z)}{f(z)}\right)$ thus [**181**] the image of $L$ under $f(z)$ has the same winding number as the one under $f(z) + \varphi(z)$.

**195.** Special case of **194**: $f(z) = ze^{\lambda - z}$, $\varphi(z) = -1$, $L$ is the unit circle. Since $ze^{\lambda - z}$ increases with $z$ on $0 \leqq z \leqq 1$ from 0 to $e^{\lambda - 1} > 1$ there is a root on this segment.

**196.** If $z = iy$, $y$ real, we have $|\lambda - iy| \geqq \lambda > 1 = |e^{-iy}|$. If $|z|$ is sufficiently large, $\Re z \geqq 0$, we have $|\lambda - z| > 1 \geqq |e^{-z}|$. Moreover $\lambda - z = 0$ has the root $z = \lambda$ in the right half-plane. Special case of **194**: $f(z) = \lambda - z$, $\varphi(z) = -e^{-z}$, $L$ a sufficiently large half-circle in $\Re z \geqq 0$ and its closing diameter. The only root is real because the modular graph is symmetric with respect to the vertical plane through the real axis.

**197.** [Cf. G. Julia: J. Math. Pures Appl. Ser. 8, Vol. 1, p. 63 (1918).] Special case of **194**: on the unit circle $|z| = 1 > |f(z)|$.

**198.** [Example of a general theorem of G. Julia: Ann. Sci. École Norm. Sup. (Paris) Ser. 3, Vol. 36, pp. 104–108 (1919).] Let $R_n$ be the rectangle with the four corners $n \pm \frac{1}{2} \pm id$, $n$ integer, $d$ fixed, $d > 0$ and write $z = x + iy = re^{i\vartheta}$. Because of I **155** we have on the boundary of $R_n$ as $n \to \infty$

$$\log |\Gamma(z)| \sim (x - \tfrac{1}{2}) \log r - y\vartheta - x,$$

so that the minimum of $|\Gamma(z)|$ on $R_n$ tends to $\infty$ as $n \to \infty$. On the other hand we find $|\sin \pi z| > c$ on the boundary of $R_n$ where $c$ is independent of $n$ and $c > 0$. Hence the minimum of

$$\left|\frac{1}{\Gamma(z)}\right| = \frac{|\sin \pi z|}{\pi}\,|\Gamma(1-z)|$$

converges on the boundary of $R_{-n}$ to $+\infty$ as $n \to \infty$. Thus we have for arbitrary $a$ and sufficiently large $n$

$$\left|\frac{1}{\Gamma(z)}\right| > |a|$$

on the boundary of $R_{-n}$ whereas at the center of $R_{-n}$ $\frac{1}{\Gamma(z)} = 0$. Apply **194** to $f(z) = \frac{1}{\Gamma(z)}$, $\varphi(z) = -a$.

**199.** Integrating by parts twice we obtain

$$zF(z) = f(0) - f(1)\cos z + \frac{1}{z}\left[f'(1)\sin z - \int_0^1 f''(t)\sin zt\,dt\right]$$
$$= f(0) - f(1)\cos z + \varphi(z).$$

We draw circles of radius $\varepsilon$ around all the zeros of the periodic function $f(0) - f(1)\cos z$, where $\varepsilon > 0$ and $2\varepsilon$ smaller than the distance between any two zeros. In the $z$-plane from which all the $\varepsilon$-disks have been lifted $\frac{\varphi(z)}{f(0) - f(1)\cos z}$ converges to 0 as $z \to \infty$. [To begin with prove this for the strip $-\pi \leqq \Re z \leqq \pi$.] With the exception of finitely many zeros $zF(z)$ has, therefore, in any disk the same number of zeros as the function $f(0) - f(1)\cos z$ [**194**], that is 1. This zero is necessarily real in the case $|f(1)| > |f(0)|$ where the disk is cut in half by the real axis; the non-real zeros of the functions that assume real values for real $z$ appear in pairs [solution **196**].

**200.** The term $z^n a^{-n^2}$ assumes its role as maximum term of the series

$$1 + \frac{z}{a} + \frac{z}{a}\cdot\frac{z}{a^3} + \frac{z}{a}\cdot\frac{z}{a^3}\cdot\frac{z}{a^5} + \cdots$$

on the circle $|z| = |a|^{2n-1}$ and abandons it on the circle $|z| = |a|^{2n+1}$ [I **117**]. To study the dominance of the maximum term in between those two circles notice the formula

$$\frac{F(z) - z^n a^{-n^2}}{z^n a^{-n^2}} = \frac{z}{a^{2n+1}} + \frac{z}{a^{2n+1}}\cdot\frac{z}{a^{2n+3}} + \frac{z}{a^{2n+1}}\cdot\frac{z}{a^{2n+3}}\cdot\frac{z}{a^{2n+5}} + \cdots$$
$$+ \frac{a^{2n-1}}{z} + \frac{a^{2n-1}}{z}\cdot\frac{a^{2n-3}}{z} + \frac{a^{2n-1}}{z}\cdot\frac{a^{2n-3}}{z}\cdot\frac{a^{2n-5}}{z} + \cdots$$

On the circle $|z| = |a|^{2n}$ the corresponding terms of the two subseries on the right hand side have the same absolute value, thus

$$\left|\frac{F(z) - z^n a^{-n^2}}{z^n a^{-n^2}}\right| < 2\left(\frac{1}{|a|} + \frac{1}{|a|}\cdot\frac{1}{|a|^3} + \frac{1}{|a|}\cdot\frac{1}{|a|^3}\cdot\frac{1}{|a|^5} + \cdots\right)$$
$$< \frac{2}{|a|}\frac{1}{1-|a|^{-3}} = \frac{2|a|^2}{|a|^3 - 1} < 1,$$

because the only positive root of the equation $z^3 - 2z^2 - 1 = 0$ is smaller than 2.5 [**16**]. Hence $F(z)$ has in the disk $|z| < |a|^{2n}$ the same number of zeros as $z^n a^{-n^2}$, namely $n$. The disk $|z| < |a|^{2n-2}$ contains, by virtue of the same proof, $n - 1$ zeros.—Cf. V **176**.

**201.** [A. Hurwitz: Math. Ann. Vol. 33, pp. 246–266 (1889).] The closed disk $D$ has $a$ as its center, lies completely in $\mathfrak{R}$ and contains no other zeros of $f(z)$ than possibly $a$. We have $|f(z)| > |f_n(z) - f(z)|$ on the boundary of $D$ when $n$ is sufficiently large. Apply **194**, $f_n(z) - f(z) = \varphi(z)$. More generally: Each subdomain of $\mathfrak{R}$ on the boundary of which there are no zeros of $f(z)$ contains exactly the same number of zeros of $f_n(z)$ as of $f(z)$ if $n$ is sufficiently large. Important for the applications!

**202.** The limit function is regular in the unit disk $|z| < 1$ [**170**]. Assume that $f(z_1) = f(z_2)$ for $z_1 \neq z_2$, $|z_1| < 1$, $|z_2| < 1$; consider the sequence $f_n(z) - f_n(z_1)$, $n = 1, 2, \ldots$ which converges to $f(z) - f(z_1)$. In a disk that has its center at $z_2$, lies completely inside the unit circle and does not contain $z_1$, $f_n(z) - f_n(z_1)$ would have to vanish for sufficiently large $n$ [**201**]: contradiction.

**203.** [**170**, **201**.]

**204.** In the case where $a$ and $d$ are integers the proposition is proved in the same way as proposition **185** because the zeros of the polynomial $a_0 z^a + a_1 z^{a+d} + a_2 z^{a+2d} + \cdots + a_n z^{a+nd}$ lie in the disk $|z| \leqq 1$ [**23**]. If $a$ and $d$ are rational $z$ has to be replaced by a suitable multiple of $z$. If $a$ and $d$ are irrational approximate these two constants by rational numbers and apply **203**.

**205.** [G. Pólya: Math. Z. Vol. 2, p. 354 (1918).] We have [II **21**]

$$\int_0^1 f(t) \cos zt\, dt = \lim_{n\to\infty} \sum_{\nu=1}^{n-1} \frac{1}{n} f\left(\frac{\nu}{n}\right) \cos\frac{\nu}{n} z \qquad [\mathbf{185}, \mathbf{203}].$$

**206.** Counter-example

$$f_n(z) = z^2 + \frac{1}{n}, \quad n = 1, 2, 3, \ldots; \quad \mathfrak{D}\colon |z| \leqq 2; \quad a = -1, \quad b = +1.$$

**206.1.** [G. Pólya, Problem: Jber. deutsch. Math. Verein. Vol. 34, 2. Abt., p. 97 (1925). Solved by R. Jungen: Jber. deutsch. Math. Verein.

Vol. 40, 2. Abt., pp. 6–7 (1931).] We may assume without loss of generality that $f_1(z), f_2(z), \ldots, f_n(z)$ are linearly independent and

$$|c_1|^2 + |c_2|^2 + \cdots + |c_n|^2 = 1$$

so that the set of coefficients $c_1, c_2, \ldots, c_n$ can be conceived as a point $c$ on the surface $\mathfrak{S}$ of the unit sphere in $2n$ dimensions. If there were no finite upper bound of the nature stated, there would exist an infinite sequence of points $c', c'', \ldots$ such that the linear combination corresponding to $c^{(m)}$ has not less than $m$ zeros in $\mathfrak{D}$. This sequence has at least one limit point $c^{(\infty)}$ on $\mathfrak{S}$, yet the linear combination corresponding to $c^{(\infty)}$ has only a finite number of zeros in $\mathfrak{D}$. Hence contradiction to the last remark of solution **201**.

**206.2.** [G. Pólya, Problem: Jber. deutsch. Math. Verein. Vol. 34, 2. Abt., p. 97 (1925). Solved by Nikola Obreschkoff: Jber. deutsch. Math. Verein. Vol. 37, 2. Abt., pp. 82–84 (1928).] Apply the argument principle to a rectangle with corners

$$-a + i\alpha, \quad a + i\alpha, \quad a + i\beta, \quad -a + i\beta$$

where $a$ is sufficiently large. Use V **75** and solution **180** in considering the horizontal sides.

**207.** From $z - w\varphi(z) = 0$ follows $1 - w\varphi'(z) = \varphi(z)\frac{dw}{dz}$. Thus Lagrange's formula $(L)$ (p. 145) for $f(z)$ implies by differentiation with respect to $w$

$$\frac{f'(z)\,\varphi(z)}{1 - w\varphi'(z)} = \sum_{n=1}^{\infty} \frac{w^{n-1}}{(n-1)!}\left[\frac{d^{n-1} f'(x)\,\varphi(x)\,[\varphi(x)]^{n-1}}{dx^{n-1}}\right]_{x=0},$$

i.e. the formula to be proved for the function $f'(z)\,\varphi(z)$. The family of admissible functions $f(z)$ is identical with the family of functions $f'(z)\,\varphi(z)$, where $f'(z)$ is the derivative of an admissible function, because $\varphi(0) \neq 0$. Thus **207** leads to Lagrange's formula $(L)$, p. 145, by integration.

**208.**

$$\frac{1}{n!}\left[\frac{d^n f(x)\,[\varphi(x)]^n}{dx^n}\right]_{x=0} = \frac{1}{2\pi i}\oint \frac{f(\zeta)\,[\varphi(\zeta)]^n}{\zeta^n}\,\frac{d\zeta}{\zeta},$$

$$\sum_{n=0}^{\infty} \frac{w^n}{n!}\left[\frac{d^n f(x)\,[\varphi(x)]^n}{dx^n}\right]_{x=0} = \frac{1}{2\pi i}\oint \frac{f(\zeta)\,d\zeta}{\zeta} \sum_{n=0}^{\infty}\left(\frac{w\varphi(\zeta)}{\zeta}\right)^n,$$

integrated along a circle around the center $\zeta = 0$ and for $w$ so small that $|\zeta| > |w\varphi(\zeta)|$ along the path of integration. Then the path of integration encloses the same number of zeros of $\zeta - w\varphi(\zeta)$ as of $\zeta$ [**194**], i.e. exactly one. Denoting this single zero by $z$ we further find

$$\frac{1}{2\pi i}\oint \frac{f(\zeta)\,d\zeta}{\zeta - w\varphi(\zeta)} = \frac{f(z)}{1 - w\varphi'(z)}.$$

**209.** [L. Euler: De serie Lambertiana, Opera Omnia, Ser. 1, Vol. 6. Leipzig and Berlin: B. G. Teubner 1921, p. 354.] In (*L*), p. 145, set $\varphi(z) = e^z$, $f(z) = z$,

$$z = w + \frac{2w^2}{2!} + \frac{3^2w^3}{3!} + \cdots + \frac{n^{n-1}w^n}{n!} + \cdots.$$

**210.** Introduce in (*L*), p. 145, $\varphi(z) = e^z$, $f(z) = e^{\alpha z}$,

$$e^{\alpha z} = 1 + \sum_{n=1}^{\infty} \frac{\alpha(\alpha+n)^{n-1}}{n!} w^n.$$

**211.** We write $x = 1 + z$, $\varphi(z) = (1+z)^\beta$, $f(z) = 1 + z$; (*L*), p. 145, yields

$$x = 1 + z = 1 + \sum_{n=1}^{\infty} \binom{\beta n}{n-1} \frac{w^n}{n}.$$

**212.** [Cf. l.c. **209**, p. 350.] Set $x = 1 + z$, $\varphi(z) = (1+z)^\beta$, $f(z) = (1+z)^\alpha$; (*L*), p. 145, implies

$$y = x^\alpha = (1+z)^\alpha = 1 + \sum_{n=1}^{\infty} \binom{\alpha+\beta n-1}{n-1} \frac{\alpha w^n}{n}.$$

**213.** We obtain for $\beta = 0$, $\beta = 1$ the binomial series, for $\beta = 2$

$$\left(\frac{1-\sqrt{1-4w}}{2w}\right)^\alpha = 1 + \alpha \sum_{n=1}^{\infty} \binom{\alpha+2n-1}{n-1} \frac{w^n}{n};$$

for $\beta = -1$ essentially the same series; for $\beta = \frac{1}{2}$

$$\left(\sqrt{1+\frac{w^2}{4}} + \frac{w}{2}\right)^{2\alpha} = 1 + \alpha \sum_{n=1}^{\infty} \binom{\alpha+\frac{n}{2}-1}{n-1} \frac{w^n}{n}.$$

Put $x = 1 + \frac{\xi}{\beta}$, $w = \frac{\omega}{\beta}$, $\alpha = a\beta$; fix $\xi$, $\omega$, $a$ and let $\beta$ increase to $+\infty$. The equation in **211** becomes

$$\omega = \xi\left(1+\frac{\xi}{\beta}\right)^{-\beta} \sim \xi e^{-\xi}.$$

**214.** By setting $\varphi(z) = e^z$, $f(z) = e^{\alpha z}$ and applying **207** we obtain

$$\sum_{n=0}^{\infty} \frac{(n+\alpha)^n w^n}{n!} = \frac{e^{\alpha z}}{1 - we^z} = \frac{e^{\alpha z}}{1-z},$$

where $z$ has the same meaning as in **209**. The radius of convergence is

$$= \lim_{n\to\infty} \frac{(n+\alpha)^n}{n!} \frac{(n+1)!}{(n+1+\alpha)^{n+1}} = e^{-1}.$$

**215.** We are dealing with

$$\sum_{n=0}^{\infty} \int_n^{n+1} \frac{t^n}{n!} e^{-\lambda t}\, dt = \sum_{n=0}^{\infty} \int_0^1 \frac{(n+\alpha)^n}{n!} e^{-\lambda(n+\alpha)}\, d\alpha.$$

The convergence of the series

$$\sum_{n=0}^{\infty} \frac{(n+\alpha)^n}{n!} e^{-\lambda n} = \frac{e^{\alpha z}}{1-z},$$

is uniform for $0 \leqq \alpha \leqq 1$, and $z$ is given by the equation $ze^{-z} = e^{-\lambda}$, $|z| < 1$ [**214, 209**]. Thus the integral in question becomes

$$= \int_0^1 \frac{e^{\alpha(z-\lambda)}}{1-z} d\alpha = \frac{e^{z-\lambda} - 1}{(1-z)(z-\lambda)} = \frac{1}{\lambda - z}.$$

We verify immediately that $\zeta = \lambda - z$ satisfies the equation $\lambda - \zeta - e^{-\zeta} = 0$; $\lambda - z$ is real and positive.

**216.** [Regarding **216–218, 225, 226**, cf. G. Pólya: Enseignement Math. Vol. 22, pp. 38–47 (1922).] We put $\varphi(z) = (1+z)^\beta$, $f(z) = (1+z)^\alpha$ and apply **207**:

$$\sum_{n=0}^{\infty} \binom{\alpha + \beta n}{n} w^n = \frac{(1+z)^\alpha}{1 - w\beta(1+z)^{\beta-1}} = \frac{x^{\alpha+1}}{(1-\beta)x + \beta},$$

where $1 + z = x$ and $x$ is the root of the equation **211** which is algebraic for rational $\beta$.

**217.** [Cf. L. Euler: Opuscula analytica, Vol. 1. Petropoli 1783, pp. 48–62.] We introduce $\varphi(z) = 1 + z + z^2$, $f(z) = 1$ and apply **207**:

$$z = \frac{1 - w - \sqrt{1 - 2w - 3w^2}}{2w},$$

$$\sum_{n=0}^{\infty} \frac{w^n}{n!} \left[\frac{d^n(1+x+x^2)^n}{dx^n}\right]_{x=0} = \frac{1}{1 - w(1+2z)} = \frac{1}{\sqrt{1 - 2w - 3w^2}}.$$

**218.** The sum of the terms in the $k$-th column to the left of the middle column is [**216**]

$$1 + \binom{k+2}{1} w + \binom{k+4}{2} w^2 + \cdots + \binom{k+2n}{n} w^n + \cdots$$

$$= \frac{1}{\sqrt{1-4w}} \left(\frac{1 - \sqrt{1-4w}}{2w}\right)^k.$$

**219.** [Cf. Jacobi: Werke, Vol. 6. Berlin: G. Reimer 1891, p. 22.] It is sufficient to consider the cases (2) and (3).

(2) Suppose $\xi \neq -1$, $\xi \neq 1$; choose some branch of $(1-\xi)^\alpha$ and of $(1+\xi)^\beta$ if $\alpha$ and $\beta$ are not integers. Introduce in **207**

$$\varphi(z) = \frac{(\xi + z)^2 - 1}{2}; \quad f(z) = (1 - \xi - z)^\alpha (1 + \xi + z)^\beta,$$

$$f(0) = (1-\xi)^\alpha (1+\xi)^\beta.$$

On the right hand side we find $(1-\xi)^\alpha (1+\xi)^\beta \sum_{n=0}^{\infty} P_n^{(\alpha,\beta)}(\xi)\, w^n$, whereas $w = \frac{z}{\varphi(z)}$ implies

$$z = \frac{1 - \xi w - \sqrt{1 - 2\xi w + w^2}}{w},$$

$$f(z) = 2^{\alpha+\beta}(1-\xi)^\alpha (1+\xi)^\beta (1 - w + \sqrt{1 - 2\xi w + w^2})^{-\alpha} \times$$
$$(1 + w + \sqrt{1 - 2\xi w + w^2})^{-\beta}, \quad 1 - w\varphi'(z) = \sqrt{1 - 2\xi w + w^2}.$$

Thus

$$\sum_{n=0}^{\infty} P_n^{(\alpha,\beta)}(\xi)\, w^n$$

$$= \frac{2^{\alpha+\beta}}{\sqrt{1 - 2\xi w + w^2}} (1 - w + \sqrt{1 - 2\xi w + w^2})^{-\alpha} (1 + w + \sqrt{1 - 2\xi w + w^2})^{-\beta}.$$

The two cases $\xi = -1$, $\xi = 1$ can be discussed directly [solution VI **98**].

(3) Assume $\xi \neq 0$ and put in **207**

$$\varphi(z) = z + \xi; \quad f(z) = e^{-(z+\xi)}(z+\xi)^\alpha, \quad f(0) = e^{-\xi}\xi^\alpha.$$

On the right hand side we have $e^{-\xi}\xi^\alpha \sum_{n=0}^{\infty} L_n^{(\alpha)}(\xi)\, w^n$; the relation $w = \frac{z}{\varphi(z)}$ yields

$$z = \frac{\xi w}{1 - w}, \quad f(z) = \frac{\xi^\alpha}{(1-w)^\alpha} e^{\frac{\xi}{w-1}}, \quad 1 - w\varphi'(z) = 1 - w,$$

therefore

$$\sum_{n=0}^{\infty} L_n^{(\alpha)}(\xi)\, w^n = \frac{1}{(1-w)^{\alpha+1}} e^{\frac{\xi w}{w-1}}.$$

The case $\xi = 0$ can be treated directly [solution VI **99**].

**220.**

$$\Delta e^{sz} = e^{sz}(e^s - 1), \quad \Delta^n e^{sz} = e^{sz}(e^s - 1)^n.$$

(1) $$(1+w)^z = 1 + \frac{z}{1} w + \frac{z(z-1)}{2!} w^2 + \cdots + \frac{z(z-1)\cdots(z-n+1)}{n!} w^n + \cdots;$$

where $w = e^s - 1$, valid for $|e^s - 1| < 1$, $z$ arbitrary.

(2) $$s = e^s - 1 - \frac{1}{2}(e^s-1)^2 + \frac{1}{3}(e^s-1)^3 - \cdots + (-1)^{n-1}\frac{1}{n}(e^s-1)^n + \cdots, \quad |e^s - 1| < 1.$$

(3) $$e^{sz} = 1 + \frac{z}{1!} se^s + \frac{z(z-2)}{2!}(se^s)^2 + \cdots + \frac{z(z-n)^{n-1}}{n!}(se^s)^n + \cdots,$$

follows from **210** with $z = -s$, $\alpha = -z$. According to II **205** we have

$$\frac{z(z-n)^{n-1}}{n!}(se^s)^n \sim (-1)^{n-1} ze^{-z}(2\pi)^{-\frac{1}{2}} n^{-\frac{3}{2}} (se^{s+1})^n,$$

hence the series converges whenever $|se^{s+1}| \leqq 1$; at any rate the formula holds in a neighbourhood of $s = 0$.

(4) in **212** put: $x = e^s$, $\beta = \frac{1}{2}$, thus $w = e^{s/2} - e^{-s/2}$, $\alpha = z$,

$$e^{sz} = 1 + \sum_{n=1}^{\infty} \frac{z}{n}\binom{z+\frac{n}{2}-1}{n-1}\left(e^{\frac{s}{2}} - e^{-\frac{s}{2}}\right)^n.$$

Replace $s$ by $-s$ and add the two formulas:

$$\frac{e^{sz}+e^{-sz}}{2} = 1 + \sum_{m=1}^{\infty} \frac{z}{2m}\binom{z+m-1}{2m-1}\left(e^{\frac{s}{2}} - e^{-\frac{s}{2}}\right)^{2m}.$$

Replace $s$ by $-s$, subtract and rearrange and then apply **216** to $x = e^s$, $\beta = \frac{1}{2}$, $w = e^{s/2} - e^{-s/2}$, $\alpha = z - \frac{1}{2}$:

$$\frac{e^{sz}-e^{-sz}}{2} = \sum_{m=1}^{\infty} \frac{1}{2}\binom{z+m-1}{2m-1}\left(e^{\frac{s}{2}} - e^{-\frac{s}{2}}\right)^{2m-2}(e^s - e^{-s}).$$

The sum of the last two formulas produces the desired expansion of $e^{sz}$. We remember Stirling's formula and write $\sin \pi z$ as an infinite product to obtain

$$\frac{z}{2m}\binom{z+m-1}{2m-1}\left(e^{\frac{s}{2}} - e^{-\frac{s}{2}}\right)^{2m} \sim -\frac{z \sin \pi z}{\sqrt{\pi}}\left(\sin\frac{is}{2}\right)^{2m} m^{-\frac{3}{2}};$$

this shows convergence for $\left|\sin\frac{is}{2}\right| < 1$; the formula is certainly correct in a neighbourhood of $s = 0$.

**221.** [With respect to the formulas (2) and (3) cf. N. H. Abel: Oeuvres, Vol. 2, Nouvelle édition. Christiania: Grøndahl & Son 1881, pp. 72—73.] Expand in **220**, $F(z) = e^{sz}$, both sides in ascending powers of $s$ and compare the coefficients of $\frac{s^k}{k!}$. We have

$$\Delta^n F(z) = \Delta^n\left(1 + \frac{s}{1!}z + \frac{s^2}{2!}z^2 + \cdots + \frac{s^k}{k!}z^k + \cdots\right) = \sum_{k=0}^{\infty}\frac{s^k}{k!}\Delta^n z^k.$$

**222.** It is sufficient to prove (1) and (2) for $F(z) = (z - w)^{-1}$; for other rational functions differentiate with respect to $w$, split them up into partial fractions and apply **221**. The formulas (1) and (2) hold for $F(z) = e^{sz}$ if $s$ is real and negative. [Solution **220**.] Multiplication with $e^{-sw}\,ds$ and integration over $-\infty < s \leqq 0$ yield (1) and (2) for

$F(z) = (z - w)^{-1}$. It is easier to discuss the range, in which the formulas are valid, in the following way: Since

$$\Delta^n (z - w)^{-1} = \frac{(-1)^n \, n!}{(z - w)(z - w + 1) \cdots (z - w + n)},$$

we have to prove

$$(1) \qquad \frac{1}{z - w} = -\frac{1}{w} - \sum_{n=0}^{\infty} \frac{z(z-1) \cdots (z-n)}{w(w-1) \cdots (w-n-1)},$$

$$(2) \qquad -\frac{1}{(z-w)^2} = -\sum_{n=0}^{\infty} \frac{n!}{(z-w)(z-w+1) \cdots (z-w+n+1)}.$$

We obtain (2) by replacing in (1) $w$ and $z$ by $w - z - 1$ and $-1$ respectively. Formula (1) is derived by a limit process from the identity

$$\frac{1}{w - z} = \frac{1}{w} + \frac{z}{w(w-1)} + \frac{z(z-1)}{w(w-1)(w-2)} + \cdots$$
$$+ \frac{z(z-1) \cdots (z-n+1)}{w(w-1)(w-2) \cdots (w-n)} + \frac{z}{w} \frac{z-1}{w-1} \cdots \frac{z-n}{w-n} \frac{1}{w-z},$$

which is easily proved by complete induction. If $w$ and $z$ are different from one another and from 0, 1, 2, 3, ... the remainder is [cf. II **31**]

$$\frac{(-z)(-z+1) \cdots (-z+n)}{n^{-z} \, n!} \, \frac{n^{-w} \, n!}{-w(-w+1) \cdots (-w+n)} \, \frac{n^{w-z}}{w-z}$$
$$\sim \frac{\Gamma(-w)}{\Gamma(-z)} \frac{n^{w-z}}{w-z}.$$

The formulas (3) and (4) cannot be valid in a non-empty region for non-entire rational functions: Otherwise the partial sums of order $n$ and $2n$ resp. would converge uniformly in any finite domain according to **255** and **254**, thus $F(z)$ would be regular everywhere: contradiction.

**223.** The identity in question is purely formal; it represents the combination of the infinitely many equations that define the quantities $\Delta^n a_k$, $k = 0, \pm 1, \pm 2, \ldots$

**224.** Since the expansion of $\frac{t}{1+t}$ does not contain a constant term the coefficient $t^n$ in the expansion of $\frac{1}{1+t} F\left(\frac{t}{1+t}\right)$ depends on $a_0, a_1, \ldots, a_n$ only. Therefore we can restrict ourselves to the case where $F(z)$ is a polynomial. Every polynomial however can be written as a linear combination of the special polynomials $(1 - z)^m$, $m = 0, 1, 2, \ldots$; besides we have for two sequences $a_k$ and $b_k$ and two constants $c_1$ and $c_2$ the linear relation $\Delta^n (c_1 a_k + c_2 b_k) = c_1 \Delta^n a_k + c_2 \Delta^n b_k$. Consequently it is sufficient to prove the statement for $F(z) = (1 - z)^m$, $m = 0, 1, 2, \ldots$ In this case $\Delta^n a_0$ is equal [**223**] to the coefficient of $z^n$ in the expansion of $(1-z)^n \cdot (1-z)^m = (1-z)^{n+m}$, i.e. $= (-1)^n \binom{n+m}{n}$. Thus

$$\sum_{n=0}^{\infty} (-1)^n \binom{n+m}{n} t^n = \frac{1}{(1+t)^{m+1}} = \frac{1}{1+t} F\left(\frac{t}{1+t}\right).$$

**225.** It is sufficient to prove the statement for $F(z) = (1-z)^m$, $m = 0, 1, 2, \ldots$ [solution **224**]. Then the quantity $\Delta^{2n} a_{-n}$ is [**223**] equal to the coefficient of $z^n$ in the expansion of

$$(1-z)^{2n} \sum_{k=-\infty}^{\infty} a_k z^k = (1-z)^{2n} \frac{F(z)+F(z^{-1})}{2} = (1-z)^{2n+m} \frac{1+(-1)^m z^{-m}}{2}$$

i.e. $= (-1)^n \binom{2n+m}{n}$. [Solution **218**.]

**226.** We put $F(z) = (1-z)^m - 1$, $m = 1, 2, 3, \ldots$ [solution **224**, **225**]. Then $\Delta^{2n} a_{-n+1} - \Delta^{2n} a_{-n-1}$ is equal [**223**] to the coefficient of $z^n$ in the expansion of

$$(1-z)^{2n} (z^{-1} - z) \sum_{k=-\infty}^{\infty} a_k z^k = (1-z)^{2n} (z^{-1} - z) \frac{F(z) - F(z^{-1})}{2}$$

$$= (1-z)^{2n+m} (z^{-1} - z) \frac{1 + (-1)^{m+1} z^{-m}}{2},$$

i.e.

$$= (-1)^n \left\{ \binom{2n+m+1}{n} - \binom{2n+m+1}{n+1} \right\}$$

$$= (-1)^{n+1} \frac{m}{n+1} \binom{2n+m+1}{n}.$$

Set $\beta = 2$, $\alpha = m$, $w = -t$ in solution **213**.

***227.**

$$\frac{\sin \pi x}{\pi} = x\left(1-\frac{x}{1}\right)\left(1+\frac{x}{1}\right)\left(1-\frac{x}{2}\right)\left(1+\frac{x}{2}\right)\cdots$$

$$\frac{\sin \pi x}{\pi x(1-x)} = \prod_{n=1}^{\infty} \left(1+\frac{x}{n}\right)\left(1-\frac{x}{n+1}\right) = \prod_{n=1}^{\infty} \left(1+\frac{x(1-x)}{n(n+1)}\right).$$

**228.** [I. Schur.] By carrying out the multiplication in the last infinite product in **227**.

**229.** Apply $(L)$, p. 145, to

$$\varphi(z) = \frac{1}{1-z}, \quad f(z) = \sin \pi z, \quad w = z(1-z).$$

We find

$$\sin \pi z = \sum_{n=1}^{\infty} \frac{A_n}{n!} w^n, \; A_n = \left[\frac{d^{n-1}(1-x)^{-n} \pi \cos \pi x}{dx^{n-1}}\right]_{x=0} \qquad [\mathbf{228}].$$

**230.** We split $f(z)$ into the real and imaginary parts, $f(re^{i\vartheta}) = U(r,\vartheta) + iV(r,\vartheta)$, $U(r,\vartheta)$, $V(r,\vartheta)$ real, $a_n = b_n + ic_n$, $b_n$, $c_n$ real. The series

$$U(r,\vartheta) = b_0 + \sum_{n=1}^{\infty} r^n (b_n \cos n\vartheta - c_n \sin n\vartheta),$$

$$V(r,\vartheta) = c_0 + \sum_{n=1}^{\infty} r^n (c_n \cos n\vartheta + b_n \sin n\vartheta)$$

are uniformly convergent for $0 \leqq \vartheta \leqq 2\pi$, hence [**117**]

$$a_n = b_n + ic_n = \frac{1}{\pi r^n} \int_0^{2\pi} U(r,\vartheta)\, e^{-in\vartheta}\, d\vartheta$$

$$= \frac{i}{\pi r^n} \int_0^{2\pi} V(r,\vartheta)\, e^{-in\vartheta}\, d\vartheta, \qquad n = 1, 2, 3, \ldots$$

**231.** Solution **230** implies

$$f(z) = a_0 + \sum_{n=1}^{\infty} (b_n + ic_n)\, z^n = \frac{1}{2\pi} \int_0^{2\pi} U(r,\vartheta) \left[1 + 2 \sum_{n=1}^{\infty} \left(\frac{ze^{-i\vartheta}}{r}\right)^n\right] d\vartheta.$$

**232.** 1st special case: $f(z)$ does not vanish in the disk $|z| \leqq r$. Apply **231** to the function $\log f(z)$, which is regular in $|z| \leqq r$.

2nd special case: $f(z) = \frac{(z-c)\,r}{r^2 - \bar{c}z}$, $|c| < r$. Since $\log |f(re^{i\vartheta})| = 0$ [**5**] the integral on the right hand side vanishes.

Any function that is regular in the disk $|z| \leqq r$ and different from 0 on the circle $|z| = r$ can be written as a product of special functions of the types 1 and 2. The condition that $f(z)$ be non-zero on the boundary $|z| = r$ can finally be dropped because both sides depend continuously on $r$.—Different solution with the help of **176**, in the manner of solution **56**.

**233.** Assume $0 < \alpha < 2\pi$ and that $Re^{i\vartheta}$, $0 < \vartheta < \alpha$, is a point of the arc in question. We use the same notation as in **231**; $U(r,\vartheta) = 0$ for $0 < \vartheta < \alpha$; $f(0)$ is supposed to be real. Two limit operations lead to

$$\lim_{r \to R-0} f(re^{i\vartheta}) = f(Re^{i\vartheta}) = \frac{1}{2\pi} \int_{\alpha}^{2\pi} U(R,\Theta)\, \frac{1 + e^{i(\vartheta - \Theta)}}{1 - e^{i(\vartheta - \Theta)}}\, d\Theta,$$

$$\Im f(Re^{i\vartheta}) = \frac{1}{2\pi} \int_{\alpha}^{2\pi} U(R,\Theta) \cot \frac{\vartheta - \Theta}{2}\, d\Theta,$$

$$\frac{d}{d\vartheta} \Im f(Re^{i\vartheta}) = -\frac{1}{2\pi} \int_{\alpha}^{2\pi} U(R,\Theta) \left(\sin \frac{\vartheta - \Theta}{2}\right)^{-2} d\Theta \leqq 0.$$

**234.** We retain the notation of **230**:

$$\frac{1}{2\pi}\int_0^{2\pi} [U(r,\vartheta)]^2\, d\vartheta = b_0^2 + \frac{1}{2}\sum_{n=1}^{\infty} r^{2n}(b_n^2 + c_n^2),$$

$$\frac{1}{2\pi}\int_0^{2\pi} [V(r,\vartheta)]^2\, d\vartheta = c_0^2 + \frac{1}{2}\sum_{n=1}^{\infty} r^{2n}(b_n^2 + c_n^2).$$

**235.** [Cf. C. Carathéodory: Rend. Circ. Mat. Palermo Vol. 32, pp. 193—217 (1911).] In the notation of **230** we have $R = 1$, $a_0 = \frac{1}{2}$ and

$$a_n = \frac{1}{\pi r^n}\int_0^{2\pi} U(r,\vartheta)\, e^{-in\vartheta}\, d\vartheta, \quad |a_n| \leqq \frac{1}{\pi r^n}\int_0^{2\pi} U(r,\vartheta)\, d\vartheta = \frac{1}{r^n},$$

$$0 < r < 1; \quad n = 1, 2, 3, \ldots$$

Let $r$ converge to 1.—Example:

$$f(z) = \frac{1}{2}\,\frac{1+z}{1-z} = \frac{1}{2} + z + z^2 + \cdots + z^n + \cdots.$$

**236.** The function

$$\frac{1}{2}\,\frac{A - f(Rz) + i\Im a_0}{A - \Re a_0} = \frac{1}{2} - \frac{1}{2}\sum_{n=1}^{\infty} \frac{a_n R^n}{A - \Re a_0} z^n$$

satisfies the conditions of **235**. Hence

$$|a_n| \leqq \frac{2(A - \Re a_0)}{R^n}, \quad \text{i.e.} \sum_{n=1}^{\infty} |a_n|\, r^n \leqq 2(A - \Re a_0)\sum_{n=1}^{\infty}\left(\frac{r}{R}\right)^n.$$

The upper bound is attained for the function given as example.

**237.** It is sufficient to prove the first equality (to obtain the second we replace $z$ by $\frac{1}{z}$). The function $\sum_{n=-\infty}^{-1} a_n z^n$ remains bounded for $|z| \geqq 1$, $\left|\sum_{n=-\infty}^{-1} a_n z^n\right| < M$. We denote by $A^*(r)$ the maximum of the real part of the entire function $\sum_{n=0}^{\infty} a_n z^n$ on the circle $|z| = r$; then we have for $r > 1$ [solution **236**]

$$A^*(r) \geqq \Re a_0 + \tfrac{1}{2}|a_n|\, r^n, \qquad n = 1, 2, 3, \ldots;$$

furthermore $A(r) \geqq A^*(r) - M$, thus

$$A(r) \geqq \Re a_0 - M + \tfrac{1}{2}|a_n|\, r^n.$$

If $a_n$ is different from 0 we conclude

$$\liminf_{r\to\infty} \frac{\log A(r)}{\log r} \geqq n.$$

There are arbitrarily large $n$'s for which $a_n \neq 0$.

**238.** [E. Landau: Arch. Math. Phys. Ser. 3, Vol. 11, pp. 32–34, (1907); cf. F. Schottky: J. reine angew. Math. Vol. 117, pp. 225–253 (1897).] It is sufficient to prove the inequality $|\Re a_1| R \leqq \frac{2}{\pi} \Delta(f)$; for if $\alpha$ is a real constant the largest oscillation of the real part of $f(e^{i\alpha} z)$ is also $\Delta(f)$ and the inequality $|\Re e^{i\alpha} a_1| R \leqq \frac{2}{\pi} \Delta(f)$ for all $\alpha$ implies $|a_1| R \leqq \frac{2}{\pi} \Delta(f)$.—Let $A$ denote the arithmetic mean of the upper and lower bounds of $\Re f(z)$ in the disk $|z| < R$. We find $|\Re f(z) - A| \leqq \frac{1}{2}\Delta(f)$ for $|z| < R$. Besides [**230**]

$$\pi r a_1 = \int_0^{2\pi} [\Re f(re^{i\vartheta})] e^{-i\vartheta} d\vartheta, \qquad 0 < r < R,$$

thus

$$\pi r \Re a_1 = \int_0^{2\pi} [\Re f(re^{i\vartheta}) - A] \cos\vartheta \, d\vartheta, \quad \pi r |\Re a_1| \leqq \frac{\Delta(f)}{2} \int_0^{2\pi} |\cos\vartheta| \, d\vartheta = 2\Delta(f).$$

Let $r$ converge to $R$.—For $R = 1$

$$f(z) = \frac{1}{2i} \log \frac{1-z}{1+z} = iz + \frac{i}{3} z^3 + \frac{i}{5} z^5 + \cdots,$$

we have

$$\Delta(f) = \frac{\pi}{2}, \quad |a_1| = 1.$$

In geometrical terms the theorem reads as follows: A disk is mapped conformally but not necessarily univalently onto a region. The width of this region in any direction is at least equal to $\frac{\pi}{2}$ times the product of the radius of the disk and the linear enlargement at the center of the disk. In the above mentioned special case the image of the disk is the vertical strip $-\frac{\pi}{4} < \Re f(z) < \frac{\pi}{4}$.

**239.** [E. Landau, O. Toeplitz: Arch. Math. Phys. Ser. 3, Vol. 11, pp. 302–307 (1907).] The least upper bound of $|f(z) - f(-z)|$ on $|z| < R$ is denoted by $D^*(f)$, then $D^*(f) \leqq D(f)$. Let $0 < r < R$.

$$4\pi r a_1 = \int_0^{2\pi} [f(re^{i\vartheta}) - f(-re^{i\vartheta})] e^{-i\vartheta} d\vartheta$$

implies $4\pi r |a_1| \leqq D^*(f) \cdot 2\pi \leqq D(f) \cdot 2\pi$; let $r$ converge to $R$. If $f(z)$ is linear, $f(z) = a_0 + a_1 z$, then $D(f) = 2 |a_1| R$. The proposition admits the following geometrical interpretation: A disk is mapped conformally but not necessarily schlicht onto a region. The maximum distance of two boundary points (diameter) of this region is at least equal to the product of the diameter of the disk and the linear enlargement at the center. In the special case mentioned the image is an open disk.

**240.** [Cf. E. Landau: Math. Z. Vol. 20, pp. 99—100 (1924).] We derive from **232** by differentiation and by setting $z = 0$ [**117**]

$$-\frac{f'(0)}{f(0)} = \sum_{\mu=1}^{m}\left(\frac{1}{c_\mu} - \frac{\bar{c}_\mu}{r^2}\right) + \frac{1}{\pi r}\int_0^{2\pi} (\log M - \log|f(re^{i\vartheta})|)\, e^{-i\vartheta}\, d\vartheta,$$

$$-\Re\frac{f'(0)}{f(0)} \leqq \sum_{\mu=1}^{m}\Re\left(\frac{1}{c_\mu} - \frac{\bar{c}_\mu}{r^2}\right) + \frac{1}{\pi r}\int_0^{2\pi} (\log M - \log|f(re^{i\vartheta})|)\, d\vartheta$$

$$= \sum_{\mu=1}^{m}\Re\left(\frac{1}{c_\mu} - \frac{\bar{c}_\mu}{r^2} - \frac{2}{r}\log\frac{r}{|c_\mu|}\right) + \frac{2}{r}\quad\frac{M}{|f(0)|}$$

[**120**]. According to the condition (2) we have $\Re c_\mu < 0$; thus for $\mu > l$

$$\Re\left(\frac{1}{c_\mu} - \frac{\bar{c}_\mu}{r^2}\right) = \frac{r^2 - |c_\mu|^2}{r^2}\,\Re\frac{1}{c_\mu} < 0.$$

For $\mu \leqq l$ we have

$$\Re\left(-\frac{\bar{c}_\mu}{r} - 2\log\frac{r}{|c_\mu|}\right) \leqq \frac{|c_\mu|}{r} + 2\log\frac{|c_\mu|}{r} < 0.$$

The inequality $x + 2\log x < 0$ holds for $0 < x \leqq \frac{2}{3}$ because the left hand side increases with $x$ and $e^{2/3}\left(\frac{2}{3}\right)^2 < 1$, i.e. $8e < 27$.

**241.** The power series in question can be written in the form

$$\sum_{n=0}^{\infty} a_n z^n = \frac{c_1}{1 - z_1 z} + \frac{c_2}{1 - z_2 z} + \cdots + \frac{c_k}{1 - z_k z} + \sum_{n=0}^{\infty} b_n z^n,$$

$|z_1| = |z_2| = \cdots = |z_k| = 1$, $\limsup_{n\to\infty} \sqrt[n]{|b_n|} < 1$, whence

$$a_n = c_1 z_1^n + c_2 z_2^n + \cdots + c_k z_k^n + b_n, \quad |b_n| < B.$$

**242.** The radius of convergence is assumed to be 1, thus $z_0\bar{z}_0 = 1$. The series can be written as

$$\sum_{n=0}^{\infty} a_n z^n = \frac{c_0 + c_1 z + \cdots + c_k z^k}{(z_0 - z)^{k+1}} + \sum_{n=0}^{\infty} b_n z^n,$$

$c_0 + c_1 z_0 + \cdots + c_k z_0^k \neq 0$, $c_k \neq 0$, $\limsup_{n\to\infty} \sqrt[n]{|b_n|} < 1$; whence for $n > k$

$$a_n = \binom{n}{k} c_k \bar{z}_0^{n+1} + \binom{n+1}{k} c_{k-1}\bar{z}_0^{n+2} + \binom{n+2}{k} c_{k-2}\bar{z}_0^{n+3} + \cdots + \binom{n+k}{k} c_0 \bar{z}_0^{n+k+1} + b_n$$

$$= \binom{n}{k}\bar{z}_0^{n+k+1}\left(\sum_{\mu=0}^{k}\frac{(n+k-\mu)(n+k-\mu-1)\cdots(n-\mu+1)}{n(n-1)\cdots(n-k+1)}\, c_\mu z_0^\mu + \frac{b_n}{\binom{n}{k}} z_0^{n+k+1}\right).$$

The expression in parentheses converges to

$$c_0 + c_1 z_0 + \cdots + c_k z_0^k \neq 0 \text{ as } n \to \infty.$$

Cf. also I **178** and II **95.3**, II **197**.

**243.** [G. Pólya: J. reine angew. Math. Vol. 151, pp. 24–25 (1921).] The power series $P(z) \sum_{n=0}^{\infty} a_n z^n = \sum_{n=0}^{\infty} b_n z^n$ can be made to satisfy the conditions of **242** by an appropriate choice of the polynomial $P(z) = c_0 + c_1 z + c_2 z^2 + \cdots + c_{q-2} z^{q-2} + z^{q-1}$. $\frac{|b_n|}{|b_{n+1}|} \to \varrho$ implies $\sqrt[n]{|b_n|} = \frac{1}{\varrho}$ [I **68**]. Furthermore

$$\begin{aligned} |b_n| &= |c_0 a_n + c_1 a_{n-1} + \cdots + c_{q-2} a_{n-q+2} + a_{n-q+1}| \\ &\leqq A_n(|c_0| + |c_1| + \cdots + |c_{q-2}| + 1). \end{aligned}$$

**244.** Let $k$ be the number of poles. We set $a_n = \alpha_n + b_n$, $\limsup_{n\to\infty} \sqrt[n]{|b_n|} = b < \frac{1}{\varrho}$, $\sum_{n=0}^{\infty} \alpha_n z^n$ rational with exactly $k$ poles on the circle of convergence $|z| = \varrho$. We choose $\varepsilon$ so small that $b + \varepsilon < \frac{1}{\varrho} - \varepsilon$. According to **243** we have

$$\max(|\alpha_n|, |\alpha_{n-1}|, \ldots, |\alpha_{n-k+1}|) > \max\left[\left(\frac{1}{\varrho} - \varepsilon\right)^n, \left(\frac{1}{\varrho} - \varepsilon\right)^{n-k+1}\right]$$

for $n$ sufficiently large; i.e. $|\alpha_{\bar{n}}| > \left(\frac{1}{\varrho} - \varepsilon\right)^{\bar{n}} > |b_{\bar{n}}|$ for at least one $\bar{n}$, $n \geqq \bar{n} \geqq n - k + 1$; thus $a_{\bar{n}} = \alpha_{\bar{n}} + b_{\bar{n}} \neq 0$. Consequently $\nu_n \geqq \frac{n}{k} - c$, $c$ independent of $n$. The proposition holds also in the case where the multiplicity of the poles is not taken into account, but the proof has to be approached differently.

**245.** [J. König: Math. Ann. Vol. 9, pp. 530–540 (1876).] If the poles are of order $k$ we have $a_n = A n^{k-1} \varrho^{-n} (\sin(n\alpha + \delta) + \varepsilon_n)$, $A, \alpha, \delta$ real, $\lim_{n\to\infty} \varepsilon_n = 0$ [solution **242**]. Assume that $A > 0$, $0 < 2\eta < \alpha < \pi - 2\eta$ and that $|\varepsilon_n| < \sin \eta$ whenever $n > N$. If the distance between $n\alpha + \delta$ and the closest multiple of $\pi$ is larger than $\eta$ then $a_n$ has the same sign as $\sin(n\alpha + \delta)$. In the case where $n > N$ and where $a_n$ does not have the same sign as $\sin(n\alpha + \delta)$ the coefficients $a_{n-1}$ and $a_{n+1}$ have certainly the same signs as $\sin((n-1)\alpha + \delta)$ and $\sin((n+1)\alpha + \delta)$ resp.; $a_{n-1}$ and $a_{n+1}$ have different signs because $-\eta < n\alpha + \delta - m\pi < \eta$ implies $-\pi + \eta < (n-1)\alpha + \delta - m\pi < -\eta$, $\eta < (n+1)\alpha + \delta - m\pi < \pi - \eta$. Therefore the number of changes of sign between $a_{n-1}$ $a_n$ $a_{n+1}$ is the same as between $\sin((n-1)\alpha+\delta)$ $\sin(n\alpha+\delta)$ $\sin((n+1)\delta+\alpha)$. Now use VIII **14**.

**246.** The radius of convergence of the series $\sum_{n=0}^{\infty} a_n z^n$ is assumed to be 1. If the series converges at some point of the circle of convergence we have $\lim_{n\to\infty} a_n = 0$, thus [I **85**]

$$\lim_{z\to 1-0} (1-z)(a_0 + a_1 z + \cdots + a_n z^n + \cdots) = \lim_{n\to\infty} \frac{a_n}{1} = 0$$

hence the point $z = 1$ cannot be a pole.

**247.** [M. Fekete: C. R. Acad. Sci. (Paris) Sér. A–B, Vol. 150, pp. 1033–1036 (1910); G. H. Hardy: Proc. Lond. Math. Soc. Ser. 2, Vol. 8, pp. 277–294 (1910).] Let

$$f(e^{-x}) = c_{-h}x^{-h} + c_{-h+1}x^{-h+1} + \cdots, \qquad h \geqq 0, \qquad c_{-h} \neq 0$$

and $\varrho$, $0 < \varrho < 1$, be so small that this series converges uniformly and absolutely for $|x| \leqq \varrho$ (disregard the terms with a negative exponent). Then we obtain, at first for $\Re s > h$,

$$\int_0^{\varrho} x^{s-1} f(e^{-x})\, dx = \sum_{n=-h}^{\infty} c_n \frac{\varrho^{s+n}}{s+n}.$$

Multiplication by $[\Gamma(s)]^{-1}$ eliminates all the poles except possibly the poles at $s = h, h-1, \ldots, 1$, if $h \geqq 1$. In this case there is a pole at $s = h$ because $c_{-h} \neq 0$.

**248.** The series $\sum_{n=1}^{\infty} e^{-\alpha\sqrt{n}}$, $\alpha > 0$, is convergent. The integral $F(u)$ converges because $\Phi(a + it)$ is bounded for all values of $t$. Term by term integration [II **115**] yields

$$F(u) = \sum_{n=0}^{\infty} \frac{a_n}{2\pi i} \int_{a-i\infty}^{a+i\infty} \frac{e^{s(\sqrt{n}-u)} + e^{-s(\sqrt{n}+u)}}{s^2}\, ds.$$

By virtue of **155** we have for $\sqrt{m-1} \leqq u \leqq \sqrt{m}$

$$F(u) = a_m(\sqrt{m} - u) + a_{m+1}(\sqrt{m+1} - u) + a_{m+2}(\sqrt{m+2} - u) + \cdots,$$

i.e. $F(u)$ is a piecewise linear function whose derivative is

$$-a_m - a_{m+1} - a_{m+2} - \cdots$$

in the interval $\sqrt{m-1} < u < \sqrt{m}$. If $\Phi(s)$ vanishes identically then $F(u) \equiv 0$ for $u > 0$, thus $a_m + a_{m+1} + a_{m+2} + \cdots = 0$ for $m = 1, 2, 3, \ldots$

**249.** We have $\Phi^{(2k+1)}(0) = 0$; $\Phi^{(2k)}(0)$ vanishes also because

$$\left(z\frac{d}{dz}\right)^k f(z) = \sum_{n=0}^{\infty} a_n n^k z^n \quad \text{and} \quad \lim_{z\to 1} \sum_{n=0}^{\infty} a_n n^k z^n = \sum_{n=0}^{\infty} a_n n^k; \quad \text{i.e. } \Phi(s) \equiv 0.$$

**250.** The function $f(z)$ described in the problem has the following properties

(1) $f(z)$ is regular for $|z| < 1$ because the integral converges absolutely when $\Re z \leqq 1$; we have

$$a_n = \int_0^\infty e^{-(x + x^\mu \cos \mu\pi)} \sin (x^\mu \sin \mu\pi) \frac{x^n}{n!}\, dx, \qquad n = 0, 1, 2, \ldots;$$

(2) $a_n$ cannot vanish for $n = 0, 1, 2, \ldots$, because

$$|e^{-(x + x^\mu \cos \mu\pi)} \sin (x^\mu \sin \mu\pi)| < e^{-x} \quad [\text{solution } \mathbf{153}];$$

(3) for $|z| < 1$, the derivatives are

$$f^{(n)}(z) = \int_0^\infty e^{-x^\mu \cos \mu\pi} \sin (x^\mu \sin \mu\pi)\, e^{-x(1-z)} x^n\, dx, \qquad n = 0, 1, 2, \ldots;$$

this integral converges absolutely and uniformly for $\Re z \leqq 1$; thus, if $z$ tends to 1 along the real axis,

$$\lim_{z\to 1} f^{(n)}(z) = \int_0^\infty e^{-x^\mu \cos \mu\pi} \sin (x^\mu \sin \mu\pi)\, x^n\, dx = 0 \qquad [\mathbf{153}];$$

(4)

$$|a_n| < \frac{1}{n!}\int_0^\infty e^{-(x + x^\mu \cos \mu\pi)} x^n\, dx < \frac{1}{n!} + \frac{1}{n!}\max_{x\geq 0} \left(e^{-(x + x^\mu \cos \mu\pi)} x^{n+2}\right) \int_1^\infty x^{-2}\, dx,$$

hence [II **222**]

$$\limsup_{n\to\infty} \frac{\log |a_n|}{n^\mu} \leqq -\cos \mu\pi < 0.$$

The use of Hamburger's function, $\exp\left(-\frac{\pi\sqrt{x} - \log x}{(\log x)^2 + \pi^2}\right) \sin\left(\frac{\sqrt{x}\log x + \pi}{(\log x)^2 + \pi^2}\right)$, given in solution **153**, instead of $e^{-x^\mu \cos \mu\pi} \sin (x^\mu \sin \mu\pi)$ shows in a similar way that in **249** $\frac{\log |a_n|}{\sqrt{n}}$ cannot be replaced even by $(\log n)^2 \frac{\log |a_n|}{\sqrt{n}}$.

**251.** [G. Pólya, Problem: Arch. Math. Phys. Ser. 3, Vol. 25, p. 337 (1917). Solved by H. Prüfer, K. Scholl: Arch. Math. Phys. Ser. 3, Vol. 28, p. 177 (1920).] We assume

$$g(z) = c_0 + \frac{c_1}{1!}(z-a) + \frac{c_2}{2!}(z-a)^2 + \cdots + \frac{c_n}{n!}(z-a)^n + \cdots$$

and that the series in question converges for $z = a$, i.e. that $c_0 + c_1 + c_2 + \cdots + c_n + \cdots$ converges, thus $|c_{m+k} + c_{m+k+1} + \cdots + c_{m+k+n}| < \varepsilon$ for $m$ sufficiently large, $k, n = 0, 1, 2, 3, \ldots$ Then

$$|g^{(m)}(z) + g^{(m+1)}(z) + \cdots + g^{(m+n)}(z)|$$
$$= \left|\sum_{k=0}^{\infty} (c_{m+k} + c_{m+1+k} + \cdots + c_{m+n+k}) \frac{(z-a)^k}{k!}\right| < \varepsilon e^{|z-a|}.$$

**252.** Write $g(z)$ as a series:

$$g(z) = a_0 + \frac{a_1}{1!}(z - z_0) + \frac{a_2}{2!}(z - z_0)^2 + \cdots + \frac{a_n}{n!}(z - z_0)^n + \cdots.$$

The sequence $|a_1|, \sqrt{|a_2|}, \ldots, \sqrt[n]{|a_n|}, \ldots$ is assumed to be bounded, $\limsup_{n\to\infty} \sqrt[n]{|a_n|} = A$. There exists a number $N$ to any given $\varepsilon$, $\varepsilon > 0$, such that $|a_n| < (A + \varepsilon)^n$ whenever $n > N$, consequently

$$g^{(n)}(z) = \left|a_n + \frac{a_{n+1}}{1!}(z - z_0) + \frac{a_{n+2}}{2!}(z - z_0)^2 + \cdots\right|$$
$$< (A+\varepsilon)^n + \frac{(A+\varepsilon)^{n+1}}{1!}|z - z_0| + \frac{(A+\varepsilon)^{n+2}}{2!}|z - z_0|^2 + \cdots$$
$$= (A+\varepsilon)^n e^{(A+\varepsilon)|z-z_0|}.$$

Hence $\limsup_{n\to\infty} \sqrt[n]{|g^{(n)}(z)|} \leq A = \limsup_{n\to\infty} \sqrt[n]{|g^{(n)}(z_0)|}$. That means the limit superior in question is at no point $z$ larger than at any other point $z_0$.

**253.** [J. Bendixson: Acta Math. Vol. 9, p. 1 (1887).] The proof proceeds along similar lines as the proof of **254**. Notice

$$Q_{n+1}(z) - Q_n(z) = \gamma_n \prod_{\nu=0}^{n}\left(1 - \frac{z}{a_\nu}\right) \sim \gamma_n \prod_{\nu=0}^{\infty}\left(1 - \frac{z}{a_\nu}\right).$$

**254.** [Cf. N. E. Nörlund: Differenzenrechnung. Berlin: Springer 1924, p. 210.] We consider the product $P_n(z)$ defined in **13**. Then we can write

$$Q_{2n+2}(z) - Q_{2n}(z) = (\gamma_n z + \delta_n) P_n(z),$$

where $\gamma_n$ and $\delta_n$ are constants. Let $a$ and $b$ be the two points of convergence. The two series $\sum_{n=0}^{\infty} A_n$ and $\sum_{n=0}^{\infty} B_n$, $(\gamma_n a + \delta_n) P_n(a) = A_n$,

$(\gamma_n b + \delta_n) P_n(b) = B_n$, converge. Since the series

$$\sum_{n=1}^{\infty} \left| \frac{P_n(z)}{P_n(\alpha)} - \frac{P_{n-1}(z)}{P_{n-1}(\alpha)} \right| = \sum_{n=1}^{\infty} \left| \frac{P_{n-1}(z)}{P_{n-1}(\alpha)} \frac{z^2 - \alpha^2}{\alpha^2 - n^2} \right|,$$

converges uniformly $\left(P_n(z) \to \frac{\sin \pi z}{\pi}\right)$ for $\alpha = a$ or $\alpha = b$ and $z$ an arbitrary point in any finite domain of the $z$-plane so do the series

$$\sum_{n=0}^{\infty} A_n \frac{P_n(z)}{P_n(a)}, \qquad \sum_{n=0}^{\infty} B_n \frac{P_n(z)}{P_n(b)}$$

[Knopp, p. 348]. We have however

$$\sum_{n=0}^{\infty} (\gamma_n z + \delta_n) P_n(z) = \frac{z-b}{a-b} \sum_{n=0}^{\infty} A_n \frac{P_n(z)}{P_n(a)} + \frac{z-a}{b-a} \sum_{n=0}^{\infty} B_n \frac{P_n(z)}{P_n(b)}.$$

**255.** According to VI **76** the polynomials are

$$Q_n(z) = c_0 + c_1 z + \frac{c_2 z(z-2)}{2!} + \cdots + \frac{c_n z(z-n)^{n-1}}{n!}.$$

Let $a$ be a point of convergence for the sequence $Q_n(z)$, $a \neq 0$, $\frac{c_n a(a-n)^{n-1}}{n!} = a_n$, $\sum_{n=0}^{\infty} a_n$ convergent. For $n > |z|$, $n > |a|$ the following product can be expanded into a power series:

$$\left(1 - \frac{a}{n}\right)^{-n+1} \left(1 - \frac{z}{n}\right)^{n-1} = e^{a-z}\left(1 + \frac{A'}{n} + \frac{A''}{n^2} + \cdots\right),$$

$A', A'', \ldots$ depend on $a$ and $z$ but not on $n$. The series with the general term

$$\left(1 - \frac{a}{n}\right)^{-n+1} \left(1 - \frac{z}{n}\right)^{n-1} - \left(1 - \frac{a}{n+1}\right)^{-n} \left(1 - \frac{z}{n+1}\right)^{n} = \frac{e^{a-z} A'}{n(n+1)} + \cdots$$

converges absolutely, therefore

$$c_0 + \sum_{n=1}^{\infty} a_n \frac{z}{a} \left(1 - \frac{a}{n}\right)^{-n+1} \left(1 - \frac{z}{n}\right)^{n-1}.$$

[Knopp, p. 348] converges too.

**256.** Cf. **285**. The theorem can be formulated more generally: The functions $f_n(z)$ are regular and different from 0 and the absolute value of each is smaller than 1 in the region $\mathfrak{R}$. If at one point $a$ of $\mathfrak{R}$ we have $\lim_{n\to\infty} f_n(a) = 0$ then $\lim_{n\to\infty} f_n(z) = 0$ everywhere in $\mathfrak{R}$; the convergence is in fact uniform in every closed subdomain of $\mathfrak{R}$. To prove this cover $\mathfrak{R}$ with appropriately overlapping disks.

**257.** [A. Harnack: Math. Ann. Vol. 35, p. 23 (1890).] We denote by $v_n(x, y)$ the harmonic conjugate of $u_n(x, y)$ and put

$$g_n(z) = e^{-u_n(x,y) - iv_n(x,y)}, \quad z = x + iy, \quad \text{and} \quad f_n(z) = g_0(z)\, g_1(z) \cdots g_n(z).$$

Suppose that $\sum_{n=0}^{\infty} u_n(x, y)$ diverges at some point $z_0 = x_0 + iy_0$ of $\mathfrak{R}$. Then we would find $\lim_{n\to\infty} f_n(z_0) = 0$, consequently [solution **256**] $\lim_{n\to\infty} f_n(z) = 0$ everywhere on $\mathfrak{R}$: contradiction.

By using III **285** to a fuller extent one proves first that the convergence is uniform in a closed disk inside $\mathfrak{R}$ and centred at the presupposed point of convergence. The region $\mathfrak{R}$ is then successively covered with appropriately overlapping disks.

**258.** Put $f_n(z) = u_n(x, y) + iv_n(x, y)$. The imaginary part is given by

$$v_n(x, y) = v_n(x_0, y_0) + \int_{x_0,y_0}^{x,y} \frac{\partial u_n(x, y)}{\partial x}\, dy - \frac{\partial u_n(x, y)}{\partial y}\, dx,$$

the path of integration is any curve in $\mathfrak{R}$ that connects the arbitrarily chosen fixed point $x_0, y_0$ with the variable point $x, y$. The integral converges for $n \to \infty$ uniformly in any subdomain of $\mathfrak{R}$ because the sequences of the partial derivatives $\frac{\partial u_n}{\partial x}, \frac{\partial u_n}{\partial y}$ converge in any subdomain of $\mathfrak{R}$ [cf. **230**]. The sequence $v_n(x, y)$ converges therefore if and only if the sequence $v_n(x_0, y_0)$ converges, in which case it converges uniformly in any subdomain of $\mathfrak{R}$.

**259.** Let $a_0, a_1, a_2, \ldots$ denote arbitrary numbers. The identity

$$a_0(1 - a_1) + a_0 a_1(1 - a_2) + a_0 a_1 a_2(1 - a_3) + \cdots$$
$$+ a_0 a_1 \cdots a_{n-1}(1 - a_n) = a_0 - a_0 a_1 a_2 \cdots a_n$$

implies

$$\sum_{n=0}^{\infty} a_0 a_1 a_2 \cdots a_n (1 - a_{n+1}) = a_0 - \lim_{n\to\infty} a_0 a_1 a_2 \cdots a_n,$$

provided the last limit exists. Put

$$a_0 = 1, \quad a_n = \frac{1}{1 + z^{2^{n-1}}}, \qquad n = 1, 2, 3, \ldots \qquad [\text{I } \mathbf{14}].$$

**260.** The power series

$$1 + \sum_{n=1}^{\infty} \frac{\alpha(\alpha + n)^{n-1}}{n!} w^n$$

has the radius of convergence $e^{-1}$ [**214**]; therefore the series in question converges in any connected region of the $x$-plane where $|xe^{-x}| < e^{-1}$ and there it represents an analytic function. The interval $0 \leqq x < 1$ as well as the interval $1 < x < \infty$ can be imbedded in such a region, one in $\mathfrak{R}_1$, the other in $\mathfrak{R}_2$, but the intervals cannot both be imbedded in the same region. According to **210** the infinite sum is equal to $e^{\alpha x}$ if $x$ is sufficiently small, thus it is equal to $e^{\alpha x}$ for $x$ in $\mathfrak{R}_1$. Suppose that $1 < x < \infty$ and that $x'$ is defined by $xe^{-x} = x'e^{-x'}$, $0 < x' < 1$. The series in question stays the same, therefore its sum is $e^{\alpha x'} \neq e^{\alpha x}$.—The series converges also for $x = 1$ [**220**, (3)], its sum is $e^{\alpha}$ according to Abel's theorem [I **86**].

**261.**

$$f_n(z) = \frac{\sum\limits_{\nu=1}^{n}\left(\left[\frac{n}{\nu}\right] - \frac{n}{\nu}\right)\nu^z}{n \sum\limits_{\nu=1}^{n} \nu^{z-1}} + 1.$$

For $\mathfrak{R}z > 0$ we have

$$\lim_{n\to\infty} f_n(z) = \lim_{n\to\infty} \frac{\sum\limits_{\nu=1}^{n}\left(\left[\frac{n}{\nu}\right] - \frac{n}{\nu}\right)\left(\frac{\nu}{n}\right)^z \frac{1}{n}}{\sum\limits_{\nu=1}^{n}\left(\frac{\nu}{n}\right)^{z-1}\frac{1}{n}} + 1 = \frac{\int\limits_0^1\left(\left[\frac{1}{x}\right] - \frac{1}{x}\right)x^z\,dx}{\int\limits_0^1 x^{z-1}\,dx} + 1$$

$$= z(z+1)^{-1}\zeta(z+1) \quad [\text{II } \mathbf{45}].$$

For $\mathfrak{R}z < 0$, $\mathfrak{R}z = x$ we obtain

$$\left|\sum_{\nu=1}^{n}\left(\left[\frac{n}{\nu}\right] - \frac{n}{\nu}\right)\nu^z\right| < \sum_{\nu=1}^{n}\nu^x < \begin{cases} An^{x+1}, & \text{when } x \neq -1, \\ A\log n, & \text{when } x = -1, \end{cases}$$

where $A$ is independent of $n$. Furthermore

$$\lim_{n\to\infty}\sum_{\nu=1}^{n}\nu^{z-1} = 1^{z-1} + 2^{z-1} + \cdots + n^{z-1} + \cdots = \zeta(1-z) \neq 0$$

[VIII **48**]. Hence in this case

$$\lim_{n\to\infty} f_n(z) = 1.$$

**262.** [G. Pólya, Problem: Arch. Math. Phys. Ser. 3, Vol. 25, p. 337 (1917). Solved by H. Prüfer: Arch. Math. Phys. Ser. 3, Vol. 28, pp. 179—180 (1920).] We put

$$\frac{z-1}{z+1} = \zeta, \qquad \frac{\varphi(z)-1}{\varphi(z)+1} = \psi(\zeta),$$

then

$$\frac{\varphi[\varphi(z)]-1}{\varphi[\varphi(z)]+1}=\psi[\psi(\zeta)], \qquad \frac{\varphi\{\varphi[\varphi(z)]\}-1}{\varphi\{\varphi[\varphi(z)]\}+1}=\psi\{\psi[\psi(\zeta)]\}, \ldots,$$

$$\psi(\zeta)=\zeta\frac{\zeta+\alpha-\beta}{1+(\alpha-\beta)\zeta}$$

and the statement becomes: the sequence

$$\psi(\zeta), \quad \psi[\psi(\zeta)], \quad \psi\{\psi[\psi(\zeta)]\}, \ldots$$

converges to 0 if $|\zeta|<1$ and converges to $\infty$ if $|\zeta|>1$, converges to 1 or diverges if $|\zeta|=1$.

Let $|\zeta|=r$, $r<1$ and denote by $M(r)$ the maximum of

$$\left|\frac{\zeta+\alpha-\beta}{1+(\alpha-\beta)\zeta}\right|$$

for $|\zeta|\leqq r$; we find $M(r)<1$ [**5**] and monotone increasing with $r$ [**267**]. The inequalities

$$|\psi(\zeta)|\leqq rM(r), \quad |\psi[\psi(\zeta)]|\leqq|\psi(\zeta)|\,M[rM(r)]\leqq r[M(r)]^2,$$
$$|\psi\{\psi[\psi(\zeta)]\}|\leqq|\psi[\psi(\zeta)]|\,M\{r[M(r)]^2\}\leqq r[M(r)]^3,$$

follow from the definition of $\psi(\zeta)$. We reason analogously if $|\zeta|>1$.

Assume finally $|\zeta|=1$. Then
$|\psi(\zeta)|=|\psi[\psi(\zeta)|=|\psi\{\psi[\psi(\zeta)]\}|=\cdots=1$. If the sequence in question converges to a limit point $\zeta_0$, $|\zeta_0|=1$, we have

$$\zeta_0=\zeta_0\frac{\zeta_0+\alpha-\beta}{1+(\alpha-\beta)\zeta_0}, \quad \text{i.e.} \quad \zeta_0=1.$$

**263.** Put $\frac{(z-1)(z-2)\cdots(z-n)}{n!}=P_n(z)$. We find on the positive imaginary axis

$$z=iy, \; y>0, \; |\sqrt{iy}\,P_n(iy)|^2=y\left(1+\frac{y^2}{1}\right)\left(1+\frac{y^2}{4}\right)\cdots\left(1+\frac{y^2}{n^2}\right)<\frac{\sin i\pi y}{i\pi}$$
$$=\frac{e^{\pi y}-e^{-\pi y}}{2\pi}.$$

For fixed $z$ we have

$$\lim_{n\to\infty}(-1)^n P_n(z)\,n^z=\frac{1}{\Gamma(1-z)}.$$

Consequently the sequence is bounded, e.g. in the half-disk $\Re z\geqq 0$, $|z|\leqq 1$. In the crescent

$$|z-n|\geqq n, \quad |z-n-1|\leqq n+1$$

the absolute value of $P_n(z)$ is larger than the moduli of $P_{n-1}(z)$, $P_{n-2}(z), \ldots, P_{n+1}(z), \ldots$ [**12**]. On the inside border of the crescent, where $|z - n| = n$, $z = 2n \cos \varphi e^{i\varphi}$, we find, if $|z| \geqq 1$, that

$$|\sqrt{z}\, P_n(z)| = \left|\frac{\sqrt{z}\,\Gamma(z)}{\Gamma(z-n)\,\Gamma(n+1)}\right| = \left|\frac{\sqrt{z}\, z^{z-\frac{1}{2}} e^{-z}}{(z-n)^{z-n-\frac{1}{2}} e^{-z+n} n^{n+\frac{1}{2}} e^{-n}}\right| e^{\psi(z,n)}$$
$$= \left|\left(\frac{z}{z-n}\right)^z\right| \left|\frac{z-n}{n}\right|^{n+\frac{1}{2}} e^{\psi(z,n)} = |(2\cos\varphi e^{-i\varphi})^{e^{i\varphi}}|^r e^{\psi(z,n)},$$

where $\psi(z, n)$ remains bounded for all pairs of values $z, n$. [For a proof of Stirling's formula to the extent as it is used here cf. e.g. Whittaker and Watson, pp. 248–253.] On the outside border of the crescent the same estimate holds for $P_{n+1}(z)$ and $P_n(z)$ because $|P_{n+1}(z)| = |P_n(z)|$ on $|z - n - 1| = n + 1$. The modulus of $P_n(z)$ increases as $z$ moves from the inner to the outer border of the crescent on a ray from the origin with slope $\varphi$, $0 \leqq \varphi \leqq \frac{\pi}{2}$, because all the factors $(z - 1), (z - 2), \ldots, (z - n)$ are increasing as can be seen by a geometric consideration.

**264.** [Cf. N. E. Nörlund, l.c. **254**, p. 214.] On the inside border of the crescent cut out by two lemniscates

$$2n^2 \cos 2\varphi \leqq r^2 \leqq 2(n+1)^2 \cos 2\varphi$$

but in the exterior of the unit circle we have the relation

$$\left|z\left(1 - \frac{z^2}{1^2}\right)\left(1 - \frac{z^2}{2^2}\right)\cdots\left(1 - \frac{z^2}{n^2}\right)\right| = \left|\frac{\Gamma(z+n+1)}{\Gamma(z-n)[\Gamma(n+1)]^2}\right|$$
$$= \left|\left(\frac{z+n}{z-n}\right)^z\right| \cdot \left|\frac{z^2 - n^2}{n^2}\right|^{n+\frac{1}{2}} e^{\psi(z,n)} = \left|\left(\frac{e^{i\varphi}\sqrt{2\cos 2\varphi} + 1}{e^{i\varphi}\sqrt{2\cos 2\varphi} - 1}\right)^{e^{i\varphi}}\right|^r e^{\psi(z,n)}$$
$$= |(e^{-i\varphi}\sqrt{2\cos 2\varphi} + e^{-2i\varphi})^{e^{i\varphi}}|^{2r} e^{\psi(z,n)},$$

where $\psi(z, n)$ is bounded [**13**, **263**].

**265.** The maximum of $\left|\left(1 + \frac{z}{n}\right)^{\frac{n}{|z|}}\right|$ along the ray $z = re^{i\varphi}$, $\varphi$ fixed, is independent of $n$; put $z = n\zeta$, $\zeta = \varrho e^{i\varphi}$. The maximum of

$$\frac{1}{|\zeta|} \log |1 + \zeta| = \cos\varphi - \frac{\varrho}{2}\cos 2\varphi + \frac{\varrho^2}{3}\cos 3\varphi - \cdots$$

is obtained by differentiation with respect to $\varrho$ and some subsequent work; it is equal to $\cos\varphi$ and is reached for $\varrho = 0$ if $-\frac{\pi}{4} \leqq \varphi \leqq \frac{\pi}{4}$. If $\frac{\pi}{4} < \varphi < \frac{7\pi}{4}$ the maximum is attained when

$$-\frac{1}{2}\log(\varrho^2 + 2\varrho\cos\varphi + 1) + \frac{\varrho^2 + \varrho\cos\varphi}{\varrho^2 + 2\varrho\cos\varphi + 1}$$
$$= \Re\left(\log\frac{1}{\zeta+1} + 1 - \frac{1}{\zeta+1}\right) = 0,$$

$|\bar{\zeta}+1|>1$ and has the value

$$\frac{1}{|\zeta|}\Re\log(\bar{\zeta}+1)=\frac{1}{|\zeta|}\Re\left(1-\frac{1}{\bar{\zeta}+1}\right).$$

Put $\frac{1}{\bar{\zeta}+1}=w$; $w$ then satisfies the equation $|we^{-w+1}|=1$; $|w|<1$, $\varphi=\arg\frac{1}{\zeta}=\arg\frac{w}{1-w}$ and the maximum in question is

$$=\left|\frac{w}{1-w}\right|\Re(1-w).$$

The discussion of the sign of the derivative can be replaced by the examination of the domain in **116**.—The appearance of convex curves in **263**—**265** is no accident; cf. G. Pólya: Math. Ann. Vol. 89, pp. 179—191 (1923). [Also Math. Z. Vol. 29, pp. 549—640 (1929).]

**266.** [**135**.]

**267.** [**135**.]

**268.** The function $f(z)=f(\zeta^{-1})$ is regular in the open disk $|\zeta|<\frac{1}{R}$ and $M(r)$ is the maximum of $|f(\zeta^{-1})|$ on the circle $|\zeta|=\frac{1}{r}$ [**266**, **267**].

**269.** The function $\frac{f(z)}{z^n}$ is regular in the "punctured" plane $|z|>0$, including the point $z=\infty$ [**268**].

**270.** [S. Bernstein: Communic. Soc. Math. Charkow, Ser. 2, Vol. 14 (1914); M. Riesz: Acta Math. Vol. 40, p. 337 (1916).] Apply **268** to $\zeta^{-n}f\left(\frac{\zeta+\zeta^{-1}}{2}\right)$ [**79**]. We find for $\zeta=r$, $r>1$, $r=a+b$,

$$\left|\frac{f\left(\frac{\zeta+\frac{1}{\zeta}}{2}\right)}{\zeta^n}\right|<\max_{|\zeta|=1}\left|\frac{f\left(\frac{\zeta+\frac{1}{\zeta}}{2}\right)}{\zeta^n}\right|\leqq M.$$

In the case $z\to\infty$ the proposition becomes: the maximum modulus of a polynomial of degree $n$ on the interval $-1\leqq z\leqq 1$ is at least equal to the absolute value of its highest coefficient multiplied by $2^{-n}$.

**271.** We may assume that the axes of $E_1$ and $E_2$ coincide with the axes of the coordinate system and that the foci are $z=\pm 1$. Let the two ellipses be the images of the circles $|z|=r_1$ and $|z|=r_2$, $1<r_1<r_2$, under the mapping $z=\frac{1}{2}\left(\zeta+\frac{1}{\zeta}\right)$, then $r_1=a_1+b_1$, $r_2=a_2+b_2$. Noting **268** we proceed as in **270**.

The extreme case in which $E_1$ degenerates into the twice covered real segment $[-1,1]$ leads to proposition **270**. If the two foci coincide we obtain two circles and the problem is the same as **269**.

**272.** We may assume without loss of generality that $f(0)$ is real and $f(0) > 0$; put $f(z) = f(\varrho e^{i\vartheta}) = U(\varrho, \vartheta) + iV(\varrho, \vartheta)$, then

$$f(0) = \frac{1}{2\pi} \int_0^{2\pi} [U(\varrho, \vartheta) + iV(\varrho, \vartheta)]\, d\vartheta$$

$$= \frac{1}{2\pi} \int_0^{2\pi} U(\varrho, \vartheta)\, d\vartheta \leqq \frac{1}{2\pi} \int_0^{2\pi} [U^2(\varrho, \vartheta) + V^2(\varrho, \vartheta)]^{\frac{1}{2}}\, d\vartheta.$$

The upper bound for $f(0)$ is reached if $V(\varrho, \vartheta) = 0$ for $0 \leqq \vartheta \leqq 2\pi$, hence $f(z) \equiv f(0)$ [**230**].

**273.** [**134**.]

**274.** The function $f(z) = \frac{\varphi(z)}{\psi(z)}$ is regular for $|z| < 1$. It is also regular on the unit circle, where $|f(z)| = 1$ unless $\psi(z) = 0$. If $z_0$ is a zero of $\psi(z)$ it is also a zero of $\varphi(z)$ and with the same multiplicity [otherwise $z = z_0$ were a zero or a pole of $f(z)$ which is impossible: at other points of the unit circle, arbitrarily close to $z_0$, we have $|f(z)| = 1$]. We drop the common factors of $\varphi(z)$ and $\psi(z)$ and so obtain the regular function $f(z)$ which is different from zero in the closed disk $|z| \leqq 1$ and whose modulus is equal to 1 on the unit circle, $|f(z)| = 1$ for $|z| = 1$, therefore [**138**] $f(z) \equiv c$, $|c| = 1$. Since $\varphi(0)$ and $\psi(0)$ are real and positive we have $c = 1$.

**275.** The absolute value $|f(z)|$ is a real continuous function in $\mathfrak{D}$; it assumes therefore its maximum in $\mathfrak{D}$. This is impossible at an inner point if $f(z)$ is not a constant [**134**].

**276.** [Cf. E. Lindelöf: Acta Soc. Sc. Fennicae, Vol. 46, No. 4, p. 6 (1915).] A rotation through $\frac{2\pi\nu}{n}$ around a point $\zeta$ maps the domain onto the domain $\mathfrak{D}_\nu$ and the set $\mathfrak{B}$ onto the set $\mathfrak{B}_\nu$, $\nu = 0, 1, 2, \ldots, n-1$, $\mathfrak{B}_0 = \mathfrak{B}$, $\mathfrak{D}_0 = \mathfrak{D}$. The intersection $\mathfrak{J}$ (largest common subdomain) of the domains $\mathfrak{D}_0, \mathfrak{D}_1, \ldots, \mathfrak{D}_{n-1}$ contains $\zeta$ as an inner point. Those inner points of $\mathfrak{J}$ that can be connected with $\xi$ by a continuous curve in the interior of $\mathfrak{J}$ form a region $\mathfrak{J}^*$. The boundary of $\mathfrak{J}^*$ consists, according to the hypothesis and the construction, of certain points of the sets $\mathfrak{B}_0, \mathfrak{B}_1, \ldots, \mathfrak{B}_{n-1}$. The absolute value of the function $f(\zeta + (z - \zeta)\,\omega^{-\nu})$ is $\leqq A$ at all the boundary points of $\mathfrak{J}^*$ [**275**] and $\leqq a$ at those boundary points of $\mathfrak{J}^*$ that belong to $\mathfrak{B}_\nu$. The absolute value of the function

$$f[\zeta + (z - \zeta)]\, f[\zeta + (z - \zeta)\,\omega^{-1}] \cdots f[\zeta + (z - \zeta)\,\omega^{-n+1}]$$

is therefore not larger than $aA^{n-1}$ at all the boundary points of $\mathfrak{J}^*$, consequently [**275**] also at the inner point $z = \zeta$.

**277.** [Cf. E. Lindelöf, l.c. **276**.] Assume $\alpha < \pi$ without loss of generality because we can consider $f(z^\beta)$, with $\beta$ suitably chosen, instead of $f(z)$. Draw a circle around an arbitrary point of the ray $\arg z = \frac{1}{2}(\alpha - \varepsilon)$ that is tangent to the ray $\arg z = \alpha$. The chord cut by this circle from the real axis is always seen from the center under the same angle. Apply **276** where $\mathfrak{D}$ is identified with the portion of the disk in the upper half-plane and $\mathfrak{B}$ with the chord on the real axis; thus $\lim f(z) = 0$ along $\arg z = \frac{1}{2}(\alpha - \varepsilon)$. By modification of the conclusion we show that the convergence $\lim f(z) = 0$ is uniform in the sector $0 \leqq \arg z \leqq \frac{1}{2}(\alpha - \varepsilon)$. Repeat the argument for the rays

$$\arg z = \frac{3}{4}(\alpha - \varepsilon), \quad \frac{7}{8}(\alpha - \varepsilon), \quad \frac{15}{16}(\alpha - \varepsilon), \ldots$$

**278.** [Cf. E. Lindelöf, l.c. **276**.] Let $R$ denote the least upper bound of $|f(z)|$ in $\mathfrak{R}$. There exists at least one point $P$ in $\mathfrak{R}$ or on the boundary of $\mathfrak{R}$ so that in the intersection of $\mathfrak{R}$ with a sufficiently small disk around $P$ the least upper bound of $|f(z)|$ is equal to $R$.

If there is no such point $P$ in $\mathfrak{R}$ then $|f(z)| < R$ in $\mathfrak{R}$, but then there exists a boundary point $P$ with the required property. According to condition (3) we have $R \leqq M$, i.e. $|f(z)| < M$ in $\mathfrak{R}$.

In case there exists at least one point $P$, $z = z_0$, of the described type, then $|f(z_0)| = R$. Along a sufficiently small circle around $z_0$ we have $|f(z)| \leqq R$, thus, according to **134**, $f(z) \equiv$ constant.

**279.** [P. Fatou: Acta Math. Vol. 30, p. 395 (1905).] Put $\omega = e^{2\pi i/n}$. If $n$ is sufficiently large then

$$\lim_{r \to 1-0} f(z)\, f(\omega z)\, f(\omega^2 z) \cdots f(\omega^{n-1} z) = 0, \qquad z = re^{i\vartheta},$$

the convergence is uniform in the unit disk, $0 \leqq \vartheta \leqq 2\pi$. The function $f(z)\, f(\omega z)\, f(\omega^2 z) \cdots f(\omega^{n-1} z)$ vanishes identically according to **278**. [This proposition is not an immediate consequence of **275**.]

**280.** [H. A. Schwarz: Gesammelte mathematische Abhandlungen, Vol. 2. Berlin: Springer 1890, pp. 110—111.] Apply **278** to the function $\frac{f(z)}{z}$ which is regular in the disk $|z| < 1$.

**281.** [E. Lindelöf: Acta Soc. Sc. Fennicae, Vol. 35, No. 7 (1908). Concerning problems **282—289** cf. P. Koebe: Math. Z. Vol. 6, p. 52 (1920), where also ample bibliography is provided.] We denote by $\zeta = \psi^{-1}(w)$ the inverse function of $w = \psi(\zeta)$. The function

$$F(\zeta) = \psi^{-1}\{f[\varphi(\zeta)]\}$$

satisfies the conditions of **280**. Hence $|F(\zeta)| \leqq \varrho$ for $|\zeta| \leqq \varrho$; there is equality only if $F(\zeta) = e^{i\alpha}\zeta$, $\alpha$ real. The inequality states that the points $F(\zeta)$ lie in the disk $|\zeta| \leqq \varrho$, that means that the points $\psi[F(\zeta)] = f[\varphi(\zeta)]$ lie in the domain $\mathfrak{s}$; $z = \varphi(\zeta)$ represents an arbitrary value in $\mathfrak{r}$. In the extreme case we have $\psi(e^{i\alpha}\zeta) = f[\varphi(\zeta)]$, i.e. $f(z) = \psi[e^{i\alpha}\varphi^{-1}(z)]$, where $\zeta = \varphi^{-1}(z)$ is the inverse function of $z = \varphi(\zeta)$, $\alpha$ is real. This is the most general function that maps $\mathfrak{R}$ univalently onto $\mathfrak{S}$ and $z = z_0$ onto $w = w_0$ [IV **86**].

**282.** [C. Carathéodory: Math. Ann. Vol. 72, p. 107 (1912).] Apply **281** to the following special case: $\mathfrak{R}$: the disk $|z| < 1$; $\mathfrak{S}$: the disk $|w| < 1$; $z_0 = 0$, $w_0 = f(0)$,

$$\varphi(\zeta) = \zeta, \qquad \psi(\zeta) = \frac{\zeta + w_0}{1 + \overline{w}_0\zeta}.$$

The subdomain $\mathfrak{r}$ is the disk $|z| \leqq \varrho$, $\mathfrak{r}$ is the image of $|\zeta| \leqq \varrho$ under the function $w = \psi(\zeta)$, thus $\mathfrak{s}$ is also a disk. The points of $\mathfrak{s}$ satisfy the relation

$$|w - w_0| = \frac{|\zeta|\,(1 - |w_0|^2)}{|1 + \overline{w}_0\zeta|} \leqq \varrho\,\frac{1 - |w_0|^2}{1 - |w_0|\,\varrho}.$$

The inequality becomes an equality only if $f(z) = \psi[e^{i\alpha}\varphi^{-1}(z)] = \psi(e^{i\alpha}z) =$ $= \dfrac{e^{i\alpha}z + w_0}{1 + \overline{w}_0 e^{i\alpha}z}$, in which case $|1 + \overline{w}_0 e^{i\alpha}z| = 1 - |w_0|\,|z|$, i.e. $\arg z = \arg w_0 - \alpha + \pi$.

**283.** Apply **281** to the following special case: $\mathfrak{R}$: disk $|z| < R$; $\mathfrak{S}$: half-plane $\Re w < A(R)$; $z_0 = 0$, $w_0 = f(0)$, $\Re w_0 = A(0)$,

$$\varphi(\zeta) = R\zeta,\ \psi(\zeta) = \frac{w_0 + [\overline{w}_0 - 2A(R)]\,\zeta}{1 - \zeta} = w_0 + [w_0 + \overline{w}_0 - 2A(R)]\,\frac{\zeta}{1 - \zeta}.$$

$\mathfrak{r}$ is the disk $|z| \leqq \varrho R = r$, $\mathfrak{s}$ is the image of $|\zeta| \leqq \varrho$ under the mapping $w = \psi(\zeta)$. The points of $\mathfrak{s}$ satisfy the inequality

$$\Re w = \Re w_0 + [w_0 + \overline{w}_0 - 2A(R)]\,\Re\frac{\zeta}{1 - \zeta} \leqq \Re w_0 - 2[\Re w_0 - A(R)]\,\frac{\varrho}{1 + \varrho}$$
$$= \frac{1 - \varrho}{1 + \varrho}\,\Re w_0 + \frac{2\varrho}{1 + \varrho}\,A(R).$$

It is an equality only if $f(z) = \psi[e^{i\alpha}\varphi^{-1}(z)] = \psi\left(e^{i\alpha}\frac{z}{R}\right)$.

**284.** The following relation [solution **283**] holds

$$|w| \leqq |w_0| + [2A(R) - w_0 - \overline{w}_0]\,\frac{\varrho}{1 - \varrho} = M(0) + \frac{2\varrho}{1 - \varrho}\,[A(R) - A(0)],$$

which is a weaker statement than **236**.

**285.** Apply **283** to $\log f(z)$:

$$\Re \log f(z) = \log |f(z)| \leqq \log M(r).$$

**286.** Proposition **285** implies

$$|f_n(z)|^2 \leqq |f_n(0)|^{2\frac{1-|z|}{1+|z|}}.$$

For $|z| \leqq \frac{1}{3}$ the exponent is $2\frac{1-|z|}{1+|z|} \geqq 1$, consequently $|f_n(z)|^2 \leqq |f_n(0)|$.

**287.** Use **281**: $\mathfrak{R}$: the disk $|z| < 1$; $\mathfrak{S}$: the half-plane $\Re w > 0$; $z_0 = 0$, $w_0 = f(0) > 0$,

$$\varphi(\zeta) = \zeta, \quad \psi(\zeta) = w_0 \frac{1+\zeta}{1-\zeta}.$$

$\mathfrak{r}$: the disk $|z| \leqq \varrho$, $\mathfrak{s}$: the disk whose boundary circle intersects the real axis orthogonally at the points $w_0 \frac{1+\varrho}{1-\varrho}$ and $w_0 \frac{1-\varrho}{1+\varrho}$. The radius of this circle is $w_0 \frac{2\varrho}{1-\varrho^2}$. The points of $\mathfrak{s}$ satisfy the inequalities

$$w_0 \frac{1-\varrho}{1+\varrho} \leqq \Re w \leqq w_0 \frac{1+\varrho}{1-\varrho}, \qquad |\Im w| \leqq w_0 \frac{2\varrho}{1-\varrho^2},$$
$$w_0 \frac{1-\varrho}{1+\varrho} \leqq |w| \leqq w_0 \frac{1+\varrho}{1-\varrho}.$$

We have equality only in the case where $f(z) = \psi[e^{i\alpha}\varphi^{-1}(z)] = \psi(e^{i\alpha}z)$, $\alpha$ real.

**288.** Special case of **281**: $\mathfrak{R}$: the disk $|z| < 1$, $\mathfrak{S}$: the vertical strip $-1 < \Re w < 1$, $z_0 = 0$, $w_0 = 0$,

$$\varphi(\zeta) = \zeta, \quad \psi(\zeta) = \frac{2}{i\pi} \log \frac{1+\zeta}{1-\zeta};$$

$\mathfrak{r}$ is the disk $|z| \leqq \varrho$. The points $|\zeta| \leqq \varrho$ are mapped by $\frac{1+\zeta}{1-\zeta}$ onto the disk that intersects the real axis orthogonally at the points $\frac{1+\varrho}{1-\varrho}$ and $\frac{1-\varrho}{1+\varrho}$. This disk lies completely in the sector whose bisector is the real axis and whose center angle is $2 \arctan \frac{2\varrho}{1-\varrho^2} = 4 \arctan \varrho$. Therefore $\mathfrak{s}$ lies completely in the strip $|\Re w| \leqq \frac{4}{\pi} \arctan \varrho$. Besides we have in $\mathfrak{s}$

$$|\Im w| = \frac{2}{\pi} \log \left|\frac{1+\zeta}{1-\zeta}\right| \leqq \frac{2}{\pi} \log \frac{1+\varrho}{1-\varrho}.$$

Equality is attained only for $f(z) = \psi[e^{i\alpha}\varphi^{-1}(z)] = \psi(e^{i\alpha}z)$, $\alpha$ real.

**289.** We may assume that $R = 1$, $\Delta = 2$, and $|\Re f(z)| < 1$ for $|z| < 1$ as in **288**; $f(0) = w_0$ however is arbitrary in the strip $-1 < \Re w < 1$. Depending on the choice of $w_0$ each disk $|z| \leqq \varrho < 1$ is mapped onto a domain that contains the range of $f(z)$. We have to consider the maximum width of these domains in the direction of the real and the imaginary axis respectively while $w_0$ moves over the entire strip. It is obviously sufficient to examine only real $w_0$-values, $-1 < w_0 < 1$. In this case

$$w = \psi(\zeta) = \frac{2}{i\pi}\log\frac{e^{\frac{i\pi w_0}{2}} + i\zeta}{1 - ie^{\frac{i\pi w_0}{2}}\zeta} = w_0 + \frac{2}{i\pi}\log\frac{1 + ie^{-\frac{i\pi w_0}{2}}\zeta}{1 - ie^{\frac{i\pi w_0}{2}}\zeta}.$$

The image of $|\zeta| = \varrho$ in the $w$-plane is convex [**318**], moreover it is symmetric with respect to the real axis because $\psi(\zeta)$ is real for real $\zeta$-values. Thus $\Re w$ assumes its maximum and its minimum for real $\zeta = \pm\varrho$. The width in the horizontal direction is therefore

$$w_0 + \frac{2}{i\pi}\log\frac{1 + ie^{-\frac{i\pi w_0}{2}}\varrho}{1 - ie^{\frac{i\pi w_0}{2}}\varrho} - \left(w_0 + \frac{2}{i\pi}\log\frac{1 - ie^{-\frac{i\pi w_0}{2}}\varrho}{1 + ie^{\frac{i\pi w_0}{2}}\varrho}\right)$$

$$= \frac{4}{\pi}\left(\arctan\frac{\varrho\cos\frac{\pi w_0}{2}}{1 + \varrho\sin\frac{\pi w_0}{2}} + \arctan\frac{\varrho\cos\frac{\pi w_0}{2}}{1 - \varrho\sin\frac{\pi w_0}{2}}\right).$$

As $w_0$ varies between $-1$ and $+1$ the derivative with respect to $w_0$ has always the same sign as $-\sin\frac{\pi w_0}{2}$. The maximum is equal to $\frac{8}{\pi}\arctan\varrho$ and it is reached for $w_0 = 0$.

The oscillation of $\Im w$ cannot be larger than twice the least upper bound of $|\Im w|$ given in **288**; proof similar.

**290.** [H. Bohr: Nyt. Tidsskr. Mat. (B) Vol. 27, pp. 73–78 (1916).] We choose $\eta$ so that $|\eta| = 1$, $\eta F(1) > 0$. Picking out the branch of $\log \eta F(z)$ that is real for $z = 1$ we set

$$w = f(z) = \frac{\log \eta F(z) - \log c}{\log \eta F(1) - \log c}.$$

We have $f(1) = 1$, $\Re f(z) > 0$. Apply **281** by identifying $\mathfrak{R}$ with $\mathfrak{T}$, $\mathfrak{S}$ with the right half-plane, $z_0$ with 1, $w_0$ with 1. The functions $z = \varphi(\zeta)$, $w = \psi(\zeta)$ are supposed to be normed in such a way that they transform real $\zeta$-values into real $z$- and $w$-values; in addition let $\varphi(1) = \infty$, $\psi(1) = \infty$ [IV **119**]. Hence

$$\psi(\zeta) = \frac{1 + \zeta}{1 - \zeta}.$$

Assume $x > 1$ and $x = \varphi(\varrho)$, $0 < \varrho < 1$. According to **281** the range of $f(z)$ in the $w$-plane is contained in the image of the disk $|\zeta| \leqq \varrho$ under the mapping $w = \psi(\zeta)$. The image of the disk is a disk whose points are at a distance of at most $\psi(\varrho)$ from the origin; i.e. $|f(x)| \leqq \psi(\varrho)$. Hence we can define $h(x) = \psi[\varphi^{-1}(x)]$.

**291.** [K. Löwner.] According to **280** we have $|f(z)| \leqq |z|$ for $|z| < 1$; hence for positive $z$, $0 < z < 1$,

$$\left|\frac{1 - f(z)}{1 - z}\right| \geqq 1.$$

By taking the limit $z \to 1$ we find

$$|f'(1)| \geqq 1.$$

Assume that $\arg f'(1) = \alpha$. A sufficiently small vector with the direction $\vartheta$, $\frac{\pi}{2} < \vartheta < \frac{3\pi}{2}$, attached to $z = 1$, points towards the inside of the unit circle and is mapped by the function $w = f(z)$, $f'(1) \neq 0$, onto an analytic curve segment through the point $w = 1$ with the direction $\vartheta + \alpha$. Since $|f(z)| < 1$ for $|z| < 1$ the direction is restricted by the condition

$$\frac{\pi}{2} < \vartheta + \alpha < \frac{3\pi}{2}$$

for any admissible $\vartheta$. This is possible for $\alpha = 0$ only. Cf. **144**.

**292.** Let $z_0$ be fixed, $|z_0| < 1$, $f(z_0) = w_0$, $|w_0| < 1$. Choose the constants $\varepsilon$ and $\eta$ so that $\varepsilon \frac{1 - z_0}{1 - \bar{z}_0} = \eta \frac{1 - w_0}{1 - \bar{w}_0} = 1$, $|\varepsilon| = |\eta| = 1$. Put

$$\varepsilon \frac{z - z_0}{1 - \bar{z}_0 z} = Z, \qquad \eta \frac{w - w_0}{1 - \bar{w}_0 w} = W.$$

The function $W = F(Z)$ defined with the help of $w = f(z)$ satisfies the hypothesis of **291**. Therefore

$$1 \leqq F'(1) = \left(\frac{dW}{dw}\right)_{w=1} \left(\frac{dw}{dz}\right)_{z=1} \left(\frac{dz}{dZ}\right)_{Z=1} = \frac{\eta(1 - |w_0|^2)}{(1 - \bar{w}_0)^2} f'(1) \frac{(1 - \bar{z}_0)^2}{\varepsilon(1 - |z_0|^2)}.$$

**293.** [G. Julia: Acta Math. Vol. 42, p. 349 (1920).] Let $z_0$ be fixed, $\Im z_0 > 0$, $f(z_0) = w_0$, $\Im w_0 > 0$. Choose the constants $\varepsilon$ and $\eta$ so that $\varepsilon \frac{a - z_0}{a - \bar{z}_0} = \eta \frac{b - w_0}{b - \bar{w}_0} = 1$, $|\varepsilon| = |\eta| = 1$. Define

$$\varepsilon \frac{z - z_0}{z - \bar{z}_0} = Z, \qquad \eta \frac{w - w_0}{w - \bar{w}_0} = W.$$

Then [**291**, **292**]

$$1 \leqq \left(\frac{dW}{dZ}\right)_{Z=1} = \left(\frac{dW}{dw}\right)_{w=b} f'(a) \left(\frac{dz}{dZ}\right)_{Z=1} = \frac{\eta(w_0 - \bar{w}_0)}{(b - \bar{w}_0)^2} f'(a) \frac{(a - \bar{z}_0)^2}{\varepsilon(z_0 - \bar{z}_0)}.$$

**294.** The function $f(z) \frac{1 - \bar{z}_1 z}{z - z_1} \cdot \frac{1 - \bar{z}_2 z}{z - z_2} \cdots \frac{1 - \bar{z}_n z}{z - z_n}$ is regular in the disk $|z| < 1$, its absolute value is smaller than $M + \varepsilon$, $\varepsilon > 0$, sufficiently close to any boundary point [**5**]. Apply **278**. – Different proof by careful application of **176**.

**295.** The function $f(z)\frac{\bar{z}_1+z}{z_1-z}\cdot\frac{\bar{z}_2+z}{z_2-z}\cdots\frac{\bar{z}_n+z}{z_n-z}$ is regular in the half-plane $\Re z>0$; its absolute value is smaller than $M+\varepsilon$, $\varepsilon>0$, sufficiently close to any boundary point [**6**]. Apply **278**.—A different proof is based on **177**. In fact, both methods go beyond the particular case of the half-plane $\Re z>0$; both can be easily adapted to a generalized proposition which relates to **294** as **281** relates to Schwarz's lemma **280**.

**296.** The function $f(z)$ is assumed to be meromorphic with the zeros $a_1, a_2, \ldots, a_m$ and the poles $b_1, b_2, \ldots, b_n$ (counted with proper multiplicity) in the disk $|z|\leqq 1$ and $|f(z)|=c>0$ for $|z|=1$. The function

$$f(z)\prod_{\mu=1}^{m}\frac{1-\bar{a}_\mu z}{z-a_\mu}\prod_{\nu=1}^{n}\frac{z-b_\nu}{1-\bar{b}_\nu z}=\varphi(z)$$

is regular and non-zero for $|z|<1$, its modulus is constant, equal to $c$, for $|z|=1$. Therefore $\varphi(z)$ is a constant [**142**].

**297.** [W. Blaschke: Sber. Naturf. Ges. Lpz. Vol. 67, p. 194 (1915).] Let $\alpha$ be real, $0\leqq\alpha<1$, $f(\alpha)\neq 0$. **294** implies that the product $\prod_{\nu=1}^{\infty}\left|\frac{z_\nu-\alpha}{1-\alpha z_\nu}\right|$ does not diverge to 0. Hence the series

$$\sum_{\nu=1}^{\infty}\left(1-\left|\frac{\alpha-z_\nu}{1-\alpha z_\nu}\right|^2\right)=\sum_{\nu=1}^{\infty}\frac{(1-\alpha^2)(1+|z_\nu|)}{|1-\alpha z_\nu|^2}(1-|z_\nu|)$$

$$\geqq\frac{1-\alpha}{1+\alpha}\sum_{\nu=1}^{\infty}(1-|z_\nu|)$$

converges.

**298.** Let $\alpha$ be real, $\alpha>1$, $f(\alpha)\neq 0$. Proposition **295** implies that the product $\prod_{\nu=1}^{\infty}\left|\frac{z_\nu-\alpha}{z_\nu+\alpha}\right|$ does not diverge to 0. Consequently the series

$$\sum_{\nu=1}^{\infty}\left(1-\left|\frac{z_\nu-\alpha}{z_\nu+\alpha}\right|^2\right)=\sum_{\nu=1}^{\infty}\frac{4\alpha}{\left|1+\frac{\alpha}{z_\nu}\right|^2}\frac{z_\nu+\bar{z}_\nu}{2z_\nu\bar{z}_\nu}\geqq\frac{4\alpha}{(1+\alpha)^2}\sum_{\nu=1}^{\infty}\Re\frac{1}{z_\nu}$$

converges.

**299.** [T. Carleman; cf. also P. Csillag: Mat. phys. lap. Vol. 26, pp. 74–80 (1917).] Let $z_0$ be an inner point of $\mathfrak{D}$ and put $|f_\nu(z_0)|=\varepsilon_\nu f_\nu(z_0)$, $\nu=1, 2, \ldots, n$. (In the case $f_\nu(z_0)=0$ we choose $\varepsilon_\nu=1$.) The function

$$F(z)=\varepsilon_1 f_1(z)+\varepsilon_2 f_2(z)+\cdots+\varepsilon_n f_n(z)$$

is regular and single-valued in $\mathfrak{D}$; $F(z)$ assumes its largest absolute value at a boundary point $z_1$ of $\mathfrak{D}$. Hence

$$\varphi(z_1)\geqq|F(z_1)|\geqq|F(z_0)|=\varphi(z_0).$$

**300.** It is more convenient to prove the statement as follows than to refine the proof of **299**: Let $z_0$ be an inner point of $\mathfrak{D}$ and the disk $|z - z_0| \leqq r$ be inside $\mathfrak{D}$. Addition of the inequalities [**272**]

$$(*) \qquad |f_\nu(z_0)| \leqq \frac{1}{2\pi}\int_0^{2\pi} |f_\nu(z_0 + re^{i\vartheta})|\, d\vartheta, \qquad \nu = 1, 2, \ldots, n$$

yields

$$\varphi(z_0) \leqq \frac{1}{2\pi}\int_0^{2\pi} \varphi(z_0 + re^{i\vartheta})\, d\vartheta.$$

If one of the summands (*) satisfies a strict inequality the sum satisfies a strict inequality; i.e. if at least one of the $f_\nu$'s is not a constant the maximum cannot be attained at an inner point $z_0$.

**301.** [G. Szegö, Problem: Jber. deutsch. Math. Verein. Vol. 32, 2. Abt., p. 16 (1923).] It is sufficient to prove that $\varphi(P)$ attains its maximum on the boundary of any *plane* domain $\mathfrak{D}$ whereby the points $P_1, P_2, \ldots, P_n$ are not necessarily in the same plane. We introduce an orthogonal coordinate system in $\mathfrak{D}$, $x$, $y$; $z = x + iy$. Now we have to show: A function of the form

$$\prod_{\nu=1}^{n} (|z - a_\nu|^2 + b_\nu^2)|,$$

where $a_\nu$ are arbitrary complex, and $b_\nu$ real, constants, $\nu = 1, 2, \ldots, n$, attains its maximum on the boundary of any domain of the $z$-plane. Work out the product [**299**].

**302.** Let $z_0$ be an inner point of $\mathfrak{D}$, where some of the functions, called $f_\mu(z)$, do not vanish and other functions, called $f_\nu(z)$, do vanish. (One or the other type may be absent.) Let the radius $r$ be so small that the disk $|z - z_0| \leqq r$ lies completely inside $\mathfrak{D}$ and does not contain any other zero of the functions $f_\nu(z)$ but $z_0$. The functions $f_\mu(z)^{p_\mu}$ are regular in this disk. According to **299** there exists a point $z_1$, $|z_1 - z_0| = r$ such that

$$\sum_\mu |f_\mu(z_1)|^{p_\mu} \geqq \sum_\mu |f_\mu(z_0)|^{p_\mu}.$$

Obviously

$$\sum_\nu |f_\nu(z_1)|^{p_\nu} \geqq 0 = \sum_\nu |f_\nu(z_0)|^{p_\nu},$$

i.e. $\varphi(z_1) \geqq \varphi(z_0)$; in fact $\varphi(z_1) > \varphi(z_0)$ if at least one $f_\mu(z)$ is not a constant or, in the other case, if at least one $f_\nu(z)$ does not vanish identically. Since $\mathfrak{D}$ is closed and $\varphi(z)$ continuous there exists a point in $\mathfrak{D}$ where the function $\varphi(z)$ assumes its maximum: it cannot be an inner point except in the particular case mentioned in the problem.

**303.** Since $\mathfrak{D}$ is closed the function $|f(z)|$, which is single-valued and continuous, attains its maximum in $\mathfrak{D}$. Proposition **134** shows that this cannot happen at an inner point of $\mathfrak{D}$ except in the case where $f(z)$ is a constant.

**304.** [J. Hadamard: Bull. Soc. Math. France Vol. 24, p. 186 (1896); O. Blumenthal: Jber. deutsch. Math. Verein. Vol. 16, p. 108 (1907); G. Faber: Math. Ann. Vol. 63, p. 549 (1907). Hurwitz-Courant, pp. 429—430; E. Hille, Vol. II, pp. 410—411.] The function $z^\alpha f(z)$ is not single-valued in the annulus $r_1 \leqq |z| \leqq r_3$, its modulus however is. Hence the maximum of $|z^\alpha f(z)|$ is either $r_1^\alpha M(r_1)$ or $r_3^\alpha M(r_3)$ [**303**]. Choose $\alpha$ so that

(*) $$r_1^\alpha M(r_1) = r_3^\alpha M(r_3).$$

Considering a specific point on the circle $|z| = r_2$ we see that

$$r_2^\alpha M(r_2) \leqq r_1^\alpha M(r_1) = r_3^\alpha M(r_3).$$

We introduce the value $\alpha$ from (*). (The condition that $f(z)$ be regular and $|f(z)|$ single-valued on the punctured disk $0 < |z| < R$ is sufficient.)

**305.** The maximum of $z^\alpha f(z)$ is reached at a point of the circle $|z| = r_2$, i.e. in the interior of the annulus $r_1 \leqq |z| \leqq r_3$, only if $z^\alpha f(z)$ is a constant.

**306.** Put $f(z) = a_0 + a_1 z + a_2 z^2 + \cdots + a_n z^n + \cdots$. The integral $I_2(r)$ becomes

$$I_2(r) = |a_0|^2 + |a_1|^2 r^2 + |a_2|^2 r^4 + \cdots + |a_n|^2 r^{2n} + \cdots = \sum_{n=0}^{\infty} p_n r^n,$$

where $p_n \geqq 0$ and at least two $p_n$'s are non-zero [II **123**].

**307.** Assume that $f(z)$ is not a constant and that none of the zeros $z_1, z_2, \ldots, z_n$ of $f(z)$ in the disk $|z| \leqq r$ coincides with $z = 0$ (for simplicity's sake assume also $f(0) = 1$). Then we have [**120**]

$$\log \mathfrak{G}(r) = n \log r - \log|z_1| - \log|z_2| - \cdots - \log|z_n|.$$

Hence the graph of $\log \mathfrak{G}(r)$, as a function of $\log r$, consists in a sequence of straight pieces with monotone increasing slopes. The change of slope for $\log r = \log r_0$ is caused by the appearance of additional zeros on the circle $|z| = r_0$. The increase in slope is equal to the number of such zeros counted with appropriate multiplicity.

**308.** [G. H. Hardy: Proc. Lond. Math. Soc. Ser. 2, Vol. 14, p. 270 (1915).] Suppose $0 < r_1 < r_2 < r_3 < R$. Define the functions $\varepsilon(\vartheta)$, $F(z)$ by the relations

$$\varepsilon(\vartheta)\, f(r_2 e^{i\vartheta}) = |f(r_2 e^{i\vartheta})|, \quad 0 \leqq \vartheta \leqq 2\pi, \quad F(z) = \frac{1}{2\pi} \int_0^{2\pi} f(z e^{i\vartheta})\, \varepsilon(\vartheta)\, d\vartheta.$$

The function $F(z)$ is regular in the disk $|z| \leqq r_3$ and its absolute value reaches the maximum on the boundary, say, at the point $r_3 e^{i\vartheta_3}$. Hence

$$I(r_2) = F(r_2) \leqq |F(r_3 e^{i\vartheta_3})| \leqq I(r_3),$$

that means $I(r)$ is not decreasing. Determine the real number $\alpha$ by the equation

$$r_1^\alpha I(r_1) = r_3^\alpha I(r_3).$$

The absolute value of the function $z^\alpha F(z)$, which is regular in the annulus $r_1 \leqq |z| \leqq r_3$, is single-valued. Hence [**303**]

$$r_2^\alpha I(r_2) = r_2^\alpha F(r_2) \leqq \max_{r_1 \leqq |z| \leqq r_3} |z^\alpha F(z)| \leqq r_1^\alpha I(r_1) = r_3^\alpha I(r_3),$$

from which the convexity property of $I(r)$ follows [**304**].

**309.** $l(r) = \int_0^{2\pi} |f'(re^{i\vartheta})| \, r \, d\vartheta$ [**308**].

**310.** Define $e^{2\pi i\nu/n} = \omega_\nu$, $\nu = 1, 2, \ldots, n$. Let $0 \leqq r_1 < r_2 < R$. There exists [**302**] a point $r_2 e^{i\vartheta_2}$, on the circle $|z| = r_2$ such that

$$\frac{1}{n} \sum_{\nu=1}^{n} |f(r_1 \omega_\nu)|^p \leqq \frac{1}{n} \sum_{\nu=1}^{n} |f(r_2 \omega_\nu e^{i\vartheta_2})|^p.$$

As $n \to \infty$ this inequality becomes

$$I_p(r_1) \leqq I_p(r_2).$$

Assume $0 < r_1 < r_2 < r_3 < R$, $\alpha$ real. The functions

$$z^{\frac{\alpha}{p}} f(\omega_1 z), \quad z^{\frac{\alpha}{p}} f(\omega_2 z), \ldots, \quad z^{\frac{\alpha}{p}} f(\omega_n z)$$

are regular in the annulus $r_1 \leqq |z| \leqq r_3$, however, only their moduli are necessarily single-valued. All the same [**303, 302**] we may conclude that the sum of the $p$-th power of their absolute values assumes its maximum on the boundary of the annulus. Applying the same arguments as in **304**, **308** and taking the limit we establish the behaviour of $I_p(r)$ with regard to convexity.—Cf. II **83** for the limit cases $p = 0$ and $p = \infty$.

**311.** We may assume that the center of $\mathfrak{K}$ is at the origin. Apply **230**. The proposition states in other words: A harmonic function that is regular in a closed disk and that vanishes on the bounding circle vanishes identically.

**312.** We denote by $u(x, y)$, $z = x + iy$, a harmonic function that is regular in the disk $(x - x_0)^2 + (y - y_0)^2 \leqq r^2$. The value of $u(x, y)$ at the center is

$$u(x_0, y_0) = \frac{1}{2\pi} \int_0^{2\pi} u(x_0 + r\cos\vartheta, y_0 + r\sin\vartheta) \, d\vartheta \quad [\mathbf{118}],$$

consequently

$$|u(x_0, y_0)| \leqq \frac{1}{2\pi} \int_0^{2\pi} |u(x_0 + r\cos\vartheta, y_0 + r\sin\vartheta)|\, d\vartheta.$$

The inequality becomes an equality if

$$\frac{1}{2\pi} \int_0^{2\pi} [|u(x_0 + r\cos\vartheta, y_0 + r\sin\vartheta)| \pm u(x_0 + r\cos\vartheta, y_0 + r\sin\vartheta)]\, d\vartheta = 0,$$

where the sign depends on whether $u(x_0, y_0) \geqq 0$ or $u(x_0, y_0) \leqq 0$. The integrand must vanish identically, i.e. $u(x, y)$ cannot change sign on the given circle ($u(x, y)$ possibly becomes 0 at some points).

**313.** Suppose that the point $x_0, y_0$ at which the maximum is reached is an interior point of $\mathfrak{D}$. Choose $r$ so small that the disk with radius $r$ and center $x_0, y_0$ lies in the interior of $\mathfrak{D}$. The equation

$$\frac{1}{2\pi} \int_0^{2\pi} [u(x_0, y_0) - u(x_0 + r\cos\vartheta, y_0 + r\sin\vartheta)]\, d\vartheta = 0 \qquad \text{[solution \textbf{312}]}$$

implies

$$u(x_0, y_0) - u(x_0 + r\cos\vartheta, y_0 + r\sin\vartheta) = 0, \quad 0 \leqq \vartheta \leqq 2\pi,$$

i.e. [**311**] $u(x, y) \equiv$ const.

**314.** Follows from **313**.

**315.** $\log|z - z_1| + \log|z - z_2| + \cdots + \log|z - z_n| = \Re \log P(z)$, which is the potential of the system of forces in question, is a harmonic function and as such it does not have a maximum nor a minimum at a regular point. Stability would require a minimum of potential energy.

**316.** Remove the finitely many exceptional points from $\mathfrak{D}$ by enclosing them in circles so small that these disks have no points in common and do not contain any point at which the function reaches its maximum in $\mathfrak{D}$. Apply **313** to the remaining domain.

**317.** According to **188** the orientation of the image of the circle $|z| = R$ is preserved. Therefore **109** can be applied. The harmonic function

$$\Re z \frac{f'(z)}{f(z)}$$

which is regular in the disk $|z| \leqq R$ [$f(z)$ being schlicht has the only and simple zero $z = 0$] is positive for $|z| = R$. Hence it is positive on any smaller concentric circle $|z| = r < R$ [**313**]. The images of the circles $|z| = r$ are star-shaped with respect to the origin according to **109**.

**318.** According to **188** the orientation of the image of the circle $|z| = R$ is preserved. Thus we can apply **108**. The harmonic function

$$\Re z \frac{f''(z)}{f'(z)} + 1,$$

which is regular on the disk $|z| \leqq R$ [$f'(z) \neq 0$ because $f(z)$ is schlicht] is positive for $|z| = R$. Hence it is positive on any smaller concentric circle $|z| = r < R$. According to **108** the images of the circles $|z| = r$ are convex. The statement is now proved in the case where the inner circle is concentric to the disk $|z| < R$.

Let the inner circle lie anywhere inside the disk $|z| < R$. We build up $w = f(z)$, the given function, by combining two functions: a linear mapping $\xi = l(z)$ of the disk $|z| \leqq R$ onto itself whereby the given inner circle is transformed into one with center at the origin $\xi = 0$ [**77**] and a second one $w = g(\xi) = f(l^{-1}(\xi))$; in fact, $w = f(z) = g(l(z))$. By considering $w = g(\xi)$ we reduce the problem to the previously discussed special case.

**319.** Proof analogous to the proof of **299**.

**320.** [Cf. A. Walther: Math. Z. Vol. 11, p. 158 (1921).] Let $u(x, y)$, $z = x + iy$, be the harmonic function in question. The function $u(x, y) + \alpha \log r$, $\alpha$ arbitrary real constant, is regular in the annulus $r_1 \leqq |z| \leqq r_3$. Its maximum there is either $A(r_1) + \alpha \log r_1$ or $A(r_3) + \alpha \log r_3$. Define $\alpha$ so that

$$A(r_1) + \alpha \log r_1 = A(r_3) + \alpha \log r_3 \qquad [\mathbf{313, 304}].$$

**321.** [Cf. A. Walther, l.c. **320**.] (1) The three circle theorem proved in **320** holds also for harmonic functions that have in the disk $|z| < R$ finitely or infinitely many isolated singularities provided their accumulation points are not interior points of the disk and that, in addition, the function tends to $-\infty$ as $z$ approaches a singular point [**316**]. Apply this generalized three circle theorem to the function

$$\Re \log f(z) = \log |f(z)|.$$

(2) Let $u(x, y)$, $z = x + iy$, be a regular harmonic function in the disk $|z| < R$ and $v(x, y)$ denote the conjugate harmonic. Now apply **304** to the function $f(z) = e^{u(x,y) + iv(x,y)}$; $|f(z)| = e^{u(x,y)}$, hence $\log M(r) = A(r)$.

**322.** [Concerning the method used and the problems **322–340** cf. in the first place E. Phragmén and E. Lindelöf: Acta Math. Vol. 31, p. 386 (1908); then P. Persson: Thèse (Uppsala 1908) and E. Lindelöf: Rend. Circ. Mat. Palermo Vol. 25, p. 228 (1908); also E. Hille, Vol. II, pp. 393–397.]

**323.** The arcs may be replaced by any continuous curves that cross the sector connecting the two rays. Their minimal distance from the origin must increase beyond all bounds.

**324.** All the estimates of solution **322** remain valid in the domain bounded by $\Gamma_1$ and $\Gamma_2$.

**325.** Cf. solution p. 168.

**326.** The function $e^{\omega z}f(z)$ satisfies the conditions (1), (2) and the modified condition (3) of **325** [final remark in the proof] however large $\omega$. Consequently

$$|f(z)| \leqq e^{-\omega x} \text{ for } \Re z = x \geqq 0.$$

Let $\omega$ increase to infinity.

**327.** Put $\arctan \frac{y}{x+1} = \psi$. Then

$$z \log(z+1) = (r\cos\vartheta + ir\sin\vartheta)\,[\tfrac{1}{2}\log(r^2 + 2r\cos\vartheta + 1) + i\psi],$$

hence for $-\frac{\pi}{2} \leqq \vartheta \leqq \frac{\pi}{2}$, $r > 1$,

$$\Re[-z\log(1+z)] = r\psi\sin\vartheta - \tfrac{1}{2}r\log(r^2 + 2r\cos\vartheta + 1)\cdot\cos\vartheta$$
$$\leqq r\frac{\pi}{2} - r\log r\,\cos\vartheta \leqq r\frac{\pi}{2}.$$

Let $0 < \beta < \frac{2\gamma}{\pi}$. The function

$$\frac{1}{C} f(z)\, e^{-\beta z\log(z+1)}$$

satisfies the conditions (1), (2) and (3) of **326** with $\alpha = 0$.—Instead of quoting **326**, **325** we could apply the ideas developed in those solutions to the function

$$f(z)\exp\left(\omega z - \beta z\log(z+1) - \varepsilon e^{-i\lambda\pi/4} z^{\lambda}\right).$$

Then we set $\varepsilon = 0$ and finally $\omega = +\infty$.

**328.** [F. Carlson: Math. Z. Vol. 11, p. 14 (1921); Thèse, Uppsala 1914.] First solution. **327** can be applied to $\frac{f(z)}{\sin\pi z}$. The existence of an inequality

$$\left|\frac{f(z)}{\sin\pi z}\right| < A' e^{B|z|}$$

(with $A' > A$) is best established first outside and then inside the circles $|z - n| = \frac{1}{2}$, $n = 0, 1, 2, \ldots$ [solution **165**].

Second solution: **178** and the fact that the terms on the left hand side are positive imply

$$\sum_{\mu=1}^{n}\left(\frac{1}{\mu}-\frac{\mu}{n^2}\right)\leq\frac{1}{2\pi}\int_1^n\left(\frac{1}{\varrho^2}-\frac{1}{n^2}\right)[2\log C+2(\pi-\gamma)\,\varrho]\,d\varrho+C',$$

where $C$ and $C'$ denote constants. The left hand side is $\sim\log n$, the right hand side $\sim\frac{\pi-\gamma}{\pi}\log n$: contradiction. This method can be generalized [F. and R. Nevanlinna, l.c. **177**].

**329.** Assume $\varepsilon>0$; the function $\varphi(z)=\frac{e^z}{[f(z)]^\varepsilon}$ is regular in the half-plane $\Re z\geq 0$. We have

$$|\varphi(z)|\leq 1 \text{ for } \Re z=0 \text{ and } |\varphi(z)|\leq 1 \text{ for } |z|=r,$$

if $r$ is so large that $\omega(r)>\frac{1}{\varepsilon}$, hence $|\varphi(z)|\leq 1$ in the entire half-plane $\Re z\geq 0$; and finally, as $\varepsilon\to 0$, $|e^z|\leq 1$: contradiction. (Borderline case of **290**: The region $\mathfrak{T}$ becomes a halfplane.)

**330.** Assume $\varepsilon>0$, $h=\varepsilon e^{-i\frac{\alpha+\beta}{2}}$, $0<\sigma<\delta$, $\sigma(\beta-\alpha)<\pi$. Apply the reasoning of **322** to the function

$$F(z)=f(z)\exp\left(-(hz)^{\frac{\pi}{\beta-\alpha}-\sigma}\right).$$

There results the conclusion: $|F(z)|\leq 1$ in the entire sector. Let $\varepsilon$ converge to 0.—The proposition could also be reduced to theorem **322** with the help of a function that maps the sector $\alpha\leq\arg z\leq\beta$ onto the sector $-\gamma\leq\arg z\leq\gamma$, $\gamma=\frac{\pi}{2}-\frac{\delta(\beta-\alpha)}{2}$, and leaves $z=0$ and $z=\infty$ unchanged.

**331.** In the special case $\alpha=-\frac{\pi}{2}$, $\beta=\frac{\pi}{2}$ the statement is weaker than **325**. The proof involves the function

$$f(z)\exp\left(-\eta e^{-i\frac{\beta+\alpha}{\beta-\alpha}\frac{\pi}{2}}z^{\frac{\pi}{\beta-\alpha}}\right).$$

Start by proving with the help of **330** that the maximum of the modulus of $f(z)$ on the bisector is not larger than 1 [**325**].

**332.** Assume $|g(-r)|\leq C$: apply **331** to the function $\frac{g(z)}{C}$ in the sector $-\pi\leq\vartheta\leq\pi$; this leads to $|g(z)|\leq C$ in the entire plane, thus $g(z)$ is a constant.

We can also examine

$$g(z)\, e^{-\eta\sqrt{z}-\varepsilon(-iz)^{\frac{3}{4}}}.$$

This function is analytic in the open sector $0 < \vartheta < \pi$ and continuous in the closed sector $0 \leqq \vartheta \leqq \pi$. Cf. **325**.—Also **325** can be applied to $C^{-1}g(z^2)$.—The function $\frac{\sin\sqrt{z}}{\sqrt{z}}$ illustrates how strong the theorem is.

**333.** Let $a < b < 1$. The absolute value of the test function $e^{e^{bz}}$

$$\left|e^{e^{b(x+iy)}}\right| = e^{e^{bx}\cos by};$$

thus on the boundary of $\mathfrak{R}$ it is $\geqq e^{\cos b\frac{\pi}{2}} > 1$. Let $l > \frac{1}{b-a}\log\frac{A}{\varepsilon\cos b\frac{\pi}{2}}$.

Then we find on the boundary of the rectangle $0 \leqq x \leqq l,\ -\frac{\pi}{2} \leqq y \leqq \frac{\pi}{2}$ the inequality

$$\left|f(z)\, e^{-\varepsilon e^{bz}}\right| < 1.$$

**334.** Assume the contrary: consider the function

$$\varphi(z) = e^{e^z}[f(z)]^{-\varepsilon}, \qquad \varepsilon > 0,\ z = x + iy,$$

in the rectangle

$$0 \leqq x \leqq x_1, \quad -\frac{\pi}{2} \leqq y \leqq \frac{\pi}{2},$$

where $x_1$ is chosen so that $\varepsilon\omega(x_1) > 1$. On the boundary of this rectangle we have

$$|\varphi(iy)| \leqq e^{\cos y}\cdot e^{-\varepsilon\omega(0)} \leqq e, \quad \left|\varphi\left(x \pm i\frac{\pi}{2}\right)\right| \leqq 1,$$

$$\left|\varphi(x_1 + iy)\right| \leqq e^{e^{x_1}[\cos y - \varepsilon\omega(x_1)]} < 1;$$

hence in the interior $|\varphi(z)| \leqq e$. As $\varepsilon \to 0$ we obtain

$$\left|e^{e^z}\right| \leqq e,$$

thus e.g. $e^e \leqq e$: contradiction. The role of the function $e^{e^z}$ is explained by **290** and **187**.

**335.** It is sufficient to consider the case hinted at in the problem [linear transformation]. Take $\varepsilon > 0$; the function

$$\varphi(z) = f(z)\prod_{\nu=1}^{n}\left(\frac{z - z_\nu}{2r}\right)^{\varepsilon}$$

is regular in the intersection $\mathfrak{R}_r$ of $\mathfrak{R}$ and the disk $|z| \leqq r$ and its modulus is single-valued in $\mathfrak{R}_r$ [**303**]. The maximum of $|f(z)|$ on the circle $|z| = r$ is denoted by $M(r)$. Taking into account the condition on $f(z)$ on the boundary of $\mathfrak{R}$ [**278**] we find that in $\mathfrak{R}_r$

$$\varphi(z) \leqq \max [M, M(r)].$$

Hence for $\varepsilon \to 0$

$$|f(z)| \leqq \max [M, M(r)].$$

If, in particular, $r'$ is also admissible [see hint] and $r' < r$, then we have

$$M(r') \leqq \max [M, M(r)].$$

On the other hand [**268**]

$$M(r) \leqq M(r').$$

Hence the alternative: Either $M(r) \geqq M$, thus $M(r) = M(r')$, $f(z) \equiv$ const. [**268**] or $M(r) < M$, $|f(z)| \leqq M$, even $|f(z)| < M$ [**278**].

**336.** We assume that $n$ boundary sections of $\mathfrak{D}$ lie on the real axis. From a variable point $z$ in the upper half-plane they are seen under the angles $\omega_1, \omega_2, \ldots, \omega_n$. Determine a regular and analytic function $\varphi_\nu(z)$ in the half-plane $\mathfrak{J}z > 0$ so that $\pi \mathfrak{R} \varphi_\nu(z) = \omega_\nu$ [**57**] and put

$$\Phi(z) = a \cdot \left(\frac{A}{a}\right)^{\varphi_1(z) + \varphi_2(z) + \cdots + \varphi_n(z)}.$$

The function $f(z)\,\Phi(z)^{-1}$ is regular and bounded in the interior, $\leqq 1$ on the boundary, of $\mathfrak{D}$ with the possible exception of $2n$ boundary points [**335**].

**337.** The function $f(e^u)$ is single-valued, regular and bounded in the half-plane $\mathfrak{R}u \leqq 0$. We have $|f(e^u)| \leqq 1$ at all the boundary points of this half-plane except at the boundary point $u = \infty$. [**335**.]

**338.** If $z = \infty$ were not a boundary point we would have [**135** suffices] necessarily $|g(z)| \leqq k$ in $\mathfrak{R}$: contradiction. Hence $z = \infty$ is a boundary point of $\mathfrak{R}$. If $g(z)$ were bounded in $\mathfrak{R}$ we could use **335** (only excluded point $z = \infty$) which would lead to $|g(z)| \leqq k$ in $\mathfrak{R}$.

**339.** There exists, according to the hypothesis, a constant $M$, $M > 0$, such that $|f(z)| < M$ in the region $\mathfrak{R}$ bounded by $\Gamma_1$ and $\Gamma_2$. Choose $R > 1$ and so large that $|f(z)| < \varepsilon$ on $\Gamma_1$ and $\Gamma_2$ outside the circle $|z| = R$. Take the branch of $\log z$ that is real for positive $z$. This branch is regular and single-valued in $\mathfrak{R}$, where $|\log z| < \log |z| + \pi$ provided that $|z| \geqq R$. The inequality

$$|M(\log R + \pi) + \varepsilon(\log z + \pi)| \geqq M(\log R + \pi) + \varepsilon(\log |z| + \pi)$$

holds and for $|z| \geqq R$ both addends on the right hand side are positive. Thus we obtain for $|z| = R$, $z$ in $\mathfrak{R}$,

$$\left|\frac{\log z}{M(\log R + \pi) + \varepsilon(\log z + \pi)} f(z)\right| < \frac{\log R + \pi}{M(\log R + \pi)} M = 1.$$

If $|z| \geqq R$, $z$ on $\Gamma_1$ or on $\Gamma_2$ we find

$$\left|\frac{\log z}{M(\log R + \pi) + \varepsilon(\log z + \pi)} f(z)\right| < \frac{\log |z| + \pi}{\varepsilon(\log |z| + \pi)} \varepsilon = 1.$$

Hence we have, by virtue of **335**, at each point $z$ in $\mathfrak{R}$ but outside the circle $|z| = R$

$$|f(z)| < \left|\frac{\varepsilon(\log z + \pi) + M(\log R + \pi)}{\log z}\right|.$$

The right hand side becomes smaller than $2\varepsilon$ when $|z|$ gets sufficiently large.

**340.** Assume, if possible, $a \neq b$ and consider two disks $D_1$ and $D_2$ in the $w$-plane, with no points in common, centered at $a$ and $b$ respectively. Outside the two disks the expression $\left|\left(w - \frac{a+b}{2}\right)^2 - \left(\frac{a-b}{2}\right)^2\right|$ has a positive minimum $= \varepsilon$. Proposition **339** applied to the function $\left(f(z) - \frac{a+b}{2}\right)^2 - \left(\frac{a-b}{2}\right)^2$ implies that the absolute value of this function is smaller than $\varepsilon$ in the region $\mathfrak{R}$ bounded by $\Gamma_1$ and $\Gamma_2$ whenever $|z| > R = R(\varepsilon)$: we consider only such points $z$. Find two points, $z_1$ on $\Gamma_1$ and $z_2$ on $\Gamma_2$, such that $w_1 = f(z_1)$ is in $D_1$ and $w_2 = f(z_2)$ is in $D_2$ and join the two points by a curve in $\mathfrak{R}$. The image in the $w$-plane of this curve leads from $D_1$ to $D_2$. Therefore there exists on it a point $w = f(z)$ for which $\left|\left(w - \frac{a+b}{2}\right)^2 - \left(\frac{a-b}{2}\right)^2\right| \geqq \varepsilon$: contradiction.

# Author Index

Numbers refer to pages. Numbers in italics refer to original contributions.

# Subject Index

Roman numerals refer to parts, Arabic numerals to problems or pages. Example: III **304** 164 refer to problem 304 of Part III on page 164.

# Die Grundlehren der mathematischen Wissenschaften

*A Series of Comprehensive Studies in Mathematics*

A Selection

Printing and binding: Druckerei Triltsch, Würzburg

CIM

M. Aigner Combinatorial Theory ISBN 978-3-540-61787-7
A. L. Besse Einstein Manifolds ISBN 978-3-540-74120-6
N. P. Bhatia, G. P. Szegő Stability Theory of Dynamical Systems ISBN 978-3-540-42748-3
J. W. S. Cassels An Introduction to the Geometry of Numbers ISBN 978-3-540-61788-4
R. Courant, F. John Introduction to Calculus and Analysis I ISBN 978-3-540-65058-4
R. Courant, F. John Introduction to Calculus and Analysis II/1 ISBN 978-3-540-66569-4
R. Courant, F. John Introduction to Calculus and Analysis II/2 ISBN 978-3-540-66570-0
P. Dembowski Finite Geometries ISBN 978-3-540-61786-0
A. Dold Lectures on Algebraic Topology ISBN 978-3-540-58660-9
J. L. Doob Classical Potential Theory and Its Probabilistic Counterpart ISBN 978-3-540-41206-9
R. S. Ellis Entropy, Large Deviations, and Statistical Mechanics ISBN 978-3-540-29059-9
H. Federer Geometric Measure Theory ISBN 978-3-540-60656-7
S. Flügge Practical Quantum Mechanics ISBN 978-3-540-65035-5
L. D. Faddeev, L. A. Takhtajan Hamiltonian Methods in the Theory of Solitons ISBN 978-3-540-69843-2
I. I. Gikhman, A. V. Skorokhod The Theory of Stochastic Processes I ISBN 978-3-540-20284-4
I. I. Gikhman, A. V. Skorokhod The Theory of Stochastic Processes II ISBN 978-3-540-20285-1
I. I. Gikhman, A. V. Skorokhod The Theory of Stochastic Processes III ISBN 978-3-540-49940-4
D. Gilbarg, N. S. Trudinger Elliptic Partial Differential Equations of Second Order ISBN 978-3-540-41160-4
H. Grauert, R. Remmert Theory of Stein Spaces ISBN 978-3-540-00373-1
H. Hasse Number Theory ISBN 978-3-540-42749-0
F. Hirzebruch Topological Methods in Algebraic Geometry ISBN 978-3-540-58663-0
L. Hörmander The Analysis of Linear Partial Differential Operators I – Distribution Theory and Fourier Analysis ISBN 978-3-540-00662-6
L. Hörmander The Analysis of Linear Partial Differential Operators II – Differential Operators with Constant Coefficients ISBN 978-3-540-22516-4
L. Hörmander The Analysis of Linear Partial Differential Operators III – Pseudo-Differential Operators ISBN 978-3-540-49937-4
L. Hörmander The Analysis of Linear Partial Differential Operators IV – Fourier Integral Operators ISBN 978-3-642-00117-8
K. Itô, H. P. McKean, Jr. Diffusion Processes and Their Sample Paths ISBN 978-3-540-60629-1
T. Kato Perturbation Theory for Linear Operators ISBN 978-3-540-58661-6
S. Kobayashi Transformation Groups in Differential Geometry ISBN 978-3-540-58659-3
K. Kodaira Complex Manifolds and Deformation of Complex Structures ISBN 978-3-540-22614-7
Th. M. Liggett Interacting Particle Systems ISBN 978-3-540-22617-8
J. Lindenstrauss, L. Tzafriri Classical Banach Spaces I and II ISBN 978-3-540-60628-4
R. C. Lyndon, P. E Schupp Combinatorial Group Theory ISBN 978-3-540-41158-1
S. Mac Lane Homology ISBN 978-3-540-58662-3
C. B. Morrey Jr. Multiple Integrals in the Calculus of Variations ISBN 978-3-540-69915-6
D. Mumford Algebraic Geometry I – Complex Projective Varieties ISBN 978-3-540-58657-9
O. T. O'Meara Introduction to Quadratic Forms ISBN 978-3-540-66564-9
G. Pólya, G. Szegő Problems and Theorems in Analysis I – Series. Integral Calculus. Theory of Functions ISBN 978-3-540-63640-3
G. Pólya, G. Szegő Problems and Theorems in Analysis II – Theory of Functions. Zeros. Polynomials. Determinants. Number Theory. Geometry ISBN 978-3-540-63686-1
W. Rudin Function Theory in the Unit Ball of $\mathbb{C}^n$ ISBN 978-3-540-68272-1
S. Sakai C*-Algebras and W*-Algebras ISBN 978-3-540-63633-5
C. L. Siegel, J. K. Moser Lectures on Celestial Mechanics ISBN 978-3-540-58656-2
T. A. Springer Jordan Algebras and Algebraic Groups ISBN 978-3-540-63632-8
D. W. Stroock, S. R. S. Varadhan Multidimensional Diffusion Processes ISBN 978-3-540-28998-2
R. R. Switzer Algebraic Topology: Homology and Homotopy ISBN 978-3-540-42750-6
A. Weil Basic Number Theory ISBN 978-3-540-58655-5
A. Weil Elliptic Functions According to Eisenstein and Kronecker ISBN 978-3-540-65036-2
K. Yosida Functional Analysis ISBN 978-3-540-58654-8
O. Zariski Algebraic Surfaces ISBN 978-3-540-58658-6